MATHEMATICS
Structure and Method

Course 2   Teacher's Edition

MARY P. DOLCIANI
EDWIN F. BECKENBACH
RICHARD G. BROWN
ROBERT B. KANE
WILLIAM WOOTON

HOUGHTON MIFFLIN COMPANY · BOSTON
Atlanta   Dallas   Geneva, Ill.   Hopewell, N.J.   Palo Alto   Toronto

Authors

MARY P. DOLCIANI Professor of Mathematics, Hunter College, City University of New York.

EDWIN F. BECKENBACH Professor of Mathematics, Emeritus, University of California, Los Angeles.

RICHARD G. BROWN Mathematics Teacher, Phillips Exeter Academy, Exeter, New Hampshire.

ROBERT B. KANE Director of Teacher Education and Head of the Department of Education, Purdue University.

WILLIAM WOOTON former Professor of Mathematics, Los Angeles Pierce College.

Editorial Advisers

ANDREW M. GLEASON Hollis Professor of Mathematics and Natural Philosophy, Harvard University.

ALBERT E. MEDER, Jr. Dean and Vice Provost and Professor of Mathematics, Emeritus, Rutgers, The State University of New Jersey.

Consultants

CAROLYN AHO Teacher Specialist, Mathematics, San Francisco Unified School District, San Francisco, California.

RUTH GREEN Mathematics Teacher, Isaac Litton Junior High School, Nashville, Tennessee.

TIMOTHY HOWELL Mathematics Coordinator, Hanover High School, Hanover, New Hampshire.

BETTY TAKESUYE Mathematics Teacher, Chaparral High School, Scottsdale, Arizona.

ISBN: 0–395–31395–3

Contents

T4

Introduction

THE LESSON

The text was designed to teach your students the computational skills and the structure of mathematics. The text is organized into twelve chapters, each chapter consisting of short sections.

9-4 Percent of Increase or Decrease

Objective To compute percents of increase and decrease.

When a quantity changes from one value to another, it either increases or decreases in value. You can compute the *percent of increase or decrease* by using the following rule:

---

rule

$$\text{percent of change} = \frac{\text{amount of change}}{\text{original amount}}$$

Example 1 Find the percent of increase from 25 to 28.

Solution The increase (amount of change) is $28 - 25 = 3$.

$$\text{percent of change} = \frac{\text{amount of change}}{\text{original amount}}$$

$$= \frac{3}{25} = \frac{12}{100} = 12\%$$

Example 2 Al used to buy a photography magazine every month for \$1.25. After subscribing to the magazine, he paid only 62.5¢ per issue. Find the percent of decrease.

Solution The decrease is $1.25 - 0.625 = 0.625$.

$$\text{percent of decrease} = \frac{0.625}{1.25}$$

$$= 0.5$$

$$= 50\%$$

A generous number of Class Exercises are provided to help you determine whether further discussion of the concept being taught is needed before going on to the exercise assignment.

Class Exercises

a. State the amount of change from the first to the second number.
b. State the percent of increase or decrease.

1. 8 to 6	**2.** 2 to 4	**3.** 2 to 5	**4.** 10 to 8
5. 10 to 11	**6.** 5 to 7	**7.** 25 to 4	**8.** 2 to 1

Exercises and problems are graded A, B, or C in order of increasing difficulty. Many exercise sets include sample exercises with complete solutions. A handy assignment guide begins on page T16 with suggestions for an average or a comprehensive course.

9. 100 to 170   **10.** 100 to 35   **11.** 8 to 9   **12.** 4 to 7

13. A $1 item is marked down 20%. The new price is then increased 20%. Is the final price $1? Explain.

14. A $1 item is marked up 50%. Then the new price is decreased 50%. Is the final price $1? Explain.

Exercises

Find the percent of increase or decrease in changing from the first number to the second. If necessary, round your answer to the nearest tenth of a percent.

A **1.** 20 to 22   **2.** 30 to 35   **3.** 40 to 33   **4.** 50 to 47

   5. 66 to 44   **6.** 70 to 98   **7.** 45 to 86   **8.** 27 to 24

   9. 18 to 51   **10.** 63 to 79   **11.** 32 to 17   **12.** 75 to 27

B **13.** 9.6 to 2.4   **14.** 12 to 8.7   **15.** 5.3 to 9.15   **16.** 0.27 to 1.59

   17. $3\frac{3}{7}$ to $1\frac{2}{7}$ **18.** $6\frac{2}{3}$ to $12\frac{1}{6}$ **19.** $3\frac{1}{2}$ to $4\frac{1}{5}$ **20.** 5 to $3\frac{1}{4}$

Problems

Solve. If necessary, round your answer to the nearest tenth of a percent

A **1.** Jeffrey Corwin's salary was increased $12 per week. If his salary was $160 per week before the raise, what percent did his salary increase?

   2. Circulation of the magazine *Sport World* increased from 40,000 copies monthly in 1974 to 150,000 copies monthly in 1979. By what percent did the circulation increase?

   3. During a sale, the price of a scarf was reduced from $4.80 to $3.60. By what percent was the price decreased?

   4. After a race, Mona's pulse rate rose from 72 beats per minute to 96 beats per minute. What percent of increase does this represent?

TESTING

Self-Tests, providing frequent opportunities for review, occur after groups of related lessons. Answers are included at the back of the book.

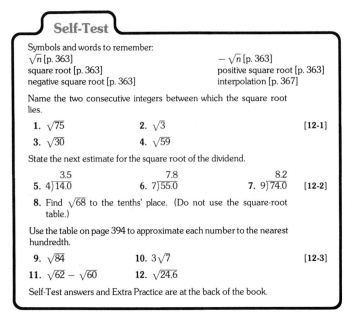

Self-Test

Symbols and words to remember:

$\sqrt{n}$ [p. 363] $-\sqrt{n}$ [p. 363]
square root [p. 363]   positive square root [p. 363]
negative square root [p. 363]   interpolation [p. 367]

Name the two consecutive integers between which the square root lies.

1. $\sqrt{75}$ **2.** $\sqrt{3}$ [12-1]

3. $\sqrt{30}$ **4.** $\sqrt{59}$

State the next estimate for the square root of the dividend.

$$\begin{array}{ccc} 3.5 & 7.8 & 8.2 \end{array}$$
5. $4)\overline{14.0}$ **6.** $7)\overline{55.0}$ **7.** $9)\overline{74.0}$ [12-2]

8. Find $\sqrt{68}$ to the tenths' place. (Do not use the square-root table.)

Use the table on page 394 to approximate each number to the nearest hundredth.

9. $\sqrt{84}$ **10.** $3\sqrt{7}$ [12-3]

11. $\sqrt{62} - \sqrt{60}$ **12.** $\sqrt{24.6}$

Self-Test answers and Extra Practice are at the back of the book.

Students who wish additional opportunity for skill mastery may refer to a bank of clearly identified exercises at the back of the book.

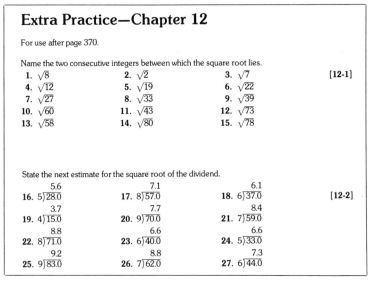

Extra Practice—Chapter 12

For use after page 370.

Name the two consecutive integers between which the square root lies.

1. $\sqrt{8}$	**2.** $\sqrt{2}$	**3.** $\sqrt{7}$	[12-1]
4. $\sqrt{12}$	**5.** $\sqrt{19}$	**6.** $\sqrt{22}$	
7. $\sqrt{27}$	**8.** $\sqrt{33}$	**9.** $\sqrt{39}$	
10. $\sqrt{60}$	**11.** $\sqrt{43}$	**12.** $\sqrt{73}$	
13. $\sqrt{58}$	**14.** $\sqrt{80}$	**15.** $\sqrt{78}$	

State the next estimate for the square root of the dividend.

5.6	7.1	6.1	
16. $5)\overline{28.0}$	**17.** $8)\overline{57.0}$	**18.** $6)\overline{37.0}$	[12-2]
3.7	7.7	8.4	
19. $4)\overline{15.0}$	**20.** $9)\overline{70.0}$	**21.** $7)\overline{59.0}$	
8.8	6.6	6.6	
22. $8)\overline{71.0}$	**23.** $6)\overline{40.0}$	**24.** $5)\overline{33.0}$	
9.2	8.8	7.3	
25. $9)\overline{83.0}$	**26.** $7)\overline{62.0}$	**27.** $6)\overline{44.0}$	

Each chapter is followed by a Chapter Review and Chapter Test. Skill Reviews provide opportunities for the students to maintain skill mastery. A Cumulative Review occurs after every three chapters, a total of four in all.

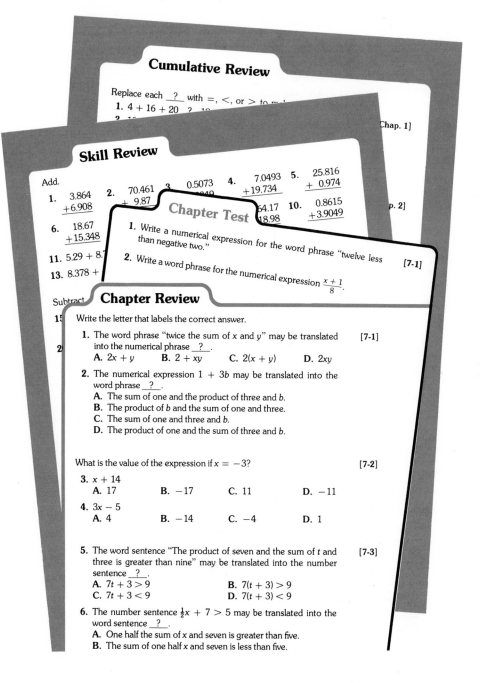

Cumulative Review

Replace each __?__ with =, <, or > to m~
1. $4 + 16 + 20$? ~

Chap. 1]

Skill Review

Add.
1. $\begin{array}{r} 3.864 \\ +6.908 \end{array}$
2. $\begin{array}{r} 70.461 \\ + 9.87 \end{array}$
3. $\begin{array}{r} 0.5073 \\ \end{array}$
4. $\begin{array}{r} 7.0493 \\ +19.734 \end{array}$
5. $\begin{array}{r} 25.816 \\ + 0.974 \end{array}$

p. 2]

6. $\begin{array}{r} 18.67 \\ +15.348 \end{array}$
64.17
18.98
10. $\begin{array}{r} 0.8615 \\ +3.9049 \end{array}$

11. $5.29 + 8.7$
13. $8.378 +$

Chapter Test

1. Write a numerical expression for the word phrase "twelve less than negative two."

2. Write a word phrase for the numerical expression $\frac{x+1}{8}$.

[7-1]

Subtract
15

2

Chapter Review

Write the letter that labels the correct answer.

1. The word phrase "twice the sum of x and y" may be translated into the numerical phrase __?__. [7-1]
   A. $2x + y$ **B.** $2 + xy$ **C.** $2(x + y)$ **D.** $2xy$

2. The numerical expression $1 + 3b$ may be translated into the word phrase __?__.
   A. The sum of one and the product of three and b.
   B. The product of b and the sum of one and three.
   C. The sum of one and three and b.
   D. The product of one and the sum of three and b.

What is the value of the expression if $x = -3$? [7-2]

3. $x + 14$
   A. 17   **B.** -17 **C.** 11   **D.** -11

4. $3x - 5$
   A. 4   **B.** -14 **C.** -4 **D.** 1

5. The word sentence "The product of seven and the sum of t and three is greater than nine" may be translated into the number sentence __?__. [7-3]
   A. $7t + 3 > 9$ **B.** $7(t + 3) > 9$
   C. $7t + 3 < 9$ **D.** $7(t + 3) < 9$

6. The number sentence $\frac{1}{2}x + 7 > 5$ may be translated into the word sentence __?__.
   A. One half the sum of x and seven is greater than five.
   B. The sum of one half x and seven is less than five.

OPTIONAL FEATURES

Many optional features like these, which the students will enjoy exploring, occur throughout the text. If you wish, you may use them as a springboard for further class discussions.

Application
Writing a Check

A person writing a **check** is generally required to write the amount to be paid both as a numeral and in words.

Research Activity Find out the correct form for a handwritten check.

Consumer Activity Find out the advantages of paying bills by check. Are there any disadvantages?

EXTRA! Computing Compound Interest

The program at the right, written in BASIC, will find the new principal based on the initial principal P, the rate R as a decimal, compounded N times a year for T years.

1. Run the program to verify the results shown on page 311.

Compute $200 at 5.5% for 3 years:

2. Compounded quarterly
3. Compounded monthly
4. Compounded daily (365 days)

```
10 PRINT "COMPOUND INTEREST:"
20 PRINT "INPUT P,R(DECIMAL),";
30 PRINT " N, AND T";
40 INPUT P,R,N,T
50 LET M=N*T
60 FOR K=1 TO M
70 LET P=P+P*R/N
80 NEXT K
90 PRINT "AMOUNT AFTER";T;
100 PRINT " YEAR(S) = $";P
110 END
```

An Architect

Julia Morgan (1872–1957) was born in San Francisco and lived most of her life near there. She received an engineering degree from the University of California at Berkeley and then turned to architecture. In 1902, she received her certificate from the École des Beaux Arts in Paris, and on returning to California, became a licensed architect. She designed more than 800 buildings, among them the Library at Mills College in Oakland, shown below, and the Hearst castle, San Simeon.

Research Activity Try to find out the names of the architects of the school buildings in your community.

Career Activity Find out the name of the nearest college of architecture and what its entrance requirements are.

More sophisticated extensions of the mathematical skills are provided in two-page features which are at the end of each chapter. None of the optional features is part of the testing program.

Random Experiments

You have learned some important ideas in probability. But how well do they work? The real test of usefulness is actually to perform some random experiment many times, and compare the results with what your knowledge of probability would lead you to expect. A computer can help us do this because it can simulate a random experiment and repeat it many times at high speed. It can also tell us thir results.

The of thro
sum of
sible s
functio

Baseball Statistics

There have been more data compiled about baseball than about any other sport, and many fans carry much of this information at the tip of their tongues. They can readily give you the answers to these:

Which team
championship

The American L

The World Seri

Let S s
below
can tu

Some baseba
For example, th
home runs duri

Ha
Ba

Sadaharu Oh o

Try to find the
statistics.

1. Which major
   number of y

2. Has any play

3. In which yea
   League batti

1. Wha

2. Wha

4. Which two p

5. Which major
   is that capa

6. Which team
   which seaso

Research Acti
and about the

House Plans

The building facing the water in the photograph at the left contains six apartments.

The drawing at the left below shows the three apartments in the left-hand half of the building, as seen from the front.

	BR.	Level 3
L.R. D.R.	L.R.	Level 2
D.R. L.R.	BR.	Level 1

KIT.　　STUDY

The apartment on the first floor (level 1) is unshaded. The two-level apartment on the second and third floors is lightly shaded, and the apartment which is entirely on the second floor is more heavily shaded.

The floor plans of the three apartments are shown below.

LEVEL 1　　LEVEL 2　　LEVEL 3

TEACHER'S ANNOTATED EDITION

The lesson commentaries, beginning on page T38, provide teaching suggestions, handy chalkboard examples, and related activities to expand, support, and extend the lessons.

For your convenience, an extra set of chapter tests begins on page T24 with answers on page T36. In addition, diagnostic tests covering eleven arithmetic skill areas begin on page T12, with answers on page T15.

T24

𝕰xtra Chapter Tests

Chapter 1

Exercises 1 and 2 refer to the table at the right.

1. What was the population of New York in 1950?

2 What was the total population of the four states in 1900?

POPULATION, in Thousands			
State Year	1800	1900	1⁹
North Carolina	478	1894	4(
New York	589	7269	14,8
Pennsylvania	602	6302	10,4
Virginia	880	1854	33

Exercises 3 and 4 refer to the bar graph at the right.

3. What is the speed of the squirrel?

4. How much faster can a giraffe run than an elephant?

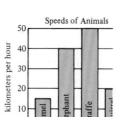

Speeds of Animals

Exercises 5 and 6 refer to the line graph at the right.

5. What is the average temperature in January?

6. In which month is the average temperature 11°C?

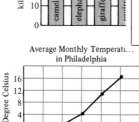

Average Monthly Temperature in Philadelphia

Exercises 7–10 refer to the table at the right.

7. Organize the data into intervals of 21–24, 25–28, 29–32, 33–36, and 37–40. Draw a histogram.

8. How many plants produced more than 28 tomatoes?

9. What is the mean of the data?

10. What is the mode of the data?

Number of Tomatoes Picked from 16 Plants			
27	24	22	29
36	34	31	27
21	39	27	21
33	24	26	27

[1-3]

[1-4]

[1-5]

users of MATHEMATICS, Structure and Method.

8-4 Scale Drawings

Teaching Suggestions
Floor plans and road maps are perhaps the most familiar examples of scale drawings to students. If possible, obtain some blueprints to show the students. Have them measure the length of, say, a wall in the blueprint and then set up a proportion using the scale to find the actual wall length.

Point out that although the text has introduced scales using arrows (1 cm → 3 m), an equality sign is commonly used. Therefore, beginning with Problem 5, the practice of using the arrow will be dropped.

Chalkboard Examples

1. A map is drawn to the scale 1 cm → 100 km. The distance between Center City and Northridge measures 4.8 cm on the map. What is the approximate actual distance between the cities? 480 km

2. A rectangular room measures 4 m × 5.5 m. If you were to make a scale drawing of the room using the scale 1 cm = 0.5 m, what would be the dimensions of the room on the drawing? 8 cm × 11 cm

Related Activities
Suggest that the students measure the lengths of a room of their house or apartment. Include in their measurements large pieces of furniture, appliances, and doorways. Have the students choose an appropriate scale and make a scale drawing of the room and its furnishings. Or, the students might be interested in making a scale drawing of their dream house with some furnishings. Students can then exchange scale drawings with a classmate and determine the actual size of the room and its furnishings.

SUPPLEMENTARY MATERIALS

The set of Progress Tests provides a convenient way to diagnose areas of difficulty and to measure the level of each student's performance.

The Solution Key provides step-by-step solutions for every exercise, problem, and activity in the text.

Diagnostic Tests in Arithmetic*

1. Whole Numbers—Addition

1. 6 +8	**2.** 2 4 +3	**3.** 6 1 +5	**4.** 54 +23	**5.** 66 +25

6. 535 +272	**7.** 662 +188	**8.** 625 118 +138	**9.** 9838 4539 1245 +9106

2. Whole Numbers—Subtraction

1. 5 −3	**2.** 12 − 5	**3.** 66 −35	**4.** 659 −408	**5.** 825 −608

6. 928 −683	**7.** 7808 −6294	**8.** 744 −485	**9.** 7440 −4659	**10.** 90,504 −45,666

3. Whole Numbers—Multiplication

1. 49 ×2	**2.** 82 ×8	**3.** 387 ×3	**4.** 86 ×30	**5.** 187 ×300

6. 4411 ×2000	**7.** 32 ×88	**8.** 532 ×18	**9.** 512 ×561	**10.** 304 ×701

4. Whole Numbers—Division

1. 7)56   **2.** 3)63   **3.** 7)175   **4.** 9)2646   **5.** 20)800

6. 43)871   **7.** 596)403,929   **8.** 62)19,293

9. Express the remainder in Exercise 8 as a fraction.

10. Express the quotient to the nearest hundredth: 6)40

5. Fractions—Basic Skills

1. Identify the figure that is divided into thirds.

A.   B.   C.

* Used by permission from INDIVIDUALIZED COMPUTATIONAL SKILLS PROGRAM, COMPUTER VERSION by Bryce R. Shaw, Miriam M. Schaefer, and Petronella M. W. Hiehle. Copyright © 1973 by Houghton Mifflin Company.

2. What is the numerator of $\frac{3}{4}$?

3. The area shaded in Figure A is represented by which fraction in Row B?

A.

B. $\frac{3}{5}; \frac{4}{5}; \frac{1}{5}; \frac{5}{3}$

4. Which of the figures represent the fraction $\frac{1}{3}$?

A.   **B.**   **C.**

5. Write the fraction represented by the set diagram.

6. Which diagram represents the fraction $\frac{3}{5}$?

A.   **B.**   **C.**

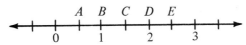

7. Which fraction represents the number 1? $\frac{3}{1}; \frac{4}{4}; \frac{5}{10}; \frac{1}{6}$

8. Which fraction represents B on the number line? $\frac{4}{1}; \frac{4}{4}; \frac{2}{4}; \frac{5}{2}$

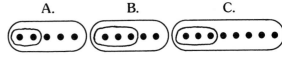

9. Write the consecutive multiples of 4: 4, __?__, __?__, __?__ .

10. Find the least common multiple of 8, 6, and 2.

11. Write 40 as the product of prime factors.

12. Find the greatest common factor of 20 and 15.

13. Find the fraction in Row Y that is equal to each fraction or mixed number in Row X.

X: (a) $\frac{1}{3}$ (b) $\frac{6}{8}$ (c) $3\frac{1}{5}$

Y: $\frac{3}{4}, \frac{8}{6}, \frac{16}{5}, \frac{3}{9}, \frac{5}{16}, \frac{1}{4}$

14. Find the lowest common denominator for the fractions $\frac{1}{2}, \frac{1}{4}, \frac{4}{5}$.

15. Find the fraction or mixed number in Row Y that is equal to each fraction or mixed number in Row X.

X: (a) $\frac{5}{3}$ (b) $1\frac{3}{6}$

Y: $\frac{2}{3}; 1\frac{2}{3}; \frac{9}{6}; \frac{6}{9}$

6. Fractions—Addition

1. $\dfrac{1}{11}$
$+\dfrac{2}{11}$

2. $4\dfrac{1}{8}$
$+3\dfrac{1}{8}$

3. $2\dfrac{3}{12}$
$+4\dfrac{11}{12}$

4. $\dfrac{1}{3}$
$+\dfrac{1}{5}$

5. $2\dfrac{5}{6}$
$+4\dfrac{2}{3}$

7. Fractions—Subtraction

1. $\dfrac{3}{4}$
$-\dfrac{2}{4}$

2. $6\dfrac{5}{8}$
$-1\dfrac{1}{8}$

3. 2
$-1\dfrac{3}{8}$

4. $4\dfrac{1}{6}$
$-1\dfrac{5}{6}$

5. $\dfrac{2}{3}$
$-\dfrac{1}{7}$

6. $13\dfrac{4}{6}$
$-3\dfrac{4}{5}$

8. Fractions—Multiplication

1. $\dfrac{1}{5} \times \dfrac{1}{3} = \underline{\quad?\quad}$

2. $\dfrac{2}{3} \times \dfrac{1}{6} = \underline{\quad?\quad}$

3. $\dfrac{3}{5} \times 15 = \underline{\quad?\quad}$

4. $3\dfrac{1}{8} \times \dfrac{7}{8} = \underline{\quad?\quad}$

5. $4\dfrac{2}{5} \times 5\dfrac{1}{4} = \underline{\quad?\quad}$

9. Fractions—Division

1. $3 \div \dfrac{1}{2} = \underline{\quad?\quad}$

2. $4 \div \dfrac{1}{2} = \underline{\quad?\quad}$

3. $\dfrac{8}{9} \div \dfrac{1}{6} = \underline{\quad?\quad}$

4. $1\dfrac{1}{5} \div \dfrac{4}{7} = \underline{\quad?\quad}$

5. $4\dfrac{5}{9} \div 2 = \underline{\quad?\quad}$

6. $5\dfrac{5}{6} \div 3\dfrac{1}{8} = \underline{\quad?\quad}$

10. Decimals

1. Write the decimal which represents "four and four thousandths."

2. $\dfrac{7}{10,000} = \underline{\quad?\quad}$ (Decimal)   **3.** $0.802 = \underline{\quad?\quad}$ (Fraction)   **4.** $4\dfrac{2}{5} = \underline{\quad?\quad}$ (Decimal)

5. Round 64.4487 to the nearest tenth.

6. $7.065 + 0.6 + 10.021 = \underline{\quad?\quad}$ **7.** $583.646 - 30.03 = \underline{\quad?\quad}$

8.   94.5
  500.85
$+\ \ 51.88$

9.   118.04
$-\ \ \ \ 9.204$

10.   4560
$\times 0.0023$

11. $0.06\overline{)25.212}$

11. Percents

1. $\dfrac{7}{5} = \dfrac{?}{100}$ **2.** $0.77 = \underline{\quad?\quad}\%$ **3.** $2.604 = \underline{\quad?\quad}\%$ **4.** $4\dfrac{2}{4} = \underline{\quad?\quad}\%$

5. $194\% = \underline{\quad?\quad}$ (Mixed Numeral) **6.** $37\dfrac{1}{2}\% = \underline{\quad?\quad}$ (Decimal) **7.** 10% of 13 $= \underline{\quad?\quad}$

8. $\dfrac{2}{7} = \underline{\quad?\quad}\%$ (Fractional Form)   **9.** $\underline{\quad?\quad}\%$ of 10 $= 2$ **10.** 60% of $\underline{\quad?\quad} = 102$

ANSWERS TO DIAGNOSTIC TESTS

1. Whole Numbers—Addition
   1. 14 **2.** 9 **3.** 12 **4.** 77 **5.** 91 **6.** 807
   7. 850 **8.** 881 **9.** 24,728

2. Whole Numbers—Subtraction
   1. 2 **2.** 7 **3.** 31 **4.** 251 **5.** 217 **6.** 245
   7. 1514 **8.** 259 **9.** 2781 **10.** 44,838

3. Whole Numbers—Multiplication
   1. 98 **2.** 656 **3.** 1161 **4.** 2580 **5.** 56,100
   6. 8,822,000 **7.** 2816 **8.** 9576 **9.** 287,232
   10. 213,104

4. Whole Numbers—Division
   1. 8 **2.** 21 **3.** 25 **4.** 294 **5.** 40 **6.** 20 R11
   7. 677 R437 **8.** 311 R11 **9.** $\dfrac{11}{62}$ **10.** 6.67

5. Fractions—Basic Skills
   1. A **2.** 3 **3.** $\dfrac{3}{5}$ **4.** B **5.** $\dfrac{2}{3}$ **6.** B **7.** $\dfrac{4}{4}$ **8.** $\dfrac{4}{4}$
   9. 8, 12, 16 **10.** 24 **11.** $2 \times 2 \times 2 \times 5$
   12. 5 **13.** (a) $\dfrac{3}{9}$ (b) $\dfrac{3}{4}$ (c) $\dfrac{16}{5}$ **14.** 20
   15. (a) $1\dfrac{2}{3}$ (b) $\dfrac{9}{6}$

6. Fractions—Addition
   1. $\dfrac{3}{11}$ **2.** $7\dfrac{1}{4}$ **3.** $7\dfrac{1}{6}$ **4.** $\dfrac{8}{15}$ **5.** $7\dfrac{1}{2}$

7. Fractions—Subtraction
   1. $\dfrac{1}{4}$ **2.** $5\dfrac{1}{2}$ **3.** $\dfrac{5}{8}$ **4.** $2\dfrac{1}{3}$ **5.** $\dfrac{11}{21}$ **6.** $9\dfrac{13}{15}$

8. Fractions—Multiplication
   1. $\dfrac{1}{15}$ **2.** $\dfrac{1}{9}$ **3.** 9 **4.** $2\dfrac{47}{64}$ **5.** $23\dfrac{1}{10}$

9. Fractions—Division
   1. 6 **2.** 8 **3.** $5\dfrac{1}{3}$ **4.** $2\dfrac{1}{10}$ **5.** $2\dfrac{5}{18}$ **6.** $1\dfrac{13}{15}$

10. Decimals
    1. 4.004 **2.** 0.0007 **3.** $\dfrac{802}{1000}$ **4.** 4.4 **5.** 64.4
    6. 17.686 **7.** 553.616 **8.** 647.23 **9.** 108.836
    10. 10.488 **11.** 420.2

11. Percents
    1. 140 **2.** 77% **3.** 260.4% **4.** 450%
    5. $1\dfrac{47}{50}$ **6.** 0.375 **7.** 1.3 **8.** $28\dfrac{4}{7}\%$ **9.** 20%
    10. 170

Management System

The following management system outlines suggested schedules and assignments for completing the course. It is intended to provide an overview of teaching and testing which we hope will make your planning easier.

For your convenience, the schedule has been divided into both a semester and a trimester program. This schedule also suggests the number of days you might spend on each chapter, but of course, you should be guided by the needs of your students.

The assignment guide outlines a suggested sequence of assignments for both an average and a comprehensive course. The average course covers all the material required to progress through the text. The comprehensive course treats the material in more depth and provides more challenging work with somewhat less drill.

| ← Semester 1 → | | | | | | ← Semester 2 → | | | | | | |
Chapter	1	2	3	4	5	6	7	8	9	10	11	12	
Number of Days	8	16	16	14	9	17	16	11	17	12	10	14	Total: 160
← Trimester 1 →				← Trimester 2 →				← Trimester 3 →				

	TEACHING		TESTING		
	Average	**Comprehensive**	**Student Text**	**Extra Tests (Teacher's Edition)**	**Progress Tests (Ancillary)**
1-1	**3–4**/all	**3–4**/all; Application			
1-2	**7**/1–8	**7**/all			
1-3	**11**/all	**11–12**/all; Application	p. 13		Test 1A
1-4	**15**/1–3	**15–16**/all; Extra!			
1-5	**18**/all; Chapter Review	**18**/all; Chapter Review; Displaying Information	p. 19		Test 1B
			p. 24	p. T24	Chapter 1 Test
	25/Skill Review	**25**/Skill Review 1–70 odd			
2-1	**29–30**/all	**29–30**/all; Extra!			
2-2	**33**/all	**33**/all			
2-3	**36**/all	**36**/1–16 odd, 17–31			

	TEACHING		TESTING		
	Average	**Comprehensive**	**Student Text**	**Extra Tests (Teacher's Edition)**	**Progress Tests (Ancillary)**
2-4	**38–39**/1–22	**38–39**/1–10 odd, 11–26	p. 39		Test 2A
2-5	**41**/all	**41**/all; Application			
2-6	**44**/1–22	**44**/1–20 odd, 21–27			
2-7	**46–47**/all; Problems 1–2	**46–47**/odd; Problems 1–2			
2-8	**49–50**/1–15; Problems 1–2	**49–50**/1–18; Problems 1–2; Application			
2-9	**53**/all; Problems 1–2	**53–54**/1–16 odd, 17–28; Problems 1–4; Application			
2-10	**56**/1–6; Chapter Review	**56**/all; Chapter Review; The Binary System	p. 57		Test 2B
			p. 62	p. T25	Chapter 2 Test
	63/Skill Review	**63**/Skill Review 1–46 odd			
3-1	**69**/1–38	**69**/1–36 odd, 37–46			
3-2	**72–73**/1–28	**72–73**/1–22 odd, 23–32; Extra!			
3-3	**76–77**/all; Problems 1–2	**76–77**/all; Problems 1–2; Application			
3-4	**79–80**/all; Problems 1–2	**79–81**/all; Problems 1–2; Extra!	p. 81		Test 3A
3-5	**83–84**/all; Problems 1–2	**83–84**/1–8 odd, 9–30; Problems 1–2; Extra!			
3-6	**86–87**/all; Problems 1–2	**86–87**/all; Problems 1–2			
3-7	**90**/1–28	**90**/all			
3-8	**92**/1–23; Problems 1–2	**92**/1–16 odd, 17–30; Problems 1–2			

TEACHING		TESTING		
Average	Comprehensive	Student Text	Extra Tests (Teacher's Edition)	Progress Tests (Ancillary)
3-9 **96**/1–20; Problems 1–2; Chapter Review	**96**/1–8 odd, 9–22; Problems 1–2; Extra!; Chapter Review; Scientific Notation	p. 97		Test 3B
		p. 102	p. T26	Chapter 3 Test
103/Cumulative Review	**103**/Cumulative Review			
				Cumulative Test
4-1 **108**/all	**108**/all			
4-2 **111**/all	**111**/1–16 odd, 17–32			
4-3 **114**/all	**114**/1–20 odd, 21–30			
4-4 **118–119**/1–18 Problems 1–2	**118–119**/1–12 odd, 13–22; Problems 1–4; Extra!	p. 120		Test 3A
4-5 **123**/all	**123**/all			
4-6 **126–127**/all Problems 1–3	**126–128**/1–24 odd, 25–28; Problems 1–5; Extra!			
4-7 **130–131**/1–20; Problems 1–2	**130–131**/all; Problems 1–2			
4-8 **134–135**/1–42; Chapter Review	**134–136**/1–26 odd, 27–45; Extra!; Chapter Review; Nonrepeating Decimals	p. 137		Test 4B
		p. 142	p. T27	Chapter 4 Test
143/Skill Review	**143**/Skill Review 1–52 odd			
5-1 **148**/all; Problems 1–2	**148**/all; Problems 1–4			
5-2 **151–152**/all; Problems 1–3	**151–152**/1–24 odd; Problems 1–5; Extra!	p. 152		Test 5A

	TEACHING		TESTING		
	Average	Comprehensive	Student Text	Extra Tests (Teacher's Edition)	Progress Tests (Ancillary)
5-3	**156–157**/all; Problems 1–5	**156–157**/1–18 odd, 19–28; Problems 3–12; Application			
5-4	**159–160**/1–10	**159–160**/1–10 odd, 11–13			
5-5	**162–163**/1–13; Chapter Review	**162–164**/all; Chapter Review; The Golden Ratio	p. 165		Test 5B
			p. 170	p. T28	Chapter 5 Test
	171/Skill Review	**171**/Skill Review 1–58 odd			
6-1	**175–176**/1–26	**175–176**/1–16 odd, 17–27; Extra!			
6-2	**179**/all	**179**/all; Application			
6-3	**183**/all	**183–184**/all; Extra!			
6-4	**186–187**/1–12	**186–188**/all; Extra!	p. 189		Test 6A
6-5	**191–192**/1–14, 15–20 odd	**191–193**/all; Extra!			
6-6	**195–196**/1–26	**195–196**/all			
6-7	**199–200**/1–15	**199–200**/all; Extra!			
6-8	**203–204**/1–16	**203–205**/all; Extra!			
6-9	**208**/all; Chapter Review	**208**/all; Chapter Review: Mirror Geometry	p. 209		Test 6B
			p. 214	p. T29	Chapter 6 Test
	215/Cumulative Review	**215**/Cumulative Review			
					Cumulative Test Mid-Year Test
7-1	**218–219**/all	**218–219**/all; Extra!			
7-2	**221**/1–24	**221**/1–20 odd, 21–28			

| | TEACHING | | TESTING | | |
	Average	Comprehensive	Student Text	Extra Tests (Teacher's Edition)	Progress Tests (Ancillary)
7-3	**223-224**/all	**223-224**/all			
7-4	**226**/1–25	**226**/1–15 odd, 16–30			
7-5	**228-229**/1–20	**228-229**/1–12 odd, 13–24; Extra!	p. 230		Test 7A
7-6	**232-233**/1–10, 11–22 odd	**232-233**/1–10 odd, 11–28; Application			
7-7	**236**/1–15	**236**/1–12 odd, 13–22			
7-8	**238-239**/1–16; Chapter Review	**238-239**/1–14 odd, 15–22; Chapter Review; Secret Codes	p. 239		Test 7B
			p. 244	p. T30	Chapter 7 Test
	245/Skill Review	**245**/Skill Review			
8-1	**248**/1–14	**248**/all			
8-2	**250-251**/1–14	**250-251**/all			
8-3	**253**/1–14	**253**/1–14 odd, 15–23	p. 254		Test 8A
8-4	**256**/1–8	**256-257**/1–6 odd, 7–13			
8-5	**259-260**/1–12	**259-260**/1–10 odd, 11–17			
8-6	**261-262**/1–6; Chapter Review	**261-262**/all; Chapter Review; Mathematics A Century Ago	p. 263		Test 8B
			p. 268	p. T31	Chapter 8 Test
	269/Skill Review	**269**/Skill Review 1–42 odd, 43–50			
9-1	**273-274**/all; Problems 1–7	**273-274**/all; Problems 1–7			
9-2	**277-278**/1–19	**277-278**/1–10 odd, 11–24			

	TEACHING		TESTING		
	Average	Comprehensive	Student Text	Extra Tests (Teacher's Edition)	Progress Tests (Ancillary)
9-3	**279–280**/1–22	**279–281**/1–12 odd, 13–28	p. 281		Test 9A
9-4	**283–284**/1–15	**283–284**/1–9 odd, 10–17			
9-5	**287–288**/1–15	**287–289**/all; Application			
9-6	**292**/all; Problems 1–4	**292**/all; Problems 1–6			
9-7	**295**/all; Problems 1–2	**295**/1–19 odd, 20–24; Problems 1–5			
9-8	**297**/1–6; Chapter Review	**297–298**/all; Extra!; Chapter Review; Accuracy of Measurement	p. 299		Test 9B
			p. 304	p. T32	Chapter 9 Test
	305/Cumulative Review	**305**/Cumulative Review			
					Cumulative Test
10-1	**308–309**/all	**308–309**/all			
10-2	**312**/all	**312**/odd			
10-3	**315**/1–12, 13–32 odd	**315–317**/1–12 odd, 13–32; Extra!	p. 317		Test 10A
10-4	**319**/all	**319**/all			
10-5	**321–322**/1–7	**321–322**/all			
10-6	**325–326**/all; Chapter Review	**325–326**/all; Chapter Review; Random Experiments	p. 327		Test 10B
			p. 332	p. T33	Chapter 10 Test
	333/Skill Review	**333**/Skill Review 1–42 odd			
11-1	**338–339**/all	**338–339**/1–16 odd, 17–28; Application			

	TEACHING		TESTING		
	Average	Comprehensive	Student Text	Extra Tests (Teacher's Edition)	Progress Tests (Ancillary)
11-2	**341–342**/1–28	**341–342**/all; Extra!	p. 343		Test 11A
11-3	**346–347**/1–16, 17–25 odd	**346–347**/1–10 odd, 11–34; Extra!			
11-4	**349–350**/all	**349–351**/all; Application			
11-5	**353**/all; Chapter Review	**353**/all; Chapter Review; Locating Points on Earth	p. 355		Test 11B
			p. 360	p. T34	Chapter 11 Test
	361/Skill Review	**361**/Skill Review 1–60 odd			
12-1	**364**/1–16, 17–28 odd	**364**/1–16 odd, 17–28			
12-2	**366**/1–16, 17–32 odd	**366**/1–12 odd, 13–32			
12-3	**368**/1–12, 13–30 odd	**368–369**/1–12 odd, 13–38; Extra!	p. 370		Test 12A
12-4	**373–374**/1–20; Problems 1–4	**373–374**/1–14 odd, 15–24; Problems 1–5; Extra!			
12-5	**377–378**/1–15; Problems 1–4	**377–378**/1–12 odd, 13–18; Problems 1–7			
12-6	**381**/all	**381**/all			
12-7	**383**/odd; Chapter Review	**383**/all; Chapter Review; John Napier and Napier's Bones	p. 385		Test 12B
			p. 390	p. T35	Chapter 12 Test
	391/Cumulative Review	**391**/Cumulative Review			
					Cumulative Test Final Test

NOTES

Extra Chapter Tests

Chapter 1

Exercises 1 and 2 refer to the table at the right.

1. What was the population of New York in 1950?

2 What was the total population of the four states in 1900?

POPULATION, in Thousands			
State \ Year	1800	1900	1950
North Carolina	478	1894	4062
New York	589	7269	14,830
Pennsylvania	602	6302	10,498
Virginia	880	1854	3319

[1-1]

Exercises 3 and 4 refer to the bar graph at the right.

3. What is the speed of the squirrel?

4. How much faster can a giraffe run than an elephant?

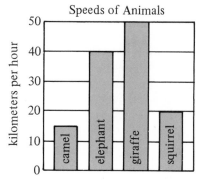

Speeds of Animals

[1-2]

Exercises 5 and 6 refer to the line graph at the right.

5. What is the average temperature in January?

6. In which month is the average temperature 11°C?

Average Monthly Temperature in Philadelphia

[1-3]

Exercises 7–10 refer to the table at the right.

7. Organize the data into intervals of 21–24, 25–28, 29–32, 33–36, and 37–40. Draw a histogram.

8. How many plants produced more than 28 tomatoes?

9. What is the mean of the data?

10. What is the mode of the data?

[1-4]

Number of Tomatoes Picked from 16 Plants			
27	24	22	29
36	34	31	27
21	39	27	21
33	24	26	27

[1-5]

Chapter 2

1. Write 10,987 in expanded form, using exponents. [2-1]

2. Write the decimal numeral for the following:
   $(4 \times 1000) + (2 \times 100) + (0 \times 10) + 3$.

3. Write 1.569 in expanded form. [2-2]

4. Write 4.004 in words.

5. Draw a number line. Graph 0.1, 0.5, 0.9, 1.2. [2-3]

6. 0.987 __?__ 0.9708
   $<, >$

Evaluate the expression for the given replacement set.

7. $31 - n$; {3, 14, 29}   8. $12n$; {6, 12, 20} [2-4]

Write a related equation. Find the value of n.

9. $n + 24 = 7$ 10. $13n = 169$ [2-5]

Replace each variable to make a true equation.

11. $9 \times n = 0$ 12. $(4 + 6) + 2 = n + (6 + 2)$ [2-6]

Add or subtract.

13. $103.9 + 32.188$ 14. $90.02 - 51.976$ [2-7]

Multiply.

15. 13.33×4.97 16. 0.08×0.2655 [2-8]

Divide. If the division is not exact, round the quotient to the nearest hundredth.

17. $197.372 \div 1.96$ 18. $4.9968 \div 0.321$ [2-9]

Solve.

19. Eight cans of beans measure 3.632 kg. What is the measure of each can? [2-10]

20. The following amounts of rain fell on Prairie City during April, May, June, and July: 8.4 cm, 7 cm, 11.9 cm, and 7.6 cm. How much rain is that in all?

Chapter 3

1. Show the graph of the following numbers on a number line: 1, ‾2, ‾3, 4.   [3-1]

2. Write the opposite of each number.

   a. ‾20   **b.** ‾1000   **c.** 64

3. On a number line, draw an arrow starting at ‾1 to represent 5.   [3-2]

4. On a number line, draw an arrow ending at ‾1 to represent ‾2.

Add.

  5. ‾715 + ‾320   **6.** ‾1.92 + ‾10.1   [3-3]

Add.

  7. ‾0.9 + 3.4   **8.** 29 + ‾92   [3-4]

Subtract.

  9. 10 − ‾2   **10.** ‾17 − 44   [3-5]

Subtract.

 11. ‾6.6 − ‾3.9   **12.** 16.9 − 32.1   [3-6]

Multiply.

 13. −8(−12)   **14.** 43(−32)   [3-7]

Multiply.

 15. −1.1(7.18)   **16.** −9.9(−4.5)   [3-8]

Divide.

 17. −13.9 ÷ (−2.78)   **18.** −19.691 ÷ 2.03   [3-9]

Chapter 4

Complete.

1. $1 \div 6 = \underline{\ ?\ }$ **2.** $3 \times \underline{\ ?\ } = \dfrac{3}{4}$ **3.** $4\dfrac{3}{5} = \dfrac{?}{5}$ [4-1]

Write each negative fraction in two other ways.

4. $-\dfrac{3}{10}$ **5.** $\dfrac{-6}{11}$ [4-2]

Complete.

6. $-1\dfrac{13}{16} = \dfrac{?}{16}$ **7.** $-\dfrac{50}{7} = \underline{\ ?\ }$ [4-3]

Reduce each fraction to lowest terms.

8. $\dfrac{75}{125}$ **9.** $-\dfrac{154}{56}$ **10.** $\dfrac{92}{108}$

Compute each product. Give your result in lowest terms.

11. $\dfrac{27}{12} \times \dfrac{8}{9}$ **12.** $-\dfrac{16}{35} \times \left(-\dfrac{19}{24}\right)$ **13.** $-9\dfrac{2}{3} \times 3\dfrac{3}{4}$ [4-4]

Replace each set fractions with equal fractions having the least common denominator (LCD).

14. $-\dfrac{4}{5}, -\dfrac{3}{2}$ **15.** $\dfrac{7}{18}, \dfrac{9}{22}$ **16.** $-\dfrac{31}{28}, -\dfrac{26}{21}, -\dfrac{5}{12}$ [4-5]

Express the sum or difference as a proper fraction in lowest terms or as a mixed number with fraction in lowest terms.

17. $\dfrac{7}{8} + \dfrac{1}{12}$ **18.** $-3\dfrac{5}{6} + \left(-4\dfrac{4}{15}\right)$ **19.** $\dfrac{19}{20} - \dfrac{19}{25}$ [4-6]

Divide. Express all results in lowest terms.

20. $-\dfrac{3}{4} \div \left(-\dfrac{8}{3}\right)$ **21.** $\dfrac{12}{17} \div \dfrac{48}{51}$ **22.** $11\dfrac{1}{10} \div \left(-1\dfrac{2}{25}\right)$ [4-7]

Express as a fraction in lowest terms.

23. 0.175 **24.** -6.88 [4-8]

Express as a terminating or repeating decimal.

25. $-\dfrac{3}{80}$ **26.** $\dfrac{32}{15}$

Chapter 5

Write the ratio as a fraction in lowest terms.

1. 20 to 56

2. 2 weeks to 10 days

[5-1]

3. Tell whether or not the proportion $\frac{6}{9} = \frac{36}{54}$ is correct.

[5-2]

4. Solve the proportion $\frac{9}{45} = \frac{n}{65}$.

Find the percent or number.

5. 6 is 12% of what number?

6. What is 52% of 250?

[5-3]

Solve.

7. Last year Sandi Rhodes earned $23,575 from her home-decorating service. This year she earned 8% less. How much did she earn this year?

[5-4]

8. Center City had a population of 75,250 ten years ago. The population is presently 70,735. What was the percent of decrease in the population in the past ten years?

9. The Chin family spends an average of $55 per week on food. The yearly family income is $17,875. What percent of their yearly income is spent on food?

[5-5]

10. Lawn mowers are on sale at a discount of 25% of their original price of $287.00. What is the sale price of a lawn mower?

Chapter 6

1. Which has two endpoints, a line or a line segment? [6-1]

2. How many endpoints does a ray have?

Complete.

3. If the diameter of a circle is 91 cm, the circumference [6-2]
   is __?__. $\left(\text{Use } \pi \approx \dfrac{22}{7}.\right)$

4. If the circumference of a circle is 94.2 m, the radius
   is __?__. (Use $\pi \approx 3.14$.)

5. The measure of a(n) __?__ angle is between 90° and 180°. [6-3]

6. The sum of two supplementary angles is __?__.

7. If two lines form a right angle, they are said to be __?__.

8. Two lines are cut by a transversal. If the lines form [6-4]
   equal corresponding angles with the transversal, the
   lines are __?__.

9. Line *l* is parallel to line *m*. We abbreviate this by
   writing __?__.

10. Two parallel lines are cut by a third line. Two alternate
    interior angles measure 65° each. The other two measure
    __?__ each.

11. An obtuse triangle has __?__ acute angles. [6-5]

12. A(n) __?__ triangle has at least two equal sides.

13. If one angle of a rhombus measures 100°, the angle [6-6]
    opposite it measures __?__.

14. A parallelogram with four right angles is called a __?__.

15. A __?__ is a four-sided polygon. [6-7]

16. If the perimeter of an equilateral hexagon is 96 cm,
    the length of one side is __?__.

17. $\triangle GHI \cong$ __?__. [6-8]

18. The triangles are congruent
    by __?__.

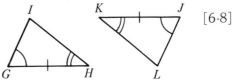

19. Corresponding __?__ of similar polygons are equal. [6-9]

20. If two polygons are congruent, are they also similar?

Chapter 7

1. Write a variable expression for the word phrase "the sum of sixteen and a number n." [7-1]

2. Write a word phrase for the numerical expression $4(8 + x)$.

Evaluate each expression if $x = 4$.

3. $9(9 - x)$ 4. $\dfrac{7}{x + 10}$ [7-2]

5. Translate into a number sentence: "The product when y is multiplied by negative two is less than or equal to zero."   [7-3]

6. Write a word sentence for the number sentence $3t + 6 > 10$.

Tell whether or not the given number is a solution of the given open sentence.

7. $\dfrac{n}{4} < 0;\ -1$ 8. $x + 6 = -4;\ -2$ [7-4]

Graph the solution set. The replacement set for x is $\{-2, -1, 0, 1, 2\}$.

9. $x + 1 \geq 0$ 10. $2 - x = 0$ [7-5]

Use one of the properties of equalities to form a true sentence.

11. If $y = -5$, then $-4y = \underline{\ ?\ }$.   [7-6]

12. If $t = 4$, then $t - 8 = \underline{\ ?\ }$.

Solve each equation and check your solution.

13. $6x = 8 + 5 - 1$ 14. $x + 7 = 17$ [7-7]

Solve. The replacement set for x is {the integers}.

15. $x + 10 > 20$ 16. $5x \leq 45$ [7-8]

Chapter 8

Write two practical problems that are expressed by the given
equation. For the first, use a simple, direct question.

1. $x + 9 = 17$ 2. $4x = 32$ [8-1]

Write an equation which can be used to solve the problem.

3. Find two consecutive integers whose sum is 67.   [8-2]

4. The perimeter of a rectangle is 32 cm. The length is
2 cm longer than the width. Find the length and width
of the rectangle.

Solve.

5. The unequal side of an isosceles triangle is 7 cm shorter   [8-3]
than the two equal sides. The perimeter of the triangle
is 41 cm. Find the length of each side.

6. One number is 9 more than twice another number. The sum
of the two numbers is 33. What are the numbers?

7. A map is drawn to the scale 1 cm = 20 km. If the distance   [8-4]
between Newburgh and Forest Haven measures 3.5 cm on the
map, what is the approximate actual distance between the
two towns?

8. A scale drawing of a rectangular table top has scale
1 cm = 0.25 m. If the table top measures 1.75 m by 2 m,
find the dimensions on the scale drawing.

9. If you invest $3000 compounded semiannually at 6%, how much   [8-5]
would you have at the end of one year?

10. Sammy Wong paid $247 interest on two loans in one year.
If one loan was for $950 at 12%, for how much was the
other loan at 9.5%?

11. For each kilogram of mass, a bicycle rider expends about   [8-6]
730 J of energy per kilometer. If a 68 kg rider pedals
a 10 kg bicycle 15 km, how many joules are expended?

12. Lee County will spend 2% of its annual budget on public
transportation and 0.9% to remodel the library, a total of
$9,007,400. Five years ago, $235,750 was spent to build a
new addition onto the library. What is the annual budget
of Lee County?

Chapter 9

For each parallelogram, give the area and perimeter.   [9-1]

1.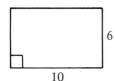

2.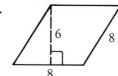

3. The area of a triangle measures 42 cm². If the base   [9-2]
measures 12 cm, what is the measure of its height?

4. The height of a trapezoid measures 10 cm. Its bases are
16 cm and 19.5 cm long. What is the area of the trapezoid?

5. What is the area of a circle with radius 3.5? (Use $\pi \approx 3.14$.)   [9-3]

6. If a circle has area 154 cm², what is its radius? $\left(\text{Use } \pi \approx \dfrac{22}{7}.\right)$

7. $6 \text{ m}^3 = \underline{} \text{ cm}^3$ [9-4]

8. Find the volume of a rectangular prism with length 17 cm,
width 8 cm, and height 10 cm.

9. The base of a pyramid is a square with sides 9 cm. If the   [9-5]
height of the pyramid is 12 cm, what is its volume?

10. A cone and a cylinder have equal radii and equal heights.
If the volume of the cone is 90 cm³, what is the volume
of the cylinder?

11. Find the lateral area and the total surface area (in terms   [9-6]
of π) of a cylinder with height 11.5 cm and radius 8 cm.

12. Each edge of a cube measures 6 cm. What is the total
surface area of the cube?

13. The mass of 1 L of milk measures 1.03 kg. What is the   [9-7]
mass of 500 mL of milk?

14. The mass of 1 cm³ of ice measures 0.9 g. What is the
mass of 2000 cm³ of ice?

15. Find the area (in terms of π) of a sphere with radius 8.5.   [9-8]

16. Find the volume (in terms of π) of a sphere with diameter 12.

Chapter 10

1. If a card is drawn at random from a standard deck of 52 playing cards, how many outcomes are possible? [10-1]

2. A letter from the alphabet is chosen at random. The probability of choosing the letter a is __?__.

3. An impossible event has probability __?__. [10-2]

4. A card is drawn at random from a standard deck of 52 playing cards. The probability that it is not a number card is __?__.

5. The probability that the toss of an ordinary die will turn up showing a number of dots which is a multiple of 2 is __?__.

6. Thirteen slips of paper are labeled with the letters of the word *parallelogram*. If you choose a slip at random, the probability that you will draw the letter l or r is __?__.

7. The probability of an event is $\frac{1}{6}$. The odds against it are __?__. [10-3]

8. You toss a dime and a nickel. The odds in favor of exactly one coin landing heads up are __?__.

9. The Trojans have won 15 of the 24 games played against the Vikings. The probability that the Vikings will win the next game against the Trojans is __?__. [10-4]

10. Of the 75 buses passing a checkpoint, 27 carried less than 40 passengers, the rest more than 40 passengers. The probability that the next bus will carry more than 40 passengers is __?__.

11. A legislator sends the voters in his district a questionaire to be completed and returned. Will the experiment produce a random sample? [10-5]

12. A random survey of 50 households in Oakview indicates that 22 own at least one luxury car. Of the 400 households in Oakview, about how many own a luxury car?

13. The number cards less than 7 from a standard deck of 52 playing cards are shuffled together. If V is the face value of a card chosen at random from these cards, what is the expected value of V? [10-6]

14. Sixteen discs are placed in a box. Five are marked with the number 1, five with the number 2, three with the number 3, and three with the number 4. If a disc is chosen at random, the expected value of the score is __?__.

Chapter 11

1. We associate the second component of an ordered pair of numbers with the __?__-axis.   [11-1]

2. The positive direction on the horizontal axis is to the __?__.

3. The vertices of a parallelogram have coordinates $(1, -4)$, $(-2, -4)$, $(-2, 3)$, and $(1, 3)$. The area of this figure equals __?__.

4. The y-coordinate of a point P is called the __?__ of P.   [11-2]

5. If $x < 0$ and $y > 0$, then (x, y) are the coordinates of a point in Quadrant __?__.

6. If a point has abscissa 0, it lies on the __?__.

7. If (a, b) are the coordinates of a point in Quadrant IV, then $(a, -b)$ are the coordinates of a point in Quadrant __?__.

8. Is $(-4, -2)$ a solution of the equation $x - y = -6$?   [11-3]

9. A solution of the equation $4x + 3y = 6$ is (__?__, 6).

10. A solution of the equation $x + 6y = -6$ whose graph lies on the x-axis is __?__.

11. The coordinates of the solutions of the equation $x + y = 0$ lie in Quadrants __?__ and __?__.

12. The graph of a linear equation in two variables is always a __?__ in the plane.   [11-4]

13. The equation $5x - y = 25$ intersects the y-axis at the point whose coordinates are __?__.

14. The graph of $x = 6$ is parallel to the __?__-axis.

15. The graph of $2x - y = 8$ does *not* pass through Quadrant __?__.

16. The graphs of two linear equations are perpendicular. The system of equations has __?__ solution(s).   [11-5]

17. The system of equations $x - y = 3$ and $3x - 3y = 8$ has no solutions. The graphs of the equations are __?__.

18. Is $(1, -1)$ a solution of the system? $x + 5y = -4$
$2x + 3y = -1$

Chapter 12

1. $-\sqrt{100} = \underline{\quad ? \quad}$ [12-1]

2. $\sqrt{20}$ lies between which two consecutive integers?

3. Using the divide and average method, $\sqrt{67} = \underline{\quad ? \quad}$ [12-2]
to the tenths' place.

4. Using the divide and average method, $\sqrt{92.3} = \underline{\quad ? \quad}$
to the tenths' place.

5. Using the table on page 394, $5\sqrt{28} = \underline{\quad ? \quad}$ to the   [12-3]
nearest hundredth.

6. Using the table on page 394, and interpolating,
$\sqrt{41.7} = \underline{\quad ? \quad}$ to the nearest hundredth.

7. A triangle has sides of lengths 4, 8, and 9. Is it a   [12-4]
right triangle?

8. The diagonal of a rectangle measures 20 cm. One side
measures 12 cm. What is the length of the other side?

9. How long is the shorter leg of a 30°-60° right triangle   [12-5]
whose hypotenuse is 11 cm long?

10. Use the table on page 394 to approximate
the value of x to the nearest tenth.

11. $\tan B = \underline{\quad ? \quad}$ [12-6]

12. $\cos 30° = \underline{\quad ? \quad}$

13. Using the table on page 395, find   [12-7]
the length of y to the nearest whole
number.

14. To the nearest degree, $\angle B = \underline{\quad ? \quad}$.

ANSWERS TO EXTRA CHAPTER TESTS

Chapter 1
1. 14,830,000   **2.** 17,319,000   **3.** 20 km/h
4. 10 km/h faster   **5.** 0°C   **6.** April

7.

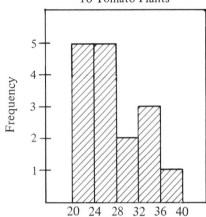

Tomatoes Picked from
16 Tomato Plants

8. 6   **9.** 28   **10.** 27

Chapter 2
1. $(1 \times 10^4) + (9 \times 10^2) + (8 \times 10^1) + 7$
2. 4203
3. $1 + (5 \times 0.1) + (6 \times 0.01) + (9 \times 0.001)$
4. Four and four-thousandths

5.

6. $>$ **7.** 28, 17, 2   **8.** 72, 144, 240
9. $n = 7 - 24$; $n = -17$
10. $n = 169 \div 13$; $n = 13$ **11.** $n = 0$
12. $n = 4$ **13.** 136.088   **14.** 38.044
15. 66.2501   **16.** 0.02124   **17.** 100.7
18. 15.57   **19.** 0.454 kg   **20.** 34.9 cm

Chapter 3
1.

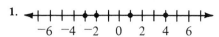

2. a. 20   **b.** 1000   **c.** $^-64$

3.

4.

5. $^-1035$ **6.** $^-12.02$ **7.** 2.5   **8.** $^-63$
9. 12   **10.** $^-61$ **11.** $^-2.7$ **12.** $^-15.2$
13. 96   **14.** -1376 **15.** -7.898
16. 44.55   **17.** 5   **18.** -9.7

Chapter 4
1. $\dfrac{1}{6}$ **2.** $\dfrac{1}{4}$ **3.** 23   **4.** $\dfrac{-3}{10}, \dfrac{3}{-10}$
5. $-\dfrac{6}{11}, \dfrac{6}{-11}$ **6.** -29 **7.** $-7\dfrac{1}{7}$ **8.** $\dfrac{3}{5}$
9. $-\dfrac{11}{4}$ **10.** $\dfrac{23}{27}$ **11.** 2   **12.** $\dfrac{38}{105}$
13. $-36\dfrac{1}{4}$ **14.** $-\dfrac{8}{10}, -\dfrac{15}{10}$ **15.** $\dfrac{77}{198}, \dfrac{81}{198}$
16. $-\dfrac{93}{84}, -\dfrac{104}{84}, -\dfrac{35}{84}$ **17.** $\dfrac{23}{24}$ **18.** $-8\dfrac{1}{10}$
19. $\dfrac{19}{100}$ **20.** $\dfrac{9}{32}$ **21.** $\dfrac{3}{4}$ **22.** $10\dfrac{5}{18}$
23. $\dfrac{7}{40}$ **24.** $-6\dfrac{22}{25}$ **25.** -0.0375 **26.** $2.1\overline{3}$

Chapter 5
1. $\dfrac{5}{14}$ **2.** $\dfrac{7}{5}$ **3.** Yes   **4.** $n = 13$
5. 50   **6.** 130   **7.** \$21,689   **8.** 6%
9. 16%   **10.** \$215.25

Chapter 6

1. line segment **2.** 1 **3.** 286 cm
4. 15 m **5.** obtuse **6.** 180°
7. perpendicular **8.** parallel **9.** $l \parallel m$
10. 115° **11.** 2 **12.** isosceles **13.** 100°
14. rectangle **15.** quadrilateral **16.** 16 cm
17. $\triangle JKL$ **18.** ASA **19.** angles **20.** Yes

Chapter 7

1. $16 + n$ **2.** Four times the sum of 8 and a number x.

3. 45 **4.** $\dfrac{1}{2}$ **5.** $-2y \le 0$ **6.** The sum of three times a number n and 6 is greater than 10.
7. Yes **8.** No

9.
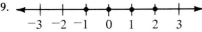

10.

11. 20 **12.** -4 **13.** {2} **14.** {10}
15. $\{x > 10\}$ **16.** $\{x \le 9\}$

Chapter 8

1. a. What number increased by 9 gives 17?
b. Answers may vary. For example, Jane needs $17 for a new calculator. She has saved $9. How much more does she need?
2. a. What number multiplied by 4 gives 32?
b. Answers may vary. For example, it costs Brendon $32 to commute to work each month. What is the average weekly cost?
3. $x + (x + 1) = 67$
4. $x + x + (x + 2) + (x + 2) = 32$
5. 16 cm, 16 cm, 9 cm **6.** 8 and 25 **7.** 70 km
8. 7 cm by 8 cm **9.** $3182.70 **10.** $1400
11. 854,100 J **12.** $310,600,000

Chapter 9

1. 60; 32 **2.** 48; 32 **3.** 7 cm
4. 177.5 cm² **5.** 38.465 **6.** 7 cm
7. 216,000,000 **8.** 1360 cm³ **9.** 324 cm³
10. 270 cm³ **11.** 184π cm²; 312π cm²
12. 216 cm² **13.** 0.515 kg **14.** 1800 g
15. 289π **16.** 288π

Chapter 10

1. 52 **2.** $\dfrac{1}{26}$ **3.** 0 **4.** $\dfrac{4}{13}$ **5.** $\dfrac{1}{2}$ **6.** $\dfrac{5}{13}$

7. 5 to 1 **8.** 1 to 1 **9.** $\dfrac{3}{8}$ **10.** $\dfrac{16}{25}$

11. No **12.** 176 **13.** 4 **14.** $2\dfrac{1}{4}$

Chapter 11

1. y **2.** right **3.** 21 **4.** ordinate **5.** II
6. y-axis **7.** I **8.** No **9.** -3
10. $(-6, 0)$ **11.** II and IV **12.** line
13. $(0, -25)$ **14.** y **15.** II **16.** 1
17. parallel **18.** Yes

Chapter 12

1. -10 **2.** 4 and 5 **3.** 8.2 **4.** 9.6
5. 26.46 **6.** 6.46 **7.** No **8.** 16 cm

9. 5.5 cm **10.** 9.9 **11.** $\dfrac{3}{4}$ **12.** $\dfrac{\sqrt{3}}{2}$

13. 28 **14.** 26°

Lesson Commentary

1-1 Reading Tables and Charts

Teaching Suggestions

Numerical information is frequently displayed in tables and charts. You need only glance through the morning newspaper or your favorite magazine to find examples. To motivate class discussions of tables and charts and their uses, cut out or copy a few tables from magazines, newspapers, or books. Try to choose tables from different subject areas. Ask students what types of information are shown in tables, why the writers used them, and why tables might be more effective than writing the information in sentence form.

Chalkboard Examples

Nutritive Value of Foods

Food	Calcium (mg)	Iron (mg)
Milk (454 g)	288	0.1
Liver (56.7 g)	6	5.0
Peas (454 g)	37	2.9

1. Which food is the best source of iron?
   liver
2. Which food is the best source of calcium?
   milk
3. An egg contains 10 mg less calcium and 1.8 mg less iron than 454 g of peas. How many milligrams of each mineral are in an egg? 27 mg calcium; 1.1 mg iron

Related Activities

Almanacs and reference books use tables in order to present large amounts of information as completely as possible. Have each student copy a table or chart from a reference book and write several questions about the information. Have a classmate refer to the table to answer the questions.

1-2 Bar Graphs

Teaching Suggestions

As you discuss the tables and bar graphs in the exposition with the students, ask what advantages and disadvantages tables and bar graphs have over one another. Although students are familiar with bar graphs, some may not really understand exactly how one is constructed. Reproduce a simple graph on the chalkboard. Explain each step: how to choose the intervals, how to draw and label the axes, and finally how to construct the bars.

Chalkboard Examples

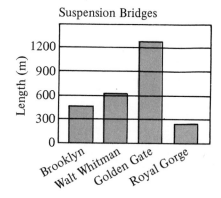

Suspension Bridges

1. Which bridge is shortest? Royal Gorge

2. Which bridge is half as long as the Golden Gate Bridge? Walt Whitman

3. About how long is the Brooklyn Bridge? 500 m

4. Can you tell exact lengths from the graph? no

Related Activities

Inquire about the material students are studying in other subjects, for example, social studies or science. Illustrate how an idea can be more easily conveyed through the use of bar graphs than through the use of tables of data.

1-3 Line Graphs

Teaching Suggestions

A broken-line graph highlights changes in data, or trends. For example, simply by looking at the first broken-line graph in the text, you have a general impression of the weather in Ottawa from April to November. Point out the use of a broken line on the horizontal axis in the graph for Exercises 1–7 to indicate a gap in the numbering for 0 to 90. This same technique is used in the graph below on the vertical axis to show a gap in the numbering for 0 to 50.

Chalkboard Examples

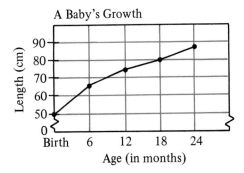

A Baby's Growth

1. How long was the baby at birth? 50 cm

2. When was the baby 80 cm long? 18 months

3. During which 6-month period did the length increase the most? birth to 6 months

4. About how much did the length increase from 1 year to 2 years? 13 cm (Answers may vary.)

Related Activities

Have the students take a survey of the class to determine favorite television programs, rock groups, or sports. Some may wish to record the outdoor temperature at 9:00 A.M. and 3:30 P.M. each Monday for two months. The students can collect their data and make graphs. Which kind of graph, a bar graph or a line graph, would be more useful for their data?

1-4 Making a Histogram

Teaching Suggestions

Bar graphs and line graphs were already familiar to students from earlier courses. Histograms are probably a new topic. It may, therefore, be advisable to work slowly through the example in the text with the students. Write the heights of some of the students on the chalkboard. Have the class help you make a frequency distribution table and histogram for the data.

Chalkboard Examples

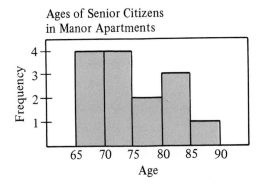

Ages of Senior Citizens in Manor Apartments

1. How many people are between 86 and 90 years old? 1

2. Which intervals have the same number of people? 66–70, 71–75

3. How many people are between 71 and 80 years old? 6

4. How many people are 77 years old? Can't tell

Related Activities

The students might enjoy performing the experiment of estimating the number of beans in a jar. (See "Extra! Another Broken-line Graph.") Other objects, such as paper clips, buttons, or raisins, can be substituted for the beans. Have the students make a frequency distribution table, histogram, and frequency polygon for their estimates.

1-5 Using Numbers to Describe Data

Teaching Suggestions

In this chapter, we are concerned with interpreting data. Bar graphs, line graphs, and histograms provide a visual summary of a set of data. Data may also be interpreted through numerical descriptions. These descriptions include the range, mean, median, and mode.

Most of the student difficulties with these concepts result from mixing up the terms mean, median, and mode and from careless computational errors. Emphasize that the first task is to organize the numbers in the set of data in increasing order. Once the numbers are arranged, it is relatively easy to find the range, mode, and median. The mode and sometimes the median require no computation. The range is simply a subtraction exercise. The mean is the only lengthly computation, involving both addition and division.

Chalkboard Examples

Arrange the data in order from least to greatest. Find the range, mean, median, and mode.

1. 9, 7, 4, 8, 7 5, 7, 7, 7

2. 18, 23, 21, 21, 17, 20 6, 20, $20\frac{1}{2}$, 21

3. 13, 9, 9, 12, 15, 13 13 6, 12, 13, 13

Related Activities

The students might enjoy this data-gathering activity. Each student tries ten free throws on the basketball court and notes the number of baskets made. The data for all students are recorded. The students then find the range, mean, median, and mode for this set of data.

2-1 Powers of Ten

Teaching Suggestions

Emphasize that 10^6 does not mean that 10 is multiplied by itself six times. Rather, the exponent 6 signifies the number of factors of 10, not the number of operations. You might show that there are actually five multiplications involved in finding 10^6. In addition, point out that the number of zeros in the product, 1,000,000, is the same as the exponent.

Remind the students that 10^2 and 10^3 are usually read "ten squared" and "ten cubed" respectively, instead of "ten to the second power" and "ten to the third power."

Chalkboard Examples

Express as a single power of 10.

1. $10^2 \times 10^1$ 10^3 **2.** $10^2 \times 10^2$ 10^4

3. $10^4 \times 10^2$ 10^6 **4.** $10^3 \times 10^3$ 10^6

5. Write the answers in Exercises 1–4 as decimal numerals 1000; 10,000; 1,000,000; 1,000,000

Write in expanded form, using exponents.

6. 172 $(1 \times 10^2) + (7 \times 10^1) + 2$

7. 2713 $(2 \times 10^3) + (7 \times 10^2) + (1 \times 10^1) + 3$

Related Activities

Give the students copies of puzzles like the one below. Have them follow the instruction, expressing each expanded form as a decimal numeral before performing any operations.

1. Write the decimal numeral for $(3 \times 10^3) + (7 \times 10^2) + (5 \times 10^1) + 1$. 3751

2. Add $(9 \times 10^4) + (4 \times 10^2) + (2 \times 10^1) + 9$. $3751 + 90,429 = 94,180$

3. Subtract $(4 \times 10^3) + (1 \times 10^2) + (8 \times 10^1)$. $94,180 - 4180 = 90,000$

4. Multiply by 10^4. Write your answer as a power of 10 and as a decimal numeral. $9 \times 10^8 = 900,000,000$

2-2 Decimals

Teaching Suggestions

When discussing the table of fractions and decimals, point out that the exponent in the fraction is the same as the number of digits, not zeros, to the right of the decimal point. In the case of whole numbers, the exponent is the same as the number of zeros in the number, for example, $10^3 = 1000$.

"And" should be used for the decimal point when reading decimals. Students sometimes incorrectly use "and" when reading whole numbers such as 105. You might also mention that in other countries, other devices are used instead of a decimal point. In Great Britain, a raised dot is used; in Germany, a comma is used.

Chalkboard Examples

Read each number.

1. 67.7 sixty-seven and seven tenths
2. 70.077 seventy and seventy-seven thousandths
3. 0.0099 ninety-nine ten-thousandths

Write in expanded form.

4. 9.04 $9 + (4 \times 0.01)$
5. 909.44 $(9 \times 100) + 9 + (4 \times 0.1) +$ (4×0.01)
6. 0.4994 $(4 \times 0.1) + (9 \times 0.01) +$ $(9 \times 0.001) + (4 \times 0.0001)$

Related Activities

Give the students the scrambled expanded forms below and have them write the decimal numeral.

1. $(4 \times 1000) + (4 \times 0.01) + (4 \times 10) + 4$
   4044.04
2. $(9 \times 0.01) + (9 \times 10) + (7 \times 0.1) + 7$
   97.79
3. $(6 \times 0.1) + (6 \times 0.0001) + 6$
   6.6006
4. $100 + (1 \times 0.001) + (1 \times 0.01)$
   100.011
5. $(7 \times 0.001) + (8 \times 0.01) + (6 \times 0.00001)$
   0.08706

2-3 The Number Line

Teaching Suggestions

Although the students are familiar with number lines, the terminology can sometimes be confusing. Stress that graphs are points, and coordinates are numbers. Explain that consecutive whole numbers are whole numbers in order without skipping, for example, 0, 1, 2, 3 or 7, 8, 9, 10.

Comparing decimals in Exercises 17–31, especially those with similar digits, is often difficult for students. Suggest that anyone having trouble arrange the decimals vertically with the decimal point aligned, as shown below. Look for the first place value in which the digits differ.

$$
\begin{array}{l}
4.392 \\
4.328 \\
\;\;\uparrow
\end{array}
$$
The digits first differ in the hundredths' place. Since $9 > 2$, then $4.392 > 4.328$.

Chalkboard Examples

Draw a number line and graph the numbers.

1. 3, 5, 6, 7

2. 0.4, 0.7, 0.2, 0.1

Write $>$ or $<$.

3. 1 $\underline{\;>\;}$ 0   4. 9 $\underline{\;<\;}$ 11   5. 21 $\underline{\;>\;}$ 12

6. 1.1 $\underline{\;>\;}$ 1.01   7. 0.06 $\underline{\;<\;}$ 0.6 $\underline{\;<\;}$ 6.0

Related Activities

The number lines pictured in this lesson have all shown the origin and used intervals of 1 or 0.1. Have the students graph the numbers below on the number lines shown.

1. 100, 125, 250, 350

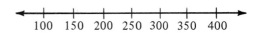

2. 0.4, 0.5, 0.6, 0.9

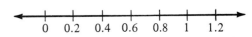

2-4 Equations and Variables

Teaching Suggestions
Make certain that the students understand the difference between an equation and an expression. Point out that an equation is formed by two expressions united by the symbol "=."

You might point out to the students that evaluating an expression can involve more than one operation, as in Exercises 21–26.

Chalkboard Examples
Is the equation true or false?

1. $6 \times 4 = 26$ F
2. $13 - 13 = 0$ T
3. $27 + 24 = 41$ F
4. $92 - 8$ Not an equation

Evaluate the expression if the replacement set is {9, 15, 27}.

5. $7 + x$ 6. $4x$ 7. $31 - x$
    16, 22, 34   36, 60, 108   22, 16, 4

Related Activities
Copy over the puzzle below. Ask the students to evaluate the expression and write the result in the puzzle, one digit per square.

ACROSS
a. $4n$; $n = 13$
b. $209 - x$; $x = 93$
c. $100 \div y$; $y = 5$
d. $44 + m$; $m = 45$
e. $17 + x$; $x = 17$
f. $238 \div y$; $y = 34$

DOWN
a. $100 - n$; $n = 50$
b. $10 + y$; $y = 9$
c. $x + 111$; $x = 179$
d. $7m$; $m = 12$
e. $3(x + 10)$; $x = 3$
f. $988 - 2n$; $n = 100$

a 5	2	b 1	1	6	
c 2	0	d 8	9	e 3	4	f 7
9	■	4	■	9	■	8
0					8

2-5 Inverse Operations

Teaching Suggestions
The concept of inverse operations is not new to the students. They have worked with inverse operations ever since they began to learn the basic addition/subtraction and multiplication/division facts. First review the use of inverse operations in basic facts. Then write the same facts with a variable for one addend or factor, as shown below.

a. $4 + 5 = 9$ $n + 5 = 9$
    $4 = 9 - 5$ $n = 9 - 5 = 4$

b. $8 \times 2 = 16$ $8 \times y = 16$
    $2 = 16 \div 8$ $y = 16 \div 8 = 2$

Chalkboard Examples
Complete.

1. If $n + 9 = 13$, then $n = 13 - \underline{9}$ and $n = \underline{4}$.
2. If $x + 1 = 7$, then $x = 7 - \underline{1}$ and $x = \underline{6}$.
3. If $y + 16 = 20$, then $\underline{y = 20 - 16}$ and $y = 4$.
4. If $9 \times n = 81$, then $n = 81 \div \underline{9}$ and $n = \underline{9}$.
5. If $6 \times x = 18$, then $x = 18 \div \underline{6}$ and $x = \underline{3}$.
6. If $10 \times y = 90$, then $\underline{y = 90 \div 10}$ and $y = 9$.

Related Activities
Ask the students to write a related subtraction equation and then a related division equation to find the value of n in the equations below. First, discuss the worked-out example at the left below.

EXAMPLE: | Find the value of n.

$(2 \times n) + 4 = 6$ 1. $(3 \times n) + 1 = 7$
   $(2 \times n) = 6 - 4$ 2. $(2 \times n) + 5 = 13$
   $(2 \times n) = 2$ 3. $(4 \times n) + 3 = 19$
       $n = 2 \div 2$ 4. $(5 \times n) + 2 = 17$
       $n = 1$ 5. $(6 \times n) + 1 = 31$

2-6 Properties of Addition and Multiplication

Teaching Suggestions
The students have previously learned and applied all the properties in this lesson, although some may have forgotten. Discuss the properties one at a time, asking the students to give an example of each.

Some students may confuse the Associative Property and the Distributive Property since both properties contain parentheses and look similar. Point out that the full name is the Distributive Property of Multiplication over Addition and that the property involves two different operations.

Chalkboard Examples
Use a property of addition or multiplication to correct each equation.

1. $2 + 4 = 4 \times 2$ 2. $7 \times (4 + 3) = 11 + 10$
   $2 + 4 = 4 + 2$ $7 \times (4 + 3) = 28 + 21$
3. $5 \times 0 = 5$ 4. $6 \times 1 = 1$
   $5 \times 0 = 0$ $6 \times 1 = 6$
5. $9 + 0 = 0$ 6. $(5 \times 5) \times 2 = 5 \times (5 + 2)$
   $9 + 0 = 9$ $(5 \times 5) \times 2 = 5 \times (5 \times 2)$

Related Activities
Suppose we invent a new set of numbers. There are three symbols in our set, ★, ▲, and ✛, and one operation shown by ●. The table below shows the basic facts for our new set of numbers. Have the students use the table to complete the statements.

1. ★ ● ▲ = $\underline{▲}$
   ▲ ● ★ = $\underline{▲}$

●	★	▲	✛
★	★	▲	✛
▲	▲	✛	★
✛	✛	★	▲

2. ★ ● ✛ = $\underline{✛}$
   ✛ ● ★ = $\underline{✛}$

3. ▲ ● ✛ = $\underline{★}$
   ✛ ● ▲ = $\underline{★}$

4. Is the operation commutative? yes

2-7 Adding and Subtracting Decimals

Teaching Suggestions
Students who add and subtract well with whole numbers should have little difficulty with decimals. The decimal points must be aligned when writing additions and subtractions vertically. Point out that adding zeros on the right is a way of expressing decimal fractions with a common denominator and makes addition and subtraction of decimals easier to perform.

Chalkboard Examples
Compute the sum or difference.

1. 1.7
   +3.8
   ‾‾‾‾
    5.5

2. 11.07
   + 9.729
   ‾‾‾‾‾‾
    20.799

3.   7.3
    20.02
   +19.297
   ‾‾‾‾‾‾
    46.617

4. 2.34
   −1.82
   ‾‾‾‾
    0.52

5. 19.9
   −10.975
   ‾‾‾‾‾‾
    8.925

6. 29
   − 7.65
   ‾‾‾‾‾
   21.35

Related Activities
Give the students copies of the addition and subtraction problems below. Allow them about five minutes to work all the problems. Tell the students to work as quickly and accurately as possible. At the end of the time alloted, read the answers. How well did each student do?

1.   2.9
      41.876
      72.9579
     199.35
       7.2917
   +   3.111
   ‾‾‾‾‾‾‾‾‾
     327.4866

2.   116.73
       9.9999
     800.01
       6.728
      22.345
   + 49
   ‾‾‾‾‾‾‾‾
    1004.8129

3.   33.8808
       9.759
     150.2
       7.6543
     300.9
   +229.85
   ‾‾‾‾‾‾‾
    732.2441

4.   200.1
    − 87.294
   ‾‾‾‾‾‾‾
    112.806

5.   506.92
    −382.9987
   ‾‾‾‾‾‾‾
    123.9213

6.   97
    −19.7201
   ‾‾‾‾‾‾
    77.2799

2-8 Multiplying Decimals

Teaching Suggestions

This section should not be very difficult for most students since multiplication of decimals is very similar to multiplication of whole numbers. Discuss the rule for finding the number of decimal places in the product with the students. Encourage them to check the placement of the decimal point in the product by rounding and estimating. Some students may require a short review of estimation.

Chalkboard Examples

Compute the product.

1. 9.35
   ×8
   ‾‾‾‾
   74.8

2. 1.96
   ×5.4
   ‾‾‾‾
   10.584

3. 3.003
   ×5.09
   ‾‾‾‾‾
   15.28527

4. Estimate to check your answers.

Related Activities

Ask the students to write the product for the problems below and then complete the rule for multiplying by a power of 10.

1. 4.54×10
   4.54×100
   4.54×1000

2. 19.2×10
   19.2×100
   19.2×1000

3. 8.795×10
   8.795×100
   8.795×1000

4. 17.7921×10
   17.7921×100
   17.7921×1000

5. To multiply by a power of 10, count the number of _zeros_ in the power of 10 and move the decimal point in the other factor to the _right_ by that many decimal places.

2-9 Dividing Decimals

Teaching Suggestions

Review multiplication of decimals by powers of 10 (See Related Activities in the previous section.). Emphasize that before we can divide with decimals, we must multiply the divisor and dividend by a power of 10 to make the divisor a counting number. A quick review of the counting numbers (whole numbers excluding zero) may be necessary. Explain that we do not include 0 as a divisor because division by 0 has no meaning. A quick review of rounding decimals might be appropriate here.

Chalkboard Examples

Divide. If the quotient is not exact, round to the nearest hundredth. Check your answer by multiplication.

1. $22\overline{)217.8}$ → 9.9
2. $1.38\overline{)10.35}$ → 7.5
3. $0.961\overline{)44.7}$ → 46.51

4. $0.09\overline{)8.4}$ → 93.33
5. $4.31\overline{)0.498}$ → 0.12
6. $6.01\overline{)789}$ → 131.28

Related Activities

Divide. Continue the division until there is a remainder of zero.

1. $4.35\overline{)34.46679}$ → 7.9234
2. $10.6\overline{)51.34852}$ → 4.8442

3. $92\overline{)343.1094}$ → 3.72945
4. $200.5\overline{)137.78761}$ → 0.68722

2-10 Solving Problems

Teaching Suggestions

Explain that learning to solve problems is the principal reason for studying mathematics. Word problems are used to convey real-life situations employing the mathematics learned.

Discuss each step in the plan for solving problems. Apply the plan to a few examples. If a student is experiencing difficulties, try to determine whether the fault lies in the lack of reading ability, weakness of interpretation, computational deficiencies, or simply a lack of confidence.

Chalkboard Examples

Solve.

1. A jar of apple juice holds 0.946 L. A large can holds 1.36 L. How much more juice is in the can? 0.414 L

2. Mike sold 33 boxes of greeting cards for $2.52 a box and 17 boxes of stationery for $3.69 a box. What were total sales for greeting cards? $83.16

3. Marcy earned $53. She worked 4 hours on Thursday and 8 hours on both Friday and Saturday. How much did she earn per hour? $2.65

4. The Tyco Building has 46 floors with 88 windows per floor. It takes a window washer 1.5 minutes to wash a window. If two washers are assigned, how many hours will it take them to complete the job? about 22.5 hours

Related Activities

Suggest a topic, for example, sales, earnings, or the weather and have the students write several word problems relating to that topic. Have each student choose a classmate with whom to exchange problems for solving.

3-1 The Extended Number Line

Teaching Suggestions

Begin this section by discussing the subject areas where positive and negative numbers are used: asset and debt, temperature above and below zero, altitude above and below sea level, and rise and fall in stock prices.

The use of the minus sign may be confusing to some of the students. Explain that the raised minus sign in $^-8$ means "negative 8," whereas the unraised minus sign in -3 means "the opposite of positive 3." In fact, use chalkboard examples to show that $-n$ may be positive or negative depending on whether n is positive or negative. Point out that another way of saying "the opposite of a number" is "the additive inverse of a number."

Chalkboard Examples

Show the graph of each of the given numbers on a number line.

1. 0   **2.** $^-2$ **3.** 3   **4.** 4   **5.** $^-5$

State the opposite of the given number. Use the number line to help you, if necessary.

6. $^-2$ 2   **7.** 0   0   **8.** $^-4$ 4   **9.** 3 $^-3$ **10.** 5 $^-5$

Related Activities

Give the students copies of some expressions like the ones below and have them state the opposite.

1. hot
   cold

2. short
   tall

3. negative 5
   positive 5

4. exit
   entrance

5. $12 withdrawal
   $12 deposit

6. 8 steps forward
   8 steps backward

7. +3.6
   −3.6

8. $25 gain
   $25 loss

9. 95 km longer
   95 km shorter

10. 7 point increase
    7 point decrease

3-2 Arrows on the Number line

Teaching Suggestions

Graph a decimal number, say 3.5, on a number line on the chalkboard. Explain that we say this is the graph of 3.5. Tell the students that there is a way to represent 3.5 on the number line using directed line segments, or arrows. Picture 3.5 on a number line using arrows beginning at the origin, at ⁻1, and at 2. Emphasize that although there is only one correct graph of 3.5 on the number line, there are many ways of representing the number using arrows. You might mention to the students that since an arrow can start at any point on the number line, the principal considerations are the length of the arrow and its direction.

Chalkboard Examples

Graph each number on a number line.

1. 1.75   2. ⁻1.1   3. ⁻2.1   4. 0.9

5. Using the number line in Exercises 1–4, draw an arrow starting at the origin to represent 2.

6. Draw an arrow starting at the origin to represent ⁻1.75.

7. Draw an arrow starting at ⁻1 to represent 2.9.

Related Activities

Give each student a copy of two tables similar to the one for Class Exercises 21–26. The tables should have headings and room for five lines of entries, but no other entries. Have each student choose two decimal numbers, one positive and one negative. Enter one number in each table in all spaces under the heading "Number Represented." Have the students list five different ways of representing each number with an arrow.

3-3 The Sum of Two Numbers of the Same Sign

Teaching Suggestions

This section will not be difficult for most students and should not require much teaching time. Draw a number line on the chalkboard and picture the sum of two positive and then two negative numbers using arrows. You may wish to recall for the students that the arrow which represents the first addend always begins at the origin.

Have the students cite some applications or situations which involve the sum of two negative numbers, for example, two consecutive withdrawals from a checking account.

Chalkboard Examples

Add. If necessary, draw a number line.

1. 9 + 8 17   2. 1.7 + 5.6 7.3   3. 13 + 0 13

4. ⁻6 + ⁻3 ⁻9   5. ⁻4.9 + ⁻10.1 ⁻15   6. ⁻19 + ⁻19 ⁻38

Related Activities

Below is a magic square. The sum of the numbers in any row, column, or diagonal is the same. Give copies of the square to the students and have them write in the missing numbers.

⁻2.6	⁻19.5	⁻20.8	⁻6.5
⁻16.9	⁻10.4	⁻9.1	⁻13
⁻11.7	⁻15.6	⁻14.3	⁻7.8
⁻18.2	⁻3.9	⁻5.2	⁻22.1

3-4 The Sum of Two Numbers of Opposite Sign

Teaching Suggestions

First, review the rules for drawing arrows to picture addition on the number line. You might illustrate the sum of two numbers of the same sign. Then illustrate the three different possibilities when finding the sum of a positive number and a negative number; that is, the sum may be positive, negative, or zero. Often, students find it easier to find the sum of two numbers of opposite signs if told to disregard the signs and subtract. The sum then has the same sign as that of the addend with the longer arrow.

Chalkboard Examples

Add.

1. $9 + 3$ 12   **2.** $9 + {}^-3$ 6   **3.** ${}^-9 + 3$ ${}^-6$

4. ${}^-9 + {}^-3$ ${}^-12$ **5.** $9 + {}^-9$ 0   **6.** ${}^-3 + 9$ 6

7. $6.2 + {}^-1.3$ **8.** ${}^-8.8 + 4.4$ **9.** ${}^-3.7 + 7.3$
    4.9 ${}^-4.4$ 3.6

Related Activities

Have the first and second students in the first row each name a number between ${}^-10$ and 10. The third student then states the sum of the two numbers. The fourth student names another number between ${}^-10$ and 10. The fifth student states the sum of that number and the previous sum. Have the class continue in this manner, each time adding a number between ${}^-10$ and 10, until everyone has had a turn.

3-5 The Difference of Two Numbers

Teaching Suggestions

The difference of two numbers can be found by finding the missing addend. It can also be found by adding the opposite, as will be discussed in the next section.

Begin by reviewing the meaning of the raised and unraised minus signs. Point out that in the stated property on page 82, the numbers a, b, and c can be positive, negative, or zero. It may help to write some subtraction exercises on the chalkboard as a sum with a missing addend, which the students can supply, as shown below.

$$ {}^-12 - {}^-6 = {}^-6 + \underline{\;?\;} = {}^-12 $$
$$ {}^-75 - 20 = 20 + \underline{\;?\;} = {}^-75 $$

Chalkboard Examples

Write two differences corresponding to the given sum.

1. $8 + {}^-3 = 5$ $5 - {}^-3 = 8$; $5 - 8 = {}^-3$

2. ${}^-19 + 4 = {}^-15$ ${}^-15 - {}^-19 = 4$;
${}^-15 - 4 = {}^-19$

3. ${}^-15 + {}^-30 = {}^-45$
${}^-45 - {}^-15 = {}^-30$; ${}^-45 - {}^-30 = {}^-15$

4. $5 + 8 = 13$ $13 - 5 = 8$; $13 - 8 = 5$

Find the given difference.

5. $16 - 12$ 4   **6.** $25 - 29$ ${}^-4$ **7.** ${}^-33 - 4$ ${}^-37$

8. ${}^-18 - {}^-9$ ${}^-9$ **9.** $4 - {}^-7$ 11 **10.** ${}^-1 - {}^-1$ 0

Related Activities

Ask the students whether they are familiar with the expression "Family of Facts." Write the following family on the chalkboard.

$$ 6 + 9 = 15 \qquad 9 + 6 = 15 $$
$$ 15 - 6 = 9 \qquad 15 - 9 = 6 $$

Have the students write a family of facts for each set of numbers.

1. ${}^-6, 13, 19$ **2.** ${}^-8, {}^-4, {}^-4$ **3.** $17, 9, 8$

4. ${}^-3, {}^-1, 2$ **5.** ${}^-4, 10, 6$ **6.** $2, {}^-2, {}^-4$

3-6 Subtraction as Adding the Opposite

Teaching Suggestions
The two arrow diagrams on page 85 show that we can subtract a number by adding its opposite. After discussing the arrow diagrams, write some subtraction exercises on the chalkboard. Have the students write each as an addition exercise, and then find the difference.

Chalkboard Examples
Express the given difference as a sum.

1. $9 - {}^-15$
   $9 + 15$

2. ${}^-4 - {}^-4$
   ${}^-4 + 4$

3. ${}^-17 - 16$
   ${}^-17 + {}^-16$

4. $3 - 13$
   $3 + {}^-13$

5. $2.5 - {}^-1.2$
   $2.5 + 1.2$

6. ${}^-9.7 - {}^-7.9$
   ${}^-9.7 + 7.9$

7. Find the difference in Exercises 1–6 by adding. 1. 24   2. 0   3. ${}^-33$ 4. ${}^-10$ 5. 3.7   6. ${}^-1.8$

Related Activities
Mention to the students that some textbooks give the following rule for subtracting two numbers. "To subtract two numbers, change the sign of the number being subtracted, and add." Ask the students why this rule works and have them complete subtraction tables like the one below.

−	⁻10	20	6	⁻14
11	21	⁻9	5	25
⁻12	⁻2	⁻32	⁻18	2
⁻13	⁻3	⁻33	⁻19	1
14	24	⁻6	8	28

3-7 The Product of Two Integers

Teaching Suggestions
The distinction between raised and unraised minus signs was helpful in teaching addition and subtraction in earlier lessons. In this section, the distinction has been dropped. Caution the students that from this point on, for notational convenience, only the unraised minus sign will be used.

Chalkboard Examples
Find the product.

1. 7(8) 56   2. 7(−8) −56   3. −7(8) −56

4. −7(−8) 56   5. −7(0) 0   6. 0(−8) 0

7. 9(−9) −81   8. −10(−4) 40   9. −13(2) −26

Related Activities
Here is a way of relating the product of positive and negative numbers to a real situation.

a. A rise of 3° an hour. In 4 hours it will be 3 × 4, or 12° warmer.
b. A drop of 3° an hour. In 4 hours it will be −3 × 4, or −12° colder.
c. A drop of 3° an hour. Four hours ago it was −3 × −4, or 12° warmer.

Have the students write other situations where the product of positive and negative numbers are related: deposits and withdrawals on a banking account or the rise and fall of stock and bond prices.

3-8 The Product of Two Numbers

Teaching Suggestions
This section will not be difficult for most students. Before discussing the examples in the text, you may wish to recall for the students the rule for placing the decimal point in the product when multiplying with decimal numbers. Then point out that the same rules of sign used to simplify products of integers are used for any numbers.

Chalkboard Examples
Multiply.

1. $0.7(0.7)$
   0.49

2. $-1.6(8)$
   -12.8

3. $10(-2.23)$
   -22.3

4. $-0.18(0.2)$
   -0.036

5. $-1.2(-2.3)$
   2.76

6. $11.1(-1.1)$
   -12.21

Related Activities
Give the students copies of the number maze below. The starting point is -5; the end point is -32.12. Begin at -5 and multiply by a number in an adjacent square. If the product is in a square adjacent to the second factor, the player moves to the square containing the product. For example, $-5 \times 2.2 = -11$, so the player moves to the square containing -11. Continue in this manner until the end is reached. Have the students draw a path connecting the numbers they used from start to finish. Students who finish the maze can make one of their own and give it to a classmate for solution.

-5	-1.1	-5.5	4	-20
2.2	-11	0.2	-2.2	-4.4
-10	-7.3	-2	14.6	-0.4
22	80.3	0.1	0.803	80.3
-176	32.14	8.03	-4	-32.12

3-9 The Quotient of Two Numbers

Teaching Suggestions
Since division can be defined in terms of multiplication, the rules of sign for multiplication are related to the rules of sign for division. Illustrate these rules by finding the quotients of some positive and negative decimals. If necessary, review how the decimal point is located in the quotient. Emphasize that division by 0 is meaningless, discussing Example 3 in detail.

Chalkboard Examples
Find the quotient.

1. $72 \div -8 \quad -9$

2. $-72 \div -9 \quad 8$

3. $-72 \div 0$
   meaningless

4. $0 \div 72 \quad 0$

5. $-17 \div 8.5 \quad -2$

6. $5.94 \div -4.4$
   -1.35

Related Activities
Have each student choose a number between -10 and 10. The students name their numbers in turn while you record them on the chalkboard in groups of five. Continue until there are at least six groups of five. Then have the students perform four operations with the five numbers in each group: add the first two, subtract the third, multiply by the fourth, and divide by the fifth. Quotients should be rounded to the nearest hundredth.

EXAMPLE: $-9, 2, 3, -4, -6$

Step 1: $-9 + 2 = -7$
Step 2: $-7 - 3 = -10$
Step 3: $-10 \times -4 = 40$
Step 4: $40 \div -6 = -6.67$, to the nearest hundredth.

4-1 Positive Fractions

Teaching Suggestions

Begin by reviewing the terms proper fraction, improper fraction, mixed number, unit fraction, and equal fraction with the students. Notice that no distinction is made in the text between fraction and fractional number.

While discussing the properties of fractions, emphasize the relationship between fractions and division. Ask the students why the condition $b \neq 0$ is needed in all of these properties. You may wish to provide extra drill on changing a mixed number to an improper fraction and vice versa.

Chalkboard Examples

Complete with a fraction.

1. $7 \times \underline{} = 1\frac{1}{7}$ **2.** $1 \div 9 = \underline{}\frac{1}{9}$

3. $14 \div 17 = \underline{}\frac{14}{17}$ **4.** $6 \times \frac{1}{9} = \underline{}\frac{6}{9}$

Write as an improper fraction.

5. $1\frac{3}{4}\ \frac{7}{4}$ **6.** $3\frac{3}{16}\ \frac{51}{16}$

Write as a whole number or mixed number.

7. $\frac{15}{5}\ 3$ **8.** $\frac{21}{8}\ 2\frac{5}{8}$

Related Activities

Interest some of your students in investigating the development and applications of fractions for the benefit of the class. Some questions to get them started might include the following.

1. Why were fractions first developed?
2. Where are fractions most frequently used?
3. In business and industry, which numbers are most frequently used in denominators?

4-2 Negative Fractions

Teaching Suggestions

The students should have little difficulty with this section on negative fractions. Explain that the properties that apply to the positive fractions also apply to the negative fractions. These positive and negative fractions together form the set of rational numbers.

Chalkboard Examples

1. Show the graphs of $\frac{-8}{3}$, $-\frac{1}{3}$, $\frac{3}{-3}$, and $-\frac{5}{3}$ on the number line.

Write each fraction in two other ways.

2. $-\frac{1}{9}\ \frac{-1}{9}, \frac{1}{-9}$ **3.** $\frac{2}{-7}\ \frac{-2}{7}, -\frac{2}{7}$

4. $\frac{-13}{12}\ -\frac{13}{12}, \frac{13}{-12}$

Related Activities

As is the case with integers, we can also picture rational numbers on the number line with directed line segments, or arrows. Remind the students that the arrow can start at any point on the number line. Have the students draw a number line and picture each arrow described below.

1. Draw arrows starting at the origin to represent **(a)** $-\frac{7}{4}$ and **(b)** $\frac{1}{4}$.

2. Draw arrows starting at the graph of $-\frac{1}{2}$ to represent **(a)** 3 and **(b)** $-1\frac{1}{2}$.

3. Draw arrows starting at the graph of $\frac{4}{4}$ to represent **(a)** $\frac{-3}{3}$ and **(b)** $\frac{5}{6}$.

4-3 Equal Fractions

Teaching Suggestions

Discuss the Equal-Fractions Rule. Stress that multiplying or dividing by $\frac{c}{c}$ is the same as multiplying or dividing by 1. Before discussing how to reduce fractions to lowest terms, have the students define common factor and greatest common factor and give examples for each.

Chalkboard Examples

Write a fraction equal to the given fraction.

1. $\frac{1}{9}$ **2.** $-\frac{5}{8}$ **3.** $\frac{10}{12}$ **4.** $-\frac{23}{18}$

Answers will vary.

Reduce each fraction to lowest terms.

5. $\frac{15}{27}$ $\frac{5}{9}$ **6.** $-\frac{49}{14}$ $-\frac{7}{2}$ **7.** $-\frac{60}{132}$ $-\frac{5}{11}$

8. $\frac{32}{4}$ 8

Related Activities

Cross-multiplication is used to determine whether two fractions are equal. You may wish to introduce this topic here in preparation for proportion in the next chapter.

Example: $\frac{2}{3} \stackrel{?}{=} \frac{26}{39}$ Solution: $2 \times 39 = 78$
$3 \times 26 = 78$

Since the products are equal, the fractions are equal.

Which fractions are equal to the first fraction?

1. $\frac{3}{4}$: $\frac{6}{9}, \frac{9}{16}, \frac{27}{36}, \frac{75}{100}, \frac{700}{1000}$ $\frac{27}{36}, \frac{75}{100}$

2. $\frac{13}{14}$: $\frac{33}{42}, \frac{65}{70}, \frac{117}{126}, \frac{146}{168}, \frac{195}{215}$ $\frac{65}{70}, \frac{117}{126}$

3. $\frac{6}{5}$: $\frac{72}{60}, \frac{110}{100}, \frac{186}{155}, \frac{300}{275}, \frac{408}{340}$ $\frac{72}{60},$
$\frac{186}{155}, \frac{408}{340}$

4. $\frac{5}{8}$: $\frac{20}{40}, \frac{40}{64}, \frac{80}{104}, \frac{120}{160}, \frac{100}{166}$ $\frac{40}{64}$

4-4 Products of Fractions

Teaching Suggestions

After discussing Examples 3–5, summarizing the material might prove helpful to students:

1. Ignore signs until the product has been computed. Then use the rule of sign for products of integers.
2. Change mixed numbers and integers to improper fractions before multiplying.
3. Try to divide by common factors before multiplying.
4. Reduce products to lowest terms.

Tell your students whether you prefer that answers be left as improper fractions in lowest terms or be written as mixed numbers.

Chalkboard Examples

Multiply.

1. $\frac{2}{9} \times (-9)$ -2 **2.** $\frac{1}{3} \times \frac{1}{3}$ $\frac{1}{9}$

3. $\left(-\frac{1}{6}\right) \times \left(-\frac{5}{12}\right)$ $\frac{5}{72}$ **4.** $-\frac{4}{7} \times \frac{9}{14}$ $-\frac{18}{49}$

5. $\left(-\frac{3}{4}\right) \times \left(-\frac{8}{15}\right)$ $\frac{2}{5}$ **6.** $4\frac{3}{5} \times 10$ 46

Related Activities

Have the students compute products like those below. After they have finished, ask them if they see any relationship between the number of negative factors and the sign of the product.

1. $\frac{1}{2} \times (-\frac{1}{2} \times \frac{1}{2})$

2. $[\frac{1}{2} \times (-\frac{1}{2})] \times [\frac{1}{2} \times \frac{1}{2}]$

3. $\frac{1}{2} \times [-\frac{1}{2} \times (-\frac{1}{2})]$

4. $[\frac{1}{2} \times (-\frac{1}{2})] \times [-\frac{1}{2} \times \frac{1}{2}]$

The product is negative if there is an odd number of factors; the product is positive if there is an even number of factors.

4-5 Least Common Denominator

Teaching Suggestions

Although the students have previously dealt with finding the least common denominator (LCD), many confuse the LCD with other concepts such as multiple, common multiple, least common multiple, prime number, and prime factorization. As you discuss the lesson, be sure the students understand these terms. Point out that a common denominator can always be found by multiplying the denominators of the given fractions but that the LCD makes calculations easier. Encourage the students to use mental arithmetic as much as possible in finding least common denominators.

Chalkboard Examples

State the LCD of the given fractions: first by finding the LCM, then by prime factorization.

1. $\frac{2}{3}, \frac{1}{9}$ 9

2. $-\frac{3}{10}, \frac{3}{4}$ 20

3. $\frac{1}{6}, \frac{1}{8}$ 24

4. $-\frac{1}{12}, -\frac{3}{8}$ 24

5. $\frac{13}{20}, \frac{13}{16}$ 80

6. $\frac{2}{15}, -\frac{3}{20}$ 60

Related Activities

Christian Goldbach, an 18th century mathematician, proposed that every even number greater than 2 may be written as the sum of two prime numbers. For example, 10 may be written as $5 + 5$ and $3 + 7$. This proposal is known as Goldbach's Conjecture. Demonstrate this conjecture to the students for the number 16. Then give them a list of ten even numbers and have them write each number as the sum of primes in as many different ways as possible.

$16 = 3 + 13 = 5 + 11$

4-6 Sums and Differences of Fractions

Teaching Suggestions

You may wish to have your students prove the two rules on page 124, mentioning that there is a distributive property of multiplication with respect to addition and subtraction.

Before calculating sums and differences with fractions, have the students complete some addition and subtraction exercises with integers. Remind them that we subtract by adding the opposite. It may be necessary to review the methods for finding a common denominator, pointing out the computational advantage of using the LCD.

Chalkboard Examples

Express the sum or difference as a proper fraction in lowest terms or as a mixed number with fraction in lowest terms.

1. $\frac{3}{10} + \frac{9}{10}$

$1\frac{1}{5}$

2. $-\frac{2}{3} + \frac{3}{7}$

$-\frac{5}{21}$

3. $-5 + 2\frac{11}{100}$

$-2\frac{89}{100}$

4. $3\frac{1}{4} - \frac{3}{4}$

$2\frac{1}{2}$

5. $-\frac{5}{6} - \frac{2}{9}$

$-1\frac{1}{18}$

6. $13\frac{7}{12} - \left(-3\frac{5}{16}\right)$

$16\frac{43}{48}$

Related Activities

Give the students copies of the magic square below and have them find the missing fractions in lowest terms.

$1\frac{1}{3}$	$\frac{1}{4}$	$-\frac{2}{3}$	$1\frac{1}{12}$
$-\frac{5}{12}$	$\frac{5}{6}$	$\frac{11}{12}$	$\frac{2}{3}$
$\frac{3}{4}$	$-\frac{1}{3}$	$\frac{7}{12}$	1
$\frac{1}{3}$	$1\frac{1}{4}$	$1\frac{1}{6}$	$-\frac{3}{4}$

4-7 Quotients of Fractions

Teaching Suggestions

The relationship between multiplication and division is used here to suggest a rule for dividing fractions. Some students may be interested in a proof of the general rule for dividing fractions which is detailed in the Related Activities below.

Point out that to divide by a fraction, we first multiply by its reciprocal, then we apply the rules of sign for division of integers. Mixed numbers should be changed to improper fractions before calculations begin.

Chalkboard Examples

State the reciprocal.

1. $-\dfrac{1}{2}$ -2 **2.** $\dfrac{4}{9}$ $\dfrac{9}{4}$ **3.** -8 $-\dfrac{1}{8}$ **4.** $2\dfrac{1}{12}$ $\dfrac{12}{25}$

Divide. Express answers in lowest terms.

5. $\dfrac{5}{8} \div \left(-\dfrac{1}{2}\right)$ $-1\dfrac{1}{4}$

6. $\dfrac{12}{29} \div \dfrac{4}{9}$ $\dfrac{27}{29}$

7. $-9\dfrac{7}{15} \div (-8)$ $1\dfrac{11}{60}$

Related Activities

A proof of the general rule for dividing fractions is given below. Have the students turn to page 129 for a statement of the rule and then have them justify each step.

Proof: **1.** If $\dfrac{a}{b} \div \dfrac{c}{d} = n$, then $\dfrac{c}{d} \times n = \dfrac{a}{b}$.

2. Since $\dfrac{c}{d} \times n$ and $\dfrac{a}{b}$ represent the same number, we can multiply each by $\dfrac{d}{c}$.

3. $\dfrac{d}{c} \times \left(\dfrac{c}{d} \times n\right) = \dfrac{d}{c} \times \dfrac{a}{b}$

4. $\left(\dfrac{d}{c} \times \dfrac{c}{d}\right) \times n = \dfrac{d}{c} \times \dfrac{a}{b}$

5. $1 \times n = \dfrac{d}{c} \times \dfrac{a}{b}$

6. $n = \dfrac{d}{c} \times \dfrac{a}{b}$ or $\dfrac{a}{b} \times \dfrac{d}{c}$

4-8 Fractions and Decimals

Teaching Suggestions

This section should be a review for most students and should require little teaching time. Explain that every rational number may be written as a terminating or a repeating decimal. Choose a fraction that you know can be written as a repeating decimal, say $\frac{5}{7}$, to work on the chalkboard. As you work the example, ask the students if they see a relationship between the remainder and the divisor. Ask them what the possible remainders are? How many digits are in the repeating block? What is the maximum number of digits possible in a repeating block?

Chalkboard Examples

Express as a fraction in lowest terms.

1. 1.8 $\dfrac{9}{5}$ **2.** -7.25 $-\dfrac{29}{4}$ **3.** 21.625 $\dfrac{173}{8}$

Express as a terminating or a repeating decimal.

4. $-\dfrac{17}{16}$ -1.0625 **5.** $-\dfrac{1}{15}$ $-0.0\overline{6}$

6. $\dfrac{13}{6}$ $2.1\overline{6}$

Related Activities

Give the students copies of exercises like the ones below. Write $<$, $>$, or $=$.

1. $\dfrac{1}{2}$ $\underline{<}$ $0.\overline{5}$ **2.** 9.375 $\underline{>}$ $\dfrac{28}{3}$

3. $\dfrac{5}{6}$ $\underline{=}$ $0.8\overline{3}$ **4.** $-\dfrac{9}{20}$ $\underline{=}$ -0.45

5. $-\dfrac{1}{8}$ $\underline{<}$ -0.0125 **6.** $0.91\overline{6}$ $\underline{>}$ $\dfrac{9}{11}$

5-1 Ratio

Teaching Suggestions

Ratio and rate are important concepts because of their widespread applications. The students can easily find examples of ratio in the newspaper, and the supermarket provides innumerable examples of rate. You might point out that although we can write a ratio in several ways, it is usually easier to work with ratios written in the form of a fraction in simplest terms. Emphasize that when forming a ratio of quantities of the same kind, the same unit of measure for each quantity must be used.

Chalkboard Examples

1. The directions say to mix 1 part polish to 4 parts water. Express the ratio of water to polish in three ways. $4 \div 1$; $4:1$; $\frac{4}{1}$

2. Maria spent $4 at the grocery store. Milk cost 75¢. Express the ratio of the cost of milk to the total bill as a fraction in lowest terms. $\frac{3}{16}$

3. Maria spent $.99 for 3 cartons of yogurt. Write a ratio to show the rate of cost per carton. What is the cost of 5 cartons? $\frac{33}{1}$; $1.65

Related Activities

Have the students find other examples of ratio. They might consider information in newspapers, particularly items in the sports and financial sections and ads of supermarket sales. Data can also be in the form of school activities, events, or enrollment. Students should write several problems about their ratios and exchange them with a classmate for solution.

5-2 Proportion

Teaching Suggestions

Students often have difficulty recognizing the extremes and means of a proportion. Point out that when a proportion is written with the ratio signs, as in $3:4 = 6:8$, the extremes are at the end and the means are in the middle.

Before solving proportions, review the method for solving simple multiplication-division equations introduced in Section 2-5. Since proportions are so useful in word problems, thoroughly discuss the four steps on page 150 and Example 3.

Chalkboard Examples

Solve the proportion.

1. $\frac{n}{36} = \frac{3}{18}$ 6

2. $\frac{5}{n} = \frac{100}{80}$ 4

3. $\frac{2.2}{11} = \frac{n}{100}$ 20

4. $\frac{48}{57} = \frac{16}{n}$ 19

Related Activities

The type of proportion discussed in this section is direct proportion. For example, a car averaging 70 km/h will travel 140 km in 2 hours. In general, we know that as time increases, the distance traveled will also increase.

The students might be interested in another type of proportion, inverse proportion. For example, a store manager must price cuts of meat for sale. The manager knows that the more he or she charges, the less consumers will be able to buy. On the other hand, as prices drop, sales increase. Have the students write other examples of inverse proportions.

5-3 Percent

Teaching Suggestions

The first example in this section illustrates ways of converting percents to fractions or decimals, and vice versa. If necessary, use additional examples to ensure that the students understand these processes. Also, include examples of conversions of percents greater than 100%, say 120%, to fractions and decimals. You might suggest that the students memorize the percents and their equal fractions and decimals in the table on page 154.

The three types of percent problems all use the same basic equation in their solution: $a \times b = c$, where a represents the percent and the multiplication sign represents the word *of*. You should point out that the percent must be changed to an equal fraction or decimal before multiplying or dividing.

Chalkboard Examples

Express as a decimal and as a fraction in lowest terms.

1. 35% $0.35, \dfrac{7}{20}$ **2.** 4% $0.04, \dfrac{1}{25}$

3. 0.5% $0.005, \dfrac{1}{200}$ **4.** 360% $3.6, \dfrac{18}{5}$

Find the percent or number.

5. What is 89% of 900? 801
6. 16 is 32% of what number? 50

7. 62 is what percent of 496? $12\dfrac{1}{2}\%$

Related Activities

Give the students copies of the cross-number puzzle at the right. The results Across can be checked by finding the results Down.

		a 8
b 1	1	
c 4	9	5
d 1	2	0
e 3	7	6	f 2

Across
b. 20% of 55 is ? .
c. 33⅓% of ? is 165.
d. ? % of 81 is 97.2.
e. 0.5% of ? is 1.88.
f. ? % of 200 is 4.

Down
a. 16⅔% of 4890 is ? .
b. ? % of 380 is 722.
c. 50% of ? is 213.
d. ? % of 7 is 1.19.
f. 1% of 200 is ? .

5-4 Percent Increase and Decrease

Teaching Suggestions

This section and Section 5-5 provide practice in applying percents to everyday situations. Percent of increase or decrease is usually difficult for many students. One of the greatest difficulties is remembering that percent of increase or decrease is the ratio of the amount of change to the *original* amount. If necessary, use more chalkboard examples to ensure understanding of this section.

Chalkboard Examples

Find the percent of increase or decrease in going from the first number to the second.

1. 6 to 7

$16\dfrac{2}{3}\%$ increase

2. 10 to 5

50% decrease

3. 20 to 18

10% decrease

4. 60 to 120

100% increase

Solve.

5. A $45 tennis racket is now on sale for $36. What is the percent of decrease? 20%

6. The Kim family bought a house for $40,000 seven years ago. Today it is worth 55% more. What is the value of the house now? $62,000

Related Activities

The students may find the following situation interesting. Pose the problem and then ask the students to explain their answers. If necessary, use actual prices and check by calculating.

The price of lettuce increased 30% after heavy rains fell during the growing season. Shortly thereafter, the price fell 30%. Is the final price the same as the old price before the wet weather? Explain.

5-5 Commissions, Discounts, Budgets

Teaching Suggestions

The topics in this section may still be relatively new to most students. Carefully discuss the terms introduced. Point out that in a circle graph, we can not only compare an amount with the total amount, but also with each other. Have the students find the sum of all the percents shown in the graph in Example 3, explaining that a complete circle graph represents 100%, or 1. It might be worthwhile to demonstrate how to construct a circle graph since many students find this difficult.

Chalkboard Examples

Solve.

1. Mrs. Vail is a real estate agent. Last year, she sold $1,250,000 worth of real estate. If her rate of commission is 2% of sales, what was her commission for the year? $25,000

2. A television set usually sells for $390. It is now on sale for $312. What is the discount rate? 20%

3. Ted spent $2.50 at the movie. This represents 20% of the money he had earned mowing lawns. How much had he earned? $12.50

Related Activities

Suggest that students investigate the commission or rates of commission of various sales positions. Which sales positions have higher commissions or rates of commission? Some students might look through newspaper advertisements to find sale prices and discount rates on items. Have the students write word problems with the information and exchange them with a classmate for solution.

6-1 Units of Length

Teaching Suggestions

Review the use of the symbol "$\approx$," emphasizing that measurements are approximations. Provide the students with metric measuring devices and have them measure some objects in the classroom in meters, centimeters, and millimeters.

Discuss the table on page 174. For those students who are having difficulties with negative and zero exponents, refer them to the Extra in Section 3-9. Finally, ask the students why we simply move the decimal point to convert from one metric unit to another.

Chalkboard Examples

Refer to the diagram to answer the following.

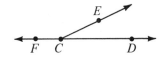

1. Name three rays.
   Rays CF, CD, CE

2. Name three segments.
   $\overline{CF}$, $\overline{CD}$, $\overline{CE}$

3. Name one line.
   Line DF

Complete.

4. 20 mm = __?__ cm 2

5. 1 km = __?__ cm 100,000

6. 3 m = __?__ mm 3000

7. 1200 m = __?__ km 1.2

Related Activities

Name five familiar objects, for example, your index finger, a pencil point, a classmate, a door, a desk. Have the students first estimate the length, or height, of each object. Ask the students which units they used for each object and why. Then actually measure each object to see how close the estimates are.

6-2 Circles

Teaching Suggestions

After discussing the parts of a circle, you might point out that the terms "radius" or "diameter" can mean the line segment itself or the length of such a line segment. If the students might enjoy finding the value of π experimentally, have them measure the circumference and diameter of two circular objects with a tape measure, for example, a bicycle wheel and the top of a wastepaper basket. Then have them write the ratio of each circumference to its diameter as a decimal. Some students may recall from "Nonrepeating Decimals" on pages 140–141 that the result, π, is an irrational number.

Chalkboard Examples

Complete.

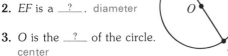

1. EO is a __?__. radius

2. EF is a __?__. diameter

3. O is the __?__ of the circle. center

Find the circumference of the circle described.

4. diameter = 16 m

$\pi \approx 3.14$

50.24 m

5. radius = 154 cm

$\pi \approx \frac{22}{7}$ 968 cm

Related Activities

Give the students copies of the circle descriptions below. Have them use a compass to draw each circle described and then find the circumference. Use $\pi \approx 3.14$.

1. radius = 35 mm
   219.8 mm

2. diameter = 8.6 cm
   27.004 cm

3. radius = 5.8 cm
   36.424 cm

4. diameter = 4.5 cm
   14.13 cm

6-3 Measuring and Constructing Angles

Teaching Suggestions

The students will already be familiar with most of the terminology for angles. You should point out that the middle letter is the vertex when naming an angle using three letters.

Be certain that every student understands how to use a protractor. The Extra! which follows the section serves as a basis for future features and should be assigned if students are not familiar with using a compass and straight-edge.

Chalkboard Examples

Answer the following.

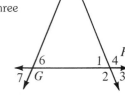

1. Name $\angle 5$ using three letters.
   $\angle FEG$ or $\angle GEF$

2. Are $\angle 2$ and $\angle 3$ complementary angles? no

3. Is $\angle 6$ an acute angle? yes

4. Name two obtuse angles.
   $\angle 2$ and $\angle 4$

5. If $\angle 1 = 65°$, then $\angle 2 = $ __?__, $\angle 3 = $ __?__, and $\angle 4 = $ __?__.
   115°, 65°, 115°

Related Activities

Give the students copies of figures which have two, three, and four lines that intersect at one point, as shown below. Have the students measure only one angle in the first diagram, two angles in the second diagram, and three in the third. The students should choose the angles for measurement so that they can find the size of all the other angles without measuring by using vertical and supplementary angles. Then have the students find the sum of all the angle measures in each diagram.

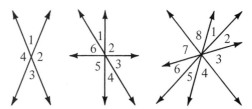

6-4 Parallel Lines

Teaching Suggestions

Begin by asking the students to explain the meaning of the term "plane." Refer to the classroom or to objects in the classroom for examples of parallel lines. For example, the top of the door is parallel to its bottom. Mention the possibility of two lines in space that do not lie in the same plane and which do not intersect. This may be illustrated by pointing out that a North-South line in the ceiling is not parallel to an East-West line in the floor although they will not meet.

Chalkboard Examples

In the figure, $l \parallel m$.

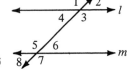

1. Name two pairs of alternate interior angles.
   $\angle 3, \angle 5; \angle 4, \angle 6$

2. Name four pairs of corresponding angles.
   $\angle 1, \angle 5; \angle 4, \angle 8; \angle 2, \angle 6; \angle 3, \angle 7$

3. $\angle 1 = 120°$. Name other angles with measure $120°$. $\angle 3, \angle 5, \angle 7$

4. If $\angle 1 = 120°$, $\angle 4 = $ __?__. $60°$

5. Name other angles with measure $60°$.
   $\angle 2, \angle 6, \angle 8$

Related Activities

Draw two parallel lines cut by a transversal on the chalkboard. Ask the students to locate the interior and exterior angles on the same side of the transversal and then the alternate exterior angles. Show that pairs of interior and exterior angles on the same side of the transversal are supplementary and that alternate exterior angles are equal.

6-5 Special Triangles

Teaching Suggestions

Note that the Class Exercises provide additional information about properties of triangles and should be carefully completed with the students. When constructing the isosceles triangles in Class Exercises 4 and 5, suggest that the students draw the equal sides first, with an acute angle between them in Exercise 4 and an obtuse angle between them in Exercise 5.

Point out that when we say to construct, we usually mean to use only a compass and straightedge. It may be necessary to explain in more detail how one uses a compass to construct an equilateral triangle in Class Exercise 6.

Chalkboard Examples

Complete.

1. An acute triangle has __?__ acute angles. 3

2. An __?__ triangle has at least two equal sides.
   isosceles

3. Two angles of a triangle measure $65°$ and $55°$. The third angle measures __?__. $60°$

4. Find the measure of each angle denoted by a question mark.
   $70°; 55°$

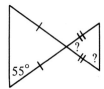

Related Activities

The results of Exercise 2 in "Extra! Paper Folding" can be obtained using a compass and straightedge. Have the students draw a large acute triangle and bisect each angle. The bisectors should meet in a point. Placing the compass on this point, the students can construct a circle that "just fits" inside the triangle.

The students can also duplicate the result of Exercise 3 by drawing the perpendicular bisectors of the sides of a triangle. Have the students complete this construction.

6-6 Special Quadrilaterals

Teaching Suggestions

The students should be familiar with most of the special quadrilaterals from earlier studies. You might draw different quadrilaterals on the chalkboard and have the students give their own definitions for each.

Point out that rhombuses, rectangles, and squares are all parallelograms. In addition, every square is a rectangle and a rhombus, but a rectangle or a rhombus may not be a square. Mention that the sum of all the angle measures for any quadrilateral is 360°.

Chalkboard Examples

Complete.

1. A trapezoid has only __?__ parallel sides. two

2. An isosceles trapezoid has one pair of equal __?__ and two pairs of equal __?__ .
   sides, angles

3. The opposite sides of a parallelogram are __?__ and __?__ . parallel, equal

4. In a rhombus, all four __?__ and the opposite __?__ are equal. sides, angles

5. A rectangle is a parallelogram with four __?__ angles. right

6. A square has __?__ equal sides and __?__ right angles. four, four

Related Activities

The diagonals of a rhombus and a rectangle are discussed in Exercises 25 and 26. What about the diagonals of a square? Since a square is both a rectangle and a rhombus, are its diagonals both equal and perpendicular? Have the students consider the diagonals of several parallelograms. Are the diagonals equal or perpendicular?

6-7 Polygons and Their Perimeters

Teaching Suggestions

The word *polygon* is of Greek origin and means "many-angled." Point out that many polygon names contain a prefix, such as "penta," which connotes a number of sides or angles. You might review some special triangles and quadrilaterals so that students can calculate the perimeters of these figures even if the length of only one or two sides is given.

Emphasize that all lengths must be in the same unit of measure before calculating the perimeter. Also mention that similar hatch marks denote parts of equal length.

Chalkboard Examples

Find the perimeter.

1. The sides of a triangle are 1 m, 93 cm, and 88 cm. 281 cm

2. Each side of a square is 6.2 cm. 24.8 cm

3. Each side of an equilateral decagon is $2\frac{1}{5}$ units. 22 units

Related Activities

Draw several four-sided figures *ABCD* and locate the midpoints *P*, *Q*, *R*, and *S* of the sides of each, as shown below.

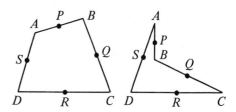

1. What kind of quadrilateral is *PQRS*? Do you think that *PQRS* will always be this kind of figure? parallelogram; yes

2. For each figure, find the perimeter of *PQRS* and compare it with *AC* + *BD*. What is the relationship? Perimeter of *PQRS* =

$$\frac{1}{2}(AC + BD)$$

6-8 Congruent Polygons

Teaching Suggestions

Some students find it difficult to list corresponding parts of congruent polygons when one figure is rotated or flipped. Suggest that the students study the congruent polygons and mark the corresponding parts which are obviously equal. The remaining corresponding parts will be easier to spot. The students can refer to the markings when naming corresponding parts. As you discuss the checks for congruent triangles, point out that the angle in SAS is the *included* angle formed by the two sides and that the side in ASA is the *included* side between the two angles.

Chalkboard Examples

Trapezoids *ABCD* and *EFGH* are congruent.

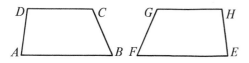

1. $AB = \underline{\ ?\ }$ EF **2.** $DA = \underline{\ ?\ }$ HE

3. $\angle C = \underline{\ ?\ }$ $\angle G$ **4.** $\angle H = \underline{\ ?\ }$ $\angle D$

5. $ABCD \cong \underline{\ ?\ }$ $EFGH$

6. Why are these triangles congruent? SAS

Related Activities

Provide the students with copies of pairs of triangles that are not congruent but have the following congruent parts:

1. Two sides and the angle opposite one of them;

2. Two angles and the side opposite one of them;

3. Three angles.

Mark the pairs of congruent parts. Ask the students to determine by measurement whether the triangles are indeed congruent.

6-9 Similar Polygons

Teaching Suggestions

Introduce the symbol "∼" (is similar to) and compare it with "≈" (is approximately) and "≅" (is congruent to). Also, some students may need a review of ratios and proportions before beginning the section. Point out that for polygons having more than three sides, corresponding angles can be equal without the sides being proportional. For example, a square and a rectangle with length twice its width have equal angles, but the sides are not proportional. Similarly, corresponding sides can be proportional without corresponding equal angles, such as a rhombus with angles 60° and 120° and a square. Both angles and sides must be considered before we can state that two polygons of more than three sides are similar.

Chalkboard Examples

Complete for the pair of similar figures.

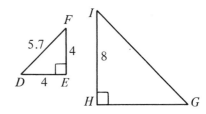

1. $\triangle DEF \sim \triangle \underline{\ ?\ }$ GHI

2. $\angle D = \underline{\ ?\ }°$ 45

3. $\angle H = \underline{\ ?\ }°$ 90   **4.** $\angle I = \underline{\ ?\ }°$ 45

5. $GH = \underline{\ ?\ }$ units 8

6. $GI = \underline{\ ?\ }$ units 11.4

Related Activities

The concept of similar triangles can be used to solve verbal problems. Have the students make a sketch for each problem showing the similar triangles involved, and then solve the problem.

1. A child 100 cm tall casts a shadow 80 cm long. At the same time, a tree casts a shadow 6 m long. How tall is the tree? 7.5 m

2. A 7 m flagpole casts a shadow 5 m long. How long a shadow does the building 28 m high cast at the same time? 20 m

7-1 Word Phrases and Numerical Expressions

Teaching Suggestions

In order to solve verbal problems, students need to be able to translate word phrases into numerical expressions, and vice versa. This translation in both directions is important and is an area where many students encounter difficulties. To further the understanding of word phrases and numerical expressions, encourage the students to create problems of the type discussed in the Extra! on page 219. The students can then choose a classmate with whom to play the game.

Review the meaning of the term <u>variable</u>, since most of the word phrases and numerical expressions that students will be working with contain variables. Point out that we usually omit the multiplication sign, ✕, when writing variable expressions.

Chalkboard Examples

Translate into a numerical expression.

1. the difference when eight is subtracted from six $6 - 8$

2. three times the sum of four and fourteen $3(4 + 14)$

3. the quotient when y is divided by three $\dfrac{y}{3}$

Translate into a word phrase.

4. $4 + 29$ the sum of four and twenty-nine

5. $-17 - x$ the difference when x is subtracted from negative seventeen

6. $-3x$ the product of negative three and x

Related Activities

Have the students perform the operations on the variables n and m, and then play the following game with a classmate. The last two digits show the age; the first two show the month born.

1. Write the number of the month born.
2. Multiply by 10.
3. Add 20.
4. Multiply by 10.
5. Add your age.
6. Add 165.
7. Subtract 165.

7-2 Evaluating Numerical Expressions

Teaching Suggestions

You might begin by evaluating some expressions which have no grouping symbols, for example, those illustrated by Example 1. Recall for the students that they have worked similar exercises in Section 2-4. You should emphasize that when an expression has a fraction bar, as in Example 3, evaluate and simplify above and below the bar first.

Many students have difficulties with expressions of the type shown in Example 4. Carefully discuss the order of operations and write some expression on the chalkboard for the students to simplify.

Chalkboard Examples

Evaluate each expression for the given value of the variable.

1. $12 + m$; $m = -9$ 3 **2.** $2x - 16$; $x = 34$ 52

3. $7(2 + n)$; $n = -2$ 7 **4.** $\dfrac{3t}{t + 8}$; $t = 1\dfrac{1}{3}$

Simplify.

5. $2 \times 2 + 12 \div 4$ 7 **6.** $-6 + 7^2 - 4 \div 4$ 42

Related Activities

Give the students copies of the cross-number puzzle at the right. Have them evaluate the expressions Across and Down and write the values in the puzzle.

a 2	b 1		c 1	2
	e 9	f 1	5	
		6		
	g 2	h 4	j 3	
i 1	0		k 3	9

ACROSS

a. $x + 13$; $x = 8$
c. $11 - x$; $x = -1$

g. $\dfrac{9^2}{x}$; $x = \dfrac{1}{3}$

i. $\dfrac{-63 + x}{x}$; $x = -7$

k. $2x + 5y$; $x = 2$, $y = 7$

DOWN

b. $-3x + 1$; $x = -6$
c. $x^2 - 1$; $x = 4$

f. $10x + 2y$; $x = 14$, $y = 12$

g. $\dfrac{400}{-5x}$; $x = -4$

j. $x(x + 8)$; $x = 3$

7-3 Word Sentences and Number Sentences

Teaching Suggestions
Write the six verb symbols on the chalkboard and have the students give numerical examples of each. As you proceed to open sentences, give more attention to inequalities since this concept is relatively new to most students. Keep an eye on those students who were having difficulties with Section 7-1 since these two sections are very similar.

Chalkboard Examples
Write a number sentence.

1. The product of four and negative three is greater than negative sixteen. $4(-3) > -16$

2. The sum of x and ten is twenty-one.
$$x + 10 = 21$$

3. The quotient when twice t is divided by eight is less than or equal to ten. $\dfrac{2t}{8} \leq 10$

Write a word sentence.

4. $-9 + 3 \neq 6$ The sum of negative nine and three is not equal to six.

5. $\dfrac{15}{r} < 0$ The quotient when fifteen is divided by r is less than zero.

Related Activities
Compound inequalities were first introduced in Section 2-3. We learned that the sentence $5 < 7 < 10$ means that 7 is between 5 and 10, or that 7 is greater than 5 and less than 10. Have the students translate the following compound inequalities.

Write a number sentence.

1. Three times the sum of n and six is between ten and thirty. $10 < 3(n + 6) < 30$

2. The quotient when x is divided by two is greater than one half and less than four.
$$\frac{1}{2} < \frac{x}{2} < 4$$

7-4 The Solution of an Open Sentence

Teaching Suggestions
In this section, we want to determine whether or not a specific value satisfies the given equation. You should point out that an open sentence is neither true nor false. Only when we replace the variable of an open sentence by a given value do we have a statement (closed sentence) which is either true or false. Also, point out that some open sentences may have more than one solution.

Chalkboard Examples
Replace the variable with the given value, and tell whether the resulting statement is true or false.

1. $n + 9 = 15$; 5 F   2. $-x + 3 \leq 0$; 3 T

3. $-5x > -10$; $\dfrac{1}{10}$ T   4. $2x + 4 \geq 10$; 1 F

Related Activities
Give the students copies of the exercises below. Have them match each equation or inequality in the left-hand column with its solution in the right-hand column.

1. $2t + 6 \leq t + 1$ a. -6

2. $\dfrac{n - 3}{3} > 3$ b. 9

3. $-27 + x = -18$ c. 15

4. $-4 - z = 0$ d. -4

5. $6t + 5 = -13$ e. 10

6. $5x \geq 50$ f. -3

7. $\dfrac{2a + 3}{7} = 3$

8. $\dfrac{r}{2} + \dfrac{1}{2} > -1$

1. d, f	2. c	3. b
4. d	5. f	6. c, e
7. b	8. b, c, e	

7-5 The Solution Set of an Open Sentence

Teaching Suggestions

In the previous section, we tested whether a given number was a solution of an open sentence. In this section, we are interested in *all* possible solutions in a given replacement set.

Most of the terms in this section have been previously introduced, although a review would help. You should particularly review the term replacement set.

Chalkboard Examples

Find the solution set of each open sentence if the replacement set of x is $\{-3, -2, -1, 0, 1, 2\}$.

1. $x \le 0$ $\{-3, -2, -1, 0\}$

2. $x + 2 = -1$ $\{-3\}$

3. $2x < 4$ $\{-3, -2, -1, 0, 1\}$

4. $6 - x > 9$ $\varnothing$

5. Graph the solution set of Exercise 1.

Related Activities

The sets of integers, rational numbers, and irrational numbers together make up the set of real numbers.

The solution set of an inequality whose replacement set is the {real numbers} is graphed on the number line as a solid line. For example, the solution set of $x > -5$ whose replacement set is {real numbers} is shown below.

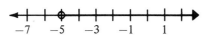

Have the students graph the solution set of each inequality, one graph for each replacement set. The replacement sets are: $\{-4, 0, 4, 8\}$, {integers}, and {real numbers}.

1. $6x - 1 > 11$ **2.** $-x > 9$

3. $5 - x < 1$ **4.** $5x > -11$

7-6 Properties of Equality

Teaching Suggestions

Up to this point, the students have solved equations using trial-and-error; that is, a variable in an open sentence is replaced by a given value and the resulting sentence is either true or false. In the next three sections, procedures will be introduced to solve open sentences.

Discuss with the students the properties on page 231. You should emphasize that when we replace a, or b, or c with a number in an equation such as $a + c = b + c$, we must replace it every time it occurs.

Chalkboard Examples

Use one of the properties of equality to form a true sentence.

1. If $t = 4$, then $t - 7 = \underline{}$. -3

2. If $x = 8$, then $4x = \underline{}$. 32

3. If $a = 40$, then $-\dfrac{1}{2}a = \underline{}$. -20

4. If $n = -40$, then $n + 10 = \underline{}$. -30

5. If $s + 9 = 13$, then $s = \underline{}$. 4

6. If $3p = 48$, then $p = \underline{}$. 16

Related Activities

Many students overlook the fact that c cannot equal zero when using the following property.

$$\text{If } ac = bc, \text{ then } a = b.$$

Show why c cannot equal zero by writing the equations below on the chalkboard. Point out that although all the calculations are correct, a contradictory statement was reached. Ask what is wrong.

$$9(16 - 7 - 9) = (9 \times 16) - (9 \times 7) - (9 \times 9)$$
$$= 144 - 63 - 81 = 0$$
$$8(16 - 7 - 9) = (8 \times 16) - (8 \times 7) - (8 \times 9)$$
$$= 128 - 56 - 72 = 0$$

Since both expressions equal zero,
$$9(16 - 7 - 9) = 8(16 - 7 - 9)$$
And, $$9 = 8$$

7-7 Solving Equations by Transformations

Teaching Suggestions
Write these equations on the chalkboard.

$$2x = 8 \qquad x + 1 = 8 \qquad 3 = 8(x - 1)$$

It should be apparent to most students that the solutions to the first two equations, but not the third, are easily attainable by inspection. Transformations are used to transform complicated equations into simpler equivalent equations. Mention that these transformations are based on the properties of equality.

The students should have little difficulty with the first two transformations. A common oversight of students when using the third transformation, the distributive property, is to multiply the first member of the expression in the parentheses but not the second by the common factor. Similarly, many students will add to or multiply one side of an equation and forget to do the same to the other side.

Chalkboard Examples
Solve. Tell which transformation is used to solve the equation. Check your solution.

1. $x + 4 + 2 = 0$ -6 **2.** $-4 + 6 = x + 2$ 0

3. $x + 9 = 17$ 8 **4.** $9x = 19$ $2\frac{1}{9}$

5. $8 - (-x) = 24$ 16 **6.** $2(x + 1) = 4$ 1

Related Activities
Have the students solve the three equations below. Ask them how the three equations are related.

1. $4x + 2x - 10 + 6 = 8(x + 1) + 2x$

2. $8x + 4x - 20 + 12 = 16(x + 1) + 4x$

3. $2x + x - 5 + 3 = 4(x + 1) + x$
   The solution to all three equations is
   -3. These are all equivalent equations. To get Equation 2, we multiply
   Equation 1 by 2; to get Equation 3, we
   divide Equation 1 by 2.

7-8 Solving Inequalities by Transformations

Teaching Suggestions
The transformations used to solve inequalities are similar to those used to solve equations. The two exceptions are transformations **b** and **f**.

Students usually find solving inequalities more difficult than solving equations, and these difficulties lie in two areas. First, the two transformations mentioned, **b** and **f**, result in a reversal of the inequality sign. Illustrate this with numerical examples. For example, if $3 < 7$, then $7 > 3$. Similarly, if $5 > 3$, then $-5 < -3$.

Another area of confusion lies in the fact that many inequalities have not one solution, but an infinite number of solutions. Point out that three introductory or trailing dots are used to show that a sequence continues in the solution set.

Chalkboard Examples
Solve. The replacement set for each variable is {the integers}.

1. $4 > x$ $\{x < 4\}$ or $\{\ldots, 1, 2, 3\}$

2. $x + 3 \geq 7$ $\{x \geq 4\}$ or $\{4, 5, 6, \ldots\}$

3. $7x < 49$ $\{x < 7\}$ or $\{\ldots, 4, 5, 6\}$

4. $-\frac{1}{3}x < -1$ $\{x > 3\}$ or $\{4, 5, 6, \ldots\}$

Related Activities
Have the students solve the inequalities below, and then answer the questions. The replacement set for each variable is {integers}.

1. $2x + 12 \geq 18$ **2.** $-8 \leq 1 - 3x$
   $\{x \geq 3\}$ $\qquad$ $\{x \leq 3\}$

3. Which number is a solution common to both inequalities in Exercises 1 and 2. $\{3\}$

4. $2x + 8 < 10$ **5.** $2x + 7 > 5 + x$
   $\{x < 1\}$ $\qquad$ $\{x > -2\}$

6. Which numbers are solutions common to both inequalities in Exercises 4 and 5. $\{-1, 0\}$

8-1 Equations into Word Problems

Teaching Suggestions

An important goal of studying mathematics is to solve practical, everyday problems using equations, a goal which most students will recognize. In this section, students will be asked to write a practical word problem for a given equation. This type of exercise may be new and difficult for many students and patience may be necessary.

The students should have little difficulty writing a simple, direct question for the given equation. However, composing a practical word problem for a given equation can be difficult for even the average student. It might help to supply for the students several subject areas or ideas for each equation as you go through the Class Exercises.

Chalkboard Examples

Write two practical word problems that are expressed by the given equation. For the first problem, use a simple, direct question.

1. $4 + 9 = x$ **2.** $19 - 12 = x$

3. $6x = 12$ **4.** $\dfrac{36}{6} = x$

Related Activities

Have the students choose one particular topic, such as shopping or taking an inventory, and write a practical problem concerning the chosen topic for each equation in the Chalkboard Examples above.

8-2 Word Problems into Equations

Teaching Suggestions

Discuss the four-step chart on page 249 with the students and illustrate how each step is applied in the Example. The students will find it very helpful if you work different types of exercises, such as the three Chalkboard Examples below, in parallel using the suggestive aids, where applicable.

Most students will have little difficulty with word problems which are simple, direct translations of an open sentence, such as Class Exercise 6. Difficulties begin to emerge when students are faced with problems which involve mathematical concepts. Therefore, it will be well worthwhile to read those problems which you intend to assign and review the mathematical terms which the students may not be too familiar with. Some of these terms might include consecutive and consecutive even integers, equilateral and isosceles triangles, and perimeter.

Chalkboard Examples

Write an equation which can be used to solve the problem. Use the suggestions listed on page 249.

1. When 15 is added to some number, the sum is 41. $x + 15 = 41$

2. Find two consecutive integers whose sum is 61. $x + (x + 1) = 61$

3. The perimeter of a rhombus is 56 cm. Find the length of each side. $4x = 56$

Related Activities

Write the problems below on the chalkboard or read them orally. Have the students write an equation which can be used to solve the problem.

1. Marcie has pennies and nickels, 15 coins in all, worth 43¢. How many pennies and how many nickles are there?
$$x + 5(15 - x) = 43$$

2. Mike opened his bank and found four times as many dimes as nickels. The dimes and nickels together total $4.95. How many dimes and how many nickels are there? $5x + 10(4x) = 495$

8-3 Solving and Checking Word Problems

Teaching Suggestions
In this section, two more steps in the problem plan are added: solving the equation and checking the answer. Begin by recalling for the students the different transformations used to solve equations. Students who need additional review can be referred to Section 7-7 on page 234. Explain that the solution should be checked in both the original equation and the word problem. Checking the solution in the word problem ensures that the correct equation was used to begin with.

Tell the students that many problems contain extraneous information and that one aspect of problem solving is to determine what information is pertinent and what is not.

Chalkboard Examples
Solve.

1. The sum when 52 is added to a number is 81. Find the number. $x + 52 = 81; x = 29$

2. The sum of two numbers is 51. One of the numbers is twice the other. Find the numbers.
   $2x = 51 - x$; The numbers are 17 and 34.

3. Nancy Gomez spent $2.32 for melon, peaches, and berries. The berries cost 13¢ more than the peaches. The peaches cost 6¢ more than the melon. She gave the clerk a $5 bill. Find the cost of each item.
   $x + (x + 6) + (x + 6 + 13) = 232$; The melon costs 69¢, peaches cost 75¢, and berries cost 88¢.

Related Activities
Many practical problems involve percents (See Chapter 5, pages 153–164.). Have the students write an equation which can be used to solve the problems below and then solve them.

1. A calculator is on sale for $60. This is 80% of the original price. What was the original price? $0.8y = 60; y = \$75$

2. Martina wants the selling price of a chair to be cost plus 15%. If the selling price is $80.50, what was the cost of the chair? $y + 0.15y = 80.50; y = \$70$

8-4 Scale Drawings

Teaching Suggestions
Floor plans and road maps are perhaps the most familiar examples of scale drawings to students. If possible, obtain some blueprints to show the students. Have them measure the length of, say, a wall in the blueprint and then set up a proportion using the scale to find the actual wall length.

Point out that although the text has introduced scales using arrows (1 cm → 3 m), an equality sign is commonly used. Therefore, beginning with Problem 5, the practice of using the arrow will be dropped.

Chalkboard Examples

1. A map is drawn to the scale 1 cm → 100 km. The distance between Center City and Northridge measures 4.8 cm on the map. What is the approximate actual distance between the cities? 480 km

2. A rectangular room measures 4 m × 5.5 m. If you were to make a scale drawing of the room using the scale 1 cm = 0.5 m, what would be the dimensions of the room on the drawing? 8 cm × 11 cm

Related Activities
Suggest that the students measure the lengths of a room of their house or apartment. Include in their measurements large pieces of furniture, appliances, and doorways. Have the students choose an appropriate scale and make a scale drawing of the room and its furnishings. Or, the students might be interested in making a scale drawing of their dream house with some furnishings. Students can then exchange scale drawings with a classmate and determine the actual size of the room and its furnishings.

8-5 Interest Problems

Teaching Suggestions

In Chapter 5, the students learned to solve problems involving percents in various business and consumer settings: commissions, discounts, and budgets. In this section, percents will be used to find interest on loans and savings.

The interest formula, $I = Prt$, should not be new to most students. As you discuss the interest formula with the students, caution them of the two areas where most errors occur: conversion of the percent to a decimal, and conversion of the time period to a decimal or a fractional part of a year. It might help to put examples like the ones below on the chalkboard for the students to complete.

$$7\frac{3}{4}\% = 0.\underline{\ ?\ } \qquad 1 \text{ yr } 9 \text{ mo} = 1.\underline{\ ?\ } \text{ yr}$$

Chalkboard Examples

Find the simple interest.

1. Borrowed $750 at 6% for 6 months $22.50

2. Lent $2000 at 9% for 3 years $570

3. If Mrs. Jacoby receives $1040 interest for a $6500 investment at the end of two years, what is the rate of interest on the investment? 8%

Related Activities

Have the students investigate the rates of interest on savings accounts in their area. Is the rate of interest the same for all savings accounts? Is the principal compounded annually, semiannually, quarterly, or daily? Have the students set up an imaginary savings account. Given the principal and the rate of interest and the number of compounding periods, have the students compute the amount of the principal at the end of two years if no deposits or withdrawals are made.

8-6 Environmental Problems

Teaching Suggestions

Many of the ideas in this section will be new to most students, and thus appear more difficult than they really are. Tell the students to read each problem carefully. You may wish to refer them back to the problem solving suggestions discussed in Sections 8-2 and 8-3.

Explain that the focus of this section is not only the mathematics involved, but also how mathematics is used to solve environmental concerns. Notice that many of the problems involve computing with large or messy numbers. You may wish to have the students use calculators to aid in calculating.

Chalkboard Examples

1. The population of the United States in 1970 was approximately 203,000,000, of which 76,000,000 lived in the suburbs. To the nearest whole percent, what percent of the population lived in the suburbs in 1970? 37%

2. A bath requires about 135 L of water; a shower uses about 95 L. The Madison Water Authority charges 20¢ per 1000 L of water. About how much money could a family of four save annually if each took a daily shower instead of a bath? $11.68

Related Activities

The chart below shows the average household usage and costs for a typical one-month billing for the given appliances.

APPLIANCE	kw · H	COST/MO
Refrigerator/freezer	94.7	$4.90
Frostless	153.4	7.90
Television—B/W	30.1	1.57
Color	41.8	2.15

How much could a household which owns a black and white television and a conventional refrigerator save in a year over a household which owns a color television and a frostless refrigerator?

9-1 Areas of Rectangles and Parallelograms

Teaching Suggestions
This section should require little teaching time since most students have already been introduced to the method for finding the area of parallelograms from an earlier course. You might begin by reviewing the different parallelograms: square, rectangle, rhombus, and parallelogram. Then illustrate how the formula is used to find the area of each figure above. You might show how the formula for finding the area of a square, $A = s^2$, can be derived from the general formula, $A = bh$.

Many students have difficulties converting between units of area. It might be a good idea to work Class Exercises 3–5 with the students and, if necessary, use additional examples to ensure understanding.

Chalkboard Examples
Find the area of each parallelogram.

1. Parallelogram: base 6 m, height 3 m 18 m²

2. Rectangle: length 2 cm, width 1.5 cm 3 cm²

3. Square: side 90 mm 8100 mm²

4. Rhombus: base 5.5 m, height 4.4 m 24.2 m²

5. Is there enough information to find the perimeter of all the figures in Exercises 1–4? Explain. There is enough information to find the perimeters in Exercises 2–4, but not in Exercise 1 where the length of only one pair of sides is known.

Related Activities
Write an equation which can be used to solve the problem. Then solve.

1. A rhombus is known to have an area of 432 square units, and one side measures 20. What is the height of the rhombus?
   $A = bh$; $432 = 20h$; $h = 21.6$

2. A rectangle has an area of 140 cm² and is 4 cm wider than it is high. What are the dimensions of the rectangle?
   $x(x + 4) = 140$; width = 14 cm; height = 10 cm

9-2 Areas of Triangles and Trapezoids

Teaching Suggestions
The formulas for finding the area of a triangle and a trapezoid are very straightforward and should give the students few difficulties. You might point out to the students that Class Exercise 13 and Exercise 24 provide an informal justification for the formulas for the area of a triangle and a trapezoid, respectively.

Chalkboard Examples
Find the area of the triangle or trapezoid.

1. Equilateral triangle: $b = 8$ cm, $h = 6.9$ cm
   27.6 cm²

2. Obtuse triangle: $b = 15$ m, $h = 12$ m
   90 m²

3. Trapezoid: $a = 9$ cm, $b = 12$ cm, $h = 6$ cm
   63 cm²

4. Isosceles trapezoid: $a = 5$ cm, $b = 6$ cm, $h = 4$ cm 22 cm²

5. Is there enough information to find the perimeters of the figures in Exercises 1–4? Which one(s)? The perimeter in Ex. 1 is 24 cm.

Related Activities
Give the students copies of the figures below. Have them find the area of each figure by first dividing into triangles, trapezoids, or parallelograms. The area of each figure is the sum of the areas of its parts.

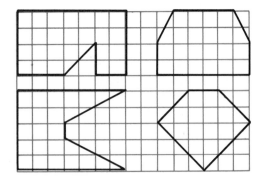

9-3 Areas of Circles

Teaching Suggestions

The students should find that determining the area of a circle to be no more difficult than finding its circumference. Reemphasize that since we are using an approximate value for π, the computed value for the area is only an approximation.

In Exercises 20–22, the students will need to determine the square root of a number in order to find the radius or diameter of the given circle. Students who are not familiar with finding square roots can refer to the Extra! on pages 127–128 for help. (More on square roots is provided in Chapter 12.)

Chalkboard Examples

Find the area of the circle described. Use $\pi \approx \dfrac{22}{7}$ for Exercises 1–2 and $\pi \approx 3.14$ for Exercises 3–4.

1. radius = 35 cm

3850 cm²

2. diameter = 7 m

$38\dfrac{1}{2}$ m²

3. radius = 6 mm

113.04 mm²

4. diameter = 40 km

1256 km²

Related Activities

Show that the formula $A = \pi r^2$ is reasonable for the area of a circle. Draw a circle with diameter 10 cm and divide it into sixteen equal parts. Cut the circle and fit the parts together to form a shape that resembles a parallelogram. Measure the base and height of the parallelogram to the nearest millimeter and find its approximate area using $A = bh$. Find the area of the circle using the formula $A = \pi r^2$. Compare.

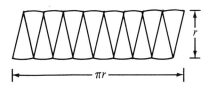

9-4 Volumes of Prisms and Cylinders

Teaching Suggestions

Have available examples of different prisms and cylinders to show the students. Use the examples to point out that each lateral edge in a right prism is perpendicular to the bases. Also point out that the bases are not only parallel, but also congruent.

The formulas for finding volume are very straightforward. You might review the formulas for finding the area of different polygons, as this is usually where many errors occur.

Remind the students that volume is measured in cubic units. Class Exercises 4–6 show the relationship between different units of volume in the metric system. Recall for the students that to compute the relationship between different units of area, we square the numbers. To compute the relationship between units of volume, we cube the numbers. For example, since 1 cm = 10 mm, $1 \text{ cm}^3 = 1000 \text{ mm}^3$.

Chalkboard Examples

Find the volume of each solid. Use $\pi \approx 3.14$.

1. A cube, each side measuring 7 units **343**

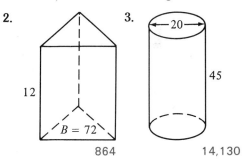

864   14,130

Related Activities

Oblique prisms, cylinders, pyramids, and cones are not discussed in this text. You might show the students some models of these irregular solids. Also, point out that cones may be noncircular. Have the students do some research to determine if there is a method for finding the surface area and volume of irregular solids.

9-5 Volumes of Pyramids and Cones

Teaching Suggestions

The formulas for the area of a pyramid and a cone are closely related to the formulas for the area of a prism and a cylinder, respectively. An informal verification of these relationships is given in the Application on page 289 and in the Related Activities below.

You might wish to point out that the perpendicular distance from the vertex to the base is called the height, but the perpendicular line segment from the vertex to the base is actually called the altitude.

Chalkboard Examples

Find the volume of the figures. Use $\pi \approx \frac{22}{7}$.

1.

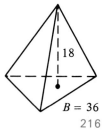

$B = 36$

216

2.

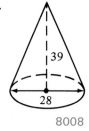

39

28

8008

Related Activities

Using the patterns below and the nine steps on page 289, show the relationship between the volume of a square pyramid and the volume of a prism. Dimensions are given in centimeters.

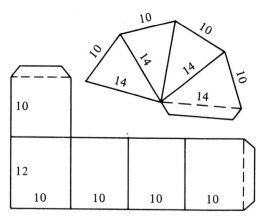

9-6 Surface Areas of Prisms and Cylinders

Teaching Suggestions

It might be a good idea to use some models of right prisms and right cylinders to illustrate the difference between lateral area and total surface area. Although the text gives clear, concise formulas for finding lateral and total surface areas, many teachers prefer to have the students find the area of each polygonal face or each circular region and then find the sum of all regions. You may wish to use this approach.

Chalkboard Examples

Find the lateral area and the total surface area of each solid. Use $\pi \approx \frac{22}{7}$.

1.

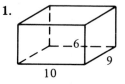

6

9

10

228; 408

2.

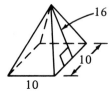

8

6

$301\frac{5}{7}$; 528

Related Activities

Have the students find the total surface areas of some pyramids, like the one at the right. Explain that we use the slant height to find the surface area of a pyramid. The slant height of a pyramid is the height of each triangular face of the pyramid. Provide pyramids with different polygonal bases.

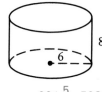

16

10

10

9-7 Mass and Liquid Capacity

Teaching Suggestions

Many students are confused as to when to use mass and when to use weight. You can usually resolve this confusion by pointing out that a baseball has the same mass whether it lies on the moon or whether it lies on Earth. However, a baseball will weigh less on the moon than on Earth because the force of gravity on the moon is less. In everyday usage, mass and weight are roughly equal, although the distinction is very real to scientists.

Discuss the relationship between units of volume and mass. Simply stated, $1000 \text{ cm}^3 = 1000 \text{ g} = 1 \text{ L}$. It might be a good idea to go over Class Exercises 1–17 before proceeding to the tables of density and the associated examples.

Chalkboard Examples

Complete.

1. $200 \text{ mL} = \underline{\ ?\ } \text{ cm}^3$
   200

2. $4 \text{ L} = \underline{\ ?\ } \text{ mL}$
   4000

3. $2.5 \text{ L} = \underline{\ ?\ } \text{ cm}^3$
   2500

4. $3.4 \text{ kg} = \underline{\ ?\ } \text{ g}$
   3400

Give the mass of each of the following. Use the tables in the text.

5. 3 L of fresh water
   3 kg

6. 800 mL of mercury
   10.88 kg

7. 2 cm^3 of gold
   38.6 g

8. 1000 cm^3 of steel
   7.7 kg

Related Activities

You can use a method called displacement to find the volume of an irregularly-shaped object for which there is no formula, such as a pair of scissors. First, fill a graduated container with enough water to cover the scissors. Note the water level reading. Place the scissors into the container and note the new water level reading. The volume of water displaced is equal to the volume of the scissors. Have the students perform this experiment. You may have to use another irregularly-shaped object if scissors are not available.

9-8 Spheres

Teaching Suggestions

Begin by comparing the definition of a sphere with that of a circle. Explain that a sphere consists of all points in space at a given distance from the center. Have the students give some examples of spheres: basketball, baseball, balloon, globe of the earth.

The formulas for the surface area and the volume of a sphere should cause little difficulties. Notice that computations will be minimal since answers to the exercises are to be expressed in terms of π.

As you are discussing the feature "Accuracy of Measurement; Significant Digits," point out that $51.5 \text{ mm} \leq XY < 52.5$ and $XY = (52 \pm 0.5) \text{ mm}$ are not exactly equivalent. This is because XY may equal $52 - 0.5$, but cannot equal $52 + 0.5$, only approach it.

Chalkboard Examples

Find the surface area and volume of a sphere with the given radius. Leave answers in terms of π.

1. radius 4 $A = 64\pi$; $V = 85\frac{1}{3}\pi$

2. radius 30 $A = 3600\pi$; $V = 36{,}000\pi$

Related Activities

The sun, planets, and their moons are all spheres. Write the diameters of the sun, the planets Mars and earth, and the earth's moon on the chalkboard.

Earth 12,700 km   Earth's moon 3500 km
Mars 6700 km   Sun 1,382,400 km

Have the students find the surface area of each body. The answers may be left in terms of π. If calculators are available, the answers should be completely worked out and expressed in scientific notation.

10-1 Experiments with Equally Likely Outcomes

Teaching Suggestions
Most students will find the topic of probability interesting. Encourage them to perform some of the experiments outlined in the examples and exercises. You should point out that all of the experiments are understood to be random experiments. Stress that a sufficiently large number of trials must be performed before the actual outcome of an experiment can approximate the probability $\frac{1}{n}$.

Chalkboard Examples
A jar contains 6 marbles of the same size: 1 red, 1 yellow, 1 black, 1 white, and 1 green. While blindfolded, you pick a marble.

1. Name the possible outcomes. You could pick any one of the six different marbles.

2. How many possible outcomes are there? 6

3. Are the outcomes equally likely? Yes

4. What is the probability of each outcome? $\frac{1}{6}$

5. If the red marble were considerably larger than the other marbles, would the outcome be equally likely? Explain. No; It would be easier to pick or avoid the red marble.

Related Activities
Ask the students to think of examples in which the outcomes of random drawings would not be equally likely. For example, why would a magician's deck of "trick" cards not result in equally likely drawings? How is this accomplished? How do "loaded" dice work?

10-2 The Probability of an Event

Teaching Suggestions
In this section, we are interested in not just one particular outcome, but in a group of favorable outcomes called an <u>event</u>. Point out some examples of events, such as the number of face cards or the number of red cards in a deck of 52 regular playing cards. Once you have gone through a few examples, the students should have little difficulty with the exercises.

Chalkboard Examples
Assume that the pointer on the board at the right is spun and can stop at random at any point but not on a division line. Find each probability.

1. p(3) $\frac{1}{4}$

2. p(number less than 4) $\frac{3}{4}$

3. p(shaded) $\frac{1}{2}$

4. p(1, 2, 3, or 4) 1

5. p(odd number) $\frac{1}{2}$

6. p(number less than 1) 0

Related Activities
Many family games make use of a board with a spinner. Have your students look through games which they have at home and bring the boards with spinners to class. Hold the spinners up before the class and have them give the probability of each described event.

10-3 Odds

Teaching Suggestions

Many students have the mistaken impression that odds and probability are the same concept. Point out that probability is a ratio of *numbers* (whole) of outcomes whereas odds is a ratio of *probabilities*. In most cases, to find the odds for or odds against an event happening, we must find the quotient of two fractions. You should emphasize that the value of a probability ranges from 0 to 1, whereas the value of odds may be greater than 1.

Chalkboard Examples

One of the cards shown at the right is picked at random. State the probability that the event below will occur and the odds in favor of the event.

1. The letter N is picked. $\frac{1}{6}$, 1 to 5

2. A vowel is picked. $\frac{1}{2}$, 1 to 1

3. The letter O is picked. $\frac{1}{3}$, 1 to 2

4. A letter in "TWO" is picked. $\frac{2}{3}$, 2 to 1

Related Activities

Use the definition of odds to derive a formula which would make computing odds easier. The formula for odds in favor of an event can be derived as follows.

$$
\text{Odds in favor} = \frac{p(\text{event occurs})}{p(\text{event does not occur})}
$$

$$
= \frac{\dfrac{f}{n}}{1 - \dfrac{f}{n}} = \frac{\dfrac{f}{n}}{\dfrac{n-f}{n}}
$$

$$
= \frac{f}{n} \div \frac{n-f}{n} = \frac{f}{n} \times \frac{n}{n-f}
$$

$$
= \frac{f}{n-f}
$$

Have the students use the formula to confirm that it actually computes the odds in favor of an event. Then have the students derive the formula for the odds against an event happening.

10-4 Estimating Probabilities

Teaching Suggestions

We now turn our attention to some practical applications of probability. In the previous sections, the probability of an outcome was computed based on the assumption that each outcome had an equally likely chance of occurring. In most real-life situations, however, the probability is actually estimated on the basis of repeated observations. These estimated probabilities are then used as standards of future performance and predictions.

Chalkboard Examples

1. Leslie has successfully completed 7 field goals out of 20 attempts. What is the probability that he will score a field goal in his next attempt? $\frac{7}{20}$

2. The temperature on the first day in May has been above 30°C 16 times in the past 20 years. What is the probability that the temperature will be below 30°C next May 1st? $\frac{1}{5}$

3. The horse *Sundance* has won 1 of his last 6 races. What are the odds in favor of his winning his next race against a similar field? 1 to 5

Related Activities

Sports events, traffic control, and the weather are some of the applications of estimated probabilities mentioned in the exercises. Discuss how estimated probabilities are used to compile mortality and accident tables so that insurance companies can fairly set their rates. Also, point out that business investments are often made on the basis of past performance. You might have the students research other areas where estimated probabilities are applied.

10-5 Estimating from a Sample

Teaching Suggestions

Bring to class some articles from newspapers or magazines which show results of television ratings, public-opinion polls, or marketing surveys. Discuss what effects, if any, a poll might have on a product.

Or, have the students imagine that they are part of a consumer-testing service that is trying to issue a report on the bacterial levels of hamburgers in fast-food restaurants. Ask the students how they might select a random sample for testing. Obviously, all hamburgers cannot be tested.

The two examples above and those in the text are a few applications of sampling. You should point out to the students that it is often very difficult to obtain a representative and yet random sample.

Chalkboard Examples

1. Every third person leaving a subway train is asked about the need for additional funds for public transportation. Would this be a random sampling of the opinions of city residents toward public transportation? No

2. The *Journal* sent a sample newspaper to 11,250 randomly selected households. Of these households, a survey of 500 show that 30 households plan to subscribe. About how many of the 11,250 households can the newspaper expect as new subscribers? 675

Related Activities

Suggest some topics of current interest to the students. Have them conduct a survey to determine the general opinion toward a particular topic. Some topics of interest might include the outcome of a school election, the current attitude toward a particular school subject, or simply the general opinion toward a particular book or teacher. Such surveys of student interests are usually well received in school papers.

10-6 Random Variables and Expected Value

Teaching Suggestions

The concepts of random variables and expected values will be new to most students. Begin by discussing how the observed average was computed in the text example. Then carefully go through the steps which led to the predicted average, or commonly referred to as the expected value. It is important to point out that the expected value of a random variable is an *average* value computed from a large number of trials.

You may wish to point out that many of the probabilities in this section are expressed as decimals or percents, a departure from the fractional representation in previous sections.

Chalkboard Examples

1. There are four slips of paper in a box, each paper has a number: 1, 2, 3, or 4. You draw a slip at random. If N is the number on the paper, what is the expected value of N? 2.5

2. There are ten new bills in a box: 4 $1-bills, 3 $2-bills, 2 $5-bills, and 1 $10-bill. You draw a bill at random. If V is the dollar value of the bill, what is the expected value of V? $3

Related Activities

Ask the students to consider a game with one die. A roll of 3, 4, or 5 wins nothing. A roll of 1 or 2 wins $2. A roll of 6 wins $4. If V is the value of the winnings, what is the expected value of V? What amount should be charged a player to play the game if the "house" wants to break even? $5.00; $5.00

Have the students invent a card game or some game of chance, complete with rules. Determine the expected value of the winnings and what should be charged a player to play the game in order for the house to break even.

11-1 Graphing an Ordered Pair of Numbers

Teaching Suggestions
Most of the students will find naming and graphing ordered pairs of numbers both simple and enjoyable. On the chalkboard, draw a rectangular number system in the plane. You might wish to point out that many textbooks refer to this number system as simply the Cartesian Plane. Point out that order is important and illustrate this by graphing, say, (5, 2), (2, 5), (5, −3), and (−3, 5). Also, graphs of points which lie on the axes are sometimes a source of confusion, so point out the distinction between (0, t) and (t, 0). Stress that each ordered pair is associated with exactly one point in the number plane.

Chalkboard Examples
Name the graph of each ordered pair.

1. (4, 2) *A*
2. (−4, −4) *E*
3. (1, −3) *F*
4. (−2, 3) *B*

Graph the point on the plane.

5. (0, 1)
6. (−3, 3)
7. (2, 4)
8. (−1, 0)

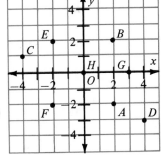

Related Activities
Extend the idea of points in a number plane to points in a number space. Now, instead of an ordered pair of the form (x, y), we have an ordered triple of the form (x, y, z) where z is a component along an axis perpendicular to both the horizontal and vertical axes. Have the students research number spaces and try to draw a graphical representation of a number space.

11-2 The Coordinates of a Point in the Plane

Teaching Suggestions
The term abscissa refers to the x-coordinate, the first number of an ordered pair; the term ordinate refers to the y-coordinate.

Point out that the quadrants are numbered counterclockwise beginning with the upper right. Students should be asked to learn the range of values for the coordinates of any point in a particular quadrant. This should be relatively simple if the students merely examine some sample points in that particular quadrant.

Chalkboard Examples
State the coordinates of the point.

1. A (2, −2)
2. B (2, 2)
3. C (−4, 1)
4. D (4, −3)
5. E (−2, 2)
6. F (−2, −2)
7. G (3, 0)
8. H (0, 0)
9. Name the points shown in Quadrant II. Are the ordinates positive? *C, E; Yes*

Related Activities
Beginning with point A, have the students name in order the coordinates of each vertex of the pentagon.

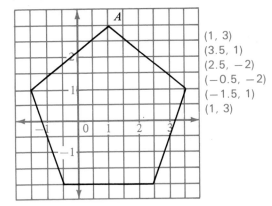

(1, 3)
(3.5, 1)
(2.5, −2)
(−0.5, −2)
(−1.5, 1)
(1, 3)

11-3 Equations in Two Variables

Teaching Suggestions

The topic of equations in two variables will be new to most students. Explain that it is usually easier to find solutions for equations in two variables if the equation is expressed in terms of one variable, that is, if the equation is solved for y in terms of x. The transformations discussed in Section 7-7 are used to find these equivalent equations.

Encourage the students to not only find integral solutions, but also decimal solutions. For example, a decimal solution of the equation $-2x + y = 3$ is $(0.3, 3.6)$.

Chalkboard Examples

Which ordered pair is a solution of the equation?

1. $x - y = 10$ $(7, 3)$ $(-13, 3)$ $\underline{(3, -7)}$

2. $4x = 2y$ $\underline{(-2, -4)}$ $(8, 4)$ $\left(\frac{1}{2}, -1\right)$

3. $2x - 3y = 1$ $\left(0, \frac{1}{3}\right)$ $(-5, 3)$ $\underline{(8, 5)}$

Solve for y in terms of x. Then find solutions of the equation which have abscissas of -4 and 2.

4. $x = y - 8$ $y = x + 8$; $(-4, 4)$, $(2, 10)$

5. $4x - 3y = 2$ $y = \dfrac{4x - 2}{3}$; $(-4, -6)$, $(2, 2)$

Related Activities

Write the equation $x + y - 2z = 5$ on the chalkboard and show them that the ordered triple $(1, 2, -1)$ is a solution of the equation. Have the students find other solutions to the equation, both integral and decimal.

11-4 The Graph of a Linear Equation

Teaching Suggestions

In Exercises 26–34, the students should have noticed that there can be an infinite number of solutions for any linear equation. Point out that although a linear equation has an infinite number of solutions, only two are needed to determine the graph of the equation. This is because one and only one line may be drawn through any two points; hence, the line is unique.

Point out that if either a or b equals zero in the general form of a linear equation, then we have a horizontal or vertical line, respectively.

Chalkboard Examples

State ordered pairs of the form $(x, 0)$ and $(0, y)$ that are solutions of the given equations.

1. $x + y = 6$
   $(6, 0)$, $(0, 6)$

2. $x - y = -2$
   $(-2, 0)$, $(0, 2)$

3. $2x + y = 5$
   $\left(\frac{5}{2}, 0\right)$, $(0, 5)$

4. $3x + 4y = 24$
   $(8, 0)$, $(0, 6)$

5. Graph the equation in Exercise 3 on a coordinate plane.

Related Activities

Write the equation $xy = 12$ on the chalkboard. Ask the students to find all possible positive whole number solutions and then all possible negative integral solutions. Graph the positive solutions on a coordinate system and point out that the graph of $xy = 12$ is not a straight line. Join the points with a smooth curve. Then graph the negative solutions and join the points with a smooth curve. Equations of the form $xy = b$ are called hyperbolas. Ask the students if the curves will ever cross the x- or y-axis. Why not?

11-5 Graphing a System of Equations

Teaching Suggestions

Go through the example in the text with the students. Emphasize that solutions obtained from graphing must be checked in both equations since only an approximate solution can be determined from looking at a graph.

Point out that there are three possible outcomes when solving a system of two linear equations by graphing: (1) the graphs can intersect in one point, (2) the graphs can be parallel, or (3) the graphs can coincide. If the graphs intersect, then the system has exactly one solution. If they are parallel, the system has no common solution. If they coincide, the system has an infinite number of common solutions.

Chalkboard Examples

Is the ordered pair of numbers a solution of the system of equations?

1. $(4, -6)$ $x + y = -2$ No
                 $x - y = 8$

2. $(3, 3)$ $2x + y = 9$ Yes
           $x - 2y = -3$

Find the solution of each system by graphing. Check your results.

                              $(-2, -1)$

3. $x - y = 12$ 4. $3x + y = -7$
   $x + y = -4$ $(4, -8)$ $2x + 2y = -6$

Related Activities

The system of linear equations (1) $2x + y = 5$ at the right can be solved by   (2) $x - y = 4$ using <u>substitution</u>.

a. Express (2) in terms of x. $y = x - 4$
b. Substitute into (1). $2x + x - 4 = 5$
c. Simplify. $3x - 4 = 5$
d. Solve for x. $3x = 9$
                         $x = 3$

e. Solve for y. $y = x - 4$
                       $= 3 - 4 = -1$
f. The solution of the system: $(3, -1)$

Have the students use substitution to solve Exercises 1–9 on page 353.

12-1 Square Roots

Teaching Suggestions

Point out that "Any nonnegative real number . . ." is just a shorter way of saying "Any positive real number or zero . . . ." Emphasize that we will primarily be concerned with the positive square root of a number, but the students should be aware that any nonnegative real number has two square roots, one positive and one negative. In fact, the symbol "$\sqrt{16}$" denotes "the positive square root of 16." The negative square root of a number is symbolized by "$-\sqrt{}$."

The students will find the exercises much easier if they are familiar with the perfect squares from 1 to, say, 225. Point out that the order of operation in the B exercises states that the expression under the radical must be simplified first. You might show how an incorrect answer can result otherwise. For example,

$$\sqrt{144 + 25} \neq \sqrt{144} + \sqrt{25} = 17,$$

but

$$\sqrt{144 + 25} = 13.$$

Chalkboard Examples

State the integer named by each symbol.

1. $\sqrt{1}$ 1   2. $\sqrt{100}$ 10   3. $-\sqrt{36}$ -6
4. $\sqrt{0}$ 0

Name the two consecutive integers between which the number lies.

5. $\sqrt{8}$ 2, 3   6. $\sqrt{95}$ 9, 10   7. $\sqrt{56}$ 7, 8

Related Activities

Any real number may be expressed as the product of three equal factors. Each factor is called a <u>cube root</u> of a real number n; that is, if a is a cube root of n, then $a^3 = n$. Moreover, if the number is positive, the cube root is positive; if the number is negative, the cube root is negative. Name some numbers between 1 and 1000, and ask the students to state the consecutive integers between which the cube root lies, provided the number is not a perfect cube.

12-2 Approximating Square Roots

Teaching Suggestions
Carefully go through the example in the text with the students. Outline the steps in the divide-and-average method. Point out that a good estimate helps to reduce the number of times Steps 1 and 2 must be performed before the quotient and divisor agree. Encourage the students to always check their answers by squaring.

Chalkboard Examples
What would you use as a first estimate for each square root? Answers will vary.

1. $\sqrt{6}$ 2. $\sqrt{18}$ 3. $\sqrt{45}$ 4. $\sqrt{111}$

State the next estimate for the square root of the dividend.

5. $2\overline{)6}$ quotient 3, answer 2.5

6. $4.1\overline{)18.0\,00}$ quotient 4.39, answer 4.25

7. $11\overline{)111.00}$ quotient 10.09, answer 10.55

Related Activities
Use the divide-and-average method to approximate $\sqrt[3]{2744}$.

1. *Estimate the root.* Use 10 as an estimate.

2. *Square the estimate.* $10^2 = 100$

3. *Divide the number by the squared estimate.*

$$2744 \div 100 \approx 27$$

4. *Find the average of the quotient above and twice the original estimate.*

$$\frac{27 + 10 + 10}{3} \approx 16$$

5. *Repeating steps 2–4, using 16 as an estimate, you will find that* $\sqrt[3]{2744} = 14$.

Notice in step 4 that a weighted average is used to approximate the cube root; a direct average is used to approximate the square root. Use the above method to approximate $\sqrt[3]{13,824}$.

12-3 Using a Square-Root Table

Teaching Suggestions
Reading a square-root table is relatively easy and the students should have little difficulty. The example in the text which illustrates the use of interpolation should be discussed on the chalkboard. It might be a good idea to work another example, say $\sqrt{5.8}$, to ensure understanding.

Chalkboard Examples
Use the table to approximate the square root.

1. $\sqrt{34}$ 2. $\sqrt{99}$ 3. $\sqrt{7}$ 4. $\sqrt{18}$
   5.831   9.95   2.646   4.243

Use interpolation to approximate each square root, to the nearest hundredth.

5. $\sqrt{26.3}$ 6. $\sqrt{84.8}$ 7. $\sqrt{60.5}$
   5.13   9.21   7.78

Related Activities
In the Extra! on page 369, the students learned that if $x > 0$ and $y > 0$, then $\sqrt{xy} = \sqrt{x} \times \sqrt{y}$. Similarly, we may reason that if $x > 0$ and $y > 0$, then $\sqrt{\dfrac{x}{y}} = \dfrac{\sqrt{x}}{\sqrt{y}}$. Thus, the positive square root of the quotient of two positive numbers is equal to the quotient of their positive square roots. For example,

$$\sqrt{\frac{3}{4}} = \frac{\sqrt{3}}{\sqrt{4}} \approx \frac{1.732}{2} = 0.866.$$

Write the exercises below on the chalkboard and have the students approximate each square root to the nearest hundredth. Use the table on page 394.

1. $\sqrt{\dfrac{8}{9}}$ 2. $\sqrt{\dfrac{16}{5}}$ 3. $\sqrt{\dfrac{11}{36}}$ 4. $\sqrt{\dfrac{19}{22}}$
   0.94   1.79   0.55   0.93

12-4 The Pythagorean Property

Teaching Suggestions

The Pythagorean Property and its converse are among the most basic and important results in geometry, both having many practical applications. Point out that the hypotenuse is always the longest side in a right triangle and that the square of the hypotenuse must equal the sum of the squares of its legs.

Carefully go through the examples in the exercises with the students. These exercises form the basis for many upcoming problems.

Chalkboard Examples

State whether or not a triangle with sides of the given lengths is a right triangle.

1. 8, 12, 15 No **2.** 12, 16, 20 Yes

The given lengths are the legs of a right triangle. Find the length of the hypotenuse of the triangle to the nearest hundredth.

3. 32, 24 40.00 **4.** 3, 5 5.83

5. 6, 6 8.49 **6.** 7, 2 7.28

Related Activities

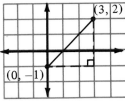

Draw a rectangular coordinate system on the chalkboard and graph the points (3, 2) and (0, −1). Point out that the Pythagorean Property can be used to find the distance between any two points on the number plane if the line segment joining the points is not parallel to either axis. Complete the triangle. The length of both legs is 3 units. Since $c^2 = 3^2 + 3^2 = 18$, c is about 4.24. Have the students graph the points below and find the distance between them, to the nearest hundredth.

1. (−4, 5), (1, −2) 8.60 **2.** (2, 0), (−6, 4) 8.94

12-5 Special Right Triangles

Teaching Suggestions

The two special right triangles discussed in this section are the isosceles right triangle and the 30°–60° right triangle. Point out that the two related properties discussed in the section form the basis from which the lengths of the sides of other similar triangles can be computed, given the length of one side. Strongly encourage the students to learn the derivations of these properties, since they will frequently encounter these in later studies in geometry and trigonometry.

Chalkboard Examples

State the values of the missing angle and lengths in each figure.

1. 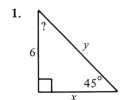 **2.**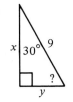

$x = 6; y = 6\sqrt{2};$ 45° | $x = 4.5\sqrt{3};$ $y = 4.5;$ 60°

Related Activities

Copy the trapezoid below on the chalkboard and label as shown. Challenge the students to find the values of the missing angles and lengths. If a hint is necessary, suggest that they draw a perpendicular line between the parallel sides forming one of the special right triangles discussed.

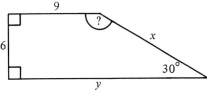

$x = 12; y = 9 + 6\sqrt{3};$ 150°

12-6 Trigonometric Ratios

Teaching Suggestions
Most of the terms in this section will be new and confusing to the students, but should become clearer as the students delve into the topic. Keep in mind that most errors which students make at this early stage result because the students are confused at the many new and different terms.

Point out the distinction between side opposite and side adjacent to an angle. Encourage the students to memorize the three trigonometric ratios on page 380 and emphasize that the value of the trigonometric ratio for an angle is the same for all similar triangles regardless of size.

Chalkboard Examples
State the value of each trigonometric ratio.

1. $\sin A \dfrac{a}{c}$ 2. $\cos A \dfrac{b}{c}$

3. $\tan A \dfrac{a}{b}$ 4. $\sin B \dfrac{b}{c}$

5. $\cos B \dfrac{a}{c}$ 6. $\tan B \dfrac{b}{a}$

7. $\sin D \dfrac{27}{29}$

8. $\cos E \dfrac{27}{29}$

9. $\tan E \dfrac{10}{27}$

Related Activities
The text points out that the value of the sine ratio approaches 1 as the size of the angle nears 90°; the value approaches 0 as the angle nears 0°. Ask the students what happens to the value of the cosine ratio as the size of the angle nears 90° or 0°. What happens to the tangent ratio as the size of the angle increases? What happens as the angle nears 0°?

12-7 Using a Trigonometric Table

Teaching Suggestions
Review the "tools" which will be used to solve right triangles: the angle measures of any triangle total 180°, the Pythagorean property, and the trigonometric ratios.

Point out that solving right triangles is similar to solving word problems. First, examine the information given. Next, formulate a plan or an equation which can be used to solve the problem. Then, carry out the operations and check the answer.

Chalkboard Examples
Use the table on page 395 to state an approximation for each ratio.

1. $\sin 75°$ 2. $\cos 6°$ 3. $\tan 44°$
    0.9659   0.9945   0.9657

Use the table to find the measure of the angle to the nearest degree.

4. $\sin A = 1.000$ 90°   5. $\cos A = 0.69$ 46°

6. $\tan A = 12.56$ 85°   7. $\sin A = 0.5$ 30°

Find the angle measure to the nearest degree and the lengths to the nearest whole number.

8. $\angle B$ 25°

9. x 12

10. y 5

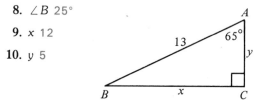

Related Activities
Solve.

1. An extension ladder 8 m long touches a house 7 m above the ground. What is the measure of the angle the ladder forms with the house? 29°

2. To straighten his tree, Mr. Weiss plans to tie a cord to the tree trunk 2 m above the ground. The other end of the cord will be tied to a stake 1.5 m from the base of the tree. Mr. Weiss has 240 cm of cord. Does he have enough? No

Optional Activities

In this next section of the Teacher's Manual, you will find twelve permission-to-reproduce pages of recreational and enrichment activities which you may use to supplement the regular course work. They are completely optional, and although they are intended for all students may be given to students who have successfully completed the regular course work. You may even wish to use them to provide a break in the normal routine between chapters. Regardless of how you use them, the activities should prove interesting and challenging to the students, and we hope you enjoy using them.

CONTENTS

Word Search of Mathematicians

Some famous mathematicians are listed below. All the names listed below appear in the matrix. They may have a vertical, horizontal, or diagonal orientation, and may be read forwards or backwards. How many names can you find?

ARCHIMEDES   ARISTOTLE   ABEL   APOLLONIUS
BERNOULLI   BOOLE   CANTOR   CAYLEY
DEDEKIND   DESCARTES   EINSTEIN   EUCLID
EULER   FERMAT   FIBONACCI   FOURIER
GALILEO   GALOIS   GAUSS   HILBERT
JACOBI   LAGRANGE   LAPLACE   LEIBNIZ
NEWTON   PASCAL   PYTHAGORAS   ZENO

Copyright © 1979 by Houghton Mifflin Company. Permission to reproduce granted to users of MATHEMATICS, Structure and Method.

```
M I V X E A D M C H C V U N T F L F V P W G U T T
C E X J L Z F L L C O R A K Z Y J U R C A A I L W
Y U U N P U E J N Q P F R L A K N T A F N M M I I
O M I C O C A N K S Y J C D B P U Z L M V J M J L
Z W I E L T M S O D T X H E T S O C Q I K P N A Y
W A C O L I W Z E H H X I S R P D L K S U C G Q O
C L C Z X Z D E I W A I M C E J L Z L T A R W Q S
Z C A Q F I W T N C G A E A B J B Y F O A U Z K E
F I N P E N C W S T O D D R L S Q T V N N I F P W
B P O G P B L E T M R V E T I Y S D G F M I Z X E
F N B U M I C M E S A O S E H D X E V J A S U Q I
B Y I D Q E S A I M S H Y S B Q S E N H C B O S E
I D F Z E L N Y N N T V J F E X V L X X K J I K R
D E D E K I N D X S F E R L R A E R G I I D G X S
Y B A D A E G F S M O R G L N P R K C A Y L E Y T
Q R K G A A V U O Q F E L O O L P I Q M H E X R R
A C T P F T A L E U C M R A U A M A S O Z C K G A
A J A A A G C D A U R L K J L T V F S T Y R Z A S
Q A K N M E F Q B P L I K E L O Z G X C O T N L S
F C G F T R S U C P L E E A I F H A B J A T W I R
E O W W J O E U R R R A R R D K G L O O Y L L L S
F B I H U S R F P Q N R C B E G W O U C O H A E J
H I J S K X V C Y T I G N E J I H I X N X L T O Z
A D G I K A Z H J S F Z J U J D E S S D T X E N G
W J E G K A B E L B F N G P H F Y N U G B M N G F
```

Do a brief biographical sketch of one of these mathematicians.

Using the Calculator

The calculator can be used to perform a large number of calculations or calculations involving large numbers both quickly and accurately. Use a calculator to help solve the problems below.

1. Write the following fractions as decimals.

$$\frac{10}{81} \qquad \frac{100}{891} \qquad \frac{1000}{8991}$$

Use the pattern to predict the decimal equivalents for these.

$$\frac{10,000}{89,991} \qquad \frac{100,000}{899,991} \qquad \frac{1,000,000}{8,999,991}$$

Check your answers using a calculator.

2. Write the decimal equivalents for the following fractions.

$$\frac{1}{111} \quad \frac{2}{111} \quad \frac{3}{111} \quad \frac{4}{111} \quad \frac{5}{111} \quad \frac{6}{111} \quad \frac{8}{111} \quad \frac{10}{111} \quad \frac{15}{111} \quad \frac{25}{111}$$

3. Find the products.

(a) 22×22

(b) 202×202

(c) 2002×2002

(d) $200,002 \times 200,002$

4. Choose a partner. Enter any number between 30 and 40 into a calculator. Players take turns subtracting 1, 2, or 3 from the number displayed. The first player to get a display of zero loses.

SAMPLE GAME

Start: Enter 31	Display: 31
Player 1: Subtracts 1	Display: 30
Player 2: Subtracts 3	Display: 27
Player 1: Subtracts 2	Display: 25
Player 2: Subtracts 1	Display: 24

AND SO ON . . .

If a player goes below zero, he or she loses. After you have played the game several times, see if you can come up with a strategy that will guarantee that you win every time.

Networks

The game of Networks can be played by two or more players. The game can begin with any number of dots although it is usually best to learn and experiment with a game involving three dots before proceeding to more.

Each player in turn connects one dot to another or to itself by drawing a curve. The curve may have any shape but must not cross itself, cross a previously drawn curve, or pass through a previously made dot. A new dot is then placed anywhere along the new curve. No more than three curves may be connected to any one dot. The winner is the last person who is able to play.

A typical game involving three dots is shown below. The dashed curves indicate the move, and the game was won in the seventh move by the first player.

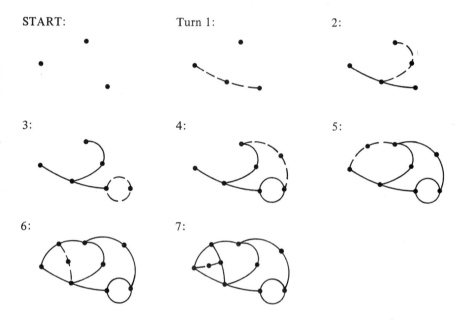

Play the game several times. See if you can come up with a strategy which guarantees a win. It has been shown that a game involving n dots must end in at most $3n - 1$ moves.

The Alphabet Wheel

The alphabet wheel at the right is a common method of coding messages. We can use the wheel to code the message "HELP, AM SINKING." into the following:

8 5 12 16 27 29 1 13 29 19 9 14 11 9 14 7 28

However, a person who knows that the alphabet wheel was used to encode the message could decode it in little time.

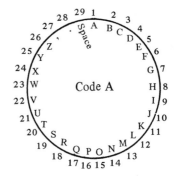

One way to make the code more difficult to break would be to rotate the alphabet, as shown at the right. The coded message above could then be coded back into an alphabetic form. Our original message now looks like this: KHOSACDPCVLQNLQJB

The following reply was coded using the method above. Decode the message and see what the reply was: WU.CVZLPPLQJB

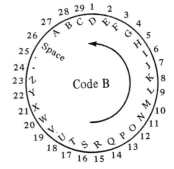

We can devise a code that is even more intricate by using Code A or B above and then adding a number to or multiplying each number by another number. For example, the message below was coded using Code B and then multiplying each number by 2. Use the spiral wheel at the right to help decode the message:

CJYPDCZDYADWEJLCWLCWVXEU

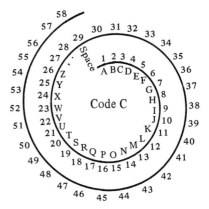

As you spend more time working with the alphabet wheel, you will see that it is not very difficult to devise a code that only a few can break. Choose a classmate. Each of you can devise a code involving the alphabet wheel. Encode a message and see if you can break each others coded message.

Cross-Number Puzzle

ACROSS

2. The reciprocal of 0.008
4. The square of 116
6. $3^2 \times 31$
7. 80% of 1240
9. 479 cubed
12. $5 + 23 \times 81 - 344$
13. $3^5 \times 5 \times 7$
15. The LCM of 6, 16, and 34
16. The repetend of $1 \div 11$
17. The cube root of 125,000
18. This palindrome contains the digits 1 and 7.
19. The GCF of 5096 and 8918
22. The sum of the first 99 counting numbers
24. $333,667 \times 1656$
27. The repetend of $25 \div 111$
28. The digits of $2 \times 3 \times 13$, reversed
29. $6(21x^4y^2 + 67)$; $x = 2$, $y = 5$
31. The decimal representation of CCIII

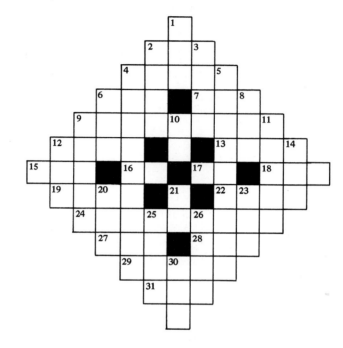

DOWN

1. The product of 12 and 52
2. The greatest prime number less than 1400
3. $58,716 \div 10.5$
4. $1001 \times 179,225$
5. The digits of 666^3 reversed
6. 178, 186, 194, __?__
8. $2^8 - 21$

DOWN (continued)

9. XXV times DCXXV
10. For $(7ab + 61) \times 13$ to equal 9906, a and b represent what digits?
11. 50% of what number is 45,376
12. The reciprocal of $0.\overline{009}$
14. 49×10.4, to the nearest whole number
20. What percent of 150 is 1128?
21. The sum of the first five counting numbers
23. The first three digits of 545×175
25. 5240, 5371, __?__
26. The first prime year in the twenty-first century
30. The number of years in eight centuries

TANGRAM Geometry

Cut out these seven shapes very carefully. Arrange all the pieces to form **(1)** a square, **(2)** an isosceles triangle, **(3)** a parallelogram, **(4)** a rectangle. Remember that all seven pieces must be used to form each figure.

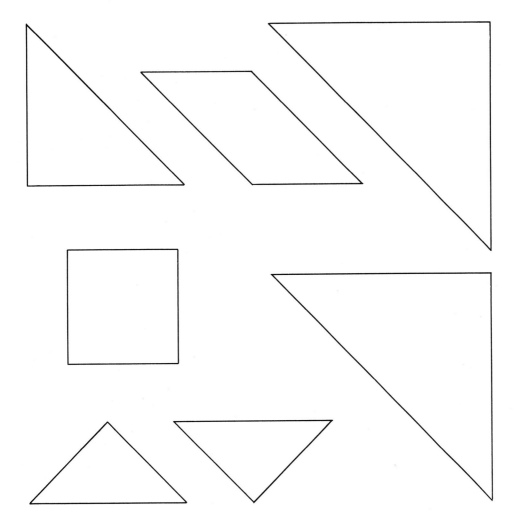

Computing with BASIC

Study this sequence of statements
from part of a computer program.

```
→50  FOR A = 1 TO 25 ┐
 60  LET S = S + A
└70  NEXT A ←
```

Line 60 is an example of a LET statement. The LET statement assigns to a variable a value or an algebraic expression. When the computer reaches line 60, the present value of S is added to the value of A and that sum is then placed back into S.

Statements 50–70 make up what is known as a FOR-NEXT loop. The FOR-NEXT loop is used to repeat an operation a defined number of times. In the example above, line 50 defines A to equal 1 through 25. Thus, the loop will be repeated 25 times. Before reading any further, try to determine what the loop was designed to do.

Earlier in the program, S was assigned the value 0. The first time around the loop, therefore, A takes on the value of 1 and S has the value 0. The sum of S and A (1) is then placed in S. The second time around the loop, A = 2 and S = 1, and the sum 3 is placed in S. The third time, A = 3 and S = 3, and the sum 6 is placed in S. The loop continues until the values defined for A are exhausted. Do you see that the loop was designed to find the sum of the first 25 counting numbers?

Study the sequence of statements below and then answer the questions.

```
 70  FOR N = 3 TO 15
 80  LET C = N * N            This is the way we write ×
                              when we use computer language.
 90  PRINT C
100  NEXT N
```

1. How many times will the loop be repeated?

2. What is the value of N the first time around the loop?

3. What is the value of C the first time around the loop?

4. What is the value of C the tenth time around the loop?

5. What will be the final value of C?

6. Explain what the loop was designed to do.

Logic Problems

Some logic problems can be solved using the method of elimination. Each clue eliminates a name or position from consideration. How many of the problems below can you solve?

1. Peg, John, and Maggie live in a three-story building, but no two on any one floor. Their ages are 5, 7, and 8, although not necessarily in that order. Peg, who is youngest, lives above the seven year old and below Maggie. Find the ages and on which floor each child lives.

2. Linda, Tim, Karen, and Gary are freshman, sophomore, junior, and senior who major in chemistry, architecture, mathematics, and business, not necessarily in this order. Pair each person according to major and class. You have the following clues:

(a) Karen and the sophomore business major room together.
(b) Tim will graduate this year.
(c) Gary had to do a term paper on Egyptian pyramids.
(d) Linda will graduate the year after Karen and the year before the architecture major.
(e) The mathematics major graduated from the same high school as Karen and Linda.

3. One of three identical triplets—Edward, Edwin, or Edmond—was seen leaving the scene of a crime. When questioned, each made the following statements:

> Edward: It wasn't me!
> Edwin: Edward is lying!
> Edmond: Edwin is lying!

Only one of the triplets was lying. Determine who lied. Can you determine who was seen leaving the scene of the crime?

4. There are three boxes. One contains two pennies; one contains two nickels; one contains a penny and a nickel. Each of the boxes is labelled PP, NN, or PN, but each is labelled incorrectly. If you are allowed to take one coin at a time out of any box, what is the smallest number of drawings needed to determine the contents of each box?

Cross-Number Puzzle

ACROSS

1. Year of the first lunar landing
3. 81×1001
8. 32% of what number is 4712?
9. $2^2 \times 3^6$
12. The GCF of 147 and 539
13. The mean of 126, 432, and 378, squared
14. 2^{11}
16. The number of primes between 1 and 100
18. The total number of four-suited poker hands possible
24. $(2.3)(2.504 \times 10^6) + 84,162$
26. The median of 77, 11, and 47
27. The year the Magna Carta was signed
31. 59°F in degree Celsius
33. In terms of greatest land area of the 50 U.S. states, Maine's rank
34. MCMIV
35. 1 century = ___?___ days
36. The product of eight times the square root of 458,329

DOWN

1. The square of 1315
2. The simple interest on a $22,910 loan at 5% for 8 years
3. CCMXIV
4. The year that Columbus discovered the Americas
5. The repetend of $7 \div 99$
6. The first and last digits of 16×547
7. The number of U.S. presidents before Lincoln

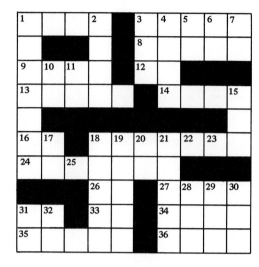

10. The largest prime number less than 100
11. $1300 loan for 2 years, Interest, $338; What is the simple interest rate?
15. The reciprocal of 0.00125
17. What percent of 1500 is 870?
18. The volume in meters of a box measuring 29 m by 800 cm by 0.101 km
19. $53 \times 35 \div \dfrac{1}{29}$
20. The measure of an angle which is supplemental to an 84° angle
21. $4352.095 \div 0.053$
22. The number of planets in the solar system
23. The number of years served by a U.S. senator for a single term
25. The number of scores referred to in the Gettysburg Address
28. The LCM of 98 and 147
29. The smallest prime between 100 and 500
30. $2^9 + 34$
31. Sum of the prime digits in 16,343,907
32. Atomic number of barium

Hexominoes

A <u>hexomino</u> is a set of six congruent squares which are connected to each other. Each square shares at least one whole side with at least one other square. Two examples are shown below.

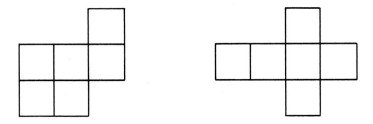

The hexomino on the right has a special property in that it can be cut out along its edges and folded to form a cube; the shape on the left cannot. Below are two hexominoes which can also be cut out to form cubes.

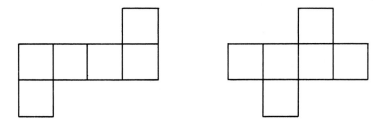

There are 35 hexomino shapes, but only eleven of those can be folded to form a cube. Using gridded paper, find all possible hexominoes. Which can be folded to form cubes?

Puzzlers

Here are some intriguing problems. How many can you solve?

1. Two cars race on an oval track. In the time trials, car A averaged 32 seconds to complete one lap, and car B averaged 36 seconds per lap. Under these conditions, and if both begin together, how long will it take car A to overtake car B.

2. A pet store owner has some perfectly healthy parakeets and dogs for sale. There are 19 heads and 52 feet altogether. How many parakeets and how many dogs are there?

3. A pencil and a notebook cost $1.10. The notebook costs $1.00 more than the pencil. How much does the pencil cost? How much does the notebook cost?

4. Hanna was given permission to go berrying in the Smiths' berry patch on the condition that she give the Smiths one-fifth of all the berries she picked. Hanna went home with exactly one kilogram of berries. How much had she picked originally?

5. The width of a river is 321 m at the point where it is spanned by a bridge. 25% of the bridge is on one side of the river, and $21\frac{1}{2}$% is on the other side of the river. How long is the bridge?

6. A light on top of a radio tower goes on at precisely 12 midnight. Thereafter, it will go on and off at regular intervals, each lasting an integral number of minutes. It was seen that at 12:09 A.M. the light was off; at 12:17 A.M. it was on; and at 12:58 A.M. it was on. Will the light be on or off at 2:00 A.M.?

7. Suppose you have to cross a river with three possessions: a fox, a goose, and a bag of corn. There is room in your boat for yourself and only one of your possessions. If you leave the fox and the goose alone together, the fox will eat the goose. If you leave the goose and the corn alone together, the goose will eat the corn. Assuming the fox will not eat the corn, how do you get yourself and your three possessions across the river?

Word Search of Mathematical Terms

All the terms below have been covered in this text. In the matrix below, they may have a vertical, horizontal, or diagonal orientation, and may be read forwards or backwards. How many of the terms can you find?

ABSCISSA	CIRCUMFERENCE	COMMUTATIVE	COMPLEMENTARY
CONGRUENT	DECIMAL	DISTRIBUTIVE	EQUILATERAL
EXPONENT	FACTOR	HISTOGRAM	HYPOTENUSE
INTEGER	INTERPOLATION	IRRATIONAL	ISOSCELES
MEAN	MEDIAN	ORDINATE	PARALLELOGRAM
PRIME	PROBABILITY	PROPORTION	QUADRILATERAL
RANDOM	RATIO	RATIONAL	RECIPROCAL
SCIENTIFIC NOTATION		TANGENT	VARIABLE

```
K E L B A I R A V X T E N H N P L K K F L Q E K F
S C I E N T I F I C N O T A T I O N H N W D X O V
N L M T R O A J G Z V D K A W M F G F J R B P F I
L A R E T A L I U Q E E S U N E T O P Y H E O T C
N P K G D W Q X E F F I J B U I V S Q V A U N H O
R W L L P L E G F A L C N Y V H D Q Y S G R E K N
E O J F U R P M S C N H I T N T E R D C E G N R G
C V T A E A O S I P O P N R E A C T O Y Z N T I R
I X Z C B V I B R R V M R O C R E Z R L W X D N U
P B K K A C I Q A X P S P S I U P M K D E W G F E
R N K H S F C T H B A J M L D T M O B V C E K X N
O B O B U M F Q U X I S F S E I R F L L E D X C T
C Q A Q F U O B A B B L I T B M Z O E A S P T B Z
A R A T I O T G D H I N I R N S E V P R T Q G P M
L A N O I T A R R I P R E T X C I N P O E I F U E
M A R G O T S I H Q O G T L Y T U D T Z R N O Y T
V M Q M T K E T X J E G F S A F M S M A N P C N F
Y E C N N C H W J T I A J T I N R D B N R C D E O
C D L P E X R A N D O M U F U D O V E W P Y D B G
Z I N E G W Z I Z C Q M Z F V J V I Z G V N T H B
U A K D N E F V F A M H X O X M T N T V R N J C Z
E N W F A C X J M O Y E X S I P C O G A Q E T Z A
B Y I W T Z Z X C D M A R G O L E L L A R A P Q S
S V S F C D V G A C D K V I S O S C E L E S B H A
V V W D E C I M A L C L A R E T A L I R D A U Q O
```

If you do not know the meaning of any of the terms above, you should go back and look up their meanings.

Answers to Optional Activities

Page T82: Word Search of Mathematicians

```
+ + + + + + + + + + + + + + + + + + + + + + + +
+ E + + + Z + + + + + + A + + + + + + + + + + +
+ + U N + + E + + + P + R + A + + + + + + + + +
+ + + C O + + N + + Y + C D + P + + + + + + + + L
+ + I + L T + + O + T + H E T + O + + + + + + A +
+ + C + + I W + E + H + I S R + + L + + + + G + +
+ + C + + Z D E I + A + M C E + + + L + + R + + +
+ + A + + I + + N + G + E A B + + + + O A + + + +
+ + N + + N + + S + O + D R L + + + + N N + + + +
+ + O + + B + + T + R + E T I + + + G + + I + + +
+ + B + + I + + E + A + S E H + + E + + + + U + +
+ + I + + E + + I + S + + S B + + + + + + + + S +
+ + F + + L + + N + + + + + E + + + + + + + + + +
D E D E K I N D + S + + + + R A + + + + + + + + +
+ + + + + + + F S + + + + + N + R + C A Y L E Y +
+ + + + + + + U O + + + + + O + P I + + + + + + +
+ C T + + + A L E U + + + + U + + A S + + + + G +
+ J A A + G + + A U R + + + L + + + S T + + + A +
+ A + N M + + + + P L I + + L + + G + C O + + L +
+ C + + T R + + + + L E E + I + + A B + A T + I +
+ O + + + O E + + + + A R R + + + L + O + L L L +
+ B + + + + R F + + + + C + + + + O + + O + + E +
+ I + + + + + + + + + + + E + + + I + + + L + O +
+ + + + + + + + + + + + + + + + + S + + + + E + +
+ + + + + A B E L + + + + + + + + + + + + + + + +
```

Page T83: Using the Calculator
1. .123456 . . .; .112233 . . .; .111222333 . . .; .11112222 . . .; .1111122222 . . .;
.111111222222 . . . **2.** 0.00$\overline{9}$; 0.01$\overline{8}$; 0.02$\overline{7}$; 0.03$\overline{6}$; 0.04$\overline{5}$; 0.05$\overline{4}$; 0.07$\overline{2}$; 0.09$\overline{0}$; 0.13$\overline{5}$;
0.2$\overline{25}$ **3. (a)** 484 **(b)** 40804 **(c)** 4008004 **(d)** 40000800004 **4.** Except when the
number is in the form $4n + 1$, the starter can force a win by grouping as follows:
$31 = 1 + 4 + 4 + 4 + 4 + 4 + 4 + 4 + 2$. The starter subtracts 2. Then, whatever
number the second player subtracts, the starter subtracts the complement of 4.

Page T84: Networks
No strategy that guarantees a win has ever been formulated,
so just have fun

Page T85: The Alphabet Wheel
TRY SWIMMING; SHAKESPEARE THIS IS NOT.

Page T86: Cross-Number Puzzle

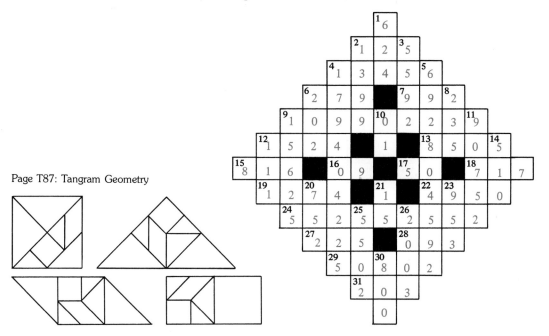

Page T87: Tangram Geometry

Page T88: Computing with BASIC
1. 13 times   **2.** 3   **3.** 9   **4.** 144   **5.** 225   **6.** The loop was designed to print the squares of the numbers from 3 to 15

Page T89: Logic Problems
1. Maggie is 8 and lives on the third floor; John is 7 and lives on the first floor; Peg is 5 and lives on the second floor.   **2.** Linda-sophomore-business; Tim-senior-mathematics; Karen-junior-chemistry; Gary-freshman-architecture   **3.** Edwin is lying; Not enough information to determine who was seen leavning the scene of the crime.   **4.** 3 drawings

Page T90: Cross-Number Puzzle

Page T91: Hexominoes
The other 8 hexominoes which can be folded to form cubes are:

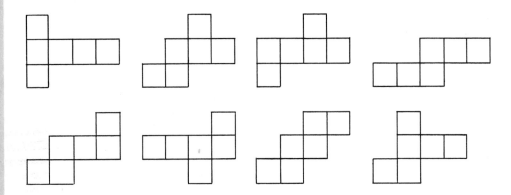

Page T92: Puzzlers
1. 4 minutes 48 seconds   **2.** 7 dogs, 12 parakeets   **3.** $.05; $1.05   **4.** 1.25 kg
5. 600 m   **6.** 7 minutes   **7.** He takes the goose across, leaves it, returns, fetches the
fox, takes it across, returns with the goose, leaves the goose, takes the grain across, returns, and brings the goose over.

Page T93: Word Search of Mathematical Terms

```
+ E L B A I R A V + + E + + + + + + + + + E + +
S C I E N T I F I C N O T A T I O N + + + + X + +
+ + + + + + + + + + + + A + + + + + + + + P + +
L A R E T A L I U Q E E S U N E T O P Y H + O + C
+ + + + + + + + + + + I + + + I + + + + + + N + O
R + + + P + E + + A + C N + + + D + + + + + E + N
E O + + + R + M S C + + I T N + + R + + + + N + G
C + T + E + O S I + O + N R E A + + O + + + T + R
I + + C + V I B + R + M + O C R E + + + + + + + U
P + + + A C I + A + P + P + I U P M + + + + + + E
R + + + S F + T + B + + + L + T M O + + + + + + N
O + + B + + + + U + I + + + E + R F L + + + + + T
C + A + + + + + + B + L + + + M + O E A + + + + +
A R A T I O + + + + I + I R + + E V P R T + + + +
L A N O I T A R R I + R E T + + I N + O E I + + +
M A R G O T S I H + + G T L Y T + + T + R N O + +
+ M + + T + + + + + E + + S A + + + + A + P C N +
+ E + + N + + + + T + + + T I N + + + + R + + E +
+ D + + E + R A N D O M U + + D O + + + + Y + + +
+ I + + G + + I + + + M + + + + + I + + + + + + +
+ A + + N + + + + + M + + + + + + + T + + + + + +
+ N + + A + + + + O + + + + + + + + + A + + + + +
+ + + + T + + + C + M A R G O L E L L A R A P + +
+ + + + + + + + + + + + + I S O S C E L E S + + +
+ + + D E C I M A L + L A R E T A L I R D A U Q +
```

MATHEMATICS
Structure and Method

Course 2

MARY P. DOLCIANI
EDWIN F. BECKENBACH
RICHARD G. BROWN
ROBERT B. KANE
WILLIAM WOOTON

HOUGHTON MIFFLIN COMPANY · BOSTON
Atlanta   Dallas   Geneva, Ill.   Hopewell, N.J.   Palo Alto   Toronto

Authors

MARY P. DOLCIANI   Professor of Mathematics, Hunter College, City University of New York.

EDWIN F. BECKENBACH   Professor of Mathematics, Emeritus, University of California, Los Angeles.

RICHARD G. BROWN   Mathematics Teacher, Phillips Exeter Academy, Exeter, New Hampshire.

ROBERT B. KANE   Director of Teacher Education and Head of the Department of Education, Purdue University.

WILLIAM WOOTON   former Professor of Mathematics, Los Angeles Pierce College.

Editorial Advisers

ANDREW M. GLEASON   Hollis Professor of Mathematics and Natural Philosophy, Harvard University.

ALBERT E. MEDER, Jr.   Dean and Vice Provost and Professor of Mathematics, Emeritus, Rutgers, The State University of New Jersey.

Consultants

CAROLYN AHO   Teacher Specialist, Mathematics, San Francisco Unified School District, San Francisco, California.

RUTH GREEN   Mathematics Teacher, Isaac Litton Junior High School, Nashville, Tennessee.

TIMOTHY HOWELL   Mathematics Coordinator, Hanover High School, Hanover, New Hampshire.

BETTY TAKESUYE   Mathematics Teacher, Chaparral High School, Scottsdale, Arizona.

Printed in U.S.A.

ISBN: 0–395–31394–5

Contents

8 Problem Solving

9 Areas and Volumes

10 Probability

SYMBOLS

		page			page
10^1, 10^4	powers of 10	27	∟	right angle	180
$\{0, 1, 2, 3, \ldots\}$	set of whole numbers	31	∥	is parallel to	185
			$\triangle ABC$	triangle ABC	190
$>$	is greater than	34	$\cong$	is congruent to	201
$<$	is less than	34	$\sim$	is similar to	206
$=$	is equal to	37	$\neq$	is not equal to	222
$^-3$	negative 3	66	$\geq$	is greater than or equal to	222
$\lvert n \rvert$	absolute value of n	80			
$6.\overline{47}$	repeating decimal	133	$\leq$	is less than or equal to	222
$\%$	percent	153			
$\overline{AB}$	segment AB	173	$\varnothing$	empty set	227
AB	length of $\overline{AB}$	174	$(6,8)$	ordered pair of numbers	336
$\approx$	is approximately equal to	174	$\sqrt{n}$	positive square root of n	363
π	pi	177			
$\angle A$	angle A	180	sin A	sin of $\angle A$	379
$30°$	30 degrees	180	cos A	cosine of $\angle A$	380
$\perp$	is perpendicular to	180	tan A	tangent of $\angle A$	380

METRIC MEASURES

PREFIXES	kilo- 1000	hecto- 100	deka- 10	deci- 0.1	centi- 0.01	milli- 0.001

LENGTH
1 millimeter (mm) = 0.001 **meter** (m)   1 m = 1000 mm
1 centimeter (cm) = 0.01 m   1 m = 100 cm
1 kilometer (km) = 1000 m   1 cm = 10 mm

MASS
1 milligram (mg) = 0.001 **gram** (g)   1 g = 1000 mg
1 kilogram (kg) = 1000 g

CAPACITY
1 milliliter (mL) = 0.001 **liter** (L)   1 L = 1000 mL
1 L = 1000 cm³

TIME
1 minute (min) = 60 seconds (s)
1 hour (h) = 60 min
1 day (d) = 24 h
1 week = 7 d
1 year = 365 d
1 leap year = 366 d

TEMPERATURE
Degrees Celsius (°C)
0°C = freezing point of water
37°C = normal body temperature
100°C = boiling point of water

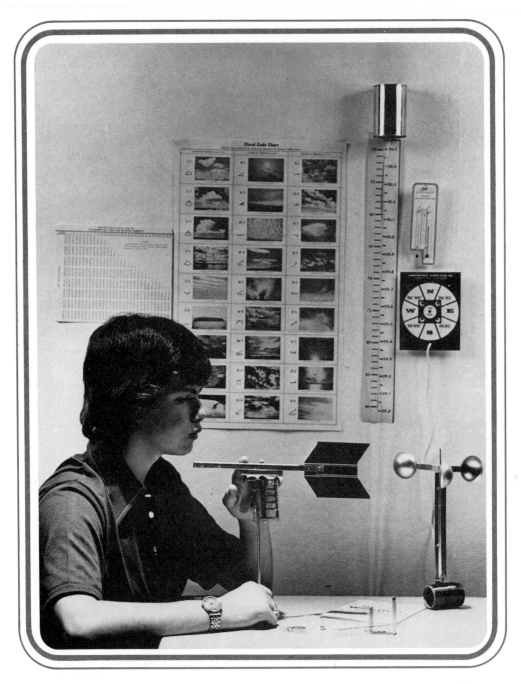

The photographs facing the chapter opening pages illustrate different aspects of students' lives. For the earlier chapters, they focus on school and home activities. For the later chapters, they broaden out to show community-related activities and, finally, careers in which some knowledge of mathematics is helpful.

1

Statistics and Graphs

1-1 Reading Tables and Charts

Objective To read and use information displayed in charts and tables.

In everyday life we see numerical facts called *data*, or *statistics*, presented in charts and tables. To make use of these data, we must be able to read such tables accurately.

For example, the table at the right uses a compact arrangement to show the distances between some European cities. As the red lines indicate, we find the distance between Budapest and London by going down the column labeled "Budapest" and finding where it meets with the row labeled "London." The distance is 1752 km. In the same way we find that the distance between London and Amsterdam is 530 km.

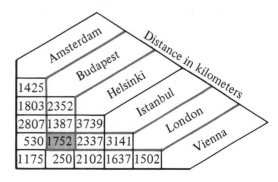

Amsterdam	Budapest	Helsinki	Istanbul	London	Vienna
1425					
1803	2352				
2807	1387	3739			
530	1752	2337	3141		
1175	250	2102	1637	1502	

Distance in kilometers

1

Some tables, like the one below, provide a variety of information. Notice that the distances of the trips around the world are given in kilometers. In the "Time" column, the "d," "h," and "m," indicate the units used are days, hours, and minutes. Study the table and tell which was the fastest trip and when it was made. Which trip covered the greatest distance? In what year was it made?

Fast Trips Around the World

Craft	Person	Distance (km)	Time			Year
			d	h	m	
train, burro, jinriksha, ship	Nellie Bly	—	72	6	11	1889
Graf Zeppelin	—	34,900	20	4	0	1929
—	Wiley Post	25,100	4	19	36	1933
Zeppelin Hindenburg	H. R. Ekins	41,300	18	11	15	1936
—	Howard Hughes	23,900	3	19	8	1938
Pan Am Clipper	Clara Adams	—	16	19	4	1939
Reynolds Bombshell	Wm. P. Odom	32,200	3	6	55	1947
—	Pamela Martin	—	3	18	59	1953
3 B-52's	—	39,100	1	21	19	1957
—	Sue Snyder	34,100	2	14	59	1960
Light plane	Robert and Joan Wallick	37,200	5	6	17	1966

Class Exercises

Using the table at the right, find the distances between these cities.

1. Honolulu and Montreal   7913

2. Melbourne and Nairobi   11,509

3. Peking and Mexico City   12,475

4. Mexico City and Honolulu   6097

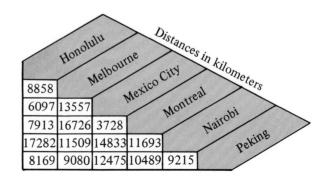

Distances in kilometers

	Honolulu	Melbourne	Mexico City	Montreal	Nairobi	Peking
	8858					
	6097	13557				
	7913	16726	3728			
	17282	11509	14833	11693		
	8169	9080	12475	10489	9215	

5. Melbourne and Peking 9080

6. Nairobi and Montreal 11,693

7. Honolulu and Melbourne 8858

8. Peking and Nairobi 9215

9. Which cities in the table are closest to each other?
Mexico City and Montreal

10. Which cities are farthest from each other?
Honolulu and Nairobi

Which is the shorter trip?

11. Melbourne to Montreal or to Mexico City?

12. Honolulu to Nairobi or to Peking?

13. Montreal to Mexico City or to Peking?

14. Nairobi to Honolulu or to Montreal?

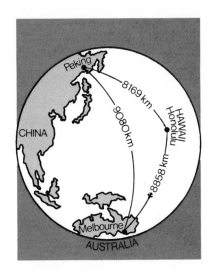

Exercises

Using the same table as for the Class Exercises,
find the total distance for each trip.

A

1. Honolulu to Melbourne to Peking 17,938

2. Melbourne to Mexico City to Peking 26,032

3. Nairobi to Peking to Honolulu 17,384

4. Nairobi to Montreal to Mexico City 15,421

5. Peking to Honolulu to Mexico City 14,266

6. Montreal to Peking to Melbourne 19,569

How much farther is it from

7. Honolulu to Melbourne than to Peking? 689

8. Mexico City to Nairobi than to Montreal? 11,105

9. Peking to Montreal than to Honolulu? 2320

10. Melbourne to Montreal than to Honolulu? 7868

How far would you travel if you made a round trip between the given cities?

11. Nairobi and Peking 18,430

12. Montreal and Mexico City 7456

13. Melbourne and Nairobi 23,018

14. Honolulu and Melbourne 17,716

How far would a jet plane travel on the trip described?

15. 5 round trips between Montreal and Honolulu 79,130

16. 3 round trips between Nairobi and Peking 55,290

17. 6 round trips between Melbourne and Mexico City 162,684

18. 7 round trips between Peking and Honolulu 114,366

A satellite used by radio hams will pass over the Denver area at the times shown in the chart at the right. How many minutes will elapse between these orbits?

Satellite Data		
Date	Time	Orbit Number
3/18	10:16 A.M.	1540
	12:11 P.M.	1541
3/20	10:10 A.M.	1565
	12:05 P.M.	1566
3/24	9:57 A.M.	1615
	11:52 A.M.	1616

B 19. 1540 and 1541   115 min

20. 1565 and 1566   115 min

21. 1615 and 1616   115 min

22. How many orbits will the satellite make between 12:11 P.M. on March 18 and 10:10 A.M. on March 20?   24 orbits

23. How much time will elapse between Orbits 1541 and 1565?   45 h 59 min

24. About how much time will the satellite take to make each orbit between Orbits 1541 and 1565? How does this compare with your answers to Exercises 19–21?   114.9 min; 0.1 min less

Application
Getting Your Money's Worth

Which is the better buy? A 400 g can of tomatoes for 60¢ or a 600 g can for 80¢? From the table below we see that if we buy the 400 g can, we are paying at a rate of $1.50 per kilogram. For the 600 g can the rate is $1.33 per kilogram. If the quality is the same and we can use the larger amount, the 600 g can is the better buy.

No. of grams in containers	Price of container in dollars				
	.20	.40	.60	.80	1.00
	Cost of food per kilogram in dollars				
200	1.00	2.00	3.00	4.00	5.00
400	.50	1.00	1.50	2.00	2.50
600	.33	.67	1.00	1.33	1.67
800	.25	.50	.75	1.00	1.25
1000	.20	.40	.60	.80	1.00

Consumer Activity Check a local supermarket to see how it displays unit prices. How would you use unit prices to get your money's worth?

1-2 Bar Graphs

Objective To read and make bar graphs.

We often use a **bar graph** to give a visual display of numerical data after they have been arranged in a table. The table and bar graph below show the number of kilojoules, kJ, of energy used up by one minute of different kinds of physical activity.

Energy Used

Activity	kJ/min	Activity	kJ/min
Walking	20	Swimming	44
Bicycling	32	Running	76

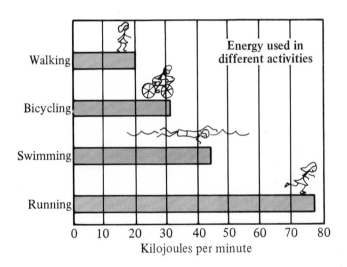

To make this graph, bars for the different activities were marked off along the vertical axis. Starting with zero, intervals of ten were marked off along the horizontal axis to represent the number of kilojoules per minute. Bars of the appropriate length were drawn for each activity. Finally, the bar graph was given its title, "Energy used in different activities."

Just by looking at the lengths of the bars, estimate about how many times more kilojoules are used in one minute of swimming than in one minute of walking, in one minute of running than in one minute of walking. We usually use bar graphs to make rough comparisons like these rather than for obtaining exact information.

The bar graph below uses different shades of color to indicate the years which the bars represent. Notice that each unit on the vertical axis represents 1000 metric tons, so that "500" on the vertical axis indicates 500,000 metric tons. A smaller unit could have been chosen to show a more exact picture.

Recyclable Materials

Year	Metric Tons (1000)		
	Iron	Aluminum	Glass
1975	260	50	250
1980 (est.)	540	100	340
1985 (est.)	1560	250	540

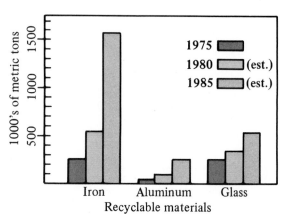

Class Exercises

From the graph at the right, tell about how much water is used for each activity.

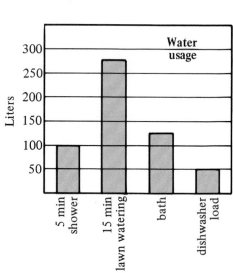

1. A 5-minute shower. 100 L

2. Watering the lawn for 15 minutes. 275 L

3. A bath. 125 L

4. A dishwasher load. 50 L

5. How much more water is used taking a bath than a 5-minute shower? 25 L

6. How much water is used watering the lawn for one hour? 1100 L

7. The Smiths washed 18 loads of dishes one week. How much water did they use? 900 L

8. How much water would 12 members of the track team use if they each took a 5-minute shower? 1200 L

9. The Browns watered their lawn for 30 minutes on each of 3 days. How much water did they use? 1650 L

Exercises

From the graph at the right, tell how far an automobile travels before stopping after the brakes are applied at the given speed.

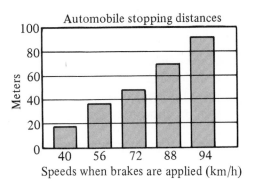

Automobile stopping distances

A
1. 40 km/h 2. 56 km/h 3. 72 km/h
   17 m   35 m   50 m
4. 88 km/h 5. 94 km/h
   70 m   90 m

B
6. What is the difference in stopping distances at 40 km/h and 72 km/h? 33 m

7. What is the difference in stopping distances at 40 km/h and 94 km/h? 73 m

8. A car traveling 72 km/h is 40 m from an intersection when the driver puts on the brakes. Will the car stop before it reaches the intersection? No

Draw a bar graph to illustrate the data in the table. Check students' graphs.

C
9.   Average Heights of Growth

Flower	Centimeters
Dahlia	56
Hollyhock	180
Larkspur	90
Marigold	30
Petunia	25
Snapdragon	40

10.   Leading Cacao-Bean-Growing Countries

Country	Kilograms of Cacao Beans
Ghana	353,000,000
Brazil	245,500,000
Nigeria	218,000,000
Ivory Coast	205,600,000
Cameroon	110,000,000

11.   Endangered Wildlife

Mammal or Bird	No. in Captivity	Mammal or Bird	No. in Captivity
Golden lion marmoset	83	Reeve's muntjac	161
Bushdog	19	Eld's deer	69
Maned wolf	75	Pere David's deer	777
Onager	142	Florida sandhill crane	109
Scimitar-horned oryx	298	Rothschild's mynah	565

1-3 Line Graphs

Objective To read and make line graphs.

The *broken-line graph* below shows the average monthly temperatures in degrees Celsius at Ottawa, Canada, from April through November. Broken-line graphs such as this are especially useful when we want to show how something changes over a period of time.

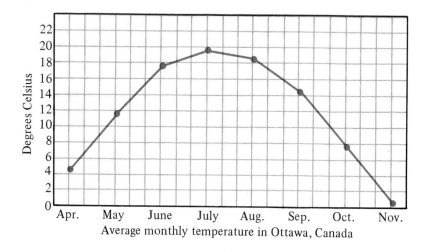

Average monthly temperature in Ottawa, Canada

Here are some things we can notice in this graph:

1. The lowest average temperature, 1°C, is recorded in November, and the highest, 20°C, in July.
2. The average temperature rises from April to July, and then falls from July to November.
3. The greatest increase in temperature from one month to the next occurs between April and May. This is an increase of 7°. Also, the line connecting the points for April and May has the steepest upward slant.
4. The least increase in temperature from one month to the next occurs between June and July. This increase is 2°. Also, the line between June and July has the least upward slant.
5. The temperature decreases the same number of degrees, 7°, from September to October as from October to November. Also, the line between September and October has the same downward slant as the line between October and November.

An almanac predicted the amount of rainfall for each month in a farm area as shown by the black figures in the table below. The actual rainfall is shown in red.

Centimeters of Rain

	Jan	Feb	Mar	Apr	May	Jun	Jul	Aug	Sept	Oct	Nov	Dec
Predicted	2	3	3	6	10	9	9	8	4	1	1	0
Actual	3	5	6	7	8	6	4	2	1	0	0	1

A clearer way to display these data is by using a double-line graph as shown below.

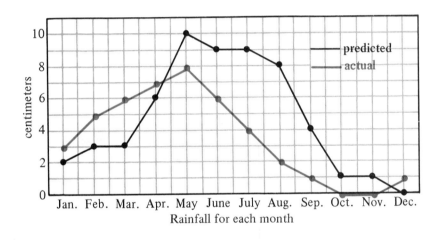

Rainfall for each month

Notice that in answering the questions in the example below, it is usually easier to use the graph than the table.

Example **a. In what months did the actual rainfall and the predicted rainfall agree most closely?**

b. By how many centimeters did they differ?

c. In those months, which was higher—the actual rainfall or the predicted rainfall?

Solution **a. In January, April, October, November, and December**

b. 1 cm

c. The actual rainfall was higher in January, April, and December. The predicted rainfall was higher in October and November.

Another kind of line graph is shown below. It uses vertical line segments to display the ranges of temperature during one day in August in several cities.

City	Temperature range in °C on August 5
Anchorage	11–18
Atlanta	21–31
Boston	19–33
Chicago	22–29
Denver	14–29
Honolulu	23–31
Las Vegas	24–42
Miami	27–32
Phoenix	29–43
San Francisco	13–18

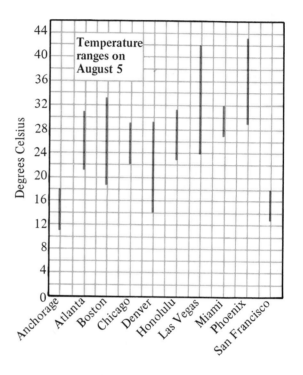

From the graph we can see at a glance which city had the highest temperature, the lowest temperature, or the greatest range in temperature.

Class Exercises

From the graph at the right, read the population of the world, approximately, for the given year. (in millions)

1. 1 250   **2.** 500 300 **3.** 1000 400

4. 1500 500 **5.** 2000 7500

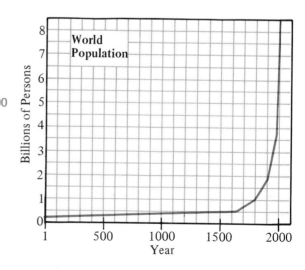

About how long did it take for the population to double from

6. 250 million to 500 million? 1500 yr

7. 500 million to 1 billion? 300 yr

8. 1 billion to 2 billion? 100 yr

Exercises

Using the graph at the right, what is the altitude, approximately, of a satellite which takes the given number of minutes to orbit Earth?

A 1. 100 2. 120
   800 km   1700 km
   3. 140 4. 160
   2600 km   3400 km

5. A communications satellite is in orbit 2000 km above Earth. About how long does it take to make one orbit? 125 min

6. A weather satellite is in orbit 1500 km above Earth. How long does it take to make one orbit? 115 min

7. A satellite was put into orbit at 3000 km at 9 A.M. When will it complete its first orbit? 11:30 A.M.

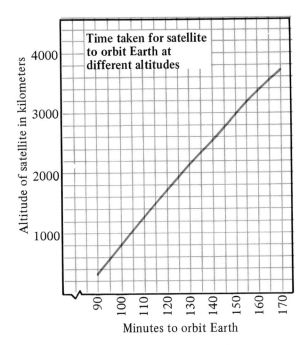

Time taken for satellite to orbit Earth at different altitudes

Altitude of satellite in kilometers

Minutes to orbit Earth

Draw a broken-line graph picturing the given data. Check students' graphs.

8. Number of Newspaper Customers

Month	Jan	Feb	Mar	Apr	May	Jun
Customers	42	45	38	45	50	53

B 9. Population Growth of Cactus City

Year	1900	1910	1920	1930	1940	1950	1960	1970	1980
Population	4000	5500	6000	7000	6800	8000	11,000	13,500	14,500

Draw a vertical line-segment graph to show the given data. Check students' graphs.

10. Temperature Ranges in Upland City

Month	Jan	Feb	Mar	Apr	May	Jun
Range in °C	2–8	4–9	7–11	8–14	12–20	14–22

Application

Wind Speed

When meteorologists talk about a "storm," they refer to weather conditions in which the wind is blowing in a range of 89–102 kilometers per hour. The graph below gives the names meteorologists use for winds of different speeds.

12	Hurricane	greater than 117
11	Violent storm	103-117
10	Storm	89-102
9	Strong gale	75-88
8	Gale	62-74
7	Near gale	50-61
6	Strong breeze	39-49
5	Fresh breeze	29-38
4	Moderate breeze	20-28
3	Gentle breeze	12-19
2	Light breeze	6-11
1	Light air	1-5
0	Calm	less than 1

Beaufort
Scale

0 10 20 30 40 50 60 70 80 90 100 110 120 130 140 150

kilometers per hour

Wind Speed

The Beaufort Scale, shown to the left of the graph, assigns the numbers 0–12 to various ranges of wind speeds. This scale was invented in 1805 by Sir Thomas Beaufort. He assigned the numbers according to the effect of various winds on sailing ships. For example, he described a wind with a force of 12 as "that which no canvas can stand."

Research Activities

1. Find out how wind speed is measured. What visible signs can be used to estimate wind speed?
2. Find out how artificial satellites are used to predict weather.
3. Find out how to make some simple instruments and set up your own weather station.

Career Activity
Find out what kind of training a meteorologist needs.

Self-Test

Words to remember:

data [p. 1] statistics [p. 1] bar graph [p. 5] broken-line graph [p. 8]

Use the table at the right for Exercises 1–4.

Layers of Earth

[1-1]

Layer	Depth below surface (km)
Crust	0–30
Mantle	30–3000
Outer Core	3000–5000
Inner Core	5000–6000

1. What is the outermost layer called? Crust

2. What is the deepest layer called? Inner Core

3. How deep is the outer core? 3000-5000 km

4. Which layer is beneath the crust? Mantle

Use the bar graph at the right for Exercises 5–8.

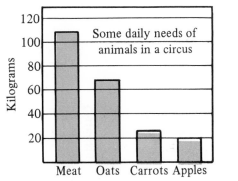

5. How much meat do the animals eat? 110 kg

6. How much oats do they eat? 70 kg

7. How much carrots and apples do they eat altogether? 45 kg

8. How much more meat do they eat than apples? 90 kg

Use the graph at the right for Exercises 9–12.

[1-3]

9. What is the range of temperature in February? 6°-22°

10. Which months have the greatest range? Feb., Mar.

11. Which has the least range? June

12. Which has the hottest temperature? May

Self-Test answers and Extra Practice are at the back of the book.

1-4 Making a Histogram

Objective To display data by means of a histogram.

Test Scores	Intervals
47]	41–50
51]	
53]	51–60
57]	
62]	
65]	61–70
70]	
71]	
72	
74	
74	
74	71–80
74	
80	
80]	
81]	
83	
83	
83	81–90
86	
88	
90]	
93]	
93	
96]	91–100
99	
99]	

A **histogram** is a type of bar graph that displays numerical data which have been organized into equal intervals. To organize the data, we first list them in increasing order as we did with the test scores at the right. Then we mark the data off into equal intervals. We used intervals of 41–50, 51–60, and so on, for the test scores. Next we count the number of scores that occur in each interval and make a table like the one at the left below. We call this a *frequency distribution*. The table shows, for example, that there was one score in the interval of 41–50, three scores in the interval of 51–60, and so on. Finally, we draw the histogram. The height of each bar shows the number of test scores which occurred in its interval. Notice that there are no spaces between the bars of a histogram.

Frequency distribution
of test scores

Interval	Frequency
41–50	1
51–60	3
61–70	3
71–80	8
81–90	7
91–100	5

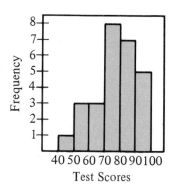

Class Exercises

Exercises 1–5 refer to the histogram at the right.

1. How many students had a test score in the interval 61–70? 3

2. In which interval were the fewest test scores made? 41-50

3. Which interval had the greatest number of test scores? 81-90

4. How many more students scored in the 91–100 interval than in the 61–70 interval? 4

5. How many students had a score of 70 or less? 6

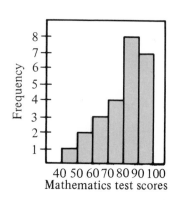

Exercises

Exercises 1–3 refer to the data given at the right.

A   **1.** Arrange the data in order from least to greatest.
     See below.
    2. Organize the data into intervals of 51–60, 61–70,
    and so on, and make a frequency distribution
    table. Check students' papers.

B   **3.** Make a histogram for the data.
     Check students' papers for histograms.

Numbers of meteors spotted
in one hour by members
of the Astronomy Club

74	65	88	89	82
53	73	94	102	91
96	56	77	90	61
107	94	106	87	65
106	80	94	99	82

Draw histograms for the following sets of data.

C   **4.** Data recorded by a marine biologist for an experiment.

Lengths of lobsters in centimeters

6	8	20	15	33	24	12	10	5	32	19	21	17
9	31	19	11	25	8	18	18	13	12	32	19	15

(Hint: Use intervals of 0–6, 7–12, and so on.)

5. Data recorded in a survey taken at Shopper's Mall.

Distances in kilometers from
shoppers' homes to the mall

18	2	5	12	7	2	13	16	4	4	15	3	10
9	4	2	3	13	19	1	20	11	5	4	6	3

(Hint: Use intervals of 0–5, 6–10 and so on.)

6. Data recorded in a frog-jumping contest. Select a reasonable interval.

Distances in centimeters jumped by frogs

333	235	350	358	338	281	345	221	279	344
247	357	305	280	211	300	213	202	298	276

1. 53, 56, 61, 65, 65, 73, 74, 77, 80, 82, 82, 87, 88, 89, 90, 91, 94, 94, 94, 96, 99, 102, 106, 106, 107

EXTRA! Another Broken-Line Graph

Sometimes we use a histogram as the basis for drawing a broken-line graph. This kind of broken-line graph is called a *frequency polygon.* Here is an example of how to draw a frequency polygon.

In an experiment, some students were asked to estimate the number of beans in a glass jar. They made a frequency distribution table of their estimates and drew a histogram as shown below.

Estimates of number of beans	Frequency
251–300	7
301–350	10
351–400	16
401–450	12

How many beans?

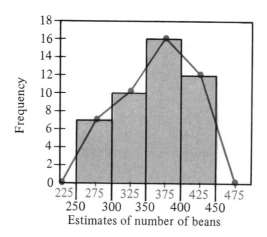

They also drew the red broken-line graph which connects the midpoints of the tops of the bars of the histogram. This graph is the frequency polygon. Notice that the frequency polygon is also connected to the horizontal axis half an interval to the left and to the right of the histogram.

Research Activity Collect some statistics about some activity that interests you, for example, football, or energy conservation. Make a frequency distribution table and draw a histogram and a frequency polygon to display on a bulletin board.

1-5 Using Numbers to Describe Data

Objective To calculate the range, mean, median, and mode.

Suppose we want to compare the number of cartons of yogurt sold daily during one month at two different stores as recorded in these tables.

Bestfood's

S	M	T	W	T	F	S
	46	53	58	47	64	
	72	64	55	73	63	
	48	39	45	50	79	
		52	51	60	64	

Eatrite's

S	M	T	W	T	F	S
	31	55	34	56	55	48
	42	50	34	34	45	62
	62	72	46	60	64	61
		34	40	43	50	

First we list each set of numbers in increasing order. The **range** of a set of numbers is the difference between the greatest and least numbers.

Range for Bestfood's $= 79 - 39 = 40$
Range for Eatrite's $\quad = 72 - 31 = 41$

The **mean,** or **average,** is the sum of the numbers divided by the number found by counting them.

Mean for Bestfood's $= \dfrac{1083}{19} = 57$

Mean for Eatrite's $\quad = \dfrac{1078}{22} = 49$

The **median** is the middle number in the set if their count is odd. If their count is even, the median is the mean of the two middle numbers.

Median for Bestfood's $= 55$
Median for Eatrite's $\quad = \dfrac{48 + 50}{2} = 49$

The **mode** is the number which occurs most frequently.

Mode for Bestfood's $= 64$
Mode for Eatrite's $\quad = 34$

Bestfood's (19 days)	Eatrite's (22 days)
39	31
45	34
46	34
47	34
48	34
50	40
51	42
52	43
53	45
55	46
58	48
60	50
63	50
64	55
64	55
64	56
72	60
73	61
79	62
1083	62
	64
	72
	1078

Note that while the range was almost the same for both stores, there were considerable differences in the other statistics:

1. The means show that Bestfood's sold 8 more cartons of yogurt per day on the average than Eatrite's.
2. The medians show that on half the days Bestfood's was open, it sold more than 55 cartons of yogurt per day. On half the days Eatrite's was open, it sold only 49 or more cartons of yogurt per day.
3. The modes show that the day's sales of yogurt most frequently made by Bestfood's was 64, and by Eatrite's was 34.

Class Exercises

Exercises 1–7 refer to the data at the right.

Number of peas in 16 different pods

6	5	2	9
12	6	5	6
11	9	6	4
7	8	10	6

1. Arrange the data in order from least to greatest.
   2, 4, 5, 5, 6, 6, 6, 6, 6, 6, 7, 8, 9, 9, 10, 11, 12
2. What is the range?   10
3. What is the sum of the data?   112
4. How many items are there in the set of data?   16
5. What is the mean?   7
6. What is the median?   6
7. What is the mode?   6

Exercises

For the data at the right, find the

B 1. range   65 cm

2. mean   62.9 cm

3. median   64 cm

4. mode   52 and 90

Centimeters of rain each month at Muddy Valley

J	F	M	A	M	J
25	36	52	48	76	85
J	A	S	O	N	D
90	90	81	77	52	43

For the data at the right, find the

5. range   6

6. mean   5

7. median   5

8. mode   4

Numbers of puppies in 13 litters

3	5	4	4	6	8	5
3	4	6	7	8	2	

Self-Test

Words to remember:

histogram [p. 14]   frequency distribution [p. 14]   range [p. 17]
mean [p. 17]   median [p. 17]   mode [p. 17]

For Exercises 1–4, use the data given at the right.

1. Organize the data into intervals of 140–144, 145–149, and so on. Make a frequency distribution table. Check students' papers for exercises 1 and 2.

2. Make a histogram.

3. Which interval has the greatest number of students? 155-159

4. How many students are at least 160 cm tall? 5

For the data at the right, find the

5. range 3   6. mean 9

7. median 9 8. mode 10

Students' heights in cm

143	157	161	155	146
153	148	144	147	155
152	169	164	142	146
156	159	168	150	163

[1-4]

Quiz scores

Sep.	9	Feb.	8
Oct.	10	Mar.	9
Nov.	10	Apr.	10
Dec.	9	May	10
Jan.	7	Jun.	8

[1-5]

Self-Test answers and Extra Practice are at the back of the book.

A Biologist

Rachel Carson (1907–1964) became interested in the environment when, as a young woman, she spent long hours observing nature. Upon completing college and graduate school she worked as an aquatic biologist. In 1941 she wrote her first book, *Under the Sea Wind,* after which she became editor-in-chief of the Fish and Wildlife Service. Her succeeding book, *The Sea Around Us,* was translated into thirty languages. Being an observant person, Carson became increasingly troubled by the consequences of technology on nature. Her last book, *Silent Spring,* published in 1962, produced a world-wide awareness of the dangers of chemical sprays. Statistics played an important role in her work.

Chapter Review

Write the letter that labels the correct answer.

Use the table below for Exercises 1–4.

Price Chart for Wearwell Tire Company

Automobile	Four-Ply	Belted	Radial
subcompact	$20.00	$23.00	$36.00
compact	$22.50	$27.00	$45.00
mid-size	$25.50	$31.50	$45.00

1. A four-ply tire for a mid-size car costs __?__   D   [1-1]
   A. $22.50   **B.** $31.50   **C.** $20.00   **D.** $25.50

2. A radial tire for a subcompact car costs __?__   A
   A. $36.00   **B.** $23.00   **C.** $45.00   **D.** $31.50

3. Four belted tires for a mid-size car cost __?__   B
   A. $102.00   **B.** $126.00   **C.** $144.00   **D.** $180.00

4. Five belted tires for a compact car cost __?__   A
   A. $135   **B.** $108   **C.** $157.50   **D.** $162

Exercises 5–8 refer to the bar graph at the right.

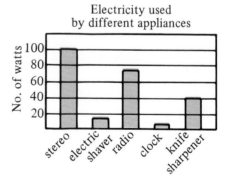

Electricity used by different appliances

5. The number of watts used by a radio is closest to __?__   B   [1-2]
   A. 55 W   **B.** 75 W
   C. 85 W   **D.** 65 W

6. A __?__ uses about 15 W.   C
   A. stereo   **B.** clock
   C. shaver   **D.** sharpener

7. A __?__ uses the least electricity.   C
   A. radio   **B.** shaver   **C.** clock   **D.** stereo

8. A __?__ uses the most electricity.   B
   A. clock   **B.** stereo   **C.** sharpener   **D.** radio

Exercises 9–12 refer to the line graph at the right.

Cost to run a standard-sized car

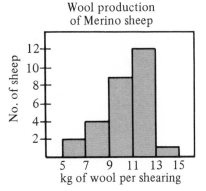

9. The cost per kilometer of running a standard-sized car in its 4th year is closest to __?__ C
   A. 9¢　　　　**B.** 10¢
   C. 11¢　　　　**D.** 12¢

[1-3]

10. It is least expensive to drive a car in its __?__ year. B
   A. 3rd　　　　**B.** 2nd　　　　**C.** 1st　　　　**D.** 4th

11. The cost of driving a car 20 km in its 1st year is __?__. C
   A. 20¢　　　　**B.** 2¢　　　　**C.** $2.00　　　　**D.** $200.00

12. The cost of driving a car 30 km in its 4th year is __?__. D
   A. $6.00　　　　**B.** $9.00　　　　**C.** 60¢　　　　**D.** $3.30

Exercises 13–14 refer to the histogram at the right.

Wool production of Merino sheep

[1-4]

13. The number of sheep producing between 9 kg and 11 kg of wool is __?__ D
   A. 12　　　　**B.** 8
   C. 11　　　　**D.** 9

14. The interval with the greatest number of sheep is __?__ C
   A. 7 kg–9 kg　　**B.** 9 kg–11 kg
   C. 11 kg–13 kg　**D.** over 13 kg

In Exercises 15–16, refer to the chart for Exercises 1–4.

15. The median price is __?__. A
   A. $27.00　　　**B.** $29.75　　　**C.** $30.25　　　**D.** $31.50

[1-5]

16. The mode is __?__. D
   A. $36.00　　　**B.** $27.00　　　**C.** $30.00　　　**D.** $45.00

Displaying Information

Many kinds of information are displayed by graphs. The type of graph used depends on the nature of the information being shown. Sometimes special visual effects are used as attention-getters.

The graph at the right uses a 3-dimensional effect to display some weather information. Notice that the numbers on the vertical scale at the left serve two purposes:

1. They refer to centimeters when reading the precipitation graph.
2. They refer to degrees Celsius when reading the temperature graph.

What was the lowest average monthly temperature? In what month did it occur? In what month was there the greatest average amount of precipitation? How many centimeters was it? 4°C; April; July; 9 cm

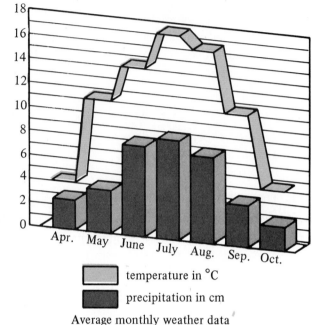

temperature in °C

precipitation in cm

Average monthly weather data for Edmonton, Alberta, Canada

The graph below uses visual symbols to indicate when total eclipses of the sun and moon will occur. This kind of graph is called a *pictograph.*

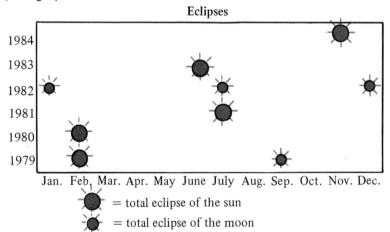

Eclipses

= total eclipse of the sun

= total eclipse of the moon

In what year does both a total eclipse of the sun and of the moon occur? Is there a total eclipse of the sun in all the years shown?

Another form of pictograph is shown below. It indicates the noise levels of various everyday situations. A *decibel* is a unit for measuring the level of sound. Wavy lines are used instead of bars because they suggest sound waves.

NOISE LEVELS

Quiet office	〰〰〰	40 decibels
Average home	〰〰〰〰	50
Vacuum cleaner	〰〰〰〰〰〰	70
Heavy truck	〰〰〰〰〰〰〰	90
Motorcycle	〰〰〰〰〰〰〰〰	100
Amplified rock music	〰〰〰〰〰〰〰〰〰	115
Jet plane (at ramp)	〰〰〰〰〰〰〰〰〰	117

Research Activity Find out about noise pollution and the effect of loud noises on one's hearing.

We can use graphs to help us plan how to spend our money.

Suppose you have $18 to spend. You like records, but you also enjoy mystery books. Records cost $6 and books cost $3 apiece. How many of each can you afford?

You can spend exactly $18 if you buy any of the following combinations:

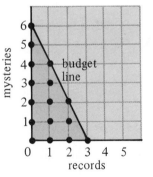

records	mysteries
0	6
1	4
2	2
3	0

The combinations can also be shown by dots on a graph. The line connecting these dots is called your budget line. You can afford any combination which is represented by dots on or below the budget line.

1. How much would 1 record and 4 mystery books cost? $18
2. How much would 1 record and 3 mystery books cost? $15
3. If you wanted to save $5 of your $18, how much would you have left to spend on mystery books and records? What combinations of books and records could you buy? $13; 1 record and 1 or 2 mysteries, or 1 to 4 mysteries

Chapter Test

Exercises 1 and 2 refer to the table below.

1. What was the least amount of peaches picked from a tree? 95 kg

2. Which row produced the most peaches? Row 3

Kilograms of peaches picked from 15 peach trees					
Row 1	120	140	95	125	190
Row 2	105	160	135	150	175
Row 3	145	200	175	180	155

[1-1]

Exercises 3 and 4 refer to the bar graph at the right.

3. What is the swimming speed of a dolphin? 40 km/h

4. How far can a pike swim in 3 hours? 30 km

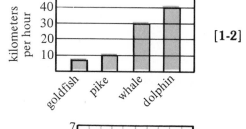

[1-2]

Exercises 5 and 6 refer to the line graph at the right which shows automobile registrations in Canada.

5. About how many automobiles were registered in 1950? 2,500,000

6. In what year were there about 3 million automobiles registered? 1952

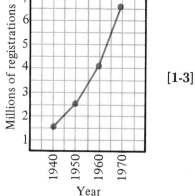

[1-3]

Exercises 7–10 refer to the table for Exercises 1 and 2.

7. Organize the data into intervals of 91–120, 121–150, 151–180, and 181–210. Draw a histogram. Check students' papers.

[1-4]

8. How many trees produced 150 kg of peaches or less? 8

9. What is the mean of the data? 150

10. What is the median of the data? 150

[1-5]

Skill Review

Perform the indicated operation.

1. $\begin{array}{r} 31 \\ +50 \\ \hline 81 \end{array}$
2. $\begin{array}{r} 72 \\ +19 \\ \hline 91 \end{array}$
3. $\begin{array}{r} 44 \\ +63 \\ \hline 107 \end{array}$
4. $\begin{array}{r} 86 \\ + 7 \\ \hline 93 \end{array}$
5. $\begin{array}{r} 65 \\ +89 \\ \hline 154 \end{array}$

6. $\begin{array}{r} 27 \\ + 4 \\ \hline 31 \end{array}$
7. $\begin{array}{r} 53 \\ +43 \\ \hline 96 \end{array}$
8. $\begin{array}{r} 90 \\ +78 \\ \hline 168 \end{array}$
9. $\begin{array}{r} 18 \\ +65 \\ \hline 83 \end{array}$
10. $\begin{array}{r} 49 \\ +72 \\ \hline 121 \end{array}$

11. $\begin{array}{r} 705 \\ +261 \\ \hline 966 \end{array}$
12. $\begin{array}{r} 634 \\ +948 \\ \hline 1582 \end{array}$
13. $\begin{array}{r} 975 \\ + 73 \\ \hline 1048 \end{array}$
14. $\begin{array}{r} 318 \\ +842 \\ \hline 1160 \end{array}$
15. $\begin{array}{r} 432 \\ +995 \\ \hline 1427 \end{array}$

16. $\begin{array}{r} 79 \\ 54 \\ +40 \\ \hline 173 \end{array}$
17. $\begin{array}{r} 31 \\ 82 \\ +36 \\ \hline 149 \end{array}$
18. $\begin{array}{r} 64 \\ 10 \\ +97 \\ \hline 171 \end{array}$
19. $\begin{array}{r} 862 \\ 473 \\ +315 \\ \hline 1650 \end{array}$
20. $\begin{array}{r} 506 \\ 994 \\ +282 \\ \hline 1782 \end{array}$

21. $\begin{array}{r} 48 \\ -36 \\ \hline 12 \end{array}$
22. $\begin{array}{r} 70 \\ -15 \\ \hline 55 \end{array}$
23. $\begin{array}{r} 39 \\ - 2 \\ \hline 37 \end{array}$
24. $\begin{array}{r} 75 \\ -45 \\ \hline 30 \end{array}$
25. $\begin{array}{r} 63 \\ -28 \\ \hline 35 \end{array}$

26. $\begin{array}{r} 36 \\ -17 \\ \hline 19 \end{array}$
27. $\begin{array}{r} 82 \\ -50 \\ \hline 32 \end{array}$
28. $\begin{array}{r} 51 \\ -29 \\ \hline 22 \end{array}$
29. $\begin{array}{r} 72 \\ - 4 \\ \hline 68 \end{array}$
30. $\begin{array}{r} 99 \\ -91 \\ \hline 8 \end{array}$

31. $\begin{array}{r} 748 \\ -273 \\ \hline 475 \end{array}$
32. $\begin{array}{r} 745 \\ -639 \\ \hline 106 \end{array}$
33. $\begin{array}{r} 640 \\ -268 \\ \hline 372 \end{array}$
34. $\begin{array}{r} 119 \\ -107 \\ \hline 12 \end{array}$
35. $\begin{array}{r} 682 \\ -488 \\ \hline 194 \end{array}$

36. $\begin{array}{r} 9817 \\ -4321 \\ \hline 5496 \end{array}$
37. $\begin{array}{r} 6257 \\ -5603 \\ \hline 654 \end{array}$
38. $\begin{array}{r} 8651 \\ - 764 \\ \hline 7887 \end{array}$
39. $\begin{array}{r} 8340 \\ -8219 \\ \hline 121 \end{array}$
40. $\begin{array}{r} 4094 \\ -1380 \\ \hline 2714 \end{array}$

41. $\begin{array}{r} 24 \\ \times 3 \\ \hline 72 \end{array}$
42. $\begin{array}{r} 81 \\ \times 7 \\ \hline 567 \end{array}$
43. $\begin{array}{r} 53 \\ \times 5 \\ \hline 265 \end{array}$
44. $\begin{array}{r} 94 \\ \times 2 \\ \hline 188 \end{array}$
45. $\begin{array}{r} 60 \\ \times 8 \\ \hline 480 \end{array}$

46. $\begin{array}{r} 71 \\ \times 49 \\ \hline 3479 \end{array}$
47. $\begin{array}{r} 23 \\ \times 60 \\ \hline 1380 \end{array}$
48. $\begin{array}{r} 54 \\ \times 76 \\ \hline 4104 \end{array}$
49. $\begin{array}{r} 63 \\ \times 13 \\ \hline 819 \end{array}$
50. $\begin{array}{r} 92 \\ \times 58 \\ \hline 5336 \end{array}$

51. $\begin{array}{r} 197 \\ \times 6 \\ \hline 1182 \end{array}$
52. $\begin{array}{r} 405 \\ \times 3 \\ \hline 1215 \end{array}$
53. $\begin{array}{r} 870 \\ \times 32 \\ \hline 27,840 \end{array}$
54. $\begin{array}{r} 458 \\ \times 84 \\ \hline 38,472 \end{array}$
55. $\begin{array}{r} 946 \\ \times 91 \\ \hline 86,086 \end{array}$

56. $6\overline{)72}$ 12
57. $4\overline{)96}$ 24
58. $9\overline{)369}$ 41
59. $3\overline{)195}$ 65
60. $5\overline{)540}$ 108

61. $2\overline{)758}$ 379
62. $7\overline{)413}$ 59
63. $8\overline{)624}$ 78
64. $4\overline{)260}$ 65
65. $3\overline{)282}$ 94

66. $67\overline{)536}$ 8
67. $29\overline{)783}$ 27
68. $56\overline{)728}$ 13
69. $13\overline{)416}$ 32
70. $48\overline{)336}$ 7

25

2 The Decimal System

2-1 Powers of Ten

Objective To identify powers of ten and write their products.

The decimal numeral for the amount of money mentioned in the newspaper headline at the right is 1,000,000. We can show 1,000,000 as a product of *factors* of 10. A **factor** is any of two or more numbers which are multiplied to form a product.

$$1{,}000{,}000 = \underbrace{10 \times 10 \times 10 \times 10 \times 10 \times 10}_{\textbf{6 factors of 10}}$$

NEW SPORTS CENTER
COSTS $1 MILLION

We say that 1,000,000 is a *power* of 10. A **power** is a product of *equal* factors. Since 1,000,000 is the product of 6 factors of 10, we say 1,000,000 is the **6th power of 10.** A simpler way to show this is

$$1{,}000{,}000 = 10^{6}$$

exponent 6

base 10

The *exponent* 6 shows the number of times the base 10 is used as a factor. The numeral 10^{6} is read "the 6th power of 10," or "10 to the 6th (power)." Notice that there are 6 zeros in the numeral 1,000,000.

The product of two powers of 10 can be found rapidly. For example

$$10^2 \times 10^3 = (10 \times 10) \times (10 \times 10 \times 10) = 10^{2+3} = 10^5$$

Note that the exponent in the product equals the sum of those in the factors, that is, $2 + 3 = 5$.

Example 1 | **Express $10^4 \times 10^5$ as a single power of 10.**

Solution | $10^4 \times 10^5 = 10^{4+5} = 10^9$

The number system we use is based on powers of ten and is called a decimal system. (The word *decimal* comes from *decem*, the Latin word for *ten*.) For example, we can show the number 3972 in expanded form as

$$3972 = (3 \times 1000) + (9 \times 100) + (7 \times 10) + 2,$$

or, using exponents, as

$$3972 = (3 \times 10^3) + (9 \times 10^2) + (7 \times 10^1) + 2.$$

Parentheses are used above to group operations which should be done first.

Example 2 | **Write in expanded form, using exponents.**
| **a. 13,599**
| **b. 8035**

Solution | **a. 13,599 $= (1 \times 10,000) + (3 \times 1000) + (5 \times 100) + (9 \times 10) + 9$**
| $= (1 \times 10^4) + (3 \times 10^3) + (5 \times 10^2) + (9 \times 10^1) + 9$
| **b. 8035 $= (8 \times 1000) + (0 \times 100) + (3 \times 10) + 5$**
| $= (8 \times 10^3) + (0 \times 10^2) + (3 \times 10^1) + 5$
| $= (8 \times 10^3) + (3 \times 10^1) + 5$

Example 3 | **Write the decimal numeral for the expanded form.**
| **a. $(2 \times 10^2) + (4 \times 10^1) + 3$**
| **b. $(3 \times 10^3) + (0 \times 10^2) + (5 \times 10^1) + 8$**

Solution | **a. 243**
| **b. 3058**

Class Exercises

Read each numeral as a power of 10.

1. 100
   10 squared

2. 100,000
   10 to the 5th

3. 1000
   10 cubed

4. 10,000,000
   10 to the 7th

5. 10,000
   10 to the 4th

Tell how many zeros there are in the decimal numeral for the power of 10.

6. 10^3 3

7. 10^6 6

8. 10^1 1

9. 10^4 4

10. 10^8 8

Complete the table below.

	Number	Decimal Numeral	Number of Zeros	Numeral with an Exponent
11.	million	1,000,000	6	? 10^6
12.	thousand	1,000	? 3	10^3
13.	hundred	100	? 2	? 10^2
14.	hundred million	100,000,000	? 8	? 10^8
15.	ten thousand	10,000	? 4	? 10^4
16.	hundred thousand	100,000	? 5	? 10^5

Exercises

Express as a single power of 10.

A 1. $10^1 \times 10^2$ 10^3 2. $10^4 \times 10^3$ 10^7 3. $10^2 \times 10^2$ 10^4 4. $10^5 \times 10^1$ 10^6 5. $10^6 \times 10^3$ 10^9

Write in expanded form using exponents.

6. 243

7. 5812

8. 62,334

9. 8049

10. 30,057

6. $(2 \times 10^2) + (4 \times 10^1) + 3$

7. $(5 \times 10^3) + (8 \times 10^2) + (1 \times 10^1) + 2$

Write the decimal numeral for the expanded form.

11. $(7 \times 1000) + (4 \times 100) + (7 \times 10) + 5$ 7475

12. $(9 \times 10,000) + (0 \times 1000) + (6 \times 100) + (4 \times 10) + 2$ 90,642

13. $(6 \times 100,000) + (3 \times 10,000) + (1 \times 1000) + (0 \times 100) + (6 \times 10) + 0$ 631,060

14. $(5 \times 10,000) + (0 \times 1000) + (0 \times 100) + (0 \times 10) + 2$ 50,002

15. $(2 \times 10^3) + (5 \times 10^2) + (7 \times 10^1) + 3$ 2573

16. $(9 \times 10^4) + (0 \times 10^3) + (3 \times 10^2) + (4 \times 10^1) + 4$ 90,344

17. $(8 \times 10^6) + (4 \times 10^5) + (0 \times 10^4) + (7 \times 10^3) + (4 \times 10^2) + (5 \times 10^1) + 9$ 8,407,459

18. $(1 \times 10^5) + (9 \times 10^4) + (1 \times 10^3) + (7 \times 10^2) + (1 \times 10^1) + 0$ 191,710

8. $(6 \times 10^4) + (2 \times 10^3) + (3 \times 10^2) + (3 \times 10^1) + 4$

9. $(8 \times 10^3) + (4 \times 10^1) + 9$

10. $(3 \times 10^4) + (5 \times 10^1) + 7$

Using some or all of the digits 8, 2, 7, 5, 4, write the number described. Each digit can be used just once in the number.

19. The largest 2-digit number   87

20. The smallest 2-digit number   24

21. The largest 3-digit number   875

22. The smallest 3-digit number   245

23. The largest 3-digit number whose second digit is 8   785

24. The smallest 3-digit number whose second digit is 8   284

25. The largest 4-digit number   8754

26. The next-to-the-largest 4-digit number   8752

27. The smallest 4-digit number   2457

28. The name for the 100th power of 10, or 10^{100}, is one *googol*. How many zeros are there in the decimal numeral for one googol?   100

EXTRA!

Powers of Bases Other than 10

We can use exponents to express powers of any number, not just 10. For example,

$2^1 = 2$ $5^1 = 5$
$2^2 = 2 \times 2 = 4$ $5^2 = 5 \times 5 = 25$
$2^3 = 2 \times 2 \times 2 = 8$ $5^3 = 5 \times 5 \times 5 = 125$
$2^4 = 2 \times 2 \times 2 \times 2 = 16$ $5^4 = 5 \times 5 \times 5 \times 5 = 625$

Calculator Activity Use a calculator for the following exercises if you wish. Operations in parentheses are to be done first.

1. a. $16^2 - 13^2$ 87   **b.** $(16 + 13) \times (16 - 13)$ 87
2. a. $45^2 - 23^2$ 1496   **b.** $(45 + 23) \times (45 - 23)$ 1496
3. a. $73^2 - 69^2$ 568   **b.** $(73 + 69) \times (73 - 69)$ 568
4. a. $124^2 - 42^2$ 13,612   **b.** $(124 + 42) \times (124 - 42)$ 13,612

Do you see a pattern? The difference of two numbers squared is the same as the product of the sum of the numbers times the difference of the numbers.

2-2 Decimals

Objective To read and write decimals.

The decimal numbers we used in the preceding section are called whole numbers. We can show the set of whole numbers like this:

$$\{0, 1, 2, \ldots\}$$

We read this "the set whose members are 0, 1, 2, and so on." The three dots show that there is no end to the list of whole numbers.

You are already familiar with another set of numbers, called fractions. For example, numbers like $\frac{1}{2}$, $\frac{4}{7}$, and so on, are fractions. A fraction whose denominator is a power of ten can be written in decimal form.

Fraction	Decimal
$\frac{3}{10}$	0.3
$\frac{5}{10}$ or $\frac{1}{2}$	0.5

decimal point

The table below shows some fractions and decimals used to name the quotient when one is divided by successive powers of ten.

Fraction	Decimal	In Words
$\frac{1}{10}$	0.1	one tenth
$\frac{1}{10^2}$ or $\frac{1}{100}$	0.01	one hundredth
$\frac{1}{10^3}$ or $\frac{1}{1000}$	0.001	one thousandth
$\frac{1}{10^4}$ or $\frac{1}{10,000}$	0.0001	one ten-thousandth
. . . .	. . . . .	. . . .

1 unit

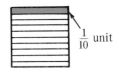

$\frac{1}{10}$ unit

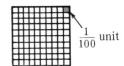

$\frac{1}{100}$ unit

Notice that as we read down the table, each succeeding decimal is equal to 0.1 of the decimal above it. That is,

$$0.01 = 0.1 \times 0.1, \quad 0.001 = 0.1 \times 0.01, \quad \text{and so on.}$$

Look at the multiplications with fractions and decimals on page 32.

$$\frac{1}{100} = \frac{1}{10} \times \frac{1}{10} \qquad \text{or} \qquad 0.01 \quad = 0.1 \times 0.1$$

$$\frac{1}{1000} = \frac{1}{10} \times \frac{1}{100} \qquad \text{or} \qquad 0.001 \ = 0.1 \times 0.01$$

$$\frac{1}{10{,}000} = \frac{1}{10} \times \frac{1}{1000} \qquad \text{or} \qquad 0.0001 = 0.1 \times 0.001$$

The chart below shows the place values for some decimals.

Place-value Chart

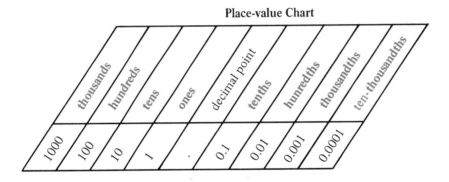

As we move to the right, place values decrease, because each new place is only one tenth of the preceding place value. The whole numbers are to the left of the decimal point and the decimals for fractions between 0 and 1 are to its right.

In the decimal 0.387,
the place value of the digit 3 is one tenth, or 0.1,
the number which 3 represents is 3 × 0.1, or 0.3.

In expanded form,
0.387 = 0.3 + 0.08 + 0.007
= (3 × 0.1) + (8 × 0.01) + (7 × 0.001)

Here are some examples of how to read decimals. Notice that, we read the decimal point as "and."

Decimal	In words
0.5	five tenths, or one half
6.32	six and thirty-two hundredths
0.387	three hundred eighty-seven thousandths
548.9	five hundred forty-eight and nine tenths

Class Exercises

Name the place value of the underlined digit.

1. 6945
hundreds

2. 83.45
ones

3. 692.01
hundredths

4. 12.336
thousandths

5. 40.6253
ten-thousandths

Read each number.

6. 5732

7. 24.6

8. 6.13

9. 1052.482

10. 34.5067

11. 0.0015

12. 8.0008

13. 0.70605

Exercises

Write each number in expanded form.

A

1. 65.4
1. (6 × 10) + 5 + (4 × 0.1)

2. 6.54

3. 0.654
2. 6 + (5 × 0.1) + (4 × 0.01)

4. 0.0654

5. 29.86
3. (6 × 0.1) + (5 × 0.01) + (4 × 0.001)

6. 10.305

7. 0.0472
4. (6 × 0.01) + (5 × 0.001) + (4 × 0 0001)

8. 9.00623

Write the decimal numeral for the expanded form.

9. (4 × 0.1) + (7 × 0.01) + (9 × 0.001) 0.479

10. (6 × 10) + 5 + (6 × 0.1) + (8 × 0.01) 65.68

11. (5 × 100) + (7 × 10) + 2 + (0 × 0.1) + (8 × 0.01) 572.08

12. (3 × 1000) + (0 × 100) × (4 × 10) + 0 + (7 × 0.1) + (0 × 0.01) + (7 × 0.001) 3040.707

13. (1 × 10) + 3 + (0 × 0.1) + (0 × 0.01) + (5 × 0.001) + (2 × 0.0001) 13.0052

14. (9 × 0.1) + (3 × 0.01) + (0 × 0.001) + (4 × 0.0001) + (5 × 0.00001) 0.93045

15. (6 × 100) + 4 + (3 × 0.1) + (7 × 0.001) 604.307

16. (2 × 10) + (3 × 0.01) + (4 × 0.0001) 20.0304

Write in words.

17. 26.5

18. 2.48

19. 0.312

20. 7.076

21. 0.0011

22. 92.072

23. 401.0906

24. 300.0036

Write the decimal.

25. fourteen and seven tenths 14.7

26. thirty-three and seven hundredths 33.07

27. five hundred and fifteen thousandths 500.015

28. seven and sixty-four ten-thousandths 7.0064

29. two hundred fifty and three hundred forty-six thousandths 250.346

30. one thousand and twenty-nine ten-thousandths 1000.0029

5. (2 × 10) + 9 + (8 × 0.1) + (6 × 0.01)

6. (1 × 10) + (3 × 0.1) + (5 × 0.001)

7. (4 × 0.01) + (7 × 0.001) + (2 × 0.0001)

8. 9 + (6 × 0.001) + (2 × 0.0001) + (3 × 0.00001)

2-3 The Number Line

Objective To graph numbers on the number line and compare them by the use of inequality symbols.

The *graphs* of the whole numbers 0 through 6 are shown on the number line below. The **graph** of a number is the point paired with

the number on a number line. The number paired with a point is called the **coordinate** of the point. Looking at the number line above we can see that:

1. The graph of 0 is called the *origin*.
2. The graphs of consecutive whole numbers are equally spaced.
3. The graph of 3 is point *A*.
4. The coordinate of point *B* is 6.
5. Coordinates increase as we move to the right.

Since *B* lies to the right of *A*,
    coordinate of *B*   coordinate of *A*
                        6 is greater than 3.
In symbols we write $6 > 3$

We can also say
                        3 is less than 6

In symbols we write $3 < 6$

We call the symbols $<$ and $>$ **inequality symbols.** To avoid confusing these symbols, think of them as arrowheads whose small ends point toward the smaller numbers.
   We can indicate that 4 is between 3 and 6 by combining two inequalities $3 < 4$ and $4 < 6$ and writing

$$3 < 4 < 6$$

Example 1 Write the sentence using an inequality symbol or symbols.

    a. 2 is less than 5

    b. 18 is greater than 7

    c. 30 is between 20 and 40

Solution   **a.** $2 < 5$

    b. $18 > 7$

    c. $20 < 30 < 40$

The number line below is marked off in tenths. On it are shown the graphs of some decimals.

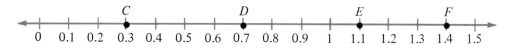

Number	Graph
0.3	point C
0.7	point D
1.1	point E
1.4	point F

In the same way we can draw number lines marked off in hundredths, thousandths, and so on, and on them show the graphs of decimals to the hundredths' and thousandths' places.

Example 2 Write the sentence using an inequality symbol or symbols.

    a. 0.2 is less than 0.8

    b. 1.3 is greater than 0.6

    c. 0.9 is between 0.5 and 1.4

    d. 3.06 is between 3.02 and 3.15

    e. 12.637 is between 12.630 and 12.640

Solution   **a.** $0.2 < 0.8$

    b. $1.3 > 0.6$

    c. $0.5 < 0.9 < 1.4$

    d. $3.02 < 3.06 < 3.15$

    e. $12.630 < 12.637 < 12.640$

Class Exercises

Use the number line below for Exercises 1–6.

1. The graph of 0.5 is point __?__. C
2. The graph of 1.1 is point __?__. D
3. The coordinate of point *B* is __?__. 0.2
4. The coordinate of point *E* is __?__. 1.4
5. Point *A* is called the __?__. origin
6. Since *D* lies to the right of *B*, 1.1 __?__ 0.2. >
    $(<,>)$

Read each sentence.

7. $2 < 8$ 8. $0.8 > 0.2$ 9. $3.2 > 0.3$ 10. $5 < 7.7$
11. $16 > 5 > 0$ 12. $0.3 < 0.5 < 1.6$ 13. $10.6 < 11 < 11.6$ 14. $1.2 > 1.02 > 1.002$

Exercises

For each exercise, draw a number line and graph the given numbers. See students' graphs.

A
1. $0, 2, 5, 6, 8$
2. $0, 0.2, 0.5, 0.6, 0.8$
3. $0.3, 0.7, 1, 1.2, 1.4$
4. $0.4, 0.6, 0.8, 1, 1.2$
5. $1.4, 0.9, 0.5, 1.2, 1.3$
6. $1.4, 0, 0.7, 1.6, 0.8$

Write the sentence using an inequality symbol.

7. 9 is less than 15. $9 < 15$
8. 14 is less than 24. $14 < 24$
9. 21 is greater than 19. $21 > 19$
10. 7 is between 4 and 10. $4 < 7 < 10$
11. 0.8 is greater than 0.5. $0.8 > 0.5$
12. 4.7 is less than 4.72. $4.7 < 4.72$
13. 6.3 is between 6.1 and 6.5. $6.1 < 6.3 < 6.5$
14. 0.5 is between 1.7 and 0.4. $1.7 > 0.5 > 0.4$
15. 1.53 is greater than 1.43. $1.53 > 1.43$
16. 9.01 is between 9.001 and 9.1. $9.001 < 9.01 < 9.1$

Replace each __?__ with $>$ or $<$ to make a true sentence.

17. 21 __?__ 16 >
18. 13 __?__ 15 <
19. 4 __?__ 9 __?__ 11 <,<
20. 0.5 __?__ 0.7 <
21. 3.8 __?__ 2.8 >
22. 5.72 __?__ 1.67 >
23. 0.08 __?__ 0.26 <
24. 12.04 __?__ 12.97 <
25. 0.26 __?__ 0.46 __?__ 0.66 <,<
26. 0.44 __?__ 0.4 >
27. 1.765 __?__ 1.76 >
28. 8.303 __?__ 8.33 <
29. 4.7 __?__ 4.76 <
30. 6.045 __?__ 6.12 <
31. 1.13 __?__ 1.093 >

2-4 Equations and Variables

Objective To use variables to represent numbers.

In Section 2-3, you learned the meaning of the inequality symbols $<$ and $>$. You are probably much more familiar with the **equality symbol,** $=$, which stands for "is equal to." When we write the *equation*

$$5 + 6 = 11$$

we mean that the *numerical expressions* $5 + 6$ and 11 name the same number. $5 + 6 = 11$ is a *true* equation. Not all equations are true; for example, $5 + 6 = 13$ is a *false* equation.

Another kind of symbol we use is called a *variable*. A **variable** is a letter which is used to stand for any member of a given set of numbers. The given set of numbers is called the **replacement set.** Each number in the set is called a **value of the variable.** The letter x is used as a variable in Example 1, below.

Example 1 | What is the sum, $x + 8$, if
| **a.** $x = 2$? **b.** $x = 9$?
Solution | **a.** $2 + 8 = 10$ **b.** $9 + 8 = 17$

In Example 1,
   2 and 9 are the values of x.
   $\{2, 9\}$ is the replacement set for x.
Replacing a variable in an expression with one of its values and then doing the indicated arithmetic is called **finding the value of the expression,** or **evaluating the expression.**

Example 2 | Evaluate $5 \times n$ if the replacement set for n is $\{3, 7, 11\}$.
Solution | $5 \times n = ?$
| $5 \times 3 = 15$
| $5 \times 7 = 35$
| $5 \times 11 = 55$

When variables are used in products, the multiplication symbol, $\times$, is often omitted. Thus we write $5n$ in place of $5 \times n$, and ab in place of $a \times b$.

Class Exercises

1. What is the difference, $20 - x$, if: **a.** $x = 11$? **b.** $x = 3$? 9; 17

2. What is the sum, $y + 5$, if: **a.** $y = 9$? **b.** $y = 67$? 14; 72

3. What is the product, $8n$, if: **a.** $n = 12$? **b.** $n = 40$? 96; 320

4. What is the quotient, $42 \div n$, if: **a.** $n = 2$? **b.** $n = 7$? 21; 6

Replace the variable with the given value and state whether the resulting equation is true or false.

Example | $9 - x = 2; x = 3$
Solution | $9 - 3 = 2$, or $6 = 2$. **False**

5. $x + 7 = 12; x = 5$ T

6. $24 - y = 20; y = 3$ F

7. $10x = 320; x = 35$ F

8. $93 - n = 6; n = 33$ F

9. $7y = 56; y = 8$ T

10. $m + 6 = 72; m = 62$ F

Exercises

Complete the table.

A

	Expression	Value of x	Value of the Expression
1.	$x + 13$	5	18 ?
2.	$x + 13$	25	38 ?
3.	$24 - x$	11	13 ?
4.	$24 - x$	22	2 ?
5.	$15x$	21	315 ?
6.	$15x$	72	1080 ?
7.	$25 + x$	67	92 ?
8.	$x - 132$	416	284 ?
9.	$58x$	39	2262 ?
10.	$217 \div x$	7	31 ?

Evaluate the given expression for the given replacement set.

11. $x + 9$; $\{3, 5, 16\}$ 12; 14; 25

12. $21 + x$; $\{4, 17, 19\}$ 25; 38; 40

13. $16n$; $\{2, 16, 18\}$ 32; 256; 288

14. $45n$; $\{8, 20, 31\}$ 360; 900; 1395

15. $37 - y$; $\{11, 15, 24\}$
26; 22; 13

16. $y - 16$; $\{74, 191, 26\}$
58; 175; 10

17. $63x$; $\{27, 11, 20\}$
1701; 693; 1260

18. $n + 184$; $\{52, 133, 326\}$
236; 317; 510

19. $120 \div x$; $\{4, 10, 60\}$
30; 12; 2

20. $144 \div y$; $\{3, 36, 9\}$
48; 4; 16

B **21.** $2x + 1$; $\{3, 7, 10\}$
7; 15; 21

22. $39 - 4x$; $\{1, 5, 9\}$
35; 19; 3

23. $5x - 4$; $\{4, 7, 9\}$
16; 31; 41

24. $3(x + 4)$; $\{6, 10, 15\}$
30; 42; 57

25. $6(x - 3)$; $\{7, 11, 21\}$
24; 48; 108

26. $21(x + 32)$; $\{2, 47, 83\}$
714; 1659; 2415

Self-Test

Symbols and words to remember:

$>$ [p. 34] $<$ [p. 34] $=$ [p. 37]
factor [p. 27] power [p. 27] exponent [p. 27]
inequalities [p. 34] equation [p. 37] expression [p. 37]
variable [p. 37] replacement set [p. 37]

1. Write 10,000 as a power of 10. 10^4 **[2-1]**

2. Express $10^3 \times 10^5$ as a single power of 10. 10^8

3. Write 40,807 in expanded form, using exponents.
$(4 \times 10^4) + (8 \times 10^2) + 7$

4. Write the decimal numeral for the following: 8691

$$(8 \times 1000) + (6 \times 100) + (9 \times 10) + 1.$$

5. Name the place value of the underlined digit in 38.0$\underline{5}$5. hundredths **[2-2]**

6. Write 7.246 in expanded form. $7 + (2 \times 0.1) + (4 \times 0.01) + (6 \times 0.001)$

7. Write the decimal for the following: 403.62

$$(4 \times 100) + (0 \times 10) + 3 + (6 \times 0.1) + (2 \times 0.01).$$

8. Write the decimal for sixty-five and nine thousandths. 65.009

9. Draw a number line. Graph 0.1, 0.7, 0.9, 1, 1.2. **[2-3]**

Replace each __?__ with $<$ or $>$ to make a true sentence.

10. 8.3 __?__ 8.03
$>$
11. 0.91 __?__ 0.916
$<$
12. 5 __?__ 5.28 __?__ 5.3
$<$ $<$

Evaluate the given expression for the given replacement set.

13. $8 + n$; $\{0, 9, 23\}$ 8; 17; 31 **14.** $x - 13$; $\{17, 53, 108\}$ 4; 40; 95 **[2-4]**

15. $12y$; $\{3, 8, 40\}$ 36; 96; 480 **16.** $t \div 24$; $\{96, 240, 408\}$ 4; 10; 17

Self-Test answers and Extra Practice are at the back of the book.

2-5 Inverse Operations

Objective To show the relationship between addition and subtraction and between multiplication and division.

Finding the difference of two numbers, or subtraction, can be looked at as the process of finding a missing addend as shown below.

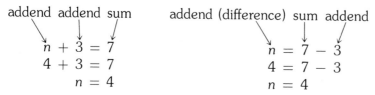

We say that adding 3 and subtracting 3 are *inverse operations*. In the same way, for every addition equation we can write a related subtraction equation. Adding a number is the inverse of subtracting the same number.

Example 1 | Write a related subtraction equation and find the value of n.

a. $n + 5 = 12$ **b.** $n + 14 = 23$

Solution | **a.** $n = 12 - 5$ **b.** $n = 23 - 14$

$n = 7$ $n = 9$

Multiplying by a number and dividing by the same number are inverse operations also. Finding the quotient of two numbers, or division, can be thought of as finding a missing factor.

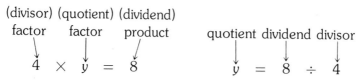

Example 2 | Write a related division equation and find the value of y.

a. $5 \times y = 15$ **b.** $7 \times y = 56$

Solution | **a.** $y = 15 \div 5$ **b.** $y = 56 \div 7$

$y = 3$ $y = 8$

Class Exercises

1. If $n + 4 = 12$, then $n = 12 - $ __?__, and $n = $ __?__. 4; 8

2. If $y + 5 = 16$, then $y = $ __?__ $- $ __?__, and $y = $ __?__. 16; 5; 11

3. If $8 \times y = 32$, then $y = 32 \div $ __?__, and $y = $ __?__. 8; 4

4. If $6 \times m = 42$, then $m = $ __?__ $\div$ __?__, and $m = $ __?__. 42; 6; 7

State a related subtraction or division equation and find the value of *n*.

5. $n + 6 = 14$
$n = 14 - 6$; 8

6. $n + 20 = 25$
$n = 25 - 20$; 5

7. $n + 2 = 12$
$n = 12 - 2$; 10

8. $n + 30 = 40$
$n = 40 - 30$; 10

9. $2 \times n = 16$
$n = 16 \div 2$; 8

10. $7 \times n = 21$
$n = 21 \div 7$; 3

11. $3 \times n = 33$
$n = 33 \div 3$; 11

12. $5 \times n = 40$
$n = 40 \div 5$; 8

Exercises

Write a related subtraction equation and find the value of *x*.

A

1. $x + 83 = 102$
$x = 102 - 83$; 19

2. $x + 47 = 96$
$x = 96 - 47$; 49

3. $x + 116 = 242$
$x = 242 - 116$; 126

4. $x + 38 = 276$
$x = 276 - 38$; 238

5. $x + 133 = 200$
$x = 200 - 133$; 67

6. $x + 88 = 217$
$x = 217 - 88$; 129

7. $x + 135 = 235$
$x = 235 - 135$; 100

8. $x + 716 = 900$
$x = 900 - 716$; 184

9. $x + 299 = 316$
$x = 316 - 299$; 17

10. $x + 308 = 803$
$x = 803 - 308$; 495

11. $x + 671 = 719$
$x = 719 - 671$; 48

12. $x + 136 = 631$
$x = 631 - 136$; 495

Write a related division equation and find the value of *y*.

13. $12 \times y = 156$
$y = 156 \div 12$; 13

14. $31 \times y = 961$
$y = 961 \div 31$; 31

15. $25 \times y = 600$
$y = 600 \div 25$; 24

16. $18 \times y = 360$
$y = 360 \div 18$; 20

17. $44 \times y = 132$
$y = 132 \div 44$; 3

18. $19 \times y = 323$
$y = 323 \div 19$; 17

19. $18 \times y = 1818$
$y = 1818 \div 18$; 101

20. $17 \times y = 289$
$y = 289 \div 17$; 17

21. $71 \times y = 781$
$y = 781 \div 71$; 11

22. $34 \times y = 1462$
$y = 1462 \div 34$; 43

23. $16 \times y = 2000$
$y = 2000 \div 16$; 125

24. $101 \times y = 7272$
$y = 7272 \div 101$; 72

Application

Calculator Activity Take your calculator and

	Example	
1. Press any 3 digits.	723	
2. Repeat the digits.	723,723	
3. Divide by 7.	?	103,389
4. Divide by 11.	?	9 399
5. Divide by 13.	?	723

What is your answer? Now multiply your answer by 1001. Explain
your result. 723; $723 \times 1001 = 723{,}723$; Dividing by 7, then by 11, then by 13 is the same
as multiplying by 1001 since $7 \times 11 \times 13 = 1001$.

2-6 Properties of Addition and Multiplication

Objective To use the properties of addition and multiplication.

Addition and multiplication follow certain basic properties as shown below.

---

properties

Commutative Property
Changing the order of the addends in a sum, or the factors in a product, does not change the sum or product.

$$7 + 3 = 3 + 7 \qquad 9 \times 5 = 5 \times 9$$

In general, for any numbers a and b,

$$a + b = b + a \qquad a \times b = b \times a$$

Associative Property
Changing the grouping of addends in a sum, or of factors in a product, does not change the sum or product.

$$(4 + 2) + 5 = 11 \qquad (2 \times 3) \times 5 = 30$$
$$4 + (2 + 5) = 11 \qquad 2 \times (3 \times 5) = 30$$

In general, for any numbers a, b, and c,

$$(a + b) + c = a + (b + c) \qquad (a \times b) \times c = a \times (b \times c)$$

The Property of Zero in Addition
The sum of zero and any given number is the given number.

$$9 + 0 = 9$$

In general, for any number a, $a + 0 = a$

The Property of Zero in Multiplication
The product of zero and any given number is zero.

$$6 \times 0 = 0$$

In general, for any given number a, $a \times 0 = 0$

properties

The Property of One in Multiplication
The product of one and any given number a is the given number.

$$2 \times 1 = 2$$

In general, for any number a, $a \times 1 = a$

The Distributive Property of Multiplication over Addition
The following computation illustrates this property.

$$3 \times (2 + 1) = 3 \times 3 \quad (3 \times 2) + (3 \times 1) = 6 + 3$$
$$= 9 \qquad\qquad\qquad\qquad = 9$$

Therefore,
$$3 \times (2 + 1) = (3 \times 2) + (3 \times 1)$$

In general, for any numbers a, b, and c,
$$a \times (b + c) = (a \times b) + (a \times c)$$

In the numerical examples of the properties shown above we have used only whole numbers, for the sake of simplicity. However, these properties hold true for all the numbers you will study in this text.

Example | Replace each variable to make a true equation.
a. $6 + 5 = 5 + n$
b. $4 \times 2 = 2 \times n = x$

Solution | **a.** $6 + 5 = 5 + n$
$6 + 5 = 5 + 6$, so $n = 6$
b. $4 \times 2 = 2 \times n = x$
$4 \times 2 = 2 \times 4 = 8$, so $n = 4$ and $x = 8$

Class Exercises

Name the property of addition or multiplication illustrated.

1. $18 \times 0 = 0$ **2.** $4 \times 8 = 8 \times 4$ **3.** $(6 + 2) + 8 = 6 + (2 + 8)$

4. $7 + 0 = 7$ **5.** $23 \times 1 = 23$ **6.** $9 \times (6 + 2) = (9 \times 6) + (9 \times 2)$

1. Prop. of Zero in Mult.   2. Commutative Prop.   3. Associative Prop.

4. Prop. of Zero in Add.   5. Prop. of One in Mult.   6. Distributive Prop.

Exercises

Use the commutative and associative properties to simplify each expression.

Example | $16 + 25 + 4$

Solution | $16 + 25 + 4 = 16 + 4 + 25$
$$= 20 + 25 = 45$$

A

1. $8 + 27 + 12$ 47

2. $17 + 44 + 3$ 64

3. $15 + 87 + 5$ 107

4. $62 + 95 + 8$ 165

5. $5 + 13 + 15 + 7$ 40

6. $6 + 12 + 24 + 8$ 50

7. $2 \times 12 \times 5$ 120

8. $15 \times 7 \times 4$ 420

9. $5 \times 52 \times 2$ 520

10. $4 \times 32 \times 5 \times 25$ 16,000

11. $75 \times 12 \times 4 \times 5$ 18,000

12. $6 \times 45 \times 15 \times 2$ 8100

Use the distributive property to simplify each expression.

Example | **a.** $(4 \times 6) + (4 \times 14)$
b. 6×58

Solution | **a.** $(4 \times 6) + (4 \times 14) = 4(6 + 14)$
$$= 4(20) = 80$$
b. $6 \times 58 = 6(50 + 8)$
$$= (6 \times 50) + (6 \times 8)$$
$$= 300 + 48 = 348$$

13. $(3 \times 7) + (3 \times 13)$ 60

14. $(5 \times 21) + (5 \times 9)$ 150

15. $(18 \times 6) + (18 \times 24)$ 540

16. $(22 \times 18) + (22 \times 2)$ 440

17. 8×43 344

18. 6×89 534

19. 4×57 228

20. 12×35 420

Use any of the properties to simplify each expression.

B

21. $44 + 19 + 2 + 16 + 81 + 58$ 220

22. $(18 + 46) + (12 + 4) + (8 \times 17) + (23 \times 8)$ 400

23. $(12 \times 7) + (56 \div 8) + (13 \times 12)$ 247

24. $5(81 \div 3) + 5(63 \div 1) - 450$ 0

25. $116 + 7(48 + 15) + (15 + 22)7 + 34$ 850

26. $(78 \times 1) + (99 \times 78) - (101 + 0 + 99)$ 7600

27. $(3 \times 10^2) + (22 \times 10^2) + (4 \times 10^3)$ 6500

28. $43(2^3 + 2) + (10 \times 27)$ 700

2-7 Adding and Subtracting Decimals

Objective To use the addition and subtraction processes with decimals.

The procedures used to add or subtract whole numbers in decimal notation extend almost without change to other decimal numbers.

Example 1 | Compute **72.8 + 6.349 + 0.76**

Solution

Step 1	Step 2	Step 3
Write the given numerals in a column with the decimal points in line.	(Not usually shown) Add as if the numerals named whole numbers.	Write the sum.

72.8	72,800	72.8
6.349	6,349	6.349
+ 0.76	+ 760	+ 0.76
	79,909	79.909

WHY
IT
WORKS

$$
\begin{array}{rcl}
72{,}800 \times 0.001 &=& 72.8 \\
6{,}349 \times 0.001 &=& 6.349 \\
760 \times 0.001 &=& 0.76 \\
\hline
79{,}909 \times 0.001 &=& 79.909
\end{array}
$$

In the same way, we can justify the subtraction process.

Example 2 | Compute **1495.76 − 321.492**

Solution

Step 1	Step 2	Step 3
Write the given numerals in a column with the decimal points in a line.	(Not usually shown) Subtract as if the numerals named whole numbers.	Write the difference.

1495.76	1,495,760	1495.760
− 321.492	− 321,492	− 321.492
	1,174,268	1174.268

Class Exercises

Compute the sum or difference.

1. 34.3 +23.5 57.8	2. 72.04 +16.7 88.74	3. 5.18 +5.605 10.785	4. 12.7 + 6.298 18.998
5. 20.347 +30.62 50.967	6. 5.2 +17.706 22.906	7. 9.73 −5.11 4.62	8. 24.868 −14.254 10.614
9. 9.043 +7.03 16.073	10. 12.827 − 2.6 10.227	11. 8.47 −3.454 5.016	12. 18.6 −13.255 5.345

Exercises

Compute the sum.

A

1. 71.5 +16.3 87.8	2. 46.53 +21.46 67.99	3. 6.297 +8.434 14.731	4. 5.06 +6.952 12.012
5. 27.954 +16.21 44.164	6. 18.6364 +13.4372 32.0736	7. 34.72 +18.6049 53.3249	8. 126.328 + 9.4078 135.7358
9. 7.805 0.46 +15.936 24.201	10. 25.6 9.372 +105.17 140.142	11. 1.6256 6.006 2.9 +1.49 12.0216	12. 24.603 18.4 9.36 + 0.216 52.579

13. 6.38 + 5.246 11.626

14. 21.329 + 16.4 37.729

15. 1.37 + 5.5 + 2.66 9.53

16. 18.205 + 11.7251 + 3.6 33.5301

17. 2.4 + 7.608 + 3.1 + 7.27 20.378

18. 37.5 + 2.0056 + 3.21 + 7.8 50.5156

Compute the difference.

19. 66.8 −26.2 40.6	20. 18.48 − 9.37 9.11	21. 3.406 −1.872 1.534	22. 18.3 −15.72 2.58
23. 21.607 − 9.85 11.757	24. 143.2 − 79.604 63.596	25. 27.205 −19.6384 7.5666	26. 19.65 − 9.3728 10.2772

Compute the difference.

27. 46.96 − 3.98 42.98

28. 53.405 − 27.3 26.105

29. 106.8 − 36.47 70.33

30. 18 − 6.52 11.48

31. 716.32 − 521.9046 194.4154

32. 137.05 − 8.6 128.45

Problems

Solve.

A **1.** Carla bought a pair of hiking boots on sale for $29.79. Later the same day, she saw the same boots in another store for $36.95. How much did she save by buying the boots on sale? **$7.16**

2. Harry set a goal of hiking 24 km each month. In May, he took hikes of 4.25 km, 6.8 km, 5.65 km, and 7.125 km. Did he reach his goal? No. He only hiked 23.825 km.

An Inventor

Did you ever wonder who invented the citizen's-band radio, or the telephone, or the walkie-talkie? Did you ever wish that you could invent a faster and better method of communication?

During his lifetime, Granville T. Woods (1856–1910) designed and acquired patents for fifty inventions ranging from an egg incubator to a steam-boiler furnace. His ability to use mathematics contributed greatly to his success.

After studying mechanical engineering at college, Woods became an engineer aboard a steamer. Later, as a steam-locomotive engineer, he patented his most ingenious invention, the Synchronous Multiplex Railway Telegraph. The purpose of his telegraph was both to prevent railway accidents by keeping each train aware of the location of nearby trains, and to enable moving trains to communicate with stations. Woods marketed his many inventions through his own company.

Research Activity Many persons have contributed to the invention and development of today's computers. Find out who some of them are and what their contributions have been.

Career Activity What career do you think you would like to follow when you finish school? Find out what important inventions are used in this field of activity.

2-8 Multiplying Decimals

Objective To use the multiplication process with decimals.

We can use the same procedure to multiply mixed numbers as we used for whole numbers. The only difference is that we need a plan for placing the decimal point in the product.

Example 1 | Compute 2.3×6.85

Solution |

Align the right-hand digits of the factors and multiply as if they were whole numbers.

The number of places to the right of the decimal point in the product is the sum of the places to the right of the decimal point in the factors.

$$
\begin{array}{r}
6.8\,5 \longleftarrow \text{2 places} \\
\times \quad 2.3 \longleftarrow \text{1 place} \\
\hline
2\,0\,5\,5 \\
13\,7\,0 \quad\quad \\
\hline
15.7\,5\,5 \longleftarrow \text{3 places}
\end{array}
$$

WHY IT WORKS |

$$
\begin{aligned}
2.3 \times 6.85 &= (23 \times 0.1)\,(685 \times 0.01) \\
&= (23 \times 685)\,(0.1 \times 0.01) \\
&= (15{,}755)\,(0.001) \\
&= 15.755
\end{aligned}
$$

Example 2 | Compute 24.7×362.008

Solution |

$$
\begin{array}{r}
36\,2.00\,8 \longleftarrow \text{3 places} \\
24.7 \longleftarrow \text{1 place} \\
\hline
253\,4\,05\,6 \\
1448\,0\,32 \quad \\
7240\,1\,6 \quad\quad \\
\hline
8941.5\,97\,6 \longleftarrow \text{4 places}
\end{array}
$$

It can be very helpful to check your answers by rounding and estimating. A rule for rounding decimal numbers can be stated as follows:

	24.7	362.008
1. Find the place to which you wish to round, and mark it with a caret (‸). Look at the digit to the right.	24.7 ‸	362.008 ‸
2. If the digit to the right is **5 or greater**, add 1 to the marked digit. If the digit to the right is **less than 5**, leave the marked digit unchanged.	2	6
3. Replace each digit to the right of the marked place with "0".	20.0	360.000

Estimate the product in Example 2: 24.7 × 362.008

$$20 \times 360 = 7200$$

Since 7200 is in rough agreement with 8941.5976, you know that the decimal point in the product is in the correct place.

Class Exercises

State the product.

1. 7 × 0.3 2.1
2. 15 × 0.2 3
3. 4 × 5.1 20.4
4. 50 × 0.4 20

5. 0.5 × 0.5 0.25
6. 0.3 × 0.9 0.27
7. 0.27 × 0.2 0.054
8. 0.4 × 0.011 0.0044

The correct digits are given for the product. State the product with the decimal point correctly inserted. Then check your answer by estimating.

9. 3.7 × 2.68; 9916 9.916
10. 0.5 × 8.65; 4325 4.325
11. 7.25 × 4.09; 296525 29.6525

12. 2.4 × 3.702; 88848 8.8848
13. 6.201 × 5.72; 3546972 35.46972
14. 0.96 × 52.303; 5021088 50.21088

Exercises

Compute the product. Check your answer by estimating.

A

1. 5.9
× 7
41.3

2. 23.8
× 42
999.6

3. 7.2
×7.7
55.44

4. 5.66
× 9.2
52.072

5. 24.4
×0.61
14.884

6. 3.807
× 6.8
25.8876

7. 13.405
× 21.3
285.5265

8. 643.21
× 1.626
1045.85946

9. 8.34 × 0.16
1.3344

10. 37.44 × 5.9
220.896

11. 0.462 × 0.93
0.42966

12. 0.012 × 5.37
0.06444

13. 0.0025 × 0.602
0.0015050

14. 0.0007 × 0.923
0.0006461

Simplify.

B **15.** $(24.6 - 9.9) \times (43.7 - 18.5)$ 370.44   **16.** $(3.01 + 14.6) \times (2.7 + 4.08)$ 119.3958

17. $0.058 \times (0.82 + 3.7 + 14.03)$ 1.07590   **18.** $6.2 \times 4.3 \times 0.08$ 2.1328

Problems

Solve.

A **1.** An oil tank holds 1050 L. If a liter of oil costs 13¢, how much does it cost to fill the tank? **$136.50**

B **2.** Janice bought 6 rolls of film at $1.29 each. Four rolls cost $2.89 each for development. For the last 2 rolls, she changed companies and was charged only $2.27 each. How much did she spend in all for her photographs? **$23.84**

Application

Estimating Electricity Costs

In trying to keep your electric bills and usage at a minimum, it is useful to know approximately what each appliance costs to run. The table gives a partial list of average costs in one community. For example, a stereo costs 0.5¢ per hour to run, or $100 \times \$0.005 = \$0.50$ for 100 hours.

Appliance	Typical wattage	Cost per hour
Air Conditioner	1500	7.9¢
Attic Fan	375	2¢
Hi-Fi/Stereo	100	0.5¢
Radio	75	0.4¢
TV (B + W)	55	0.3¢
TV (Color)	200	1.1¢

1. Which costs less to run, a black and white TV or a color TV? How much would it cost to run each for 150 hours? (B + W) TV; 45¢

2. Which costs less, listening to a radio for 50 hours or playing a stereo for 30 hours? Stereo

3. How much less does it cost to run an attic fan for 240 hours than an air conditioner for 240 hours? $14.16

Consumer Activity Find out the average costs for electricity in your community.

2-9 Dividing Decimals

Objective To use the division process with decimals.

The chart below reviews the way division relates to multiplication.

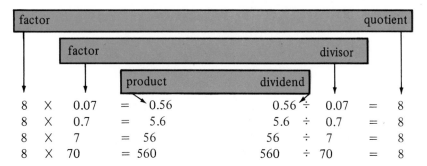

| factor | | | | | | quotient |

| | factor | | | | divisor | |

| | | product | | dividend | | |

8	×	0.07	=	0.56	0.56 ÷ 0.07	=	8
8	×	0.7	=	5.6	5.6 ÷ 0.7	=	8
8	×	7	=	56	56 ÷ 7	=	8
8	×	70	=	560	560 ÷ 70	=	8

In the column at the right above, the quotient in each case is 8, but the dividend and divisor change. We can see that multiplying the dividend and divisor by the same power of ten does not change the quotient. This fact means that we can always change the divisor into a counting number.

rule

1. **Multiply the dividend and divisor by the least power of 10 needed to produce a counting number as divisor.**
2. **In the quotient, place the decimal point directly above the decimal point in the new dividend.**
3. **Divide as with whole numbers.**

Example 1 Compute 20.35 ÷ 3.7

Solution **Estimate the quotient: 20.35 ÷ 3.7**

$$20 \div 4 = 5$$

Multiply dividend and divisor by 10 so that the divisor will be a counting number, then divide. The arrows show you the new position of the decimal point. Notice where the decimal point is placed in the quotient.

$$
\begin{array}{r}
5.5 \\
3.7\,\overline{)20.3\,5} \\
\underline{18\,5} \\
1\,8\,5 \\
\underline{1\,8\,5} \\
\end{array}
$$

→0
remainder

In the next example the division process does not lead to a remainder of zero. In a case like this, the quotient may be approximated to a nearest place value.

Example 2 | Compute 7.812 ÷ 0.43 to the nearest hundredth.

Solution | Estimate the quotient: 7.812 ÷ 0.43

$$8 \div 0.4 = 8 \times 10 \div 0.4 \times 10$$
$$= 80 \div 4 = 20$$

Compute to the thousandths' place and round to hundredths.

```
      18.167                Check:      18.1 67
0.43 )7.81 200                      ×      0.43
      4 3                              54 5 01
      3 51                            7 26 6 8
      3 44                            7.81 1 81
        7 2                          +0.00 0 19
        4 3                           7.81 2 00
        2 90
        2 58
          320
          301
           19
```

To the nearest hundredth, 7.812 ÷ 0.43 = 18.17.

Notice that the number of decimal places to the right of the decimal point in the remainder, 0.00019, is the same as the number of decimal places in the product of the divisor and quotient, 0.43 × 18.167.

Notice also that the original estimate of the quotient, 20, is in rough agreement with the computed quotient. The check by multiplication, however, is a better guarantee of accuracy.

Class Exercises

State the least power of ten which makes the divisor a counting number.

1. 0.8)168 10^1

2. 1.5)45 10^1

3. 9.25)18 10^2

4. 0.46)2.3 10^2

5. 0.012)6.06 10^3

6. 3.7)1.48 10^1

7. 8.01)16.2 10^2

8. 1.246)9.76 10^3

The correct digits are given for the quotient. Place the decimal point and state the quotient.

9.   27
  4.3)‾1.161‾ 0.27

10.   5 89
  6.2)‾365.18‾ 58.9

11.   41
  0.37)‾1.517‾ 4.1

12.   2 08
  3.52)‾732.16‾ 208

13.   157
  0.68)‾10.676‾ 15.7

14.   ˙36
  0.358)‾1.2888‾ 3.6

Exercises

a. Divide. If the division is not exact, round to the nearest hundredth.
b. Check.

A   **1.** 16.24 ÷ 5.6 2.9   **2.** 18.09 ÷ 6.7 2.7   **3.** 2.55 ÷ 0.34 7.5   **4.** 3.57 ÷ 0.42 8.5

5. 61.476 ÷ 14.1 4.36 **6.** 90.072 ÷ 32.4 2.78 **7.** 1.06172 ÷ 5.08   **8.** 0.50142 ÷ 7.32 0.0685
        0.209

9. 0.28)‾8.7553‾ 31.27 **10.** 0.46)‾0.0361‾ 0.08 **11.** 2.3)‾96.14‾ 41.8   **12.** 1.9)‾88.92‾ 46.8

13. 0.068)‾3.5156‾ 51.7 **14.** 0.049)‾1.4357‾ 29.3 **15.** 1.78)‾90.068‾ 50.6   **16.** 20.9)‾361.57‾ 17.3

17. 1.14912 ÷ 0.0057 201.6   **18.** 6.4272 ÷ 0.0103 624   **19.** 7.33 ÷ 0.8 9.16

20. 37.7 ÷ 0.6 62.83   **21.** 6.834 ÷ 0.39 17.52   **22.** 6.08 ÷ 0.92 6.61

23. 0.3321 ÷ 0.004 83.03   **24.** 0.3692 ÷ 0.058 6.37   **25.** 87.391 ÷ 0.161 542.80

26. 10.078 ÷ 5.04 2.00   **27.** 570 ÷ 74.7 7.63   **28.** 26.923 ÷ 1.001 26.90

Problems

Solve.

A   **1.** A photographer makes $295 for a 35-hour week. What is the hourly rate to the nearest cent? $8.43

2. A chemist has 558 mL of a solution and needs to prepare samples of it, each containing 23.25 mL. How many samples can be prepared? 24 samples

B   **3.** For a graduation party, the Donaldsons budgeted $25 for cold cuts. If the cold cuts cost $6.19 a kilogram, how many kilograms can they buy? How much money will they have left over? 4 kg; 24¢

4. Mario bought three packages of weather-stripping tape. Each package contained 9 m of tape. How many windows can be weather-stripped if each window needs 4.85 m? Will there be enough tape left to weather-strip a window needing 2.55 m of tape?
5 windows; Yes

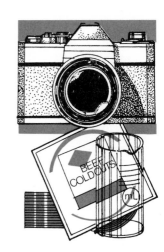

Application

The Metric System

Just like our number system, the metric system of measurement is based on powers of ten. For example, the meter, a unit of length, can be multiplied by 10, 100, 1000, and so on to form longer units of length.

$$10 \text{ meters} = 1 \text{ dekameter}$$
$$100 \text{ meters} = 1 \text{ hectometer}$$
$$1000 \text{ meters} = 1 \text{ kilometer}$$

Shorter units of length can be formed by multiplying the meter by 0.1, 0.01, 0.001, and so on.

$$0.1 \text{ meter} = 1 \text{ decimeter}$$
$$0.01 \text{ meter} = 1 \text{ centimeter}$$
$$0.001 \text{ meter} = 1 \text{ millimeter}$$

The table at the right lists some of the prefixes used in the metric system and their symbols. Using the symbol m for meter, we can write, for example,

$$1000 \text{ m} = 1 \text{ km}$$
$$0.01 \text{ m} = 1 \text{ cm}$$

and so on.

Metric Prefixes

Multiply by	Prefix	Symbol
10	deka	da
100	hecto	h
1000	kilo	k
0.1	deci	d
0.01	centi	c
0.001	milli	m

Here are some other examples of relationships in the metric system.

2000 m = 2.000 km	0.01 m = 1 cm
2500 m = 2.500 km	0.02 m = 2 cm
2530 m = 2.530 km	0.72 m = 72 cm
2538 m = 2.538 km	3.72 m = 372 cm

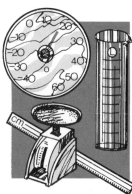

Complete the following.

1. 3000 m = __?__ km 3
2. 3764 m = __?__ km 3.764
3. 764 m = __?__ km 0.764
4. __?__ m = 0.064 km 64
5. 1 m = __?__ cm 100
6. 2.5 m = __?__ cm 250
7. 50 cm = __?__ m 0.5
8. 300 cm = __?__ m 3
9. 1 m = __?__ mm 1000
10. 1 cm = __?__ mm 10
11. 375 mm = __?__ m 0.375
12. 375 mm = __?__ cm 37.5

2-10 Solving Problems

Objective To solve simple word problems including the use of one or more arithmetic operations.

Word problems tell you how certain numbers are related and then ask a question about them. Using a plan, such as the following, is a good way to gain skill in solving problems.

Plan for Solving Problems
1. Read the problem carefully. Make sure you understand all the words used.
2. Use questions like these in planning the solution: • What is asked for? • What facts are given? • Are enough facts given? If not, what else is needed? • Are unnecessary facts given? If so, what are they? • Will a sketch or diagram help?
3. Which operation or operations can be used to solve the problem?
4. Perform the operations carefully.
5. Check your result with the facts given in the problem. Is your result reasonable?

Example 1 | During a blizzard, 64.5 cm of snow fell on Prairie City. Two weeks later another storm dropped 35.6 cm of snow. How much more snow fell in the first storm than in the second?

Solution | • The problem asks for the difference in the depths of the snowfalls.
• Given facts: 64.5 cm of snow in first storm
              35.6 cm of snow in second storm
To find how much more snow fell in the first storm, we subtract: $64.5 - 35.6 = 28.9$.
28.9 cm more snow fell in the first storm than in the second.

Example 2 | After 15 selling days, the cost of a calculator was reduced $2.99 to a sale price of $8.59. What was the original cost of the calculator?

Solution | • The problem asks for the original cost of the calculator.

• Given facts: $2.99, amount of reduction

$8.59, sale price after reduction

15 selling days, fact not needed

To find the original cost, we add:

$2.99 + $8.59 = $11.58

The original cost of the calculator was $11.58.

Problems

Solve.

A 1. A bird-feeder holds 0.5 kg of mixed bird seed. Sal filled it 3 times in one week. How much bird seed did the birds eat that week? 1.5 kg

2. During a flood, a river reached a high-water mark of 4.6 m. This was 1.4 m lower than the record high. What was the record high? 6 m

3. A microwave oven uses 1.5 kW · h of electricity in 15 minutes. How much electricity does it use in 5 minutes? 0.5 kW·h

4. A baked potato has 31 mg of vitamin C. How many milligrams of vitamin C do 8 potatoes have? 248 mg

B 5. Sue wrote five checks in May: $255, $14.63, $75, $47.59, and $82.90. She made one deposit of $624. What was her balance at the end of May if her balance at the beginning of May was $681.70? $830.58

6. A certain long distance telephone call costs $1.95 for the first 3 minutes and $.32 for each additional minute. How much does a 17-minute call cost? $6.43

7. A photograph is enlarged so that its new dimensions are 4 times its original dimensions. If the new dimensions are 19.2 cm by 25.6 cm, what were the original dimensions? 4.8 cm by 6.4 cm

8. Martin bought 0.8 kg of bananas at $.55 a kilogram, 1.5 kg of tomatoes at $1.54 a kilogram, and 18 oranges at 6 oranges for $.59. How much did Martin pay? $4.52

Self-Test

Words to remember:
inverse operations [p. 40]   commutative property [p. 42]
associative property [p. 42]   distributive property [p. 43]

Write a related equation. Find the value of n.

1. $14 + n = 21$ **2.** $n + 8 = 32$ **3.** $6n = 48$ **4.** $8 \times n = 24$ **[2-5]**
$n = 21 - 14; 7$ $n = 32 - 8; 24$ $n = 48 \div 6; 8$ $n = 24 \div 8; 3$

Replace each variable to make a true equation.

5. $(7 + 2) + 5 = 7 + (x + 5)$ 2   **6.** $24 \times n = 0$ 0   **[2-6]**

7. $3 \times (1 + 9) = (3 \times 1) + (t \times 9)$ 3   **8.** $6 \times 8 = y \times 6$ 8

Add or subtract.

9. $42.566 + 7.08$ 49.646   **10.** $3.9 + 6.141 + 8$ 18.041   **[2-7]**

11. $71.354 - 62.86$ 8.494   **12.** $25.3 - 25.297$ 0.003

Multiply.

13. 5.1×3.04 15.504   **14.** 0.72×0.85 0.6120   **[2-8]**

15. 4.96×0.003 0.01488   **16.** 6.017×2.2 13.2374

Divide. If the division is not exact, round to the nearest hundredth.

17. $120.13 \div 29.3$ 4.1   **18.** $4.758 \div 0.096$ 49.5625   **[2-9]**

19. $3.0804 \div 0.37$ 8.33   **20.** $1.8927 \div 7.01$ 0.27

Solve.

21. A swimming pool is 24 m long. Sam swam its length 3 times.   **[2-10]**
How far did he swim? 72 m

22. Rita commutes to work 1.7 km by subway and 3.8 km by bus.
How far is that in all? 5.5 km

23. The Rays planned a 2-day canoe trip of 26.8 km. They traveled
19.9 km the first day. How far did they have left to travel? 6.9 km

24. Jack bought a paperback book for $2.75. He gave the clerk a
$5 bill. How much did he receive in change? $2.25

Self-Test answers and Extra Practice are at the back of the book.

Chapter Review

Write the letter that labels the correct answer.

1. 100,000,000 may be written as __?__ . D [2-1]

   A. 10^6 B. 10^9 C. 10   D. 10^8

2. The expanded form for 519 is __?__ . B

   A. $(5 \times 1000) + (1 \times 100) + (9 \times 10)$
   B. $(5 \times 10^2) + (1 \times 10^1) + 9$
   C. $(5 \times 10^3) + (1 \times 10^2) + 9$
   D. $(5 \times 10) + 19$

3. In 384.9617, the 6 is in the __?__ place. D [2-2]

   A. hundreds'   B. thousandths'   C. tenths'   D. hundredths'

4. The decimal for $8 + (2 \times 0.1) + (0 \times 0.01) + (7 \times 0.001)$ is __?__ . D

   A. 8.27   B. 8207   C. 8.027   D. 8.207

In Exercises 5 and 6, use the number line shown.

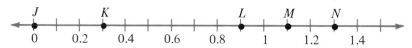

5. The graph of 1.3 is __?__ . D [2-3]

   A. the origin   B. point K C. point M D. point N

6. The coordinate of point L is __?__ . B

   A. 9   B. 0.9   C. 1   D. 1.1

What is the value of the expression when $x = 14$?

7. $x + 9$ A [2-4]

   A. 23   B. 5   C. 125   D. 104

8. $112 \div x$ C

   A. 7   B. 18   C. 8   D. 17

Find the value of n.

9. $8 + n = 16$ B [2-5]

   A. 24   B. 8   C. 2   D. 9

10. $5n = 35$ A
  A. 7   **B.** 6   **C.** 30   **D.** 40

11. $(53 + 89) + 27 = \underline{\ ?\ } + (89 + 27)$ D   [2-6]
  A. 0   **B.** 27   **C.** 89   **D.** 53

12. $23(14 + 6) = (\underline{\ ?\ } \times 14) + (\underline{\ ?\ } \times 6)$ B
  A. 20   **B.** 23   **C.** 14   **D.** 6

13. $0.692 + 5.5 = \underline{\ ?\ }$ B   [2-7]
  A. 1.242   **B.** 6.192   **C.** 5.192   **D.** 4.808

14. $47.516 - 46.903 = \underline{\ ?\ }$ D
  A. 1.613   **B.** 1.387   **C.** 1.413   **D.** 0.613

15. An estimate for 94.771×0.62 is $\underline{\ ?\ }$. A   [2-8]
  A. 50   **B.** 5.6   **C.** 5.875   **D.** 0.58

16. $4.99 \times 3 = \underline{\ ?\ }$ B
  A. 1.497   **B.** 14.97   **C.** 149.7   **D.** 1497

17. An estimate for $5.428 \div 0.92$ is $\underline{\ ?\ }$. C   [2-9]
  A. 0.05   **B.** 0.5   **C.** 5   **D.** 50

18. To the nearest hundredth, $1.4538 \div 0.017 = \underline{\ ?\ }$. D
  A. 85   **B.** 85.5   **C.** 85.51   **D.** 85.52

Solve.

19. Anna jogged 9.5 km each day for 5 days. How far did she jog   [2-10]
  in all? C
  A. 4.75 km   **B.** 74.5 km   **C.** 47.5 km   **D.** 475 km

20. A car can go 600 km on 1 tank of gas. How many kilometers
  can it go on 2.5 tanks of gas? C
  A. 1200 km   **B.** 1850 km   **C.** 1500 km   **D.** 240 km

The Binary System

The number system that we use is, of course, the decimal system. The place value of each digit in a number is a power of ten. In the *binary system,* place values are based on powers of two. The only digits used in the binary system are zero and one.

Place Value Chart

	2^6	2^5	2^4	2^3	2^2	2^1	1
. . .	64	32	16	8	4	2	1

Example 1 Write 1011_{two} as a decimal number.

Solution
$$1011_{two} = (1 \times 2^3) + (0 \times 2^2) + (1 \times 2^1) + 1$$
$$= 8 + 0 + 2 + 1$$
$$= 11$$

Example 2 Write 27 as a binary number.

Solution

Find the greatest power of 2 that is less than or equal to 27: $16 < 27 < 32$.

$27 = 2^4 + 11$
$27 = 16 + 11$

Find the greatest power of 2 that is less than or equal to 11: $8 < 11 < 16$.

$27 = 2^4 + 2^3 + 3$
$27 = 16 + 8 + 3$

Repeat this process.

$27 = 2^4 + 2^3 + 2^1 + 1$
$27 = 16 + 8 + 2 + 1$

Write the binary number:
$$(1 \times 2^4) + (1 \times 2^3) + (0 \times 2^2) + (1 \times 2^1) + 1$$
$$= 11011_{two}$$

The binary system is important in the operation of digital computers. The digits in a number are represented by switches that are either in the on or off positions. Zero is represented by a switch that is off and one is represented by a switch that is on.

Exercises

Write the binary number as a decimal number. (Answers in Base 10)

1. 101_{two} 5   **2.** 110_{two} 6   **3.** 1011_{two} 11   **4.** 10000_{two} 16   **5.** 10111_{two} 23

6. 11110_{two} **7.** 101100_{two} **8.** 111001_{two} **9.** 111111_{two} **10.** 1011100_{two}
    30   44   57   63   92

Write the decimal number as a binary number. (Answers in Base 2)

11. 7 111   **12.** 8 1000   **13.** 10 1010   **14.** 19 10011   **15.** 20 10100   **16.** 25 11001

17. 31   **18.** 64   **19.** 52   **20.** 46   **21.** 87   **22.** 100
   11111   1000000   110100   101110   1010111   1100100

23. Complete the binary addition table at the right.

+	0	1
0	0?	? 1
1	1?	? 10

Add. Check your work by writing the addition in base ten. (Answers in Base 2)

24. 1001_{two} **25.** 1111_{two} **26.** 100100_{two} **27.** 1011010_{two}
     $+1100_{two}$ $+10010_{two}$ $+110110_{two}$ $+\ 101000_{two}$
      10101   100001   1011010   10000010

28. Complete the binary multiplication table at the right.

×	0	1
0	0?	? 0
1	0?	? 1

Multiply. Check your answer by writing the multiplication in base ten. (Answers in Base 2)

29. 11_{two} **30.** 1101_{two} **31.** 10001_{two} **32.** 10101_{two}
    $\times 10_{two}$ $\times\ 100_{two}$ $\times 11010_{two}$ $\times\ 1101_{two}$
     110   110100   110111010   100010001

Research Activity Look up digital computers in an encyclopedia. Find out more about the role played by the binary system in digital computers.

Chapter Test

1. Write 1708 in expanded form, using exponents. [2-1]
   $(1 \times 10^3) + (7 \times 10^2) + 8$

2. Write the decimal numeral for the following:
   $(2 \times 10{,}000) + (9 \times 1000) + (0 \times 100) + (0 \times 10) + 5.$
   29,005

3. Write 0.5164 in expanded form. [2-2]
   $(5 \times 0.1) + (1 \times 0.01) + (6 \times 0.001) + (4 \times 0.0001)$

4. Write 20.32 in words.
   twenty and thirty-two hundredths

5. Draw a number line. Graph 0, 0.3, 0.6, 1.1. See students' graph. [2-3]

6. 12.163 ___?___ 12.5 $<$
   $<, >$

Evaluate the expression for the given replacement set

7. $n + 15$; $\{8, 16, 22\}$ 8. $900 \div n$; $\{5, 30, 225\}$ [2-4]
   23; 31; 37   180; 30; 4

Write a related equation. Find the value of n.

9. $n + 24 = 72$ 48   10. $5n = 50$ 10   [2-5]

Replace each variable to make a true equation.

11. $37 \times n = 37$ 1   12. $9 + 14 = 14 + x$ 9   [2-6]

Add or subtract.

13. $21.28 + 9.5034$ 30.7834   14. $67.7 - 43.19$ 24.51   [2-7]

Multiply.

15. 5.06×18.3 92.598   16. 0.042×0.795 0.033390   [2-8]

Divide. If the division is not exact, round the quotient to the nearest hundredth.

17. $0.31548 \div 2.96$ 0.11   18. $1737.68 \div 74.9$ 23.2   [2-9]

Solve.

19. A bee travels about 0.4 km in 1 minute. About how far does   [2-10]
    a bee travel in 5 minutes? about 2 km

20. A hiker climbed from a base camp at 1560 m above sea level
    to a hut 3680 m above sea level. How far did she climb? 2120 m

Skill Review

Add.

1. 7.3
$+1.2$
8.5

2. 4.08
$+5.96$
10.04

3. 38.571
$+\ 9.466$
48.037

4. 6.29
$+75.03$
81.32

5. 0.45
$+5.138$
5.588

6. 0.007
$+3.6$
3.607

7. 290.01
$+\ 77.4$
367.41

8. 0.8602
$+0.093$
0.9532

9. $6307.59 + 478.223$ 6785.813

10. $3.721 + 145.08$ 148.801

11. $0.0064 + 0.092$ 0.0984

12. $501.47 + 29.0033$ 530.4733

Subtract.

13. 12.4
$-\ 9.8$
2.6

14. 55.903
-54.081
1.822

15. 5.27
-2.76
2.51

16. 0.0606
-0.0039
0.0567

17. 7.05
-2
5.05

18. 38.014
-36.9
1.114

19. 50.2
-47.158
3.042

20. 0.6
$-\ 0.1105$
0.4895

21. $1.007 - 0.03$ 0.977

22. $219.38 - 54.64$ 164.74

23. $8790.72 - 606.1435$ 8184.5765

24. $5.943 - 5.8886$ 0.0544

Multiply.

25. 8.4
$\times 3.9$
32.76

26. 57.1
$\times 0.06$
3.426

27. 2.053
$\times\ 4.08$
8.37624

28. 0.904
$\times\ 0.25$
0.22600

29. 0.095
$\times 0.017$
0.001615

30. 0.0082
$\times\ 6.09$
0.049938

31. 7.33
$\times 0.01$
0.0733

32. 40.06
$\times 0.0052$
0.208312

33. 8.921×508 4531.868

34. 0.2852×0.0437 0.01246324

35. 19.74×3.601 71.08374

36. 0.952×0.0836 0.0795872

Divide. If the division is not exact, round to the nearest hundredth.

37. $28\overline{)267.4}$ 9.55   **38.** $45\overline{)1.62}$ 0.036   **39.** $6.5\overline{)20.79}$ 3.20   **40.** $0.37\overline{)0.2294}$ 0.62

41. $0.9\overline{)40}$ 44.44   **42.** $0.03\overline{)1.95}$ 65   **43.** $8.04\overline{)7.36}$ 0.92   **44.** $0.016\overline{)5.12}$ 320

45. $1.4505 \div 3.92$ 0.37

46. $32.53718 \div 0.598$ 54.41

47. $7.903 \div 61.3$ 0.13

48. $2.4 \div 0.0458$ 52.40

3 Positive and Negative Numbers

3-1 The Extended Number Line

Objective To construct an extended number line and to graph numbers on it.

When engineers want to describe the state of a rocket launching at any particular moment, they use the time at which the rocket is to be fired as their reference point, and they denote it by zero. Then they use *negative numbers* to refer to time before that point and *positive numbers* to refer to time after it.

We use positive and negative numbers to measure many *opposite* quantities, such as profit and loss, distance up and down, or points won or lost in a game. The weather thermometer at the right illustrates one such use. It pictures the idea of an extended number line in which a two-way scale extends above and below a **zero reference point.**

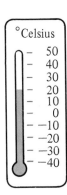

°Celsius

50
40
30
20
10
0
−10
−20
−30
−40

In Section 2-3 we graphed numbers to the right of the origin, or 0, on a horizontal number line. By extending the number line to the left of the origin we can graph both positive and negative numbers as shown in the figure below.

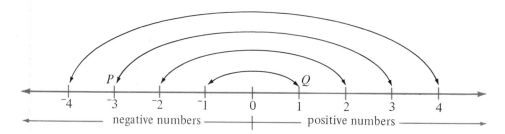

Coordinates of points to the right of 0 are positive. Those to the left are negative. The coordinate of P is ⁻3, which we read as "negative three." The coordinate of Q is 1, which we read as "positive one," or simply "one." The symbol "+1" is sometimes used for positive one, for emphasis.

Example 1 | Show the graphs of ⁻2, ⁻5, 0, and 3 on a number line.

Solution

Notice that each point on the number line has an opposite point, located the same distance from the origin but on the opposite side of it. The coordinates of such a pair of points are said to be **opposites** of one another. For example, 1 and ⁻1, 2 and ⁻2, 3 and ⁻3, 4 and ⁻4 are pairs of opposite numbers. The opposite of 0 is simply 0.

In Section 2-2 we learned that the set of whole numbers is

$$\{0, 1, 2, 3, \ldots\}.$$

The set whose members are the whole numbers and their opposites,

$$\{\ldots, ⁻3, ⁻2, ⁻1, 0, 1, 2, 3, \ldots\},$$

is called the set of **integers**.

An *unraised* minus sign before a number is frequently used to show subtraction as, for example, in $7 - 4 = 3$. Sometimes, however, an unraised minus sign is used to mean "the opposite of." For example,

$$-2 \text{ means the opposite of positive 2}$$
$$-(^-4) \text{ means the opposite of negative 4}$$

From the figure on page 66, you can see that

$$-2 = {}^-2 \quad \text{and} \quad -(^-4) = 4.$$

Example 2 | State the opposite of the given number.
  | **a.** $^-6$ **b. 18**   **c. 0**

Solution | **a. 6**   **b.** $^-18$ **c. 0**

Earlier, we used variables to stand for positive numbers and zero. We can also use variables to stand for negative numbers. Thus,

$$p = 4 \quad \text{is read} \quad p \text{ is equal to positive 4}$$
$$q = {}^-3 \quad \text{is read} \quad q \text{ is equal to negative 3}$$

We can also use an unraised minus sign with a variable. For example,

$$-m = 2 \quad \text{is read} \quad \text{the opposite of } m \text{ is equal to positive 2}$$
$$-x = {}^-5 \quad \text{is read} \quad \text{the opposite of } x \text{ is equal to negative 5}$$

Example 3 | **a.** If $m = 3$, then $-m = \underline{\ \ ?\ \ }$
  | **b.** If $n = {}^-7$, then $-n = \underline{\ \ ?\ \ }$
  | **c.** If $-p = 5$, then $p = \underline{\ \ ?\ \ }$
  | **d.** If $-q = {}^-2$, then $q = \underline{\ \ ?\ \ }$

Solution | **a.** $^-3$
  | **b. 7**
  | **c.** $^-5$
  | **d. 2**

Class Exercises

1. If east is called the positive direction, what is the negative direction? west

2. If down is the negative direction, then what is up? the positive direction

State the number implied by the given phrase.

3. 5 meters below sea level ⁻5

4. 2 minutes after the start of the game 2

5. 20 degrees below zero ⁻20

6. 3 seconds before blast-off ⁻3

State the opposite of the given number.

7. 4 ⁻4   **8.** ⁻3 3   **9.** ⁻7 7   **10.** 5 ⁻5

11. ⁻27 27   **12.** 0 0   **13.** 167 ⁻167   **14.** ⁻583 583

Supply the missing word or phrase as you read the given sentence aloud.

15. The opposite of 8 is __?__. negative eight   **16.** The opposite of ⁻9 is __?__ positive nine

17. The opposite of 0 is __?__. zero   **18.** The symbol $-(^-6)$ denotes the __?__ of ⁻6. opposite

19. If $m = 11$, then $-m =$ __?__. ⁻11   **20.** If $n = {}^-2$, then $-n =$ __?__. 2

21. If $-p = {}^-5$, then $p =$ __?__. 5   **22.** If $-q = 0$, then $q =$ __?__. 0

State the letter above the graph of the *opposite* of the given number.

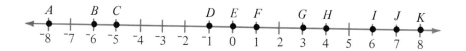

Example	**a.** ⁻8 **b.** 1
Solution	**a.** The opposite of ⁻8 is 8. The letter is *K*.
	b. The opposite of 1 is ⁻1. The letter is *D*.

23. ⁻4 H   **24.** ⁻3 G   **25.** 6 B   **26.** 5 C

27. 8 A   **28.** 1 D   **29.** ⁻6 I   **30.** 0 E

Exercises

Write as a positive or negative number.

A **1.** A loss of 4 kg ⁻4

2. A gain of 8 grade points 8

3. A profit of $46 46

4. A debt of $25 ⁻25

Copy the number line and write the coordinates of the points shown.

5.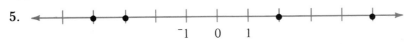

6.

Show the graph of each of the given numbers on a number line. See students' graphs.

7. 0   **8.** 1   **9.** ⁻4   **10.** ⁻1   **11.** 4   **12.** ⁻3

13. ⁻8   **14.** 7   **15.** 6   **16.** ⁻6   **17.** 5   **18.** 3

Write the opposite of the given number.

19. 6 ⁻6   **20.** 2 ⁻2   **21.** ⁻1 1   **22.** 13 ⁻13   **23.** ⁻19 19   **24.** ⁻37 37

25. 118 ⁻118   **26.** ⁻56 56   **27.** 1776 ⁻1776   **28.** ⁻2001 2001   **29.** ⁻26,475 26,475   **30.** 999,999 ⁻999,999

Show the graph of the opposite of the given number on a number line.

31. 4   **32.** ⁻3   **33.** 0   **34.** ⁻12   **35.** 11   **36.** ⁻7

See students' graphs.

Write the missing word.

B **37.** If the opposite of a certain number is a negative number, then the number itself is a __?__ number. positive

38. If the opposite of a certain number is a positive number, then the number itself is a __?__ number. negative

Write the missing number.

C **39.** If $-n = 6$, then $n = $ __?__. ⁻6

40. If $-n = 7$, then $n = $ __?__. ⁻7

41. If $-(-n) = -0.75$, then $n = $ __?__. ⁻0.75

42. If $-(-n) = 1.6$, then $n = $ __?__. 1.6

43. If $-(-n) = -n$, then $n = $ __?__. 0

44. If $n = 5$, then $-(n + 2) = $ __?__. ⁻7

45. If $-n = -7$, then $-(n - 3) = $ __?__. ⁻4

46. If $n = -8$, then $-(-n + 1) = $ __?__. ⁻9

3-2 Arrows on the Number Line

Objective To picture numbers on the number line by graphs and by directed line segments, or arrows.

The decimals like 1.5, 2.97, and 8, that you studied in Chapter 2, are called *positive decimals*. Their opposites, like ⁻1.5, ⁻2.97, and ⁻8 are *negative decimals*. The positive decimals, negative decimals, and zero, all belong to the set of decimal numbers. Notice that the integers are included in this set.

Example 1 | Graph these decimals on the number line: ⁻2.5, ⁻2, ⁻0.75, 0, 0.75, 2.

Solution

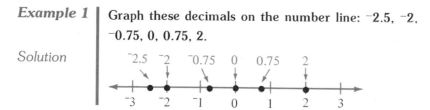

We can also use directed line segments, or arrows, to picture numbers on a number line. For example, in the figure below, the arrow directed to the right represents 2, and the arrow directed to the left represents ⁻2.5.

Any arrow directed to the left (the negative direction) represents a negative number, while any arrow directed to the right (the positive direction) represents a positive number. The *length* of the arrow is always a *positive number or zero*, even when the direction of the arrow is negative.

An arrow representing a number may have its starting point at any point on the number line, as long as it has the length and direction indicated by that number. Each of the arrows shown in color in the figure below has length 2; but since each extends to the left, each represents ⁻2.

Example 2 │ **a.** On a number line, picture an arrow that starts at 4 and ends at ⁻3.

b. By inspection, find the number represented by the arrow.

Solution │ **a.**

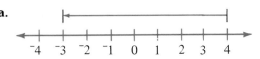

b. The arrow is 7 units long and is directed to the left. It represents ⁻7.

Class Exercises

State the letter written above the graph of the given number.

1. 3.5 F **2.** −1.5 C **3.** ⁻3 A **4.** 0 D **5.** 1.5 E **6.** ⁻2 B **7.** −(⁻4) H **8.** ⁻4 G

State the starting point of the directed arrow.

9. 0

10. 0

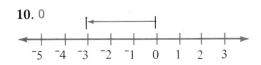

11. ⁻1

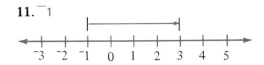

12. 2

State the endpoint of the arrow in the given exercise.

13. Exercise 9 2 **14.** Exercise 10 ⁻3 **15.** Exercise 11 3 **16.** Exercise 12 ⁻3

State the length of the arrow and the number represented by the arrow in the given exercise.

17. Exercise 9
 2 units; 2

18. Exercise 10
 3 units; ⁻3

19. Exercise 11
 4 units; 4

20. Exercise 12
 5 units; ⁻5

Tell how you would complete the table.

	Coordinate of Starting Point	Coordinate of Endpoint	Length of Arrow	Number Represented
21.	2	5	? 3	? 3
22.	2	−4	? 6	? ⁻6
23.	−1	? ⁻5	? 4	−4
24.	−4	? ⁻1	? 3	3
25.	−3	? 2	? 5	5
26.	4	? ⁻2	? 6	−6

Exercises   For solutions to Exercises 1-22, check students' graphs.

Graph the numbers on a number line.

A   **1.** 2, −1.5, 0, −10, 5.4, −1.2   **2.** −4, 6.2, 7.3, 0, −3.5, −2.5

   3. −1.1, −12.4, −2, −3.0, 0.7   **4.** −3.9, −9, 1, 2.8, 1.4, −0.5

Draw an arrow starting at the origin to represent the given number.

   5. 2   **6.** 5   **7.** −3   **8.** −4   **9.** −3.5   **10.** 4.5

Draw an arrow starting at the graph of −3 to represent the given number.

   11. 4   **12.** 3   **13.** −2   **14.** −1   **15.** 2.5   **16.** −1.5

Draw an arrow with endpoint at the graph of 2 to represent the given number.

   17. 3   **18.** 4   **19.** −2.5   **20.** −1.25   **21.** −3   **22.** 1.75

Copy and complete.

	Coordinate of Starting Point	Coordinate of Endpoint	Length of Arrow	Number Represented
23.	2	? 5	? 3	3
24.	−4	? ⁻1	? 3	3
25.	−2	−5	? 3	? ⁻3
26.	3	−2	? 5	? ⁻5

Copy and complete.

B 27.	−6.25	? ⁻2.5	? 3.75	3.75
28.	4.58	? 1.97	? 2.61	−2.61
29.	−3.125	? ⁻9.875	? 6.75	−6.75
30.	? 10.1	7.25	? 2.85	−2.85
C 31.	? ⁻0.4	−0.75	? 0.35	−0.35
32.	? 3.99	2.53	? 1.46	−1.46

EXTRA! Order on the Number Line

We can use the number line to help us put decimals in order. For example, if we wish to arrange 2.6, ⁻1.2, 0, ⁻1.4, and 3.2 in increasing order from least to greatest, we can first graph them on a number line.

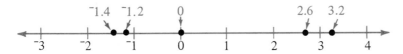

Then we list them as they appear from left to right:

$$-1.4, -1.2, 0, 2.6, 3.2$$

They are now listed in order from least to greatest.
   On the number line above, notice that, for example,
   ⁻3 < ⁻2 (⁻3 is less than ⁻2)
while
   3 > 2 (3 is greater than 2).

Graph the numbers on a number line and then list them in order from least to greatest.

1. 3, 4, ⁻1, 0, ⁻5   **2.** ⁻2, 2, 6, ⁻5, 3   **3.** 0, ⁻3, ⁻4, ⁻1, 2
    ⁻5, ⁻1, 0, 3, 4   ⁻5, ⁻2, 2, 3, 6   ⁻4, ⁻3, ⁻1, 0, 2
4. 2.5, 2, ⁻1.5, 3   **5.** ⁻4, −2.5, 0, 3.5   **6.** 1, ⁻0.5, ⁻1, 0.5
    ⁻1.5, 2, 2.5, 3   ⁻4, ⁻2.5, 0, 3.5   ⁻1, ⁻0.5, 0.5, 1

Replace each ? with < or >.

7. 5 ? 1 >   **8.** ⁻5 ? ⁻1 <   **9.** 5 ? ⁻1 >   **10.** ⁻5 ? 1 <

11. 2 ? ⁻4 >   **12.** ⁻3 ? 0 <   **13.** 6 ? ⁻6 >   **14.** ⁻7 ? 3 <

3-3 The Sum of Two Numbers of the Same Sign

Objective To find the sum of two numbers when both are positive or both are negative, and to picture the addition on a number line.

We can picture the addition of two numbers on a number line by three arrows: an arrow for each addend and an arrow for the sum. The first arrow starts at the origin. The second arrow starts where the first arrow ends. The arrow for the sum *starts* where the *first* arrow starts and *ends* where the *second* arrow ends.

The arrows in the diagram below show an addition in which both addends are positive: $2 + 3 = 5$.

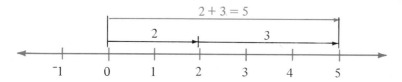

Since the addends and their sum are positive, all three of the arrows are directed to the right. This suggests the following rule:

rule

The sum of two positive numbers is a positive number.

If one of the addends is 0, then the arrow for the sum is identical with that for the other addend. For example, $4 + 0 = 4$, and also $0 + 0 = 0$.

Example 1 | Find the sum of 1.8 and 3.6.

Solution | The length of the arrow for the sum is $1.8 + 3.6$, or 5.4. Since both addends are positive, the sum is positive. Hence, $1.8 + 3.6 = 5.4$.

The diagram below shows an addition in which both addends are negative: $^-3 + ^-1 = ^-4$.

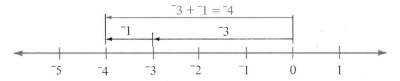

Since the addends and their sum are negative, all three of the arrows are directed to the left.

rule

The sum of two negative numbers is a negative number.

Example 2 | Find the sum of $^-2.5$ and $^-3.5$.

Solution | The length of the arrow for the sum is $2.5 + 3.5$, or 6. Since each of the addends is negative, the sum is negative. Therefore, $^-2.5 + ^-3.5 = ^-6$.

Class Exercises

Look at the diagram at the right.

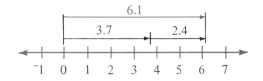

1. The first addend is __?__. 3.7

2. The second addend is __?__. 2.4

3. The sum is __?__. 6.1

4. Name the starting point and the endpoint of the arrow for the first addend; for the second addend; for the sum. 0, 3.7; 3.7, 6.1; 0, 6.1

5–8. Repeat Exercises 1–4 for the diagram at the right.
5. $^-3.3$ 6. $^-2.7$
7. $^-6$
8. 0, $^-3.3$; $^-3.3$, $^-6$; 0, $^-6$

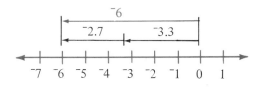

State the first addend, the second addend, and the sum. Then read the
equation as a number sentence.

9.
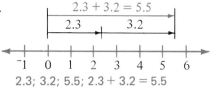

2.3; 3.2; 5.5; 2.3 + 3.2 = 5.5

10.
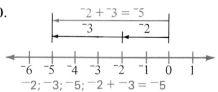

⁻2; ⁻3; ⁻5; ⁻2 + ⁻3 = ⁻5

11.
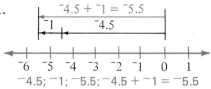

⁻4.5; ⁻1; ⁻5.5; ⁻4.5 + ⁻1 = ⁻5.5

12.
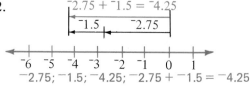

⁻2.75; ⁻1.5; ⁻4.25; ⁻2.75 + ⁻1.5 = ⁻4.25

13. In arrow addition of positive numbers, the arrows are __?__ (never, some-
    times, always) directed to the right. always

14. In arrow addition of negative numbers, the arrows are always directed
    to the __?__ . left

15. If two addends are positive, their sum is __?__ (never, sometimes, always)
    positive. always

16. If two addends are negative, their sum is always __?__ . negative

State the sum.

17. −2.8 + −3.1 ⁻5.9   18. −10 + −2.4 ⁻12.4   19. −2 + 0 ⁻2   20. 0 + −3.7 ⁻3.7

21. 0.4 + 0.2 0.6   22. 0.6 + 1.4 2   23. −0.3 + −0.6 ⁻0.9   24. −2.7 + −0.2 ⁻2.9

Exercises

Draw a number line and find the sums of the given numbers by arrow addition. See students' graphs.

A   1. 4 + 5   2. 2 + 3   3. ⁻4 + ⁻5   4. ⁻2 + ⁻3

    5. −16 + −14   6. −20 + −35   7. −2.5 + −4.5   8. −3.75 + −4.25

Write the sum.

 9. 3.8 + 8.3 12.1   10. 4.9 + 7.6 12.5   11. −1.7 + −1.7 ⁻3.4   12. −2.5 + −2.6 ⁻5.1

13. −78 + −69 ⁻147   14. −64 + −87 ⁻151   15. −148 + −256 ⁻404   16. −409 + −208 ⁻617

17. −0.8 + −0.5 ⁻1.3   18. −2.4 + −2.2 ⁻4.6   19. −2.4 + −4.6 ⁻7   20. −5.2 + −2.8 ⁻8

Supply the missing addend.

B **21.** $^-7 + \underline{\quad?\quad} = {}^-15$ $^-8$ **22.** $^-9 + \underline{\quad?\quad} = {}^-23$ $^-14$ **23.** $^-47 + \underline{\quad?\quad} = {}^-93$ $^-46$

24. $\underline{\quad?\quad} + {}^-6 = {}^-13$ $^-7$ **25.** $\underline{\quad?\quad} + {}^-24 = {}^-41$ $^-17$ **26.** $\underline{\quad?\quad} + {}^-128 = {}^-257$ $^-129$

Problems

Solve.

A **1.** Some research scientists drilled 92 m below the surface of the Ross Ice Shelf in Antarctica when their drill broke. After fixing it, they drilled 206 m deeper. How far below the surface had they then drilled? 298 m

2. A parachutist jumped from an airplane flying at an altitude of 1100 m, dropped 200 m in the first 25 seconds, and then dropped 350 m in the next 35 seconds. How many meters did the parachutist drop in 60 seconds? 550 m

Application

Ups and Downs

The Rocky Ledge Inn has 15 floors. It is located on a steep hill overlooking the ocean and must be entered at the sixth floor from the top. To number the floors, the management uses the positive and negative integers from 5 to −9, with the entrance floor or lobby assigned the number 0. The floors above the lobby are assigned positive integers, and those below the lobby are assigned negative integers.

1. If an elevator rises two floors from the lobby and then rises two more floors, it is then located at the floor assigned the integer $\underline{\quad?\quad}$. 4

2. This floor is $\underline{\quad?\quad}$ (above/below) the lobby. above

3. If the elevator drops two floors from the lobby and then drops an additional four floors, it is located at the floor assigned the integer $\underline{\quad?\quad}$. $^-6$

4. This floor is $\underline{\quad?\quad}$ (above/below) the lobby. below

3-4 The Sum of Two Numbers of Opposite Sign

Objective To find the sum of two numbers when one is positive and the other is negative, and to picture the addition on a number line.

When the addend with the longer arrow is positive, the sum is positive.

$$3 + {}^-2 = 1$$

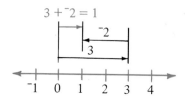

When the addend with the longer arrow is negative, the sum is negative.

$$1 + {}^-5 = {}^-4$$

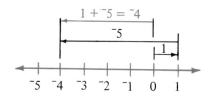

When the addends have arrows of equal length, the sum is 0.

$$3 + {}^-3 = 0$$

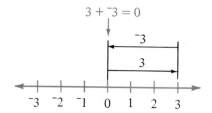

Notice that, in each case, the length of the arrow for the sum is the *difference* of the lengths of the two arrows for the addends. The sum has the same sign as that of the addend with the longer arrow.

Example | Find the sum.

a. ⁻6 and 4   b. 2.6 and ⁻1.2

Solution | a. The length of the arrow for the sum is the difference $6 - 4$, or 2. Since the addend with the longer arrow is ⁻6, the sum is negative. Thus, $^-6 + 4 = {}^-2$.

b. The arrows have lengths of 2.6 and 1.2. The difference is $2.6 - 1.2$, or 1.4. Since the addend with the longer arrow is 2.6, the sum is positive. Hence, $2.6 + {}^-1.2 = 1.4$.

With practice, you should become able to abbreviate this process and simply write, for example,

$$8 + {}^-5 = 3, \quad 2.6 + {}^-5.8 = {}^-3.2, \quad {}^-7 + 7 = 0$$

Class Exercises

1. Positive numbers are represented by arrows directed to the __?__ . right

2. Negative numbers are represented by arrows directed to the __?__ . left

3. The length of the arrow for the sum of two numbers of opposite sign is the __?__ of the lengths of the arrows for the addends. difference

4. The sum of two numbers of opposite sign has the same sign as the addend with the __?__ arrow. longer

State the first addend, the second addend, and the sum. Then read the equation as a number sentence.

5. $5.5 + {}^-3.2 = 2.3$

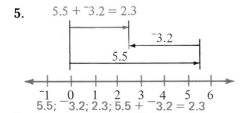

5.5; ${}^-3.2$; 2.3; $5.5 + {}^-3.2 = 2.3$

6. $2 + {}^-3 = {}^-1$

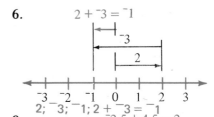

2; ${}^-3$; ${}^-1$; $2 + {}^-3 = {}^-1$

7. ${}^-5 + 3 = {}^-2$

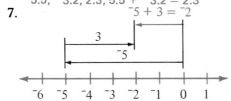

${}^-5$; 3; ${}^-2$; ${}^-5 + 3 = {}^-2$

8. ${}^-2.5 + 4.5 = 2$

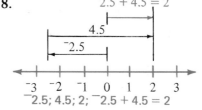

${}^-2.5$; 4.5; 2; ${}^-2.5 + 4.5 = 2$

Exercises

Draw a number line and find the sum by arrow addition. Check students' graphs.

A 1. $5 + {}^-2$ 2. $4 + {}^-3$ 3. $2 + {}^-3$ 4. $3 + {}^-5$

5. ${}^-2 + 4$ 6. ${}^-6 + 4$ 7. ${}^-3 + 3$ 8. $4 + {}^-4$

Write the sum.

9. $8.2 + {}^-3.2$ 5   10. $7.4 + {}^-2.3$ 5.1   11. ${}^-8.7 + 5.2$ ${}^-3.5$ 12. ${}^-2.4 + 1.3$ ${}^-1.1$

13. $28 + {}^-25$ 3   14. $17 + {}^-23$ ${}^-6$ 15. $42 + {}^-54$ ${}^-12$ 16. ${}^-248 + 236$ ${}^-12$

17. ${}^-8.3 + 5.6$ ${}^-2.7$ 18. $8.2 + {}^-2.4$ 5.8   19. $1.7 + {}^-2.3$ ${}^-0.6$ 20. ${}^-4.5 + 2.2$ ${}^-2.3$

Supply the missing addend.

B 21. $^-5 + \underline{\quad?\quad} = ^-3$ 2

22. $^-2 + \underline{\quad?\quad} = 4$ 6

23. $3 + \underline{\quad?\quad} = ^-4$ $^-7$

24. $7 + \underline{\quad?\quad} = 4$ $^-3$

25. $4.3 + \underline{\quad?\quad} = ^-1.6$ $^-5.9$

26. $^-3.25 + \underline{\quad?\quad} = 1.75$ 5

Problems

Solve.

A 1. After take-off, an airliner rose steadily to an elevation of 11,000 m, then descended 4000 m to avoid a storm, and next rose 3000 m. What was its elevation at that time? 10,000 m

2. Because of a grain shortage, the price of a loaf of bread increased by 2 cents. When grain again became plentiful, the price decreased by 4 cents, but later it again rose by 5 cents because of inflation. If the original price was 46 cents per loaf, what was the final price? 49¢

EXTRA! Absolute Value

On page 70 we stated that the length of a directed line segment is always a positive number or zero, even when the direction of the arrow is negative. This number is called the **absolute value** of the number represented by the arrow.

A directed line segment representing 3 has a length 3, so the absolute value of 3 is 3.

A directed line segment representing $^-5$ has a length 5, so the absolute value of $^-5$ is 5.

The symbol we use to indicate "the absolute value of the number n" is $|n|$. Thus we can write

$$n = 9, \quad \text{and} \quad |9| = 9$$
$$n = ^-1.2, \quad \text{and} \quad |^-1.2| = 1.2$$
$$n = 0, \quad \text{and} \quad |0| = 0$$

We use absolute value to find the sum of two numbers. For example, find $^-5 + 2$.

$|^-5| = 5$ and $|2| = 2$

The absolute value of the sum, $|^-5 + 2|$, is $5 - 2$, or 3.

Since $|^-5| > |2|$, the sign of the sum is negative.

So, $^-5 + 2 = ^-3$.

Use absolute value, as shown on page 80, to find the sum.

1. $8 + {}^-7$ 1

2. ${}^-8 + 7$ ${}^-1$

3. $5 + {}^-5$ 0

4. $32 + {}^-40$ ${}^-8$

5. ${}^-32 + 40$ 8

6. $58 + {}^-106$ ${}^-48$

Self-Test

Words to remember:
positive number [p. 66]
opposite numbers [p. 66]
directed line segment [p. 70]
See students' graphs.

negative number [p. 66]
integers [p. 66]
arrow [p. 70]

Show the graph of the opposite of the number on a number line.

1. 21

2. 701

[3-1]

Write the missing number.

3. If $n = {}^-6$, then $-n = \underline{\quad?\quad}$ 6

4. If $-n = ({}^-1.2)$, then $n = \underline{\quad?\quad}$ 1.2

Graph the numbers on a number line. See students' graphs.

5. ${}^-3.5$ and 0.7

6. 3.5 and ${}^-0.7$

[3-2]

7. Draw an arrow starting at the graph of ${}^-5$ to represent the number 7. See students' graphs.

8. Draw an arrow with endpoint at the graph of ${}^-3$ to represent the number 4. See students' graphs.

Complete.

9. ${}^-3.6 + {}^-1.8 = \underline{\quad?\quad}$ ${}^-5.4$

10. $1.6 + 5.9 = \underline{\quad?\quad}$ 7.5

[3-3]

11. $\underline{\quad?\quad} + {}^-4.2 = {}^-4.8$ ${}^-0.6$

12. $17.76 + \underline{\quad?\quad} = 19.75$ 1.99

13. ${}^-6 + 3 = \underline{\quad?\quad}$ ${}^-3$

14. $5.1 + {}^-1.3 = \underline{\quad?\quad}$ 3.8

[3-4]

15. $4 + \underline{\quad?\quad} = 0$ ${}^-4$

16. ${}^-10.25 + 6.54 = \underline{\quad?\quad}$ ${}^-3.71$

Self-Test answers and Extra Practice are at the back of the book.

3-5 The Difference of Two Numbers

Objective To find the difference of two numbers by finding a missing addend.

We learned in Section 2-5 that for *every* subtraction equation we can write a related addition equation. Thus, finding the difference of two numbers can be looked at as the process of finding a missing addend.

The arrow diagram below pictures a sum, 3, and one of its addends, ⁻2. The dashed arrow for the missing addend, labeled d, represents the difference of 3 and ⁻2, namely 5.

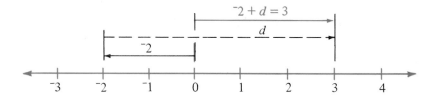

To indicate that 5 is the difference of 3 and ⁻2, we write

$$3 - {}^-2 = 5.$$

From the arrow diagram, you can also see another difference, namely,

$$3 - 5 = {}^-2.$$

Each of these subtraction facts corresponds to the addition fact that

$$^-2 + 5 = 3.$$

Thus, we have the following property:

---

property

For any numbers a, b, and c, if

$$a + b = c,$$

then it is also true that

$$c - a = b, \text{ and } c - b = a.$$

Example | Write two differences corresponding to the sum

$$4 + {}^-6 = {}^-2$$

$$(a + b = c)$$

Solution | $${}^-2 - 4 = {}^-6 \qquad \text{and} \qquad {}^-2 - {}^-6 = 4$$

$$(c - a = b) \qquad\qquad\qquad (c - b = a)$$

Class Exercises

For the given arrow diagram, state the value of the missing addend.

1.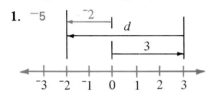

2.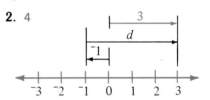

In the given exercise, express *d* as a difference.

3. Exercise 1 $^-2 - 3 = {}^-5$

4. Exercise 2 $3 - {}^-1 = 4$

Find the value of the given expression.

5. $12 - 6$ 6

6. $^-12 - {}^-6$ $^-6$

7. $12 - {}^-6$ 18

8. $^-12 - 6$ $^-18$

9. $6 - 12$ $^-6$

10. $^-6 - {}^-12$ 6

11. $6 - {}^-12$ 18

12. $^-6 - 12$ $^-18$

State two differences corresponding to the given sum. See below

13. $25 + {}^-15 = 10$

14. $3 + {}^-12 = {}^-9$

15. $^-4 + 5 = 1$

16. $^-7 + 3 = {}^-4$

17. $5 + 10 = 15$

18. $4 + 6 = 10$

19. $^-4 + {}^-7 = {}^-11$

20. $^-2 + {}^-3 = {}^-5$

21. Use examples to explain why the difference of two whole numbers is not always a whole number. $3 - 7 = {}^-4$; $7 - 12 = {}^-5$ are some examples.

22. Use examples to explain why the difference of two integers must always be an integer. $10 - 3 = 7$; $5 - 11 = {}^-6$; $1 - {}^-5 = 6$ are some examples.

Exercises

Picture the given difference on a number line. See students' graphs.

A

1. $6 - 8$

2. $4 - 7$

3. $^-6 - {}^-4$

4. $^-4 - {}^-6$

5. $3 - {}^-2$

6. $^-4 - 2$

7. $^-5 - 0$

8. $0 - {}^-5$

13. $10 - 25 = {}^-15$; $10 - {}^-15 = 25$

14. $^-9 - {}^-12 = 3$; $^-9 - 3 = {}^-12$

15. $1 - 5 = {}^-4$; $1 - {}^-4 = 5$

16. $^-4 - 3 = {}^-7$; $^-4 - {}^-7 = 3$

17. $15 - 10 = 5$; $15 - 5 = 10$

18. $10 - 6 = 4$; $10 - 4 = 6$

19. $^-11 - {}^-7 = {}^-4$; $^-11 - {}^-4 = {}^-7$

20. $^-5 - {}^-3 = {}^-2$; $^-5 - {}^-2 = {}^-3$

Find the given difference.

9. $20 - 14$ 6

10. $12 - 1$ 11

11. $^-52 - {}^-15$ $^-37$

12. $^-75 - 20$ $^-95$

13. $12 - 20$ $^-8$

14. $^-32 - {}^-21$ $^-11$

15. $^-1 - {}^-34$ 33

16. $68 - 52$ 16

B **17.** $^-8 - {}^-16$ 8

18. $35 - 43$ $^-8$

19. $^-8 - {}^-42$ 34

20. $23 - {}^-133$ 156

21. $^-15 - {}^-7$ $^-8$

22. $3 - {}^-17$ 20

23. $60 - {}^-100$ 160

24. $^-88 - 90$ $^-178$

Write two differences corresponding to the given sum.

25. $22 + {}^-6 = 16$
$16 - 22 = {}^-6; 16 - {}^-6 = 22$
28. $^-20 + {}^-12 = {}^-32$
$^-32 - {}^-20 = {}^-12;$
$^-32 - {}^-12 = {}^-20$

26. $9 + 6 = 15$
$15 - 9 = 6; 15 - 6 = 9$
29. $^-36 + 15 = {}^-21$
$^-21 - {}^-36 = 15;$
$^-21 - 15 = {}^-36$

27. $^-13 + 29 = 16$
$16 - {}^-13 = 29; 16 - 29 = {}^-13$
30. $^-76 + 23 = {}^-53$
$^-53 - {}^-76 = 23;$
$^-53 - 23 = {}^-76$

Problems

Solve.

A **1.** Two stages of a rocket burn for a total of 114.5 seconds. If the first stage burns for 86.8 seconds, how long does the second stage burn? 27.7 seconds

2. On Judy's paper route during June, her papers cost her $72.50 and she collected $68.75. What was her profit or loss for the month? $3.75 loss

EXTRA! Calculating Quickly

Show your friends you can add faster than a calculator.

A. Choose a 6-digit number and write it down.

B. Ask a friend to choose a number.

C. You choose, making sums of 9 with the digits of your friend's number.

D. Ask your friend to choose another number.

E. You choose again, making sums of 9.

F. While your friend uses a calculator to find the sum, you quickly write it down. It is your original number minus 2, plus 2 million. (2 is the number of choices your friend made.)

Example

730,598 **(A)**
343,716 **(B)**
656,283 **(C)**
174,662 **(D)**
825,337 **(E)**

2,730,596 **(F)**

$8 - 2$

$2 \times 1,000,000$

3-6 Subtraction as Adding the Opposite

Objective To find the difference of two numbers by adding the opposite of the second to the first.

At the left below is an arrow diagram for the difference $6 - 2$. Compare this with the figure at the right showing the sum of 6 and *the opposite* of 2.

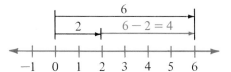 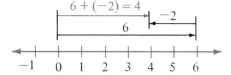

You can see that the arrow for the difference $6 - 2$ has the same length and direction as the arrow for the sum $6 + (-2)$. That is:

$$6 - 2 = 6 + (-2)$$

subtract   add the opposite

(Remember that the unraised minus sign is used to denote *subtraction* as well as the *opposite*.)

In general, we can say:

To subtract a number, add its opposite.

---

property

For any numbers a and b,

$$a - b = a + (-b).$$

We learned how to add any two numbers (positive, negative, or zero) in Sections 3-3 and 3-4. The simplest way to find the *difference* of any two numbers is to rewrite it as a sum and then add. The example which follows on the next page illustrates this method of finding differences.

Example	**Find the difference by first expressing it as a sum.**
	a. ⁻5 – ⁻8 **b.** ⁻5 – 6
	c. ⁻4.7 – 2.8 **d.** 7 – ⁻8
Solution	**a.** ⁻5 – ⁻8 = ⁻5 + (– ⁻8) = ⁻5 + 8 = 3
	b. ⁻5 – 6 = ⁻5 + (– 6) = ⁻5 + ⁻6 = ⁻11
	c. ⁻4.7 – 2.8 = ⁻4.7 + (– 2.8) = ⁻4.7 + ⁻2.8 = ⁻7.5
	d. 7 – ⁻8 = 7 + (– ⁻8) = 7 + 8 = 15

Class Exercises

Express the given difference as a sum.

1. 7 – 4 7 + ⁻4 **2.** 8 – 2 8 + ⁻2 **3.** ⁻10 – 5 ⁻10 + ⁻5 **4.** ⁻3 – ⁻8 ⁻3 + 8

5. 1.8 – 4.7 **6.** ⁻2.3 – ⁻1.7 **7.** 0.6 – ⁻1.3 **8.** ⁻4.5 – 2.4
  1.8 + ⁻4.7 ⁻2.3 + 1.7 0.6 + 1.3 ⁻4.5 + ⁻2.4

Express the given sum as a difference.

9. 4.9 + ⁻2.8 **10.** ⁻3.2 + ⁻1.5 **11.** 0.8 + 0.2 **12.** ⁻3 + 2
  4.9 – 2.8 ⁻3.2 – 1.5 0.8 – ⁻0.2 ⁻3 – ⁻2

13. Subtracting 6 is the same as adding __?__ ⁻6

14. Subtracting ⁻2 is the same as adding __?__ 2

Find the difference in the given exercise

15. Exercise 1 3 **16.** Exercise 2 6 **17.** Exercise 3 ⁻15 **18.** Exercise 4 5

Exercises

Write the given difference as a sum.

A **1.** 7.5 – ⁻1.3 **2.** ⁻2.1 – ⁻5.1 **3.** 1.3 – 3.4 **4.** ⁻21.1 – 4.2
    7.5 + 1.3 ⁻2.1 + 5.1 1.3 + ⁻3.4 ⁻21.1 + ⁻4.2
  5. 33.3 – 46.5 **6.** 18.9 – ⁻25.6 **7.** 3.56 – ⁻8.73 **8.** 40.03 – 19.97
    33.3 + ⁻46.5 18.9 + 25.6 3.56 + 8.73 40.03 + ⁻19.97

Find the given difference.

9. 23 – 18 5 **10.** 12 – 18 ⁻6 **11.** ⁻3 – 17 ⁻20 **12.** 10 – ⁻12 22

13. 7.4 – ⁻2.5 9.9 **14.** 3.5 – ⁻4.5 8 **15.** ⁻1.8 – 2.7 ⁻4.5 **16.** ⁻0.02 – 0.31 ⁻0.33

17. 13.15 – 2.27 10.88 **18.** 17.3 – 2.4 14.9 **19.** ⁻3.5 – 9.5 ⁻13 **20.** ⁻4.7 – ⁻5.8 1.1

21. 66.30 – ⁻32.12 **22.** 50.05 – 65.11 **23.** 97.44 – ⁻2.01 **24.** 0.75 – 26.15 ⁻25.4
    98.42 ⁻15.06 99.45

Problems

Solve.

A

1. At the beginning of October, Jean Johnson had a balance of $538.27 in her checking account. During the month, she made a deposit of $203.80 and wrote checks for $112.42, $89.17, $209.06, and $250.00. What was the balance in her account at the end of the month? $81.42

2. During a week, the stock of United Environmental Control Corporation had the following daily changes in price: Monday, up $3; Tuesday, up $4; Wednesday, down $5; Thursday, up $1; Friday, down $9. What was the change in price of the stock for the week? down $6

A Nutritionist

Do you know what kinds of vitamins and minerals your body needs? Do you know what kinds of foods contain these vitamins and minerals?

As a nutritionist, Hsien Wu (1893–1959) often dealt with the above questions. His research on food composition and the effects of eating habits on health made him the foremost nutrition scientist in China. Perhaps Wu's most notable discovery was his development of a system of analyzing blood for research and testing. His methods enabled scientists to measure the content of blood with only 10 mL samples. Hsien Wu wrote over 150 research papers which brought him international recognition and an honorary membership on the Standing Advisory Committee on Nutrition of the Food and Agriculture Organization of the United Nations, among other memberships.

Career Activity Besides doing research, some nutritionists coordinate nutrition programs as part of public health programs. Find out if there is a nutrition program at your school or in your neighborhood. How is the program organized?

3-7 The Product of Two Integers

Objective To find the product of two integers by applying the rules of sign for multiplication.

On page 67, we saw that, for instance, $-2 = {}^-2$, that is, the opposite of two is negative two.

Notice! From now on, we shall use an unraised minus sign in denoting:

a negative number: -2 can stand for *negative 2*
the opposite of a number: -2 can stand for *the opposite of 2*
subtraction: -2, as in $4 - 2$, means *subtract 2*

In earlier work with whole numbers we have seen that if all the addends in a given sum are the *same* whole number, we can use multiplication to find the sum. For example, to add

$$3 + 3 + 3 + 3 + 3 = 15$$

we can, instead, multiply

$$5 \times 3 = 15$$

This is true also if all the addends are the same *integer*.

Example 1 | Find the product $2(-3)$ with the aid of an arrow diagram.

Solution

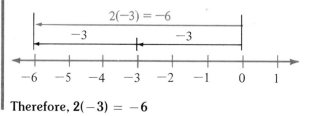

Therefore, $2(-3) = -6$

We cannot picture a product like

$$-2(3) \text{ or } -2(-3)$$

on a number line, because in each case the first factor indicates that a *negative* number of arrows of equal length must be drawn. Nevertheless, if we assume that the distributive property holds for all integers, then we can decide what such a product must be.

$$(4 + -2)(3) = (4)(3) + (-2)(3)$$
$$(2)(3) = 12 \quad + (-2)(3)$$
$$6 = 12 \quad + (-2)(3)$$

But
$$6 = 12 \quad + -6$$

so
$$(-2)(3) \text{ must be equal to } -6.$$

---
rule

The product of a negative integer and a positive integer is a negative integer.

In the same way

$$(4 + -2)(-3) = (4)(-3) + (-2)(-3)$$
$$2(-3) = -12 \quad + (-2)(-3)$$
$$-6 = -12 \quad + (-2)(-3)$$

But
$$-6 = -12 \quad + 6$$

so
$$(-2)(-3) \text{ must be equal to } 6.$$

---
rule

The product of two negative integers is a positive integer.

The following table summarizes the facts about products of integers.

Two Factors	Product	Example
both positive	positive	$3(2) = 6$
both negative	positive	$-3(-2) = 6$
of opposite sign	negative	$-3(2) = -6$
one or both 0	0	$-3(0) = 0$

Example 2 | Find the product.
           | **a.** $3(5)$ **b.** $-3(5)$ **c.** $-3(-5)$ **d.** $0(-5)$

Solution | **a.** 15   **b.** -15 **c.** 15   **d.** 0

Class Exercises

Express the given sum as a product.

1. $5 + 5 + 5$ 3(5)

2. $0 + 0 + 0 + 0$ 4(0)

3. $(-6) + (-6)$ 2(−6)

4. $(-1) + (-1) + (-1)$ 3(−1)

5. $(-3.5) + (-3.5)$ 2(−3.5)

6. $(-4.8) + (-4.8)$ 2(−4.8)

7. $(-8.2) + (-8.2)$ 2(−8.2)

8. $(-2) + (-2) + (-2)$ 3(−2)

Express the given product as a sum.

9. $3(-4)$
$(-4) + (-4) + (-4)$

10. $5(0)$
$0 + 0 + 0 + 0 + 0$

11. $6(-1)$
$(-1) + (-1) + (-1) + (-1) + (-1) + (-1)$

12. $4(-12)$
$(-12) + (-12) + (-12) + ($

Find the product.

13. $5(10)$ 50

14. $5(-10)$ −50

15. $-5(10)$ −50

16. $-5(-10)$ 50

17. $-2(-3)$ 6

18. $-1(-1)$ 1

19. $1(-7)$ −7

20. $-1(7)$ −7

Exercises

Draw an arrow diagram picturing the given product. See students' graphs.

A   **1.** $3(-4) = -12$ **2.** $5(-3) = -15$ **3.** $4(-6) = -24$ **4.** $2(-5) = -10$

Find the product.

5. $8(-9)$ −72

6. $7(-6)$ −42

7. $6(-8)$ −48

8. $10(-100)$ −1000

9. $-11(12)$ −132

10. $-13(10)$ −130

11. $-16(16)$ −256

12. $-40(28)$ −1120

13. $-4(-3)$ 12

14. $-3(-5)(-2)$ −30

15. $-20(-30)$ 600

16. $-25(-15)(4)$ 1500

17. $-42(-40)$ 1680

18. $-50(-50)$ 2500

19. $-100(24)$ −2400

20. $25(-25)$ −625

B   **21.** $-27(42 + 28)$ −1890

22. $-15(37 + 13)$ −750

23. $-20(40) + 60(-10)$ −1400

24. $-15(12) + 50(-20)$ −1180

25. $-42(17 - 10)$ −294

26. $-52(28 - 14)$ −728

27. $-33(8 - 15)$ 231

28. $-75(13 - 34)$ 1575

C   **29.** $[-14 + 18 + (-27)][-22 + 14 + (-12)]$ 460

30. $[-50 - 17 + 13][-43 + 14 + (-21)]$ 2700

3-8 The Product of Two Numbers

Objective To find the product of any two given numbers by applying the rules of sign for multiplication.

You can use the same rules of sign you used with integers to simplify expressions for products of any numbers.

Example 1 | Multiply $-4.21(1.6)$

Solution | **First, note that one factor is positive and one negative, so the product is negative. Then multiply 4.21(1.6).**

$$\begin{array}{r} 4.2\,1 \\ \underline{1.6} \\ 2\,5\,2\,6 \\ \underline{4\,2\,1} \\ 6.7\,3\,6 \end{array}$$

Since the product is negative,
$$-4.21(1.6) = -6.736$$

Example 2 | Multiply $-0.7(-0.6)$

Solution | **Since both factors are negative, the product is positive. Working with positive numbers only, you have:**
$$0.7(0.6) = 0.42$$
Since the given product is positive, you therefore have
$$-0.7(-0.6) = 0.42$$

Example 3 | Multiply $5.25(-3)$

Solution | **Note that the product is negative.**
Multiply: 5.25(3) = 15.75
Therefore, 5.25(-3) = -15.75

Class Exercises

Multiply.

1. $-3(2)$ -6

2. $4(-10)$ -40

3. $-2(1.5)$ -3

4. $-3(1.2)$ -3.6

5. $2(-14)$ -28

6. $-10(6)$ -60

7. $3(-12)$ -36

8. $15(-2)$ -30

9. $-1.2(-1.2)$ 1.44

10. $-0.8(-0.2)$ 0.16

11. $-0.4(-0.4)$ 0.16

12. $-1.5(-0.5)$ 0.75

13. $-0.5(-0.5)$ 0.25

14. $-0.4(-0.3)$ 0.12

15. $-1.5(0.5)$ -0.75

16. $-2.5(4)$ -10

Exercises

A 1. $7.4(-1.3)$ -9.62 2. $8.3(-2.6)$ -21.58 3. $-5.6(7.1)$ -39.76 4. $-4.8(1.2)$ -5.76

5. $-3.4(-5.6)$ 19.04 6. $-9.2(-7.3)$ 67.16 7. $-10.1(-8.9)$ 89.89 8. $-12.3(-4.3)$ 52.89

9. $2.8(-5.7)$ -15.96 10. $4.5(-2.3)$ -10.35 11. $-5.3(2.7)$ -14.31 12. $-6.5(3.3)$ -21.45

13. $-5.6(-3.2)$ 17.92 14. $-3.4(-8.9)$ 30.26 15. $-1.5(-1.4)$ 2.1 16. $-3.6(-5.2)$ 18.72

17. $-4.15(1.15)$
    -4.7725
18. $2.14(-7.30)$
    -15.622
19. $-2.25(-3.12)$ 7.02 20. $6.28(2.16)$ 13.5648

B 21. $4(3)(-3)$ -36 22. $-15(-2)(4)$ 120

23. $1.2(-4.8)(-3.1)$ 17.856 24. $-7.6(2.3)(-4.1)$ 71.668

25. $-4.5(-1.3)(-0.23)$ -1.3455 26. $-6.8(4.1)(-1.1)$ 30.668

C 27. $(4.02 - 6.7)(-2.1 + 1.2)$ 2.412 28. $(52.8 - 107.3)(86.1 - 92.7)$ 359.7

29. $(7.1 - 9.4)(4.2 - 8.5)$ 9.89 30. $[-6.7 - (-4.53)][-5.1 + 2.12]$ 6.4666

Problems

Solve.

A 1. Mary had an allowance of $10 for her lunch at school during the month. She spent 45¢ for lunch each of the 22 school days. How much money did she have left at the end of the month? 10¢

2. In Summit City 78 cm of snow fell on Sunday. It melted an average of 5.8 cm each day. How much snow remained six days later? 43.2 cm

EXTRA! Zero and Negative Exponents

On page 28 we noted that we can find the product of two powers of 10 by adding the exponents. For example,

$$10^2 \times 10^3 = 10^{2+3} = 10^5$$

Thus we have the following property:

> **property**
>
> For all positive integers a and b, $10^a \times 10^b = 10^{a+b}$

Does the symbol 10^0 stand for a number? If it does, and if the property on page 92 is to apply to this power, then we must have

$$10^1 \times 10^0 = 10^{1+0} = 10$$
$$10^2 \times 10^0 = 10^{2+0} = 10^2$$

and so on. That is,

$$10^a \times 10^0 = 10^a$$

Since multiplying a given power of 10 by 10^0 produces the given power of 10, 10^0 must be equal to 1. Thus we can write

$$10^0 = 1$$

If the property on page 92 is also to apply to *negative* exponents, we must have

$$10^1 \times 10^{-1} = 10^{1+(-1)} = 10^0 = 1$$
$$10^2 \times 10^{-2} = 10^{2+(-2)} = 10^0 = 1$$

and so on. Since we know that

$$10^1 \times \frac{1}{10} = 1 \quad \text{and} \quad 10^2 \times \frac{1}{10^2} = 1$$

we then must have

$$10^{-1} = \frac{1}{10} \quad \text{and} \quad 10^{-2} = \frac{1}{10^2}$$

Thus we can write

$$10^{-a} = \frac{1}{10^a}$$

Copy and complete the following table.

	Fraction	Decimal	10^a	
1.	$\frac{1}{10}$	0.1	_?_	10^{-1}
2.	$\frac{1}{10^2}$	_?_	10^{-2}	0.01
3.	_?_	0.001	_?_	$\frac{1}{10^3}$; 10^{-3}
4.	_?_	_?_	10^{-4}	$\frac{1}{10^4}$; 0.0001

3-9 The Quotient of Two Numbers

Objective To find quotients of numbers by applying the rules of sign for multiplication.

In Section 2-5 we learned that division can be thought of as the process of finding a missing factor. You can use this fact, together with the rules of sign for multiplication, in finding quotients.

Example 1	**Find the quotient $q = 15 \div (-5)$.**
Solution	**If $q = 15 \div (-5)$, then $q \times (-5) = 15$. Since the product, 15, is positive and one of its factors, -5, is negative, you know that the other factor, q, must be negative. Thus, $q = -3$.**
Example 2	**Find the quotient $0 \div (-4)$.**
Solution	**If $q = 0 \div (-4)$, then $q \times (-4) = 0$. Since the product is 0, at least one of the factors must be 0, so $q = 0$. Hence $0 \div (-4) = 0$.**
Example 3	**If possible, find the quotient $(-4) \div 0$.**
Solution	**If $q = (-4) \div 0$, then $q \times 0 = -4$. But the product of 0 and any number is 0, so there is no such value q. We cannot divide by 0.**

From examples such as the ones above, we have the following facts about quotients:

Dividend and divisor	Quotient	Example
both positive	positive	$6 \div 3 = 2$
both negative	positive	$-6 \div (-3) = 2$
of opposite sign	negative	$-6 \div 3 = -2$ $6 \div (-3) = -2$
dividend 0 divisor not 0	0	$0 \div 3 = 0$

Example 4 | Divide 126 by -9.

Solution | Apply the division process to the positive numbers 126 and 9. Since 126 is positive and -9 is negative, their quotient is negative. Therefore, $126 \div (-9) = -14$.

$$\begin{array}{r} 14 \\ 9\overline{)126} \\ 9 \\ \hline 36 \\ 36 \\ \hline 0 \end{array}$$

Example 5 | Divide -6.58 by -1.4.

Solution | Apply the division process to positive numbers. Since -6.58 and -1.4 are both negative, their quotient is positive. Therefore, $(-6.58) \div (-1.4) = 4.7$.

$$\begin{array}{r} 4.7 \\ 1.4\overline{)6.58} \\ 56 \\ \hline 98 \\ 98 \\ \hline 0 \end{array}$$

Example 6 | What is one third of -21?

Solution | Divide 21 by 3:

$$21 \div 3 = 7$$

Since -21 is negative and 3 is positive, their quotient is negative. Therefore, $-21 \div 3 = -7$, or one third of -21 is -7.

Class Exercises

Find the quotient. If the expression is meaningless, so state.

1. $-8 \div 2$ $\quad -4$
2. $-9 \div 3$ $\quad -3$
3. $10 \div (-5)$ $\quad -2$
4. $12 \div (-4)$ $\quad -3$
5. $-4 \div (-2)$ $\quad 2$
6. $-25 \div 5$ $\quad -5$
7. $-25 \div (-5)$ $\quad 5$
8. $25 \div (-5)$ $\quad -5$
9. $0 \div (-6)$ $\quad 0$
10. $2 \div 0$
    meaningless
11. $0 \div 0$
    meaningless
12. $0 \div 4$ $\quad 0$

Exercises

Find the quotient.

A
  1. $-42 \div 7$ -6 **2.** $-56 \div 8$ -7 **3.** $-85 \div (-5)$ 17 **4.** $-132 \div (-12)$ 11

  5. $-1.6 \div 2$ -0.8 **6.** $-2.8 \div 4$ **7.** $6.3 \div (-7)$ -0.9 **8.** $8.4 \div (-7)$ -1.2

                                                 -0.7

  9. $0 \div 12$ 0 **10.** $0 \div (-25)$ 0 **11.** $3.23 \div (-1.9)$ -1.7 **12.** $22.1 \div (-1.3)$ -17

  13. $-12.768 \div (-0.48)$ 26.6 **14.** $-0.31705 \div (-8.5)$ 0.0373

  15. $-24.984 \div 720$ -0.0347 **16.** $5.544 \div (-3.6)$ -1.54

Simplify.

B
  17. $[32.3 \times (-1.84)] \div 2.8$ -21.23 **18.** $[4.36 \times (-1.84)] \div (-14.4)$ 0.56

  19. $(-134.86 \div 110) \div (-0.36)$ 3.41 **20.** $(-10.1144 \div 470) \div (-0.04)$ 0.538

C
  21. $(-0.7)(4.2)(-3.8) \div (0.02)(-7.1)(2.3)$ -34.21

  22. $(5.6)(-2.1)(0.36) \div (-4.9)(0.008)(49.7)$ 2.17

Problems

Solve.

B
  1. John had a week's allowance of $3.00 for school lunches. He spent the same amount on each school day and had 50¢ left over. How much did he spend each day?   50¢

  2. A farmer had 360 kg of fertilizer. He spread 0.8 of it evenly on 3 fields of the same size. How much did he spread on each field?   96 kg

EXTRA!   Comparing Calculators

Does your calculator round off quotients? To find out, divide 5 by 3. If your answer is 1.6666 . . . 7, then your calculator rounds. If your answer is 1.6666 . . . , then your calculator "cuts off," or "truncates" any digits not shown on the display.

Two students used different calculators to do the following multiplication problem: $0.0000008 \times 0.2 \times 2$. One student got an answer of 0.0000004 and the other got an answer of 0.0000002. Can you explain the difference in answers?

Self-Test

Find the given difference.

1. $7 - 15$ ⁻8

2. $-2 - {}^-6$ 4 [3-5]

Write two differences corresponding to

3. $9 + 2 = 11$
$11 - 9 = 2; \; 11 - 2 = 9$

4. $-37 + 29 = -8$
$-8 - {}^-37 = 29; \; {}^-8 - 29 = {}^-37$

Write the given difference as a sum.

5. $1.2 - 3.6$
$1.2 + ({}^-3.6)$

6. $8.72 - {}^-13.51$
$8.72 + 13.51$ [3-6]

Find the difference.

7. $6.5 - 1.8 = \underline{\ ?\ }$ 4.7

8. $6.12 - {}^-8.47$ 14.59

Find the product.

9. $2(-3.1)$ ⁻6.2

10. $-81(-5)$ 405 [3-7]

11. $-16(-12 + 12)$ 0

12. $(-14 - 3)(6 - 9)$ 51

Find the product.

13. $0.6(-1.5)$ ⁻0.9

14. $(-3.7)(-5.1)$ 18.87 [3-8]

15. $3.12(-7.04)$ ⁻21.9648

16. $-9.55(-7.34)$ 70.097

Simplify.

17. $-24 \div 6$ ⁻4

18. $13.68 \div (-1.8)$ ⁻7.6 [3-9]

19. $(-73.44 \div 5.1) \div (-3.2)$ 4.5

20. $(0.6)(-4.2) \div (-6)(-0.7)$ ⁻0.6

Self-Test answers and Extra Practice are at the back of the book.

Chapter Review

Write the letter that labels the correct answer.

1. If $t = 0$, then $-t$ is __?__ C   **[3-1]**
    A. a positive number   **B.** a negative number
    C. zero   **D.** 0.1

2. The opposite of $^{-}12$ is __?__ A

    A. 12   **B.** $^{-}12$ **C.** 0   **D.** $\dfrac{-1}{12}$

Use the diagram below for Exercises 3–4.

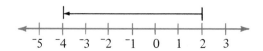

3. The coordinate of the starting point of the arrow is __?__ C   **[3-2]**
    A. $^{-}4$ **B.** $^{-}2$ **C.** 2   **D.** $^{-}6$

4. The coordinate of the endpoint of the arrow is __?__ A
    A. $^{-}4$ **B.** 2   **C.** 6   **D.** $^{-}6$

Use the diagram for Exercises 5–6.

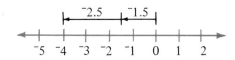

5. The diagram shows the sum __?__ B   **[3-3]**
    A. $1.5 + 2.5$ **B.** $^{-}1.5 + ^{-}2.5$ **C.** $1.5 + ^{-}2.5$ **D.** $^{-}1.5 + 2.5$

6. The sum of the addends is __?__ B
    A. 4   **B.** $^{-}4$ **C.** $^{-}3$ **D.** $^{-}1$

7. $47 + ^{-}19 =$ __?__ A   **[3-4]**
    A. 28   **B.** $^{-}28$ **C.** 66   **D.** $^{-}66$

8. $6.3 + ^{-}2.9 =$ __?__ C
    A. 9.2   **B.** $^{-}9.2$ **C.** 3.4   **D.** $^{-}3.4$

9. $^-8 - 30 =$ ___?___ D   [3-5]
    A. 22   B. $^-22$ C. 38   D. $^-38$

10. $16 - ^-22 =$ ___?___ A
    A. 38   B. $^-38$ C. 6   D. $^-6$

The given difference expressed as a sum in Exercises 11–12 is ___?___

11. $^-5.1 - ^-4.9 =$ ___?___ A   [3-6]
    A. $^-5.1 + 4.9$ B. $^-5.1 + ^-4.9$
    C. $5.1 + ^-4.9$ D. $5.1 + 4.9$

12. $0.86 - ^-0.11 =$ ___?___ A
    A. $0.86 + 0.11$ B. $^-0.86 + ^-0.11$
    C. $^-0.86 + 0.11$ D. $0.86 + ^-0.11$

13. $8(-2) =$ ___?___ D   [3-7]
    A. 4   B. -4 C. 16   D. -16

14. $-7(-35) =$ ___?___ A
    A. 245   B. -245 C. 5   D. -5

15. $-2.9(-3) =$ ___?___ C   [3-8]
    A. 87   B. -87 C. 8.7   D. -8.7

16. $0.01(-6.45) =$ ___?___ D
    A. 0.645   B. -0.645 C. 0.0645   D. -0.0645

17. $6.298 \div -0.94 =$ ___?___ D   [3-9]
    A. 0.67   B. -0.67 C. 6.7   D. -6.7

18. $-0.03787 \div -0.007 =$ ___?___ A
    A. 5.41   B. -5.41 C. 54.1   D. -54.1

Scientific Notation

Scientists often have to work with very large or extremely small quantities from their observations. For example, the diameter of a giant star is about 2,700,000,000 km, while that of a hydrogen atom is approximately $\frac{1}{100,000,000}$ cm.

Scientists have adopted a standard method for abbreviating numbers known as scientific notation. It makes use of positive exponents to show large numbers and negative exponents to show small numbers. Negative exponents were explained on page 93.

Notice that

$$4800 = 4.8 \times 1000 = 4.8 \times 10^3$$

$$0.000507 = 5.07 \times \frac{1}{10,000} = 5.07 \times 10^{-4}$$

$$835.62 = 8.3562 \times 100 = 8.3562 \times 10^2$$

rule

To express any positive number in scientific notation, you write it as the product of a power of ten and a number between 1 and 10.

Here are some things to know about scientific notation:

Decimal Number	Scientific Notation	Comments
2600	2.6×10^3	Read, "Two point six times ten to the third"
0.0000394	3.94×10^{-5}	Read, "Three point nine four times ten to the minus fifth"
7.46	7.46	The second factor, $10^0 = 1$, can be omitted.
1,000,000	10^6, or 1×10^6	The first factor, 1, can be omitted.
3 or −3	3 or −3	The notation for any integer from −9 to 9 remains unchanged.

To express a positive number in scientific notation, we can follow these steps:

1. Shift the decimal point to *standard position,* that is, to just after the first nonzero digit.
2. Then multiply by 10^n if you shifted the decimal point n places to the left, or by 10^{-n} if you shifted the decimal point n places to the right.

Number	Scientific notation
3407.2	3.4072×10^3
0.0951	9.51×10^{-2}

Express in scientific notation.

1. 3,460,092
3.460092×10^6

2. 2001
2.001×10^3

3. 0.00659
6.59×10^{-3}

4. 99.44
9.944×10^1

5. 385
3.85×10^2

6. 0.1728
1.728×10^{-1}

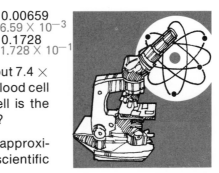

7. The diameter of a red blood cell is about 7.4×10^{-4} cm, and the diameter of a white blood cell is about 0.0015 cm. Which blood cell is the larger and by about how many times?
The white blood cell is two times larger.

8. The diameter of a hydrogen atom is approximately $\frac{1}{100,000,000}$ cm. Express this in scientific notation. 10^{-8} cm

9. Light travels at a speed of about 3×10^5 km per second. Express this as a decimal numeral. 300,000 km per second

10. The sun has a mass of 1.99×10^{33} g. Express this as a decimal numeral. 199 followed by 31 zeros

11. Earth has a mass of 5.98×10^{27} g. Express this as a decimal numeral. 598 followed by 25 zeros

12. What is the difference between the mass of the sun and Earth's mass? 1.989994×10^{33}

13. The sun has a radius of 69,600,000,000 cm. Express this in scientific notation. 6.96×10^{10}

14. Earth has a radius of 637,100,000 cm. Express this in scientific notation. 6.371×10^8

Chapter Test

1. Show the graph of the following numbers on a number line: [3-1]
4, ⁻3, 0, ⁻1. _See students' graph._

2. Write the opposite of each number.
 a. 38   **b.** ⁻12 12 **c.** 0 0
 ⁻38

3. On a number line, draw an arrow starting at the origin to repre- [3-2]
sent ⁻2. _See students' graph._

4. On a number line, draw an arrow ending at ⁻3 to represent 4.
See students' graph.

Add.

5. ⁻3.6 + ⁻2.5 ⁻6.1   **6.** ⁻640 + ⁻920 ⁻1560   [3-3]

Add.

7. 0.8 + ⁻1.1 ⁻0.3   **8.** ⁻73 + 79 6   [3-4]

Subtract.

9. ⁻9 − 7 ⁻16   **10.** ⁻52 − ⁻64 12   [3-5]

Subtract.

11. 23.5 − 30.6 ⁻7.1   **12.** 4.3 − ⁻5.7 10   [3-6]

Multiply.

13. −8(24) ⁻192   **14.** −30(−39) 1170   [3-7]

Multiply.

15. 9.6(−1.5) ⁻14.4   **16.** −5.02(−4.7) 23.594   [3-8]

Divide.

17. 5.1057 ÷ (−6.1) ⁻0.837   **18.** −33.88 ÷ (−0.8) 42.35   [3-9]

Cumulative Review

Exercises 1–4 refer to the graph at the right.

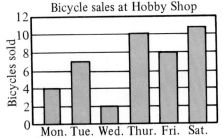

Bicycle sales at Hobby Shop

[Chap. 1]

1. Which day had the fewest sales? Wed.

2. Which day had the most sales? Sat.

3. On which day were 8 bicycles sold? Fri.

4. How many bicycles were sold in all during the week? 42

For the data at the right, find the

Heights in cm of members of the tennis club
160 162 160 165 162
163 155 162 162 159

5. range 10   **6.** mean 161
7. median 162   **8.** mode 162

Use one of the symbols < or > to make a true sentence.

9. 0.9 _?_ 0.4 >   **10.** 1.5 _?_ 2.3 <   **11.** 6.5 _?_ 5.6 [Chap. 2] >

12. 3.4 _?_ 0.4 >   **13.** 9.9 _?_ 6.5 >   **14.** 0.3 _?_ 9.0 <

Compute the sum or difference.

15. 63.05
      +20.06
      83.11

16. 198.361
      + 47.555
      245.916

17. 89.54
      −16.45
      73.09

18. 133.73
      − 92.08
      41.65

Compute the product or quotient.

19. 38.02
      × 0.6
      22.812

20. 165.09
      × 3.12
      515.0808

21. 18)53.6 2.98

22. 3.5)0.54 0.15

Compute the sum or difference.

23. 7.6 + ⁻9.2
      ⁻1.6

24. ⁻0.5 + 12.2
      11.7

25. ⁻3.4 − ⁻12.5   [Chap. 3]
      9.1

Compute the product or quotient.

26. −4.23(3.5)
      −14.805

27. −12.3(−5.4)
      66.42

28. 18.2 ÷ (−0.91)
      −20

4 Rational Numbers

4-1 Positive Fractions

Objective To use terms and properties concerning positive fractions.

The table at the right reviews some important terms concerning fractions. As with decimals, fractions can be graphed on the number line. The figure below shows the graphs of several positive fractions.

proper fraction
$\dfrac{3}{5}$ ←—numerator ←—denominator numerator less than denominator
improper fraction
$\dfrac{7}{2}$ numerator greater than or equal to denominator
mixed number
$3\dfrac{1}{2}$ whole number and a fraction

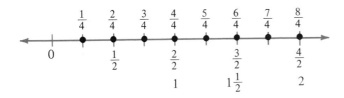

In this figure, the fractions $\frac{2}{4}$ and $\frac{1}{2}$ have the same graph. Fractions which have the same graph are called equal fractions. Thus, $\frac{2}{4}$ and $\frac{1}{2}$ are equal fractions. Can you find three other sets of equal fractions in the figure above?

105

Calculations involving fractions depend on several important patterns. For instance, some basic properties of $\frac{1}{5}$ are as follows:

$$\underbrace{\frac{1}{5} + \frac{1}{5} + \frac{1}{5} + \frac{1}{5} + \frac{1}{5}}_{\text{5 addends}} = 1 \qquad 5 \times \frac{1}{5} = 1 \qquad 1 \div 5 = \frac{1}{5}$$

Fractions such as $\frac{1}{5}$, which have the numerator 1, are called **unit fractions**. All unit fractions follow the same patterns.

properties

Properties of $\frac{1}{b}$

If b is a positive integer ($b \neq 0$), then $\frac{1}{b}$ represents the unique (one and only) number such that

$$\underbrace{\frac{1}{b} + \frac{1}{b} + \cdots + \frac{1}{b}}_{b \text{ addends}} = 1 \qquad b \times \frac{1}{b} = 1 \qquad 1 \div b = \frac{1}{b}$$

Fractions with numerators other than 1 have similar properties:

$$\frac{3}{7} = \frac{1}{7} + \frac{1}{7} + \frac{1}{7} \qquad \frac{5}{9} = 5 \times \frac{1}{9} \qquad 31 \div 40 = \frac{31}{40}$$

properties

Properties of $\frac{a}{b}$

If a and b are integers ($b \neq 0$), then $\frac{a}{b}$ represents the unique number such that

$$\underbrace{\frac{1}{b} + \frac{1}{b} + \cdots + \frac{1}{b}}_{a \text{ addends}} = \frac{a}{b} \qquad a \times \frac{1}{b} = \frac{a}{b} \qquad a \div b = \frac{a}{b}$$

The fact that $a \div b = \frac{a}{b}$ enables us to write whole numbers in fractional form:

$$\text{Since } 28 \div 7 = 4, \ \frac{28}{7} = 4.$$

$$\text{Since } 100 \div 5 = 20, \ \frac{100}{5} = 20.$$

$$\text{Since } 8 \div 1 = 8, \ \frac{8}{1} = 8.$$

$$\text{Since } 6 \div 6 = 1, \ \frac{6}{6} = 1.$$

We use fractional forms of whole numbers when we change mixed numbers to improper fractions:

$$2\frac{5}{6} = \frac{2}{1} + \frac{5}{6} = \frac{2 \times 6}{1 \times 6} + \frac{5}{6} = \frac{(2 \times 6) + 5}{6} = \frac{17}{6}$$

$$7\frac{1}{4} = \frac{7}{1} + \frac{1}{4} = \frac{7 \times 4}{1 \times 4} + \frac{1}{4} = \frac{(7 \times 4) + 1}{4} = \frac{29}{4}$$

or improper fractions to mixed numbers:

$$\frac{18}{7} = \frac{(7 \times 2) + 4}{7} = \frac{7 \times 2}{7} + \frac{4}{7} = \frac{2}{1} + \frac{4}{7} = 2\frac{4}{7}$$

$$\frac{106}{9} = \frac{(9 \times 11) + 7}{9} = \frac{9 \times 11}{9} + \frac{7}{9} = \frac{11}{1} + \frac{7}{9} = 11\frac{7}{9}$$

Class Exercises

State the fractions whose graphs are shown in Exercises 1–8.

Example

Solution **A is the graph of $\frac{2}{7}$ and B is the graph of $\frac{6}{7}$.**

1. 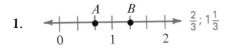 $\frac{2}{3}; 1\frac{1}{3}$

2.  $1; 1\frac{3}{4}$

3.  $1\frac{1}{2}; 3$

4.  $\frac{1}{2}; \frac{7}{8}$

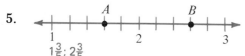

5.

$$1\frac{3}{5}; 2\frac{3}{5}$$

6.

$$2\frac{2}{3}; 4\frac{1}{3}$$

7.

$$11\frac{1}{2}; 13$$

8.

$$7\frac{1}{4}; 8\frac{3}{4}$$

Exercises

Complete.

Example $5 \times \underline{\quad?\quad} = 1$

Solution $5 \times \frac{1}{5} = 1$

A

1. $\underline{\quad?\quad} \times \frac{1}{6} = 1$ 6

2. $7 \times \frac{1}{7} = \underline{\quad?\quad}$ 1

3. $\frac{1}{4} \times \underline{\quad?\quad} = \frac{3}{4}$ 3

4. $\frac{1}{5} \times 5 = \underline{\quad?\quad}$ 1

5. $\underline{\quad?\quad} \times 3 = 1$ $\frac{1}{3}$

6. $\underline{\quad?\quad} \times 5 = 1$ $\frac{1}{5}$

7. $3 \div 7 = \underline{\quad?\quad}$ $\frac{3}{7}$

8. $6 \div 5 = \underline{\quad?\quad}$ $\frac{6}{5}$

Example $2\frac{2}{3} = \frac{?}{3}$

Solution $2\frac{2}{3} = \frac{(3 \times 2) + 2}{3} = \frac{6 + 2}{3} = \frac{8}{3}$

9. $1\frac{2}{5} = \frac{?}{5}$ 7

10. $3\frac{1}{2} = \frac{?}{2}$ 7

11. $4\frac{3}{4} = \frac{?}{4}$ 19

12. $2\frac{5}{7} = \frac{?}{7}$ 19

13. $1\frac{7}{10} = \frac{?}{10}$ 17

14. $3\frac{13}{20} = \frac{?}{20}$ 73

For each improper fraction, give an equal whole number or mixed number.

Example $\frac{21}{5}$

Solution $\frac{21}{5} = \frac{(4 \times 5) + 1}{5} = 4\frac{1}{5}$

15. $\frac{17}{3}$ $5\frac{2}{3}$

16. $\frac{15}{7}$ $2\frac{1}{7}$

17. $\frac{31}{10}$ $3\frac{1}{10}$

18. $\frac{22}{11}$ 2

19. $\frac{70}{5}$ 14

20. $\frac{41}{4}$ $10\frac{1}{4}$

4-2 Negative Fractions

Objective To use terms and properties concerning negative fractions.

The figure below shows the graphs of several negative fractions.

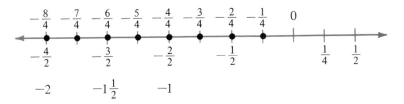

Do you see that $-\frac{2}{4}$ and $-\frac{1}{2}$ are equal fractions?

Patterns for negative fractions are the same as those for positive fractions:

$$\left(-\frac{1}{5}\right) + \left(-\frac{1}{5}\right) + \left(-\frac{1}{5}\right) + \left(-\frac{1}{5}\right) + \left(-\frac{1}{5}\right) = -1,$$

$$5 \times \left(-\frac{1}{5}\right) = -1.$$

In Section 3-9, we saw that the quotient of a positive decimal and a negative decimal is a negative decimal. Since $1 \div 5 = \frac{1}{5}$, it seems reasonable to assume that

$$1 \div (-5) = \frac{1}{-5} = -\frac{1}{5}, \quad \text{and} \quad (-1) \div 5 = \frac{-1}{5} = -\frac{1}{5}.$$

That is,

$$\frac{1}{-5} = \frac{-1}{5} = -\frac{1}{5}.$$

Example 1 | **Express $-\frac{1}{7}$ in two other ways.**

Solution | $-\frac{1}{7} = \frac{1}{-7} = \frac{-1}{7}$

Negative fractions may have numerators other than 1:

$$-\frac{5}{7} = \frac{-5}{7} = \frac{5}{-7} \quad \text{and} \quad -\frac{3}{10} = \frac{-3}{10} = \frac{3}{-10}$$

The fact that $\frac{a}{b} = a \times \frac{1}{b}$ enables us to write

$$-\frac{5}{8} = \frac{5}{-8} = 5 \times \left(\frac{1}{-8}\right) = 5 \times \left(-\frac{1}{8}\right).$$

Example 2 | **Express $-\frac{7}{3}$ in two other ways.**

Solution | $-\frac{7}{3} = -7 \times \frac{1}{3} = 7 \times \left(-\frac{1}{3}\right)$

A negative mixed number may be expressed as a negative improper fraction:

$$-3\frac{1}{3} = -\left[\frac{(3 \times 3) + 1}{3}\right] = -\frac{10}{3} \qquad\qquad -5\frac{3}{4} = -\left[\frac{(4 \times 5) + 3}{4}\right] = -\frac{23}{4}$$

Also, a negative improper fraction may be expressed as a negative mixed number:

$$-\frac{32}{5} = -\left(\frac{30 + 2}{5}\right) = -6\frac{2}{5} \qquad\qquad -\frac{24}{17} = -\left(\frac{17 + 7}{17}\right) = -1\frac{7}{17}$$

Since we now have numerators and denominators that are integers, rather than simply whole numbers, we are working with a new set of numbers called *rational numbers*. Any number that can be represented by a fraction $\frac{a}{b}$, where a and b are integers and b is not 0, is a **rational number.**

Class Exercises

State the fractions whose graphs are shown.

Example |
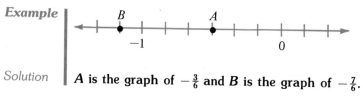

Solution | **A is the graph of $-\frac{3}{6}$ and B is the graph of $-\frac{7}{6}$.**

1.
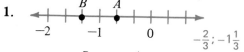
$-\frac{2}{3}; -1\frac{1}{3}$

2. $-\frac{1}{4}; -1\frac{3}{4}$

3. $-1\frac{1}{2}; -3$

4. $-\frac{1}{3}; -\frac{7}{9}$

5. $-1\frac{3}{4}; -2\frac{1}{2}$

6. $-2\frac{2}{3}; -4$

7.

$-9\frac{1}{2}; -11$

8. $-6\frac{3}{5}; -7\frac{2}{5}$

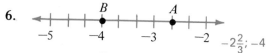

Exercises

Write each fraction in two other ways.

Example $\dfrac{1}{-6}$

Solution $\dfrac{1}{-6} = \dfrac{-1}{6} = -\dfrac{1}{6}$

A **1.** $-\dfrac{1}{8}$ $\dfrac{-1}{8}; \dfrac{1}{-8}$

2. $-\dfrac{1}{10}$ $\dfrac{-1}{10}; \dfrac{1}{-10}$

3. $\dfrac{-1}{12}$ $-\dfrac{1}{12}; \dfrac{1}{-12}$

4. $\dfrac{-1}{11}$ $-\dfrac{1}{11}; \dfrac{1}{-11}$

5. $\dfrac{1}{-3}$ $-\dfrac{1}{3}; \dfrac{-1}{3}$

6. $\dfrac{1}{-5}$ $-\dfrac{1}{5}; \dfrac{-1}{5}$

7. $-\dfrac{1}{25}$ $\dfrac{-1}{25}; \dfrac{1}{-25}$

8. $\dfrac{-1}{100}$ $-\dfrac{1}{100}; \dfrac{1}{-100}$

Write each fraction in two other ways.

Example $\dfrac{-5}{11}$

Solution $\dfrac{-5}{11} = \dfrac{5}{-11} = -\dfrac{5}{11}$

9. $\dfrac{3}{-17}$ $-\dfrac{3}{17}; \dfrac{-3}{17}$

10. $\dfrac{-5}{23}$ $-\dfrac{5}{23}; \dfrac{5}{-23}$

11. $-\dfrac{7}{17}$ $\dfrac{-7}{17}; \dfrac{7}{-17}$

12. $\dfrac{3}{-22}$ $-\dfrac{3}{22}; \dfrac{-3}{22}$

13. $\dfrac{-11}{3}$ $-\dfrac{11}{3}; \dfrac{11}{-3}$

14. $-\dfrac{15}{7}$ $\dfrac{-15}{7}; \dfrac{15}{-7}$

15. $\dfrac{-5}{2}$ $-\dfrac{5}{2}; \dfrac{5}{-2}$

16. $\dfrac{31}{-6}$ $-\dfrac{31}{6}; \dfrac{-31}{6}$

Write each mixed number as an improper fraction.

Example $-5\dfrac{1}{8}$

Solution $-5\dfrac{1}{8} = -\left(\dfrac{(8 \times 5) + 1}{8}\right) = -\dfrac{41}{8}$

17. $-3\dfrac{1}{3}$ $-\dfrac{10}{3}$

18. $-4\dfrac{2}{5}$ $-\dfrac{22}{5}$

19. $-5\dfrac{7}{8}$ $-\dfrac{47}{8}$

20. $-5\dfrac{5}{9}$ $-\dfrac{50}{9}$

21. $-8\dfrac{3}{7}$ $-\dfrac{59}{7}$

22. $-10\dfrac{2}{9}$ $-\dfrac{92}{9}$

23. $-11\dfrac{5}{8}$ $-\dfrac{93}{8}$

24. $-20\dfrac{3}{5}$ $-\dfrac{103}{5}$

Write each improper fraction as a mixed number.

Example $-\dfrac{27}{5}$

Solution $-\dfrac{27}{5} = -\left(\dfrac{25 + 2}{5}\right) = -5\dfrac{2}{5}$

25. $-\dfrac{8}{5}$ $-1\dfrac{3}{5}$

26. $-\dfrac{12}{7}$ $-1\dfrac{5}{7}$

27. $-\dfrac{18}{11}$ $-1\dfrac{7}{11}$

28. $-\dfrac{24}{5}$ $-4\dfrac{4}{5}$

29. $-\dfrac{32}{7}$ $-4\dfrac{4}{7}$

30. $-\dfrac{37}{6}$ $-6\dfrac{1}{6}$

31. $-\dfrac{53}{5}$ $-10\dfrac{3}{5}$

32. $-\dfrac{29}{10}$ $-2\dfrac{9}{10}$

4-3 Equal Fractions

Objective To use the equal-fractions rule.

In Sections 4-1 and 4-2 we saw that equal fractions have the same graph on the number line. We can calculate fractions equal to a given fraction using the following rule:

> ## rule
>
> ### Equal-Fractions Rule
>
> **For all integers a, b, and c ($b \neq 0$, $c \neq 0$),**
>
> $$\frac{a}{b} = \frac{a \times c}{b \times c} \quad \text{and} \quad \frac{a}{b} = \frac{a \div c}{b \div c}$$

Example 1 | Find three fractions equal to the given fraction.

a. $\frac{5}{7}$ b. $\frac{12}{36}$ c. $-\frac{1}{8}$

Solution | Many solutions are possible. For instance:

a. $\frac{5}{7} = \frac{5 \times 2}{7 \times 2} = \frac{10}{14}$; $\frac{5}{7} = \frac{5 \times 3}{7 \times 3} = \frac{15}{21}$; $\frac{5}{7} = \frac{5 \times 12}{7 \times 12} = \frac{60}{84}$

b. $\frac{12}{36} = \frac{12 \div 2}{36 \div 2} = \frac{6}{18}$; $\frac{12}{36} = \frac{12 \div 12}{36 \div 12} = \frac{1}{3}$; $\frac{12}{36} = \frac{12 \times 5}{36 \times 5} = \frac{60}{180}$

c. $-\frac{1}{8} = -\frac{1 \times 7}{8 \times 7} = -\frac{7}{56}$; $-\frac{1}{8} = -\frac{1 \times 100}{8 \times 100} = -\frac{100}{800}$;

$-\frac{1}{8} = -\frac{1 \times 125}{8 \times 125} = -\frac{125}{1000}$

If the numerator and denominator of a fraction have a common factor, then we can use the equal-fractions rule to *reduce* the fraction to an equal fraction:

$$\frac{40}{100} = \frac{40 \div 2}{100 \div 2} = \frac{20}{50}$$

$$\downarrow$$

$$\frac{20}{50} = \frac{20 \div 5}{50 \div 5} = \frac{4}{10}$$

$$\downarrow$$

$$\frac{4}{10} = \frac{4 \div 2}{10 \div 2} = \frac{2}{5} \quad \text{STOP!}$$

We can no longer reduce a fraction when the numerator and denominator have no common factor except 1. We say that the fraction is in *lowest terms*. In the preceding example, $\frac{40}{100}, \frac{20}{50}, \frac{4}{10}$, and $\frac{2}{5}$ are all equal, but only $\frac{2}{5}$ is in lowest terms.

Example 2 | Reduce $\frac{42}{28}$ to lowest terms.

Solution

$$\frac{42}{28} = \frac{42 \div 2}{28 \div 2} = \frac{21}{14}; \quad \frac{21}{14} = \frac{21 \div 7}{14 \div 7} = \frac{3}{2}$$

We often indicate the reduction process by using slashes:

$$\frac{\overset{\overset{3}{\cancel{21}}}{\cancel{42}}}{\underset{\underset{2}{\cancel{14}}}{\cancel{28}}} = \frac{3}{2}$$

The black slashes indicate "divide by 2."
The red slashes indicate "divide by 7."

We can reduce a fraction to lowest terms in a single step by dividing both members by their *greatest* common factor (GCF).

Example 3 | Reduce the fraction to lowest terms.

a. $\frac{42}{28}$ b. $-\frac{90}{100}$

Solution

a. The greatest common factor of 42 and 28 is 14.

$$\frac{42}{28} = \frac{42 \div 14}{28 \div 14} = \frac{3}{2}, \quad \text{or} \quad \frac{42}{28} = \frac{\overset{3}{\cancel{42}}}{\underset{2}{\cancel{28}}} = \frac{3}{2}$$

b. The GCF of 90 and 100 is 10.

$$-\frac{90}{100} = -\frac{90 \div 10}{100 \div 10} = -\frac{9}{10},$$

$$\text{or} \quad -\frac{90}{100} = -\frac{\overset{9}{\cancel{90}}}{\underset{10}{\cancel{100}}} = -\frac{9}{10}$$

In some cases, the GCF is easy to find and the fraction may be reduced to lowest terms in one step. For other fractions, it is easier to reduce in several steps.

Example 4 | Reduce $\frac{84}{140}$ to lowest terms.

Solution | $\dfrac{84 \div 2}{140 \div 2} = \dfrac{42 \div 2}{70 \div 2} = \dfrac{21 \div 7}{35 \div 7} = \dfrac{3}{5}$

Class Exercises

State two fractions that are equal to the given fraction. Answers to Ex. 1-10 may vary.

1. $\dfrac{1}{6}$ $\dfrac{2}{12}; \dfrac{5}{30}$
2. $\dfrac{2}{3}$ $\dfrac{6}{9}; \dfrac{8}{12}$
3. $-\dfrac{3}{4}$ $-\dfrac{6}{8}; -\dfrac{9}{12}$
4. $-\dfrac{5}{6}$ $-\dfrac{15}{18}; -\dfrac{20}{24}$
5. $\dfrac{2}{7}$ $\dfrac{4}{14}; \dfrac{12}{42}$

6. $\dfrac{3}{2}$ $\dfrac{12}{8}; \dfrac{9}{6}$
7. $-\dfrac{5}{3}$ $-\dfrac{25}{15}; -\dfrac{30}{18}$
8. $-\dfrac{8}{5}$ $-\dfrac{16}{10}; -\dfrac{32}{20}$
9. $\dfrac{10}{3}$ $\dfrac{30}{9}; \dfrac{50}{15}$
10. $\dfrac{20}{7}$ $\dfrac{80}{28}; \dfrac{60}{21}$

Reduce to lowest terms.

11. $\dfrac{3}{6}$ $\dfrac{1}{2}$
12. $\dfrac{5}{15}$ $\dfrac{1}{3}$
13. $-\dfrac{2}{10}$ $-\dfrac{1}{5}$

14. $-\dfrac{12}{3}$ -4
15. $\dfrac{15}{20}$ $\dfrac{3}{4}$
16. $\dfrac{6}{9}$ $\dfrac{2}{3}$

Exercises

Give two fractions equal to the given fraction. Answers to Ex. 1-10 may vary.

A

1. $\dfrac{13}{14}$ $\dfrac{26}{28}; \dfrac{52}{56}$
2. $\dfrac{7}{25}$ $\dfrac{21}{75}; \dfrac{70}{250}$
3. $-\dfrac{11}{7}$ $-\dfrac{55}{35}; -\dfrac{33}{21}$
4. $-\dfrac{21}{13}$ $-\dfrac{84}{52}; -\dfrac{105}{65}$
5. $\dfrac{12}{25}$ $\dfrac{36}{75}; \dfrac{60}{125}$

6. $-\dfrac{7}{30}$ $-\dfrac{21}{90}; -\dfrac{35}{150}$
7. $-\dfrac{21}{29}$ $-\dfrac{42}{58}; -\dfrac{84}{116}$
8. $\dfrac{13}{31}$ $\dfrac{39}{93}; \dfrac{52}{124}$
9. $\dfrac{37}{10}$ $\dfrac{74}{20}; \dfrac{370}{100}$
10. $-\dfrac{29}{41}$ $-\dfrac{116}{164}; -\dfrac{58}{82}$

Reduce each fraction to lowest terms.

11. $\dfrac{12}{36}$ $\dfrac{1}{3}$
12. $\dfrac{10}{35}$ $\dfrac{2}{7}$
13. $-\dfrac{14}{28}$ $-\dfrac{1}{2}$
14. $-\dfrac{26}{39}$ $-\dfrac{2}{3}$
15. $\dfrac{40}{80}$ $\dfrac{1}{2}$

16. $\dfrac{70}{100}$ $\dfrac{7}{10}$
17. $\dfrac{22}{55}$ $\dfrac{2}{5}$
18. $\dfrac{30}{45}$ $\dfrac{2}{3}$
19. $-\dfrac{21}{14}$ $-\dfrac{3}{2}$
20. $-\dfrac{36}{9}$ -4

21. $-\dfrac{48}{36}$ $-\dfrac{4}{3}$
22. $-\dfrac{64}{36}$ $-\dfrac{16}{9}$
23. $\dfrac{63}{84}$ $\dfrac{3}{4}$
24. $\dfrac{24}{148}$ $\dfrac{6}{37}$
25. $\dfrac{70}{105}$ $\dfrac{2}{3}$

B
26. $\dfrac{126}{210}$ $\dfrac{3}{5}$
27. $-\dfrac{48}{72}$ $-\dfrac{2}{3}$
28. $-\dfrac{54}{63}$ $-\dfrac{6}{7}$
29. $\dfrac{150}{210}$ $\dfrac{5}{7}$
30. $\dfrac{126}{252}$ $\dfrac{1}{2}$

4-4 Products of Fractions

Objective To use properties and rules of fractions to find products of fractions.

We can use the properties of numbers listed in Chapter 2 along with those introduced in this chapter to find products of fractions.

By recalling the rules of sign for products from Section 3-8, we can first concentrate on products of positive fractions, and then make adjustments for negative fractions as needed.

Example 1 | Compute.

a. $\frac{5}{3} \times 3$ **b.** $-\frac{3}{7} \times 7$

Solution | **a.** $\frac{5}{3} \times 3 = \left(5 \times \frac{1}{3}\right) \times 3 = 5\left(3 \times \frac{1}{3}\right) = 5(1) = 5$

Property of $\frac{a}{b}$　　Commutative and Associative Properties　　Property of $\frac{1}{b}$

b. $-\frac{3}{7} \times 7$　　**Working with positive numbers only:**

$\frac{3}{7} \times 7 = \left(3 \times \frac{1}{7}\right) \times 7 = 3\left(7 \times \frac{1}{7}\right) = 3(1) = 3$

Because one of the two factors is negative, the answer is -3.

Example 2 | Compute $(5 \times 4) \times \left(\frac{1}{5} \times \frac{1}{4}\right)$.

Solution | $(5 \times 4) \times \left(\frac{1}{5} \times \frac{1}{4}\right) = \left(5 \times \frac{1}{5}\right) \times \left(4 \times \frac{1}{4}\right) = 1 \times 1 = 1$

Commutative and Associative Properties　　Property of $\frac{1}{b}$

Examples 1 and 2 suggest additional rules which simplify multiplication of fractions. From Example 1 we have the following rule.

> ## rule
>
> For all integers a and b $(b \neq 0)$,
>
> $$\frac{a}{b} \times b = a$$

A closer look at Example 2 reveals a very important property of unit fractions:

$$(5 \times 4) \times \left(\frac{1}{5} \times \frac{1}{4}\right) = 20 \times \left(\frac{1}{5} \times \frac{1}{4}\right) = 1 \qquad \text{(Result of Example 2)}$$

Also,

$$20 \times \frac{1}{20} = 1 \qquad \left(\text{Property of } \frac{1}{b}\right)$$

Thus, we see that

$$\frac{1}{5} \times \frac{1}{4} = \frac{1}{20}, \quad \text{or} \quad \frac{1}{5} \times \frac{1}{4} = \frac{1}{5 \times 4}.$$

> ## rule
>
> **Unit-Fraction Rule**
>
> For all integers a and b $(a \neq 0, b \neq 0)$,
>
> $$\frac{1}{a} \times \frac{1}{b} = \frac{1}{ab}$$

The unit-fraction rule and the property of $\frac{a}{b}$ can be used to derive the rule for multiplying any two fractions.

> ## rule
>
> For all integers a, b, c, and d $(b \neq 0, d \neq 0)$,
>
> $$\frac{a}{b} \times \frac{c}{d} = \frac{ac}{bd}$$

Example 3 | Compute each product.

a. $\frac{1}{2} \times \frac{5}{6}$ **b.** $\frac{2}{5} \times \frac{3}{5}$ **c.** $-\frac{9}{2} \times \frac{3}{4}$ **d.** $\left(-\frac{3}{5}\right) \times \left(-\frac{3}{8}\right)$

Solution

a. $\frac{1}{2} \times \frac{5}{6} = \frac{1 \times 5}{2 \times 6} = \frac{5}{12}$ **b.** $\frac{2}{5} \times \frac{3}{5} = \frac{2 \times 3}{5 \times 5} = \frac{6}{25}$

c. $-\frac{9}{2} \times \frac{3}{4}$ Note that the product is negative.

$\frac{9}{2} \times \frac{3}{4} = \frac{9 \times 3}{2 \times 4} = \frac{27}{8}$ Answer: $-\frac{27}{8}$

d. $\left(-\frac{3}{5}\right) \times \left(-\frac{3}{8}\right)$ Note that the product is positive.

$\frac{3}{5} \times \frac{3}{8} = \frac{3 \times 3}{5 \times 8} = \frac{9}{40}$ Answer: $\frac{9}{40}$

When mixed numbers are involved in products, we first change each mixed number to an improper fraction and then multiply.

Example 4 | Compute each product.

a. $4\frac{1}{8} \times 1\frac{1}{2}$ **b.** $-2\frac{2}{7} \times 3\frac{1}{3}$

Solution

a. $4\frac{1}{8} \times 1\frac{1}{2} = \frac{33}{8} \times \frac{3}{2} = \frac{99}{16} = 6\frac{3}{16}$

b. $-2\frac{2}{7} \times 3\frac{1}{3}$ Note that the product is negative.

$2\frac{2}{7} \times 3\frac{1}{3} = \frac{16}{7} \times \frac{10}{3} = \frac{160}{21} = 7\frac{13}{21}$ Answer: $-7\frac{13}{21}$

Whenever a product results in a fraction that is not in lowest terms, let us agree to reduce the fraction to lowest terms.

Example 5 | $\frac{3}{4} \times \frac{6}{15}$

Solution

$\frac{3}{4} \times \frac{6}{15} = \frac{3 \times 6}{4 \times 15} = \frac{\overset{3}{\cancel{18}}}{\underset{10}{\cancel{60}}} = \frac{3}{10}$

Note that in Example 5 we could have divided by common factors *before* multiplying. Thus,

$$\frac{3}{4} \times \frac{6}{15} = \frac{\overset{1}{\cancel{3}}}{\underset{2}{\cancel{4}}} \times \frac{\overset{3}{\cancel{6}}}{\underset{5}{\cancel{15}}} = \frac{1 \times 3}{2 \times 5} = \frac{3}{10}.$$

Class Exercises

Complete.

Example | $\underline{\quad?\quad} \times (-4) = 5$

Solution | $-\dfrac{5}{4} \times (-4) = 5$

1. $\dfrac{3}{5} \times 5 = \underline{\quad?\quad}$ 3

2. $\dfrac{2}{3} \times \underline{\quad?\quad} = 2$ 3

3. $\underline{\quad?\quad} \times (-6) = 7$ $-\dfrac{7}{6}$

4. $-\dfrac{11}{13} \times \underline{\quad?\quad} = -11$ 13

5. $\underline{\quad?\quad} \times \dfrac{3}{5} = -3$ -5

6. $\dfrac{12}{17} \times 17 = \underline{\quad?\quad}$ 12

7. $\dfrac{1}{2} \times \dfrac{3}{11} = \underline{\quad?\quad}$ $\dfrac{3}{22}$

8. $\dfrac{2}{3} \times \dfrac{4}{7} = \underline{\quad?\quad}$ $\dfrac{8}{21}$

9. $\dfrac{1}{2} \times \dfrac{7}{8} = \underline{\quad?\quad}$ $\dfrac{7}{16}$

10. $\dfrac{1}{3} \times \dfrac{5}{6} = \underline{\quad?\quad}$ $\dfrac{5}{18}$

11. $\left(-\dfrac{2}{5}\right) \times \dfrac{1}{7} = \underline{\quad?\quad}$ $-\dfrac{2}{35}$

12. $\left(-\dfrac{3}{8}\right) \times \left(-\dfrac{5}{7}\right) = \underline{\quad?\quad}$ $\dfrac{15}{56}$

Exercises

Compute each product. Give your result in lowest terms.

A

1. $\dfrac{4}{5} \times \dfrac{2}{7}$ $\dfrac{8}{35}$

2. $\dfrac{3}{7} \times \dfrac{6}{11}$ $\dfrac{18}{77}$

3. $-\dfrac{1}{4} \times \dfrac{2}{3}$ $-\dfrac{1}{6}$

4. $-\dfrac{3}{2} \times \dfrac{5}{6}$ $-1\dfrac{1}{4}$

5. $\left(-\dfrac{7}{8}\right) \times \left(-\dfrac{4}{9}\right)$ $\dfrac{7}{18}$

6. $\left(-\dfrac{8}{15}\right) \times \left(-\dfrac{5}{16}\right)$ $\dfrac{1}{6}$

7. $\dfrac{7}{9} \times 0$ 0

8. $0 \times \left(-\dfrac{5}{4}\right)$ 0

9. $-\dfrac{8}{6} \times \dfrac{12}{4}$ -4

10. $\dfrac{6}{10} \times \left(-\dfrac{5}{3}\right)$ -1

11. $\dfrac{5}{16} \times \left(-\dfrac{4}{15}\right)$ $-\dfrac{1}{12}$

12. $\dfrac{31}{32} \times \dfrac{16}{7}$ $2\dfrac{3}{14}$

13. $8 \times -\dfrac{3}{4}$ -6

14. $-\dfrac{7}{8} \times (-16)$ 14

15. $2\dfrac{1}{3} \times -6$ -14

16. $4\dfrac{1}{5} \times 10$ 42

17. $-3\dfrac{1}{8} \times \left(-4\dfrac{1}{5}\right)$ $13\dfrac{1}{8}$

18. $4\dfrac{2}{7} \times 5\dfrac{1}{6}$ $22\dfrac{1}{7}$

B

19. $-\dfrac{3}{16} \times \left[\dfrac{2}{15} \times \left(-\dfrac{4}{3}\right)\right]$ $\dfrac{1}{30}$

20. $-\dfrac{11}{32} \times \left(-\dfrac{8}{7} \times \dfrac{2}{33}\right)$ $\dfrac{1}{42}$

21. $3\dfrac{1}{8} \times \left(-\dfrac{11}{4} \times 2\dfrac{1}{10}\right)$ $18\dfrac{3}{64}$

22. $4\dfrac{3}{4} \times \left[-\dfrac{8}{15} \times \left(-3\dfrac{1}{4}\right)\right]$ $8\dfrac{7}{30}$

Problems

Solve.

B 1. Of 24 people who took a driving test, $\frac{7}{8}$ passed. Of those who passed, $\frac{2}{7}$ had failed the test previously. How many of those who passed had failed the test before? 6

2. Laura's usual work week is $37\frac{1}{2}$ hours long. Last week she worked $\frac{3}{5}$ as long as usual. How many hours did she work last week? $22\frac{1}{2}$ hours

3. A restaurant uses $13\frac{1}{3}$ dozen eggs a day. How many dozen eggs will it use in $4\frac{1}{2}$ days? 60 dozen

4. Only $\frac{1}{5}$ of the downtown workers drive to work. Of those who do not drive, $\frac{3}{16}$ ride bicycles to work. What fraction of the workers ride bicycles to work? $\frac{3}{20}$

EXTRA! Using Halves

An old method of multiplying two numbers makes use of doubling and halving. For example, to multiply 162 × 21, form two columns as shown below. Then double the numbers in the left column, and halve the numbers in the right column, dropping any remainders. Continue until you reach the number 1 in the right column. The product of 162 and 21 is the sum of the numbers in the left column that are opposite odd numbers in the right column.

This is often called the *Russian peasant* method of multiplication.

Double	Halve
162	21 (odd)
324	10
648	5 (odd)
1296	2
2592	1 (odd)

$$
\begin{array}{r}
162 \\
648 \\
\underline{2592} \\
\end{array}
$$

162 × 21 = 3402

Self-Test

Words to remember:
numerator [p. 105]
denominator [p. 105]
proper fraction [p. 105]
improper fraction [p. 105]
mixed number [p. 105]

equal fractions [p. 105]
unit fractions [p. 106]
rational number [p. 110]
reduce a fraction [p. 112]
fraction in lowest terms [p. 113]

Complete.

1. $9 \times \underline{\quad?\quad} = 1 \quad \frac{1}{9}$

2. $5\frac{5}{6} = \frac{?}{6} \quad 35$ [4-1]

3. $4 \div 7 = \underline{\quad?\quad} \quad \frac{4}{7}$

4. $\frac{39}{5} = \underline{\quad?\quad} \quad 7\frac{4}{5}$

Write each negative fraction in two other ways.

5. $-\frac{2}{7} \quad \frac{-2}{7}; \frac{2}{-7}$

6. $\frac{-1}{4} \quad -\frac{1}{4}; \frac{1}{-4}$ [4-2]

Complete.

7. $-4\frac{5}{8} = \frac{?}{8} \quad -37$

8. $-\frac{32}{5} = \underline{\quad?\quad} \quad -6\frac{2}{5}$

9. Give two fractions equal to $-\frac{11}{15}$. $-\frac{22}{30}; -\frac{44}{60}$ [4-3]
   Answers to Ex. 9 may vary.

Reduce each fraction to lowest terms.

10. $\frac{50}{34} \quad \frac{25}{17}$

11. $-\frac{42}{189} \quad -\frac{2}{9}$

12. $\frac{75}{135} \quad \frac{5}{9}$

Compute each product. Give your result in lowest terms.

13. $-\frac{3}{4} \times \frac{2}{5} \quad -\frac{3}{10}$

14. $\frac{21}{25} \times \frac{20}{56} \quad \frac{3}{10}$ [4-4]

15. $-\left(\frac{7}{18}\right) \times \left(-\frac{36}{49}\right) \quad \frac{2}{7}$

16. $4\frac{1}{16} \times 6\frac{2}{5} \quad 26$

Self-Test answers and Extra Practice are at the back of the book.

120

4-5 Least Common Denominator

Objective To calculate the least common multiple (LCM) of two or more whole numbers and the least common denominator (LCD) of two or more fractions.

You may recall from previous work that a **multiple** of a whole number is any product of that number and a whole number. For example,

0, 2, 4, 6, 8, and so on are multiples of 2
0, 7, 14, 21, 28, and so on are multiples of 7

In calculations involving more than one fraction, it is sometimes necessary to replace fractions with equal ones so that all have the same denominator. We say that the fractions have a *common denominator*. A **common denominator** is a *common multiple* of the denominators of the fractions.

Example 1 | Replace the fractions with equal fractions having a common denominator.

$$\frac{1}{2}, \; -\frac{3}{4}, \frac{1}{6}$$

Solution

$$\frac{1}{2} = \frac{1 \times 4 \times 6}{2 \times 4 \times 6} = \frac{24}{48}$$

$$-\frac{3}{4} = -\frac{3 \times 2 \times 6}{4 \times 2 \times 6} = -\frac{36}{48}$$

$$\frac{1}{6} = \frac{1 \times 2 \times 4}{6 \times 2 \times 4} = \frac{8}{48}$$

common denominator
$$= 2 \times 4 \times 6$$

Example 1 above suggests that a common denominator can always be found by taking the *product* of all the denominators. It is often convenient, however, to find the *least common denominator (LCD)*. This number is the *least common multiple (LCM) of the denominators of the fractions.*

Example 2 | Find the least common multiple (LCM) of 10, 8, and 5.

Solution | **Method 1**

multiples of 10: 10, 20, 30, ⑩ , 50, 60, 70, 80, . . .
multiples of 8: 8, 16, 24, 32, ⑩ , 48, 56, 64, . . .
multiples of 5: 5, 10, 15, 20, 25, 30, 35, ⑩ , . . .
The LCM is 40.

Method 2

This method makes use of *prime factorizations*. A **prime** is a whole number greater than 0 having only two factors, itself and 1. For example, 2, 3, 5, 7, and 11 are primes.

prime factorization of 10: $\qquad$ 2×5

prime factorization of 8: $\quad$ 2 × 2 × 2

prime factorization of 5: $\qquad\qquad\qquad$ 5

The LCM must have $\quad$ three factors $\quad$ one factor

$\qquad\qquad\qquad\qquad$ of 2 $\qquad\qquad$ of 5

Hence, the LCM is (2 × 2 × 2) × 5 = 40.

Example 3 | Find the least common denominator of the fractions. Then calculate equal fractions having this LCD.

a. $\frac{5}{6}$, $-\frac{9}{8}$ $\qquad$ **b.** $\frac{1}{2}$, $\frac{3}{4}$, $\frac{1}{6}$

Solution

a. The LCD is the least common multiple of 6 and 8.

6, 12, 18, $\boxed{24}$
8, 16, $\boxed{24}$
The LCD is 24.

$$\frac{5}{6} = \frac{5 \times 4}{6 \times 4} = \frac{20}{24} \qquad -\frac{9}{8} = -\frac{9 \times 3}{8 \times 3} = -\frac{27}{24}$$

b. The LCD is the least common multiple of 2, 4, and 6.

2, 4, 6, 8, 10, $\boxed{12}$
4, 8, $\boxed{12}$
6, $\boxed{12}$
The LCD is 12.

$$\frac{1}{2} = \frac{1 \times 6}{2 \times 6} = \frac{6}{12} \qquad \frac{3}{4} = \frac{3 \times 3}{4 \times 3} = \frac{9}{12} \qquad \frac{1}{6} = \frac{1 \times 2}{6 \times 2} = \frac{2}{12}$$

Class Exercises

State the LCD of the given fractions.

1. $\frac{1}{2}$, $\frac{3}{4}$ $\quad$ 4

2. $-\frac{1}{2}$, $\frac{5}{8}$ $\quad$ 8

3. $\frac{2}{3}$, $\frac{5}{12}$ $\quad$ 12

4. $\frac{1}{8}$, $\frac{3}{24}$ $\quad$ 24

5. $-\frac{2}{3}$, $\frac{1}{2}$ $\quad$ 6

6. $\frac{1}{3}$, $\frac{1}{4}$ $\quad$ 12

7. $-\frac{1}{2}$, $-\frac{5}{7}$ $\quad$ 14

8. $-\frac{3}{8}$, $\frac{5}{3}$ $\quad$ 24

9. Examine the factorizations of 20 and 42 below.

$$20 = 2 \times 2 \times 5 \qquad 42 = 2 \times 3 \times 7$$

The LCM of 20 and 42 must have:

a. __?__ factor(s) of 2. 2 **b.** __?__ factor(s) of 3. 1
c. __?__ factor(s) of 5. 1 **d.** __?__ factor(s) of 7. 1

10. The LCM of 20 and 42 is __?__. 420

11. a. Find the LCM of 20 and 42 by listing sets of multiples. See students' papers.
    b. Does your answer agree with that for Exercise 10? yes

12. Find the LCM of 30 and 36. 180   **13.** Find the LCM of 35 and 8. 280

Exercises

Replace each set of fractions with equal fractions having the least common denominator (LCD).

A **1.** $\dfrac{2}{3}, \dfrac{5}{9}$ $\dfrac{6}{9}, \dfrac{5}{9}$ **2.** $\dfrac{3}{2}, \dfrac{7}{6}$ $\dfrac{9}{6}, \dfrac{7}{6}$ **3.** $-\dfrac{3}{5}, \dfrac{7}{20}$ $-\dfrac{12}{20}, \dfrac{7}{20}$ **4.** $-\dfrac{4}{7}, -\dfrac{18}{49}$

5. $\dfrac{5}{3}, \dfrac{7}{5}$ $\dfrac{25}{15}, \dfrac{21}{15}$ **6.** $\dfrac{3}{5}, \dfrac{3}{7}$ $\dfrac{21}{35}, \dfrac{15}{35}$ **7.** $-\dfrac{8}{3}, \dfrac{9}{11}$ $-\dfrac{88}{33}, \dfrac{27}{33}$ **8.** $-\dfrac{5}{7}, -\dfrac{12}{17}$

9. $\dfrac{7}{3}, \dfrac{3}{4}$ $\dfrac{28}{12}, \dfrac{9}{12}$ **10.** $\dfrac{5}{4}, \dfrac{8}{5}$ $\dfrac{25}{20}, \dfrac{32}{20}$ **11.** $-\dfrac{3}{34}, -\dfrac{5}{39}$ **12.** $-\dfrac{5}{24}, \dfrac{8}{21}$

13. $\dfrac{5}{54}, \dfrac{17}{42}$ $\dfrac{35}{378}, \dfrac{153}{378}$ **14.** $\dfrac{8}{45}, \dfrac{11}{75}$ $\dfrac{40}{225}, \dfrac{33}{225}$ **15.** $\dfrac{17}{32}, -\dfrac{3}{128}$ $\dfrac{68}{128}, -\dfrac{3}{128}$ **16.** $\dfrac{7}{87}, \dfrac{4}{375}$

Example $\dfrac{3}{4}, \dfrac{1}{6}, \dfrac{3}{30}$ **4.** $-\dfrac{28}{49}; -\dfrac{18}{49}$ **8.** $-\dfrac{85}{119}; -\dfrac{84}{119}$ **11.** $-\dfrac{117}{1326}; -\dfrac{170}{1326}$

Solution **Find the LCM of 4, 6, and 30 by factoring:**   **12.** $-\dfrac{35}{117}; \dfrac{64}{168}$

$$4 = 2 \times 2$$
$$6 = 2 \times \bigm| 3$$
$$30 = 2 \times \bigm| 3 \times 5$$

two factors of 2   one factor of 3   one factor of 5

16. $\dfrac{875}{10,875}; \dfrac{116}{10,875}$

The LCM is $2 \times 2 \times 3 \times 5 = 60$

$$\dfrac{3}{4} = \dfrac{3 \times 15}{4 \times 15} = \dfrac{45}{60}, \quad \dfrac{1}{6} = \dfrac{1 \times 10}{6 \times 10} = \dfrac{10}{60}, \quad \dfrac{3}{30} = \dfrac{3 \times 2}{30 \times 2} = \dfrac{6}{60}$$

17. $\dfrac{1}{2}, \dfrac{3}{4}, \dfrac{7}{8}$ **18.** $-\dfrac{2}{3}, \dfrac{3}{4}, -\dfrac{1}{20}$ **19.** $\dfrac{3}{16}, \dfrac{5}{18}, -\dfrac{1}{24}$ **20.** $\dfrac{5}{12}, \dfrac{3}{16}, \dfrac{7}{24}$

$\dfrac{4}{8}, \dfrac{6}{8}, \dfrac{7}{8}$ $-\dfrac{40}{60}, \dfrac{45}{60}, -\dfrac{3}{60}$ $\dfrac{27}{144}, \dfrac{40}{144}, -\dfrac{6}{144}$ $\dfrac{20}{48}, \dfrac{9}{48}, \dfrac{14}{48}$

B **21.** $\dfrac{2}{15}, -\dfrac{5}{12}, -\dfrac{4}{35}$ **22.** $\dfrac{7}{18}, \dfrac{5}{28}, -\dfrac{3}{56}$ **23.** $\dfrac{7}{30}, \dfrac{3}{35}, \dfrac{29}{60}$ **24.** $\dfrac{11}{24}, -\dfrac{19}{54}, \dfrac{13}{72}$

$\dfrac{56}{420}, -\dfrac{175}{420}, -\dfrac{48}{420}$ $\dfrac{196}{504}, \dfrac{90}{504}, -\dfrac{27}{504}$ $\dfrac{98}{420}, \dfrac{36}{420}, \dfrac{203}{420}$ $\dfrac{99}{216}; -\dfrac{76}{216}, \dfrac{39}{216}$

4-6 Sums and Differences of Fractions

Objective To apply the rules for adding and subtracting fractions and mixed numbers.

If two or more fractions have the same denominator, sums and differences may be calculated easily. The following rules can be justified using a property of $\frac{a}{b}$ and the distributive property.

---

rules

For all integers a, b, and c ($c \neq 0$),

1. $\dfrac{a}{c} + \dfrac{b}{c} = \dfrac{a+b}{c}$

2. $\dfrac{a}{c} - \dfrac{b}{c} = \dfrac{a-b}{c}$

Example 1 | Calculate and reduce to lowest terms.

a. $\dfrac{3}{8} + \dfrac{1}{8}$ **b.** $\dfrac{3}{8} - \dfrac{1}{8}$ **c.** $\dfrac{5}{9} - \dfrac{1}{9} + \dfrac{2}{9}$ **d.** $-\dfrac{2}{5} + \dfrac{3}{5}$

Solution

a. $\dfrac{3}{8} + \dfrac{1}{8} = \dfrac{3+1}{8} = \dfrac{\overset{1}{\cancel{4}}}{\underset{2}{\cancel{8}}} = \dfrac{1}{2}$

b. $\dfrac{3}{8} - \dfrac{1}{8} = \dfrac{3-1}{8} = \dfrac{\overset{1}{\cancel{2}}}{\underset{4}{\cancel{8}}} = \dfrac{1}{4}$

c. $\dfrac{5}{9} - \dfrac{1}{9} + \dfrac{2}{9} = \dfrac{5-1+2}{9} = \dfrac{\overset{2}{\cancel{6}}}{\underset{3}{\cancel{9}}} = \dfrac{2}{3}$

d. $-\dfrac{2}{5} + \dfrac{3}{5} = \dfrac{-2}{5} + \dfrac{3}{5} = \dfrac{-2+3}{5} = \dfrac{1}{5}$

These rules can also be used to add mixed numbers whose fractions have the same denominator.

Example 2 | Calculate and reduce to lowest terms.

a. $5\frac{3}{5} + 2\frac{4}{5}$ **b.** $5\frac{3}{5} - 2\frac{4}{5}$ **c.** $2\frac{1}{3} - \left(-1\frac{2}{3}\right)$

Solution | **Method 1: Convert to improper fractions.**

a. $5\frac{3}{5} + 2\frac{4}{5} = \frac{28}{5} + \frac{14}{5} = \frac{42}{5}$, or $8\frac{2}{5}$

b. $5\frac{3}{5} - 2\frac{4}{5} = \frac{28}{5} - \frac{14}{5} = \frac{14}{5}$, or $2\frac{4}{5}$

c. $2\frac{1}{3} - \left(-1\frac{2}{3}\right) = \frac{7}{3} - \left(-\frac{5}{3}\right) = \frac{7}{3} + \frac{5}{3} = \frac{\overset{4}{\cancel{12}}}{\underset{1}{\cancel{3}}} = 4$

Method 2: Add or subtract in columns.

a. $\quad 5\frac{3}{5}$ **b.** $\quad 5\frac{3}{5} = \quad 4\frac{8}{5}$ **c.** $\quad 2\frac{1}{3} = \quad 2\frac{1}{3}$

$\quad +2\frac{4}{5}$ $\quad -2\frac{4}{5} = -2\frac{4}{5}$ $\quad -\left(-1\frac{2}{3}\right) = +1\frac{2}{3}$

$\quad 7\frac{7}{5} = 8\frac{2}{5}$ $\qquad\qquad\quad 2\frac{4}{5}$ $\qquad\qquad\quad 3\frac{3}{3} = 4$

If two or more fractions have different denominators, we must first find equal fractions with a common denominator. Then we can add or subtract according to the rules on page 124.

Example 3 | Calculate and reduce to lowest terms.

a. $\frac{3}{4} + \frac{5}{6}$ **b.** $\frac{7}{8} - \frac{2}{5} + \frac{1}{5}$

Solution | **a.** The LCD of $\frac{3}{4}$ and $\frac{5}{6}$ is 12.

$\frac{3}{4} = \frac{3 \times 3}{4 \times 3} = \frac{9}{12}$ $\qquad \frac{3}{4} + \frac{5}{6}$

$\qquad\qquad\qquad\qquad\qquad \downarrow \quad\searrow$

$\frac{5}{6} = \frac{5 \times 2}{6 \times 2} = \frac{10}{12}$ $\qquad \frac{9}{12} + \frac{10}{12} = \frac{19}{12}$, or $1\frac{7}{12}$

b. The LCD of $\frac{7}{8}, \frac{2}{5}$, and $\frac{1}{5}$ is 40.

$\frac{7}{8} = \frac{7 \times 5}{8 \times 5} = \frac{35}{40}$ $\qquad \frac{7}{8} - \frac{2}{5} + \frac{1}{5}$

$\frac{2}{5} = \frac{2 \times 8}{5 \times 8} = \frac{16}{40}$ $\qquad \downarrow \quad\searrow \quad\searrow$

$\frac{1}{5} = \frac{1 \times 8}{5 \times 8} = \frac{8}{40}$ $\qquad \frac{35}{40} - \frac{16}{40} + \frac{8}{40} = \frac{27}{40}$

Class Exercises

Express the sum or difference in lowest terms.

1. $\frac{3}{7} + \frac{1}{7}$ $\frac{4}{7}$

2. $\frac{5}{13} + \frac{7}{13}$ $\frac{12}{13}$

3. $\frac{6}{11} - \frac{1}{11}$ $\frac{5}{11}$

4. $\frac{16}{17} - \frac{5}{17}$ $\frac{11}{17}$

5. $\frac{3}{4} + \frac{1}{4}$ 1

6. $-\frac{3}{6} + \frac{1}{6}$ $-\frac{1}{3}$

7. $-\frac{3}{4} - \frac{3}{4}$ $-1\frac{1}{2}$

8. $\frac{4}{3} - \frac{2}{3}$ $\frac{2}{3}$

9. $\frac{5}{2} + \frac{3}{2} - \left(-\frac{1}{2}\right)$ $4\frac{1}{2}$

10. $\frac{7}{3} - \frac{1}{3} + \left(-\frac{2}{3}\right)$ $1\frac{1}{3}$

11. $2\frac{5}{16} - 1\frac{1}{16}$ $1\frac{1}{4}$

12. $3\frac{7}{8} - 2\frac{3}{8}$ $1\frac{1}{2}$

Exercises

Express the sum or difference as a proper fraction in lowest terms or as a mixed number with fraction in lowest terms.

A

1. $\frac{3}{8} + \frac{5}{6}$ $1\frac{5}{24}$

2. $\frac{5}{9} + \frac{1}{6}$ $\frac{13}{18}$

3. $\frac{7}{8} + \frac{1}{12}$ $\frac{23}{24}$

4. $\frac{3}{4} + \frac{4}{5}$ $1\frac{11}{20}$

5. $\frac{5}{12} - \frac{3}{16}$ $\frac{11}{48}$

6. $\frac{11}{12} - \frac{11}{18}$ $\frac{11}{36}$

7. $\frac{7}{48} - \frac{3}{40}$ $\frac{17}{240}$

8. $\frac{16}{35} - \frac{3}{14}$ $\frac{17}{70}$

9. $\frac{11}{48} + \left(-\frac{18}{64}\right)$ $-\frac{5}{96}$

10. $\frac{17}{27} + \left(-\frac{5}{24}\right)$ $\frac{91}{216}$

11. $-\frac{15}{32} - \frac{3}{40}$ $-\frac{87}{160}$

12. $-\frac{12}{15} - \frac{5}{21}$ $-1\frac{4}{105}$

13. $1\frac{1}{4} + \frac{3}{8}$ $1\frac{5}{8}$

14. $-2\frac{3}{4} + 1\frac{5}{18}$ $-1\frac{17}{36}$

15. $-3\frac{5}{12} + 1\frac{15}{16}$ $-1\frac{23}{48}$

16. $3\frac{5}{8} + 1\frac{2}{7}$ $4\frac{51}{56}$

17. $4\frac{5}{12} - 2\frac{15}{16}$ $1\frac{23}{48}$

18. $2\frac{5}{35} - 1\frac{5}{14}$ $\frac{11}{14}$

19. $2\frac{3}{8} - \frac{7}{12}$ $1\frac{19}{24}$

20. $4\frac{11}{14} - \left(-\frac{20}{21}\right)$ $5\frac{31}{42}$

21. $12\frac{3}{7} + \left(-8\frac{11}{42}\right)$ $4\frac{1}{6}$

22. $10\frac{9}{48} + 8\frac{3}{20}$ $18\frac{27}{80}$

23. $21\frac{11}{16} - 12\frac{1}{2}$ $9\frac{3}{16}$

24. $15\frac{5}{8} - 11\frac{13}{30}$ $4\frac{23}{120}$

B

25. $\frac{1}{2} + \frac{2}{3} + \frac{3}{4}$ $1\frac{11}{12}$

26. $\frac{3}{8} + \frac{7}{15} - \frac{5}{18}$ $\frac{203}{360}$

27. $6\frac{1}{2} + 5\frac{3}{4} - 2\frac{1}{3}$ $9\frac{11}{12}$

28. $\frac{7}{8} + 2\frac{2}{3} - 1\frac{5}{12}$ $2\frac{1}{8}$

Problems

Solve.

A 1. Nancy and Gary paddled upstream for $2\frac{7}{15}$ hours. Their return trip took $1\frac{2}{3}$ hours. How much faster was the trip downstream? $\frac{4}{5}$ of an hour

B 2. Anthony has a part-time job. One week he worked $2\frac{1}{3}$ hours Monday, $3\frac{1}{2}$ hours Tuesday, $1\frac{3}{4}$ hours Wednesday, $4\frac{5}{6}$ hours Thursday, $2\frac{2}{3}$ hours Friday, and $7\frac{2}{3}$ hours Saturday. How many hours did he work that week? $22\frac{3}{4}$ hours

C 3. The west side of town is divided into three precincts. About $\frac{1}{12}$ of the voters live in the first precinct, about $\frac{3}{20}$ of the voters live in the second precinct, and about $\frac{4}{25}$ of the voters live in the third precinct. What part of the town's voters live on the west side of town? $\frac{59}{150}$

4. It takes Lisa $\frac{4}{5}$ of a minute to play the first half of a march on her trumpet. The second half takes $\frac{7}{16}$ of a minute. How long does it take to play the march if both halves are played twice? $2\frac{19}{40}$ min

5. A teacher assigned a class a set of mathematics exercises. After two days $\frac{3}{7}$ of the class had completed them. After another day an additional $\frac{4}{9}$ had finished. What part of the class still had not completed the exercises? $\frac{8}{63}$

EXTRA! Fractions and Exponents

Just as we have used exponents to show powers of 10, so can we use exponents to show powers of other positive numbers. Thus we may have, for example

$$2^2 = 2 \times 2 = 4 \qquad 2^3 = 2 \times 2 \times 2 = 8$$
$$3^2 = 3 \times 3 = 9 \qquad 3^3 = 3 \times 3 \times 3 = 27$$
$$4^1 = 4 \qquad 4^3 = 4 \times 4 \times 4 = 64$$

We saw in Chapter 2 that we could find the product of two powers of 10 by adding exponents. Thus,

$$10^2 \times 10^3 = 10^{2+3} = 10^5$$

In general,

$$10^a \times 10^b = 10^{a+b}$$

In the same way, using 4 as a base, we can see that

$$4^1 \times 4^3 = 4 \times (4 \times 4 \times 4) = 4^{1+3} = 4^4$$

or

$$4^a \times 4^b = 4^{a+b}$$

In Chapter 3, we explored the use of zero and negative integers for exponents. Can we extend the use of exponents to fractions? For example, can we find a meaning for $4^{\frac{1}{2}}$?

If $4^{\frac{1}{2}}$ names a number, and if it is true that

$$4^{\frac{1}{2}} \times 4^{\frac{1}{2}} = 4^{\frac{1}{2}+\frac{1}{2}} = 4^1 = 4$$

then $4^{\frac{1}{2}}$ must be one of two equal factors of 4; that is,

$$\left(4^{\frac{1}{2}}\right)^2 = 4$$

This means that $4^{\frac{1}{2}}$ must be equal to 2 or to -2, since these are the only numbers whose squares are 4. Because we do not wish to use the same symbol for two different numbers, we agree that $4^{\frac{1}{2}}$ stands for 2, and we call $4^{\frac{1}{2}}$, or 2, a **square root** of 4. (Square roots will be studied in more detail in Chapter 12.) In general, if a stands for zero or a positive number, then

$$\left(a^{\frac{1}{2}}\right)^2 = a$$

We say that

$$a^{\frac{1}{2}} \text{ is the square root of } a$$

or

$$a \text{ is the square of } a^{\frac{1}{2}}.$$

To find the value of $16^{\frac{1}{2}}$, we may ask, "What positive number multiplied by itself equals 16?" Since $4 \times 4 = 16$,

$$16^{\frac{1}{2}} = 4$$

Find the indicated square root.

1. $100^{\frac{1}{2}}$ 10   **2.** $4^{\frac{1}{2}}$ 2   **3.** $25^{\frac{1}{2}}$ 5   **4.** $9^{\frac{1}{2}}$ 3   **5.** $81^{\frac{1}{2}}$ 9   **6.** $49^{\frac{1}{2}}$ 7

7. Explore the use of fractions other than one half as exponents.

For example, what meaning would you give to $8^{\frac{1}{3}}$? the cube root of 8

4-7 Quotients of Fractions

Objective To use a rule for dividing fractions.

To find a rule for dividing fractions, we use the relationship between multiplication and division:

$$\frac{3}{5} \div \frac{2}{7} = n \quad \text{means} \quad n \times \frac{2}{7} = \frac{3}{5}$$

It is easy to verify that substituting $\left(\frac{3}{5} \times \frac{7}{2}\right)$ for n makes the multiplication sentence true.

$$\left(\frac{3}{5} \times \frac{7}{2}\right) \times \frac{2}{7} = \frac{3 \times \overset{1}{\cancel{7}} \times \overset{1}{\cancel{2}}}{5 \times \underset{1}{\cancel{2}} \times \underset{1}{\cancel{7}}} = \frac{3}{5}$$

In general:

---
rule

For all integers a, b, c, and d ($b \neq 0$, $c \neq 0$, $d \neq 0$),

$$\frac{a}{b} \div \frac{c}{d} = \frac{a}{b} \times \frac{d}{c}$$

Example 1 | Calculate $\frac{8}{5} \div \frac{4}{5}$ and reduce to lowest terms.

Solution | $\dfrac{8}{5} \div \dfrac{4}{5} = \dfrac{\overset{2}{\cancel{8}}}{\cancel{5}} \times \dfrac{\overset{1}{\cancel{5}}}{\cancel{4}} = \dfrac{2}{1} = \mathbf{2}$

Example 2 | Calculate $-\frac{5}{6} \div 2\frac{1}{2}$ and reduce to lowest terms.

Solution | Convert $2\frac{1}{2}$ to an improper fraction: $2\frac{1}{2} = \frac{5}{2}$

$$-\frac{5}{6} \div 2\frac{1}{2} = -\frac{5}{6} \div \frac{5}{2} = -\dfrac{\overset{1}{\cancel{5}}}{\underset{3}{\cancel{6}}} \times \dfrac{\overset{1}{\cancel{2}}}{\underset{1}{\cancel{5}}} = -\frac{1}{3}$$

We may state the rule for dividing fractions as follows:

---
rule

To divide by a fraction $\frac{c}{d}$, multiply by $\frac{d}{c}$.

The rule for division works because the product of $\frac{c}{d}$ and $\frac{d}{c}$ is 1. We note this special relationship by calling $\frac{c}{d}$ and $\frac{d}{c}$ **reciprocals.** For all rational numbers a and b, if $a \times b = 1$, then a is the *reciprocal* of b and b is the *reciprocal* of a. A property of $\frac{1}{b}$, namely, $b \times \frac{1}{b} = 1$, tells us that $\frac{1}{b}$ and b are reciprocals.

Example 3 Find the reciprocal of each number.

a. $\frac{3}{7}$ **b.** $-\frac{19}{12}$ **c.** -8 **d.** $\frac{1}{8}$

Solution

a. $\frac{7}{3}$ is the reciprocal of $\frac{3}{7}$.   **b.** $-\frac{12}{19}$ is the reciprocal of $-\frac{19}{12}$.

c. $-\frac{1}{8}$ is the reciprocal of -8.   **d.** 8 is the reciprocal of $\frac{1}{8}$.

Of course, 0 has no reciprocal, and division by 0 is impossible.

Class Exercises

State the reciprocal of each given rational number.

1. 6 $\quad \frac{1}{6}$ **2.** $\frac{1}{9}$ $\quad$ 9   **3.** -3 $\quad -\frac{1}{3}$ **4.** $-\frac{2}{7}$ $\quad -\frac{7}{2}$ **5.** $\frac{3}{8}$ $\quad \frac{8}{3}$ **6.** $\frac{5}{2}$ $\quad \frac{2}{5}$ **7.** $-\frac{21}{11}$ $\quad -\frac{11}{21}$ **8.** $\frac{11}{15}$ $\quad \frac{15}{11}$

Complete.

9. $\frac{2}{3} \div \frac{1}{5} = \frac{2}{3} \times$ ___?___ $\quad$ 5

10. $\frac{1}{4} \div \frac{3}{8} = \frac{1}{4} \times$ ___?___ $\quad \frac{8}{3}$

11. $-\frac{7}{2} \div \frac{2}{3} =$ ___?___ $\times \frac{3}{2}$ $\quad -\frac{7}{2}$

12. $-\frac{3}{5} \div \left(-\frac{1}{5}\right) = -\frac{3}{5} \times$ ___?___ $\quad -5$

13. $\frac{3}{8} \div$ ___?___ $= \frac{3}{8} \times \frac{7}{5}$ $\quad \frac{5}{7}$

14. $\frac{2}{5} \div$ ___?___ $= \frac{2}{5} \times \frac{3}{4}$ $\quad \frac{4}{3}$

Exercises

Divide. Express all results in lowest terms.

A

1. $\frac{3}{5} \div \frac{6}{15}$ $\quad 1\frac{1}{2}$ **2.** $\frac{4}{7} \div \frac{3}{14}$ $\quad 2\frac{2}{3}$ **3.** $-\frac{4}{5} \div \left(-\frac{4}{5}\right)$ $\quad 1$ **4.** $-\frac{7}{16} \div \frac{7}{16}$ $\quad -1$

5. $-\frac{5}{16} \div \frac{11}{8}$ $\quad -\frac{5}{22}$ **6.** $-\frac{3}{32} \div \left(-\frac{9}{16}\right)$ $\quad \frac{1}{6}$ **7.** $-8 \div \frac{16}{21}$ $\quad -10\frac{1}{2}$ **8.** $\frac{14}{15} \div (-7)$ $\quad -\frac{2}{15}$

9. $\frac{24}{20} \div \frac{18}{15}$ $\quad 1$ **10.** $-\frac{32}{20} \div \left(-\frac{8}{30}\right)$ $\quad 6$ **11.** $2\frac{1}{5} \div 5\frac{1}{2}$ $\quad \frac{2}{5}$ **12.** $4\frac{1}{3} \div 3\frac{1}{4}$ $\quad 1\frac{1}{3}$

13. $-2\frac{3}{4} \div 3\frac{1}{8}$ $\quad -\frac{22}{25}$ **14.** $2\frac{3}{5} \div \left(-1\frac{11}{15}\right)$ $\quad -1\frac{1}{2}$ **15.** $3\frac{2}{9} \div 29$ $\quad \frac{1}{9}$ **16.** $-4\frac{4}{5} \div (-12)$ $\quad \frac{2}{5}$

17. $12 \div 2\frac{2}{5}$ $\quad 5$ **18.** $-15 \div 2\frac{1}{2}$ $\quad -6$ **19.** $-4\frac{2}{7} \div \left(-1\frac{1}{14}\right)$ $\quad 4$ **20.** $-1\frac{7}{15} \div 3\frac{2}{3}$ $\quad -\frac{2}{5}$

B **21.** $\left(-\frac{18}{5} \div \frac{9}{35}\right) \times \frac{3}{7}$ -6 **22.** $\frac{12}{25} \times \left[-\frac{5}{7} \div \left(-\frac{9}{14}\right)\right]$ $\frac{8}{15}$ **23.** $\left(-1\frac{2}{3} \times 4\frac{2}{3}\right) \div 6\frac{1}{9}$

$-1\frac{3}{11}$

Problems

Solve.

A **1.** A gardener uses $8\frac{1}{3}$ packets of seeds in planting $3\frac{3}{4}$ rows of cabbage. How many packets of seeds are needed for 1 row of cabbage? $2\frac{2}{9}$ packets

2. A stamping machine can stamp 250 envelopes in $1\frac{2}{3}$ minutes. How many envelopes can the machine stamp in 1 minute? 150 envelopes

An Educator

Every year when Charlotte Hawkins Brown (1882–1961) was a child, she would visit her native North Carolina with her parents. As she grew older, she became particularly concerned about the poor conditions of the schools in her hometown. With the help of Alice F. Palmer, Brown attended the Salem Normal School and Wellesley College, where she studied to become a teacher. The opportunity for Brown to assist her native region came, and she accepted the position to teach in a small school near Sedalia, North Carolina. The school, however, was forced to close due to lack of funds. Brown, aware of the neighborhood's urgent need, decided to remain in Sedalia and establish her own school. In two years, she raised enough money to begin the construction of what was later to become Palmer Memorial Institute.

Research Activity Find out when the first public school was opened in this country. What subjects were taught? What is meant by a "one-room school"?

Career Activity In some places, teachers must be certified to teach a certain grade level. Find out what it means to be "certified." What types of certification are needed in your school?

4-8 Fractions and Decimals

Objective To find equal fractions and decimals.

Any decimal can be transformed into an equal fraction or mixed number as shown in Example 1, below.

Example 1 | Convert 1.85 to a mixed number.

Solution | $1.85 = \frac{185}{100} = \frac{185 \div 5}{100 \div 5} = \frac{37}{20}$, or $1\frac{17}{20}$

A property of $\frac{a}{b}$, namely $\frac{a}{b} = a \div b$, tells us that we can convert a fraction into a decimal by dividing.

Example 2 | Convert $\frac{15}{40}$ to a decimal.

Solution

$$
\begin{array}{r}
0.375 \\
40\overline{)15.000} \\
\underline{12\ 0} \\
3\ 00 \\
\underline{2\ 80} \\
200 \\
\underline{200} \\
0
\end{array}
$$

So, $\frac{15}{40} = 0.375$

In Example 2, we reach a remainder of 0 and the division process ends. We say that $\frac{15}{40}$ is equal to the *terminating decimal* 0.375. Every terminating decimal is equal to a fraction that has a power of 10 as its denominator. For example, $0.375 = \frac{375}{10^3}$.

Sometimes when we convert a fraction into a decimal, the division process does not end.

Example 3 | Convert $\frac{5}{33}$ to a decimal.

$$
\begin{array}{r}
0.1515 \\
33\overline{)5.0000} \\
3\,3 \\
\hline
1\,70 \\
1\,65 \\
\hline
50 \\
33 \\
\hline
170 \\
165 \\
\hline
5
\end{array}
$$

Solution

So, $\frac{5}{33} = 0.1515 \ldots$

In Example 3, we see that $\frac{5}{33}$ is equal to the *repeating decimal* 0.1515 . . . . We may write this decimal as $0.\overline{15}$, where the bar indicates the **repetend** 15.

Examples 2 and 3 represent a general property of rational numbers:

property

Every rational number can be represented either by a terminating decimal number or by a repeating decimal number.

There is a simple test we can make to determine whether the decimal for a fraction will terminate or repeat:

1. Express the fraction in lowest terms.
2. If the denominator then has only powers of 2 or 5, the decimal for the fraction terminates.
3. If the denominator includes powers of any numbers other than 2 or 5, the decimal form repeats.

Class Exercises

Complete.

1. $81 = \frac{81}{?}$ 1

2. $23.6 = \frac{236}{?}$ 10

3. $7.07 = \frac{707}{?}$ 100

4. $1.309 = \frac{1309}{?}$ 1000

State as a fraction in which the numerator is an integer and the denominator is a power of ten.

Example	**56.1**
Solution	$\dfrac{561}{10}$

5. 23.11 $\dfrac{2311}{100}$ **6.** 5.213 $\dfrac{5213}{1000}$ **7.** -0.021 $-\dfrac{21}{1000}$ **8.** -1.0032 $-\dfrac{10032}{10000}$

State the terminating decimal for each fraction.

9. $\dfrac{1}{10}$ 0.1   **10.** $\dfrac{3}{10}$ 0.3   **11.** $\dfrac{1}{4}$ 0.25   **12.** $\dfrac{1}{2}$ 0.5

13. $-\dfrac{3}{4}$ -0.75 **14.** $-\dfrac{1}{5}$ -0.2 **15.** $-2\dfrac{2}{5}$ -2.4 **16.** $-\dfrac{9}{10}$ -0.9

State the repetend for each decimal.

17. $\dfrac{1}{6} = 0.1\overline{6}$ 6   **18.** $\dfrac{1}{18} = 0.0\overline{5}$ 5   **19.** $\dfrac{3}{11} = 0.\overline{27}$ 27

20. $\dfrac{3}{7} = 0.\overline{428571}$ 428571   **21.** $\dfrac{12}{11} = 1.\overline{09}$ 09   **22.** $\dfrac{1}{13} = 0.\overline{076923}$ 076923

Exercises

For each decimal, write a fraction in which the numerator is an integer and the denominator is a power of 10.

A

1. 3.5 $\dfrac{35}{10}$ **2.** 6.25 $\dfrac{625}{100}$ **3.** -0.45 $-\dfrac{45}{100}$ **4.** -1.54 $-\dfrac{154}{100}$

5. 2.023 $\dfrac{2023}{1000}$ **6.** 0.0011 $\dfrac{11}{10000}$ **7.** 201.3 $\dfrac{2013}{10}$ **8.** 402.112 $\dfrac{402,112}{1000}$

Express as a fraction in lowest terms.

9. 2.75 $2\dfrac{3}{4}$ **10.** 6.125 $6\dfrac{1}{8}$ **11.** -0.375 $-\dfrac{3}{8}$ **12.** -1.025 $-1\dfrac{1}{40}$

13. 1.625 $1\dfrac{5}{8}$ **14.** 2.213 $2\dfrac{213}{1000}$ **15.** 0.416 $\dfrac{52}{125}$ **16.** 0.036 $\dfrac{9}{250}$

Find a terminating decimal for the given fraction.

17. $\dfrac{5}{4}$ 1.25   **18.** $\dfrac{7}{2}$ 3.5   **19.** $\dfrac{3}{20}$ 0.15   **20.** $\dfrac{6}{25}$ 0.24   **21.** $\dfrac{3}{8}$ 0.375

22. $\dfrac{7}{8}$ 0.875   **23.** $\dfrac{5}{16}$ 0.3125   **24.** $\dfrac{4}{40}$ 0.1   **25.** $-\dfrac{1}{80}$ -0.0125 **26.** $-\dfrac{6}{125}$ -0.048

Find a repeating decimal for the given fraction. Use a bar to show the repetend.

27. $\frac{4}{3}$ $1.\overline{3}$ **28.** $\frac{7}{6}$ $1.1\overline{6}$ **29.** $\frac{5}{9}$ $0.\overline{5}$ **30.** $\frac{7}{12}$ $0.58\overline{3}$ **31.** $-\frac{7}{11}$ $-0.\overline{63}$

32. $-\frac{3}{22}$ $-0.1\overline{36}$ **33.** $\frac{5}{7}$ $0.\overline{714285}$ **34.** $\frac{8}{21}$ $0.\overline{380952}$ **35.** $\frac{1}{13}$ $0.\overline{076923}$ **36.** $\frac{5}{13}$ $0.\overline{384615}$

Tell whether the decimal for the fraction is terminating or repeating.

Example	$\frac{12}{15}$
Solution	In lowest terms, $\frac{12}{15} = \frac{4}{5}$. The denominator has no factors other than 2 or 5, so the decimal will terminate.

37. $\frac{2}{7}$ **38.** $\frac{1}{4}$ **39.** $\frac{5}{12}$ **40.** $\frac{3}{16}$ **41.** $\frac{8}{25}$ **42.** $\frac{9}{15}$

repeating   terminating   repeating   terminating   terminating   terminating

B **43.** Divide 1 by the given integer and state the number of digits to the right of the decimal point in the decimal that results.
   a. 2 1
   b. 2×2, or 4 2
   c. $2 \times 2 \times 2$, or 8 3
   d. $2 \times 2 \times 2 \times 2$, or 16 4

44. Study your results in Exercise 43 and guess the number of digits to the right of the decimal point in the terminating decimal for

$$\frac{1}{2 \times 2 \times 2 \times 2 \times 2 \times 2}. 6$$

Check your guess by dividing 1 by 64. 0.015625

C **45.** A terminating decimal for $\frac{1}{4}$ is 0.25. Suppose that in dividing 1 by 4, there was a failure to notice that $20 \div 4 = 5$ and $40 \div 4 = 10$ so that the steps shown at the right were taken. We now have a repeating decimal for $\frac{1}{4}$, namely, $0.24\overline{9}$. Do you think that every terminating decimal has an equivalent repeating decimal with repetend $\overline{9}$? Try to explain.   Yes. Whenever we underestimate the quotient by 1 at the last step in the division, the divisor appears as the remainder. If the underestimating continues, the quotient of 9 keeps repeating.

$$
\begin{array}{r}
0.249\\
4\overline{)1.000}\\
\underline{8}\\
20\\
\underline{16}\\
40\\
\underline{36}\\
4
\end{array}
$$

EXTRA! Fractions for Repeating Decimals

We have seen that every terminating decimal can be converted into a fraction. Thus, every terminating decimal represents a rational number.

Does every *repeating* decimal represent a rational number? The answer is "Yes!" The example below shows a method for finding the rational number represented by a repeating decimal.

Example	Find the fraction in lowest terms that represents $2.\overline{18}$.
Solution	Suppose first that there is such a fraction naming a rational number n. Then

$$n = 2.\overline{18} = 2.18\overline{18}.$$

Since there are 2 digits in the repetend 18, we multiply both members of $n = 2.18\overline{18}$ by 10^2, or 100, getting

$$100n = 218.\overline{18}$$
$$\text{Subtract:} \quad n = 2.\overline{18}$$
$$99n = 216$$

$$n = \frac{216}{99} = \frac{24 \times 9}{11 \times 9} = \frac{24}{11}.$$

Thus, if there *is* a rational number n represented by $2.\overline{18}$, then $n = \frac{24}{11}$. The division below shows that, in fact, $\frac{24}{11} = 2.\overline{18}$. Therefore, $2.\overline{18} = \frac{24}{11}$.

$$
\begin{array}{r}
2.18 \\
11\overline{)24.00} \\
22 \\
\hline
2\,0 \\
1\,1 \\
\hline
90 \\
88 \\
\hline
2
\end{array}
$$

Find a fraction in lowest terms for the repeating decimal.

1. $0.\overline{4}$ $\frac{4}{9}$
2. $0.\overline{7}$ $\frac{7}{9}$
3. $2.\overline{1}$ $\frac{19}{9}$
4. $1.\overline{3}$ $\frac{4}{3}$
5. $1.\overline{21}$ $\frac{40}{33}$
6. $2.\overline{42}$ $\frac{80}{33}$
7. $0.2\overline{3}$ $\frac{7}{30}$
8. $5.2\overline{13}$ $\frac{5161}{990}$

Words to remember:

least common denominator (LCD) [p. 121]   reciprocals [p. 130]
prime factorization [p. 122]

Replace each set of fractions with equal fractions having the least common denominator (LCD).

1. $\dfrac{4}{5}, \dfrac{2}{9}$ $\dfrac{36}{45}, \dfrac{10}{45}$

$-\dfrac{25}{30}; -\dfrac{2}{30}$ **2.** $-\dfrac{5}{6}, -\dfrac{1}{15}$ [4-5]

3. $\dfrac{7}{18}, \dfrac{5}{12}, \dfrac{1}{15}$ $\dfrac{70}{180}; \dfrac{75}{180}; \dfrac{12}{180}$

4. $\dfrac{3}{4}, \dfrac{4}{7}, -\dfrac{1}{6}$
$\dfrac{63}{84}; \dfrac{48}{84}; -\dfrac{14}{84}$

Express the sum or difference as a proper fraction in lowest terms or as a mixed number with fraction in lowest terms.

5. $\dfrac{5}{8} + \left(-\dfrac{1}{6}\right)$ $\dfrac{11}{24}$

6. $-\dfrac{8}{15} - \dfrac{7}{40}$ $-\dfrac{17}{24}$ [4-6]

7. $3\dfrac{5}{6} - 2\dfrac{7}{10}$ $1\dfrac{2}{15}$

8. $\dfrac{1}{4} + 3\dfrac{2}{9} + 2\dfrac{5}{12}$ $5\dfrac{8}{9}$

Divide. Express all results in lowest terms.

9. $\dfrac{3}{7} \div \dfrac{6}{21}$ $1\dfrac{1}{2}$

10. $-\dfrac{11}{12} \div \left(-\dfrac{7}{8}\right)$ $1\dfrac{1}{21}$ [4-7]

11. $-12 \div 3\dfrac{3}{4}$ $-3\dfrac{1}{5}$

12. $1\dfrac{1}{6} \div \left(-4\dfrac{2}{3}\right)$ $-\dfrac{1}{4}$

Express as a fraction in lowest terms.
13. 1.78 $1\dfrac{39}{50}$

14. -0.325 $-\dfrac{13}{40}$ [4-8]

Express as a terminating or repeating decimal.

15. $\dfrac{23}{24}$ $0.958\overline{3}$

16. $-\dfrac{258}{125}$ -2.064

Self-Test answers and Extra Practice are at the back of the book.

Chapter Review

Write the letter that labels the correct answer.

1. ___?___ $\times 4 = 1$ D   [4-1]

    A. 1   **B.** -3 **C.** -4 **D.** $\frac{1}{4}$

2. $1\frac{3}{20} = \frac{?}{20}$ A

    A. 23   **B.** 61   **C.** 13   **D.** 203

3. $\frac{-2}{3} = $ ___?___   C   [4-2]

    A. $\frac{3}{-2}$ **B.** $\frac{2}{3}$ **C.** $\frac{2}{-3}$ **D.** -1

4. $-\frac{23}{7} = $ ___?___   C

    A. $-2\frac{3}{7}$ **B.** $\frac{23}{7}$ **C.** $-3\frac{2}{7}$ **D.** -16

5. $\frac{2}{5} = $ ___?___   C   [4-3]

    A. $\frac{4}{25}$ **B.** $\frac{5}{15}$ **C.** $\frac{12}{30}$ **D.** $-\frac{8}{20}$

6. $-\frac{56}{80}$ in lowest terms is ___?___ .   D

    A. $\frac{7}{10}$ **B.** $\frac{28}{40}$ **C.** $-\frac{28}{40}$ **D.** $-\frac{7}{10}$

7. In lowest terms, $\left(-\frac{3}{8}\right) \times \left(-\frac{1}{5}\right) = $ ___?___ .   B   [4-4]

    A. $-\frac{3}{13}$ **B.** $\frac{3}{40}$ **C.** $\frac{15}{8}$ **D.** $\frac{8}{15}$

8. In lowest terms, $-12 \times \frac{5}{6} = $ ___?___ .   C

    A. $-\frac{5}{72}$ **B.** 10   **C.** -10 **D.** $-\frac{60}{6}$

Name the set of equal fractions having the least common denominator (LCD).

9. $-\dfrac{5}{7}, \dfrac{15}{28}$ D

[4-5]

A. $-\dfrac{33}{35}, \dfrac{22}{35}$ **B.** $-\dfrac{140}{196}, \dfrac{105}{196}$ **C.** $-\dfrac{40}{56}, \dfrac{30}{56}$ **D.** $-\dfrac{20}{28}, \dfrac{15}{28}$

10. $\dfrac{8}{3}, \dfrac{4}{5}$ A

A. $\dfrac{40}{15}, \dfrac{12}{15}$ **B.** $\dfrac{20}{15}, \dfrac{14}{15}$ **C.** $\dfrac{13}{8}, \dfrac{7}{8}$ **D.** $\dfrac{80}{30}, \dfrac{24}{30}$

11. $\dfrac{5}{2} + \dfrac{2}{3} = \underline{\quad?\quad}$ A

[4-6]

A. $\dfrac{19}{6}$ **B.** $\dfrac{7}{6}$ **C.** $\dfrac{7}{5}$ **D.** $\dfrac{19}{12}$

12. $3\dfrac{1}{8} - 2\dfrac{7}{10} = \underline{\quad?\quad}$ B

A. $5\dfrac{33}{40}$ **B.** $\dfrac{17}{40}$ **C.** $1\dfrac{17}{40}$ **D.** $1\dfrac{23}{40}$

13. In lowest terms, $\dfrac{2}{3} \div \dfrac{7}{9} = \underline{\quad?\quad}$.   B

[4-7]

A. $\dfrac{14}{27}$ **B.** $\dfrac{6}{7}$ **C.** $\dfrac{18}{21}$ **D.** $\dfrac{7}{6}$

14. In lowest terms, $-5\dfrac{2}{5} \div \left(-\dfrac{9}{20}\right) = \underline{\quad?\quad}$.   A

A. 12   **B.** $\dfrac{243}{100}$ **C.** $-\dfrac{243}{100}$ **D.** -12

15. In lowest terms, $1.375 = \underline{\quad?\quad}$.   D

[4-8]

A. $1\dfrac{75}{200}$ **B.** $\dfrac{1375}{1000}$ **C.** $1\dfrac{1}{375}$ **D.** $1\dfrac{3}{8}$

16. In decimal form, $-\dfrac{13}{6} = \underline{\quad?\quad}$.   C

A. $-0.21\overline{6}$ **B.** $-2.\overline{16}$ **C.** $-2.1\overline{6}$ **D.** 2.16

Nonrepeating Decimals

We have seen that every rational number is represented either by a terminating decimal or by a repeating decimal. But what about decimals that are nonterminating and nonrepeating? An example of such a decimal is

$$0.535533555333 \ldots ,$$

in which the digits after the decimal point are first one 5 and one 3, then two 5's and two 3's, and so on. It is neither terminating nor repeating. We know, then, that it does not represent a rational number.

Does the decimal expression given above represent any number n? If so, then n must be:

greater than 0.5 . . . . . but less than 0.6,
greater than 0.53 . . . . but less than 0.54,
greater than 0.535 . . . . but less than 0.536,

and so on.

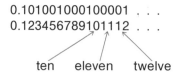

The figure indicates that the sequence of graph points is closing in from both sides to some point P. We say that P is the graph of the *irrational* number 0.535533555333 . . . . An **irrational number** is a number represented by a nonterminating, nonrepeating decimal. The set of rational numbers together with the set of irrational numbers is called the set of **real numbers.**

It is easy to find other nonterminating and nonrepeating decimal numerals. Here are two:

0.101001000100001 . . .
0.123456789101112 . . .

ten   eleven   twelve

Research Activity

1. Write a decimal number which, when added to 0.535533555333 . . . will make a sum of 0.777777777777 . . . , or 0.$\overline{7}$. Is the number you wrote rational or irrational? Is the sum rational or irrational? 0.242244222444...

2. Use a pattern of your own making to write a decimal for an irrational number between 0 and 1. Write another decimal which when added to your first decimal will make a sum of $0.\overline{9}$. Are the addends rational or irrational? Is the sum rational or irrational? Answers may vary.

3. Experiment with finding the sums of other pairs of irrational numbers. Can you come to any conclusions about the sum of two irrational numbers? Is the sum always rational, always irrational, or may it be either one or the other?

Not all irrational numbers have decimals with digits arranged in a pattern. A famous irrational number with which you may be familiar is π (pi), the quotient of the circumference of a circle and its diameter. Its decimal representation begins 3.14159, but there is no simple rule for continuing the sequence of the digits.

From antiquity, mathematicians have searched for methods to determine better and better approximations to the value of π. The Greek mathematician, Archimedes, who lived in the 3rd century B.C., calculated that π was greater than $3\frac{10}{71}$ and less than $3\frac{1}{7}$.

In more recent times, some formulas have been developed to express π. In 1655, an English mathematician, John Wallis, found the following:

$$\pi = 2\left(\frac{2}{1} \cdot \frac{2}{3} \cdot \frac{4}{3} \cdot \frac{4}{5} \cdot \frac{6}{5} \cdot \frac{6}{7} \cdot \frac{8}{7} \cdot \frac{8}{9} \cdot \ \cdots \ \right)$$

The three dots at the right inside the parentheses indicate that the pattern continues without end.

Another nice formula for π was developed by a German mathematician, Gottfried Wilhelm von Leibniz, in 1674:

$$\pi = 4\left(1 - \frac{1}{3} + \frac{1}{5} - \frac{1}{7} + \frac{1}{9} - \frac{1}{11} + \ \cdots \ \right)$$

Another irrational number is

$$\sqrt{2} = 1.41428 \ . \ . \ . \ \text{(read ''the positive square root of 2''),}$$

the positive number whose square is 2. (In the Extra! on page 128 we used the exponent $\frac{1}{2}$ to indicate square root, thus $\sqrt{2} = 2^{\frac{1}{2}}$.) Many square roots of positive integers are irrational; in fact, the square root of a positive integer is either another positive integer or it is an irrational number.

Chapter Test

Complete.

1. $\underline{}\times\frac{1}{7}=1$ 7   **2.** $2\div5=\underline{}$ $\frac{2}{5}$ **3.** $2\frac{3}{8}=\frac{?}{8}$ 19   **[4-1]**

Write each negative fraction in two other ways.

4. $\frac{3}{-4}$ $\frac{-3}{4};-\frac{3}{4}$ **5.** $-\frac{29}{100}$ $\frac{-29}{100};\frac{29}{-100}$ **[4-2]**

Complete.

6. $-6\frac{2}{9}=\frac{?}{9}$ -56 **7.** $-\frac{59}{8}=\underline{}$ $-7\frac{3}{8}$

Reduce each fraction to lowest terms.

8. $\frac{60}{48}$ $\frac{5}{4}$ **9.** $\frac{378}{630}$ $\frac{3}{5}$ **10.** $-\frac{147}{168}$ $-\frac{7}{8}$ **[4-3]**

Compute each product. Give your result in lowest terms.

11. $\frac{7}{5}\times\left(-\frac{20}{21}\right)$ $-1\frac{1}{3}$ **12.** $-1\frac{1}{5}\times\left(-7\frac{3}{11}\right)$ $8\frac{8}{11}$ **13.** $\frac{15}{36}\times\frac{2}{33}$ $\frac{5}{198}$ **[4-4]**

Replace each set of fractions with equal fractions having the least common denominator (LCD).

14. $\frac{2}{3},\frac{5}{4}$ $\frac{8}{12};\frac{15}{12}$ **15.** $\frac{7}{33},-\frac{4}{275}$ $\frac{175}{825};-\frac{12}{825}$ **16.** $\frac{1}{3},\frac{2}{15},-\frac{22}{27}$ **[4-5]**

$\frac{45}{135};\frac{18}{135};-\frac{110}{135}$

Express the sum or difference as a proper fraction in lowest terms or as a mixed number with fraction in lowest terms.

17. $\frac{5}{6}+\frac{7}{15}$ $1\frac{3}{10}$ **18.** $-\frac{11}{16}+\frac{17}{20}$ $\frac{13}{80}$ **19.** $5\frac{5}{12}-2\frac{7}{18}$ $3\frac{1}{36}$ **[4-6]**

Divide. Express all results in lowest terms.

20. $\frac{9}{10}\div\left(-\frac{9}{10}\right)$ -1 **21.** $-1\frac{5}{27}\div\left(-1\frac{7}{9}\right)$ $\frac{2}{3}$ **22.** $-\frac{6}{19}\div\frac{9}{38}$ $-1\frac{1}{3}$ **[4-7]**

Express as a fraction in lowest terms.

23. -4.125 $-4\frac{1}{8}$ **24.** 0.628 $\frac{157}{250}$ **[4-8]**

Express as a terminating or repeating decimal.

25. $-\frac{55}{12}$ $-4.58\overline{3}$ **26.** $\frac{21}{32}$ 0.65625

Skill Review

Add. Give your result in lowest terms.

1. $\frac{3}{5} + \frac{1}{5}$ $\frac{4}{5}$ **2.** $-\frac{3}{8} + \frac{2}{8}$ $-\frac{1}{8}$ **3.** $\frac{1}{5} + \frac{2}{5}$ $\frac{3}{5}$ **4.** $-\frac{1}{4} + \frac{2}{4}$ $\frac{1}{4}$

5. $2\frac{1}{3} + 3\frac{1}{3}$ $5\frac{2}{3}$ **6.** $-5\frac{2}{7} + 3\frac{3}{7}$ $-1\frac{6}{7}$ **7.** $12\frac{5}{6} + \left(-8\frac{4}{6}\right)$ $4\frac{1}{6}$ **8.** $25\frac{3}{8} + 6\frac{3}{8}$ $31\frac{3}{4}$

9. $\frac{1}{2} + \frac{2}{3}$ $1\frac{1}{6}$ **10.** $\frac{1}{3} + \left(-\frac{1}{4}\right)$ $\frac{1}{12}$ **11.** $-\frac{2}{5} + \left(-\frac{1}{2}\right)$ $-\frac{9}{10}$ **12.** $\frac{1}{10} + \frac{3}{5}$ $\frac{7}{10}$

13. $7\frac{1}{4} + 2\frac{1}{2}$ **14.** $12\frac{2}{3} + 11\frac{3}{4}$ **15.** $-6\frac{2}{5} + 3\frac{1}{3}$ **16.** $34\frac{1}{2} + \left(-12\frac{5}{6}\right)$

$9\frac{3}{4}$ $24\frac{5}{12}$ $-3\frac{1}{15}$ $21\frac{2}{3}$

Subtract. Give your result in lowest terms.

17. $\frac{5}{6} - \frac{2}{6}$ $\frac{1}{2}$ **18.** $\frac{7}{8} - \frac{3}{8}$ $\frac{1}{2}$ **19.** $\frac{9}{10} - \frac{7}{10}$ $\frac{1}{5}$ **20.** $-\frac{4}{5} - \frac{3}{5}$ $-1\frac{2}{5}$

21. $8\frac{3}{4} - 5\frac{1}{4}$ **22.** $-2\frac{1}{7} - 3\frac{2}{7}$ **23.** $8\frac{1}{3} - 9\frac{2}{3}$ **24.** $12\frac{5}{6} - \left(-2\frac{1}{6}\right)$

$3\frac{1}{2}$ $-5\frac{3}{7}$ $-1\frac{1}{3}$ 15

25. $\frac{3}{4} - \frac{1}{8}$ **26.** $-\frac{1}{5} - \frac{3}{10}$ **27.** $\frac{8}{9} - \left(-\frac{1}{3}\right)$ **28.** $\frac{1}{2} - \frac{5}{8}$

$\frac{5}{8}$ $-\frac{1}{2}$ $1\frac{2}{9}$ $-\frac{1}{8}$

Multiply. Give your result in lowest terms.

29. $\frac{3}{5} \times 5$ 3 **30.** $7 \times \frac{5}{7}$ 5 **31.** $-\frac{3}{8} \times 8$ -3 **32.** $-4 \times \frac{1}{4}$ -1

33. $\frac{1}{3} \times \frac{1}{4}$ $\frac{1}{12}$ **34.** $\frac{3}{8} \times \frac{1}{2}$ $\frac{3}{16}$ **35.** $-\frac{4}{5} \times \frac{5}{6}$ $-\frac{2}{3}$ **36.** $-\frac{2}{3} \times \left(-\frac{5}{8}\right)$ $\frac{5}{12}$

37. $5\frac{1}{2} \times 4\frac{2}{3}$ **38.** $-8\frac{1}{8} \times 3\frac{1}{5}$ **39.** $12\frac{1}{2} \times \left(-5\frac{1}{3}\right)$ **40.** $9\frac{3}{4} \times 5\frac{2}{9}$

$25\frac{2}{3}$ -26 $-66\frac{2}{3}$ $50\frac{11}{12}$

Divide. Give your result in lowest terms.

41. $3 \div \frac{1}{2}$ 6 **42.** $\frac{3}{4} \div 2$ $\frac{3}{8}$ **43.** $-5 \div \frac{1}{8}$ -40 **44.** $-\frac{1}{4} \div (-5)$ $\frac{1}{20}$

45. $\frac{2}{3} \div \frac{1}{2}$ $1\frac{1}{3}$ **46.** $\frac{5}{8} \div \frac{1}{4}$ $2\frac{1}{2}$ **47.** $\frac{2}{9} \div \left(-\frac{2}{3}\right)$ $-\frac{1}{3}$ **48.** $-\frac{4}{5} \div \frac{8}{15}$ $-1\frac{1}{2}$

49. $5\frac{5}{6} \div 2\frac{1}{3}$ **50.** $-8 \div 3\frac{3}{7}$ **51.** $10 \div \left(-4\frac{3}{8}\right)$ **52.** $-12\frac{1}{2} \div \left(-1\frac{7}{8}\right)$

$2\frac{1}{2}$ $-2\frac{1}{3}$ $-2\frac{2}{7}$ $6\frac{2}{3}$

5

Ratio, Proportion, Percent

5-1 Ratio

Objective To compare two numbers by use of a ratio expressed as a fraction, as a division, or with a ratio sign.

The **ratio** of a number c to another number d (d cannot be zero) is the quotient of the first number divided by the second, or $\frac{c}{d}$. A ratio compares one quantity with another.

We can write a ratio in several ways. For example, the ratio 5 to 10 can be expressed as follows:

$$\frac{5}{10} \qquad\qquad 5 \div 10 \qquad\qquad 5:10$$

as a fraction · with a division sign · with a ratio sign

We may express a ratio in lowest terms:

$$5 \text{ to } 10 = 1 \text{ to } 2 \qquad \frac{5}{10} = \frac{1}{2} \qquad 5 \div 10 = 1 \div 2 \qquad 5:10 = 1:2$$

145

Example 1

There are 12 boys and 8 girls in the Art Club. Express in three ways the ratio of
a. boys to girls, and
b. girls to club members.
Then give a fraction in lowest terms for each ratio.

Solution

a. boys $\longrightarrow \dfrac{12}{8}$, or $12 \div 8$, or $12:8$

In lowest terms, $\dfrac{12}{8} = \dfrac{3}{2}$.

b. $\dfrac{\text{girls} \;\rightarrow\; 8}{\text{club members} \;\rightarrow\; 20}$, or $8 \div 20$, or $8:20$.

In lowest terms, $\dfrac{8}{20} = \dfrac{2}{5}$.

When forming a ratio of quantities of the same kind, be sure to use the same unit of measure for each quantity.

Example 2

Express the ratio of 20 g to 3 kg as a fraction in lowest terms.

Solution

$3 \text{ kg} = 3000 \text{ g}$

$\dfrac{20}{3000} = \dfrac{1}{150}$

A ratio can be used to compare quantities of different kinds. Such a ratio is sometimes called a **rate**. We say the roses are selling at a rate of 41¢ for 2 or 41¢ per pair. The word **per** indicates division.

We may write the rate as a ratio:

$\dfrac{41}{2}$, or $41 \div 2$, or $41:2$.

ROSES
2 for 41¢

Example 3

Blueberries are selling at the rate of 89¢ for 2 boxes.
a. Write this information as a ratio.
b. Write a ratio to show the cost of 6 boxes of blueberries; of 10 boxes.

Solution

a. 89¢ for 2 boxes; $\frac{89}{2}$

b. $\frac{3 \times 89}{3 \times 2} = \frac{267}{6}$; 6 boxes cost $2.67

$\frac{5 \times 89}{5 \times 2} = \frac{445}{10}$; 10 boxes cost $4.45

It is often useful to show a rate as a ratio per unit.

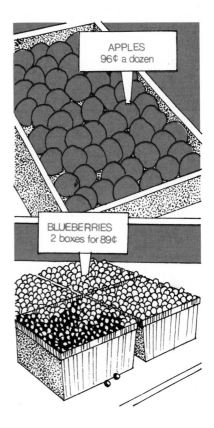

APPLES
96¢ a dozen

BLUEBERRIES
2 boxes for 89¢

Example 4

Apples cost 96¢ per dozen.
a. Write this rate as a ratio.
b. Write a ratio showing the rate of cost per *one* apple.
c. What is the cost of 5 apples? 8 apples? 14 apples?

Solution

a. 96¢ per 12 apples; $\frac{96}{12}$

b. $\frac{96 \div 12}{12 \div 12} = \frac{8}{1}$; one apple costs 8¢

c. 5×8¢ $= 40$¢
8×8¢ $= 64$¢
14×8¢ $= 112$¢, or $1.12

Class Exercises

Compare the two quantities by stating the ratio of the first to the second.

1. One dozen and 20. 12 to 20

2. A pair and a half dozen. 2 to 6

3. Number of days in a week and number of days in June. 7 to 30

4. Number of states in the United States today and number when George Washington became president. 50 to 13

5. Number of legs on a snake and number of legs on a chicken. 0 to 2

6. Number of hours in a Monday and number of hours in a week. 24 to 168

7. Number of windows in your classroom and number of doors in your classroom.   Answer will vary according to your classroom.

8. Number of elephants in your classroom and number of students in your classroom.   Answer will vary; for example, 0 to 23.

Write each ratio three ways.

$\frac{1}{4}$; $1 \div 4$; $1 : 4$

9. 2 to 1 $\frac{2}{1}$; $2 \div 1$; $2 : 1$ **10.** 3 to 4 $\frac{3}{4}$; $3 \div 4$; $3 : 4$ **11.** 5 to 7 $\frac{5}{7}$; $5 \div 7$; $5 : 7$ **12.** 1 to 4

13. 9 to 10 $\frac{9}{10}$; $9 \div 10$; $9 : 10$ **14.** 5 to 6 $\frac{5}{6}$; $5 \div 6$; $5 : 6$ **15.** 2 to 3 $\frac{2}{3}$; $2 \div 3$; $2 : 3$ **16.** 7 to 9

$\frac{7}{9}$; $7 \div 9$; $7 : 9$

Exercises

Write each ratio as a fraction in lowest terms.

A   **1.** 10 to 4 $\frac{5}{2}$ **2.** 5 to 20 $\frac{1}{4}$ **3.** $\frac{8}{30}$ $\frac{4}{15}$ **4.** $45 \div 9$ $\frac{5}{1}$

5. $7 \div 21$ $\frac{1}{3}$ **6.** $51:17$ $\frac{3}{1}$ **7.** $25 \div 5$ $\frac{5}{1}$ **8.** $\frac{3}{4}$ to $\frac{1}{3}$ $\frac{9}{4}$

9. 3 dollars to 78 cents $\frac{50}{13}$ **10.** 8 months to 2 years $\frac{1}{3}$

11. 50 minutes to 1 hour $\frac{5}{6}$ **12.** 2 kL to 75 L $\frac{80}{3}$

13. 14 cm to 1 m $\frac{7}{50}$ **14.** 700 m to 2 km $\frac{7}{20}$

Platt's Pet Palace has 24 puppies and 16 kittens for sale. Compare each pair of quantities by writing a ratio as a fraction in lowest terms.

15. Number of puppies to number of kittens. $\frac{3}{2}$

16. Number of kittens to total of kittens and puppies. $\frac{2}{5}$

17. Total of kittens and puppies to number of puppies. $\frac{5}{3}$

Problems

Solve.

A   **1.** Jessie bought 3 cans of cat food for 66¢. Write a ratio to show the rate of cost per can. What is the cost of 6 cans? 9 cans? 21 cans? $1.32; $1.98; $4.62

B   **2.** A toy car travels 6 m per second. Write a ratio to show the rate of travel per minute. How far does the car travel in 2 minutes? 2.5 minutes? 5 seconds?
720 m; 900 m; 30 m

   3. In a recent election the ratio of "Yes" votes to "No" votes on a zoning amendment was 4 to 3. What fraction of the votes were "Yes" votes? "No" votes? More than half the votes were required for passage. Was the amendment passed? $\frac{4}{7}$; $\frac{3}{7}$; Yes

C   **4.** A Ferris wheel turns 9 times in 3 minutes. Write a ratio to show how long it takes to make one turn. $\frac{20}{1}$

5-2 Proportion

Objective To set up and solve proportions.

A **proportion** is a statement of equality of two ratios such as

$$\frac{3}{4} = \frac{18}{24} \quad \text{or} \quad 3{:}4 = 18{:}24$$

with "means" pointing to the 4 and 18, and "extremes" pointing to the 3 and 24.

The numbers 3, 4, 18, and 24 are called the **terms** of the proportion. The numbers 3 and 24 are called the **extremes** of the proportion, and the numbers 4 and 18 are called the **means**. From what we know about fractions, we have:

$$\frac{3}{4} = \frac{18}{24}$$

$$96 \times \frac{3}{4} = \frac{18}{24} \times 96$$

$$\overset{24}{\cancel{96}} \times \frac{3}{\cancel{4}} = \frac{18}{\cancel{24}} \times \overset{4}{\cancel{96}}$$

$$\text{extremes} \longrightarrow 24 \times 3 = 18 \times 4 \longleftarrow \text{means}$$
$$72 = 72$$

Thus we see that the product of the extremes is equal to the product of the means. This property is true for any proportion, that is,

property

If $\frac{a}{b} = \frac{c}{d}$, where b and d are not zero, then $ad = bc$.

We can use this property to help determine whether the two ratios in a proportion are indeed equal.

Example 1	Are these proportions correct?
	a. $\frac{9}{16} = \frac{17}{31}$ **b.** $\frac{8}{25} = \frac{4}{12.5}$
Solution	**a.** Since $9 \times 31 = 279$ and $16 \times 17 = 272$, the proportion is not correct.
	b. Since $25 \times 4 = 100$ and $8 \times 12.5 = 100$, the proportion is correct.

We can also use the property just stated to find one term of a proportion when we know the other three terms. This is called *solving the proportion*.

Example 2 Solve $\frac{10}{n} = \frac{60}{24}$ for *n*.

Solution Since $\frac{10}{n} = \frac{60}{24}$, we have.

$$n \times 60 = 10 \times 24$$
$$60n = 240$$
$$n = 4$$

We can sometimes use proportions to solve word problems involving ratios or rates. The following steps are helpful.

1. Determine what quantity is to be found.
2. Determine whether the given facts involve any ratios or rates.
3. Write the ratios or rates involved as a proportion.
4. Solve the proportion.

Example 3 Maria spends 17 hours in a 2-week period practicing on the guitar. How many hours does she practice in 5 weeks?

Solution

1. The problem asks for the number of hours practiced in 5 weeks.

2. The given facts include the rate of 17 h in (per) 2 weeks, and the unknown number of hours in 5 weeks.

3. The proportion is $\frac{17}{2} = \frac{n}{5}$.

4. $2n = 5 \times 17$

$$n = \frac{85}{2}$$

$$n = 42\frac{1}{2}$$

Therefore, in 5 weeks Maria practices $42\frac{1}{2}$ hours.

Class Exercises

Name the means and the extremes of the proportion.

1. $2{:}3 = 4{:}6$
3 and 4; 2 and 6

2. $1{:}5 = 3{:}15$
5 and 3; 1 and 15

3. $4{:}9 = 8{:}18$
9 and 8; 4 and 18

4. $7{:}21 = 3{:}9$
21 and 3; 7 and 9

5. $\frac{1}{2} = \frac{2}{4}$
2 and 2; 1 and 4

6. $\frac{12}{16} = \frac{3}{4}$
16 and 3; 12 and 4

7. $\frac{b}{3} = \frac{45}{27}$
3 and 45; b and 27

8. $\frac{n}{6} = \frac{9}{54}$
6 and 9; n and 54

9. $\frac{13}{19.5} = \frac{r}{1.5}$
19.5 and r; 13 and 1.5

10. $\frac{4.8}{12} = \frac{s}{3}$
12 and s; 4.8 and 3

11. $\frac{f}{g} = \frac{j}{k}$
g and j; f and k

12. $\frac{x}{y} = \frac{v}{w}$
y and v; x and w

Write a proportion for the situation described.

13. 5 m in one hour; 15 m in h hours $\frac{5}{1} = \frac{15}{h}$

14. 4 L of paint for 160 m²; 0.25 L for x m² $\frac{4}{160} = \frac{0.25}{x}$

15. 480 km on one tank of gas; 960 km on x tanks.
$\frac{480}{1} = \frac{960}{x}$

16. 49¢ for 2 lemons; $1.47 for n lemons
$\frac{49}{2} = \frac{147}{n}$

Exercises

Use the property of proportions to tell whether or not the proportion is correct.

A

1. $\frac{2}{3} = \frac{18}{27}$ Yes

2. $\frac{8}{5} = \frac{24}{15}$ Yes

3. $\frac{3}{12} = \frac{4}{14}$ No

4. $\frac{6}{5} = \frac{42}{35}$ Yes

5. $\frac{50}{32} = \frac{26}{16}$ No

6. $\frac{34}{15} = \frac{17}{8}$ No

7. $\frac{70}{90} = \frac{17}{22}$ No

8. $\frac{0}{112} = \frac{0}{65}$ Yes

Solve the proportion.

9. $\frac{n}{7} = \frac{15}{21}$ 5

10. $\frac{18}{12} = \frac{6}{n}$ 4

11. $\frac{n}{8} = \frac{2}{32}$ $\frac{1}{2}$

12. $\frac{24}{3} = \frac{n}{12}$ 96

13. $\frac{n}{21} = \frac{9}{27}$ 7

14. $\frac{19}{n} = \frac{57}{3}$ 1

15. $\frac{16}{3} = \frac{n}{12}$ 64

16. $\frac{n}{3.4} = \frac{2}{17}$ 4

17. $\frac{13}{42} = \frac{65}{n}$ 210

18. $\frac{3}{7} = \frac{n}{200}$ $85\frac{5}{7}$

19. $\frac{35}{100} = \frac{7}{n}$ 20

20. $\frac{1.21}{11} = \frac{5.5}{n}$ 5

21. $\frac{2}{n} = \frac{12.5}{2.5}$ $\frac{2}{5}$

22. $\frac{9.6}{1.6} = \frac{n}{3}$ 18

23. $\frac{2.7}{n} = \frac{8.1}{4.5}$ 1.5

24. $\frac{n}{1.7} = \frac{5.1}{6.8}$ 1.275

EXTRA! Two Teasers

1. If 9 sopranos can sing 3 songs in 12 minutes, how long will it take 27 sopranos to sing the same 3 songs? 12 minutes

2. If a cat and a half catches a mouse and a half in a day and a half, how long will it take 6 cats to catch 6 mice? 1 day

Problems

Solve.

A **1.** Art can type 75 words in 2 minutes. At this rate how many words can he type in 12 minutes? 450 words

2. Mr. Roberts mixes 3 bags of sand with every 2 bags of cement. How many bags of cement should he mix with 8 bags of sand? $5\frac{1}{3}$ bags of cement

3. A fruit punch contains 5 parts of apple juice to 2 parts of cranberry juice. If 12 L of apple juice are used, how much cranberry juice should be used? 4.8 L

B **4.** A recipe calls for 2 spoonfuls of butter for each egg used. One egg is used for every 3 servings to be made. How much butter is needed for 6 servings? 4 spoonfuls

5. The Athletic Department bought a dozen cans of tennis balls at $2.50 each, and 3 tennis rackets. If the ratio of the price of one can of balls to the price of one racket was 1:14, what was the total cost? $135

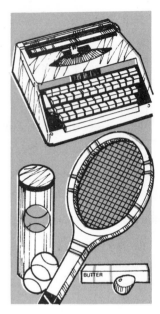

Self-Test

Words to remember:
ratio [p. 145] rate [p. 146] proportion [p. 149]
extremes [p. 149] means [p. 149]

Write each ratio as a fraction in lowest terms.

1. 3 to 12 $\frac{1}{4}$ **2.** $\frac{9}{15}$ $\frac{3}{5}$ **3.** 30 cm to 5 m $\frac{3}{50}$ **4.** 10 km to 50 m $\frac{200}{1}$ [5-1]

Tell whether or not the proportion is correct.

5. $\frac{2}{3} = \frac{5}{6}$ No **6.** $\frac{4}{14} = \frac{2}{7}$ Yes [5-2]

Solve the proportion.

7. $\frac{n}{3} = \frac{12}{9}$ 4 **8.** $\frac{21}{n} = \frac{9}{12}$ 28

Self-Test answers and Extra Practice are at the back of the book.

5-3 Percent

Objective To express percents as fractions and as decimals, and to solve problems involving percent.

The word **percent** (symbol: %) means divided by 100, or hundredths. For example, 85% means 85 divided by 100, which is the same as the ratio 85:100, or $\frac{85}{100}$. Another way to name 85% is to use a decimal.

$$85\% = \frac{85}{100} = 0.85.$$

Similarly, $3\% = \frac{3}{100} = 0.03$. Note that $1.00 = \frac{100}{100} = 100\%$; that is, 100% is another name for 1. Percents greater than 100% name numbers greater than 1; for example, $246\% = \frac{246}{100} = 2.46$.

Example 1 | Copy and complete the table. Give each fraction in lowest terms.

	Percent	Fraction	Decimal
a.	70%	?	?
b.	?	$\frac{7}{8}$	?
c.	?	?	0.025
d.	0.2%	?	?

Solution

a. $70\% = \frac{70}{100} = \frac{7}{10} = 0.7$

b. $\frac{7}{8} = \frac{n}{100}$

$8n = 7 \times 100$

$n = 700 \div 8 = 87.5$

$\frac{7}{8} = \frac{87.5}{100} = 87.5\% = 0.875$

c. $0.025 = \frac{25}{1000} = \frac{2.5}{100} = 2.5\%$

$\frac{25}{1000} = \frac{1}{40}$

d. $0.2\% = \frac{0.2}{100} = \frac{2}{1000} = 0.002$

$\frac{2}{1000} = \frac{1}{500}$

The following table shows some of the more commonly used percents and the fractions and decimals equal to them.

Percent	$12\frac{1}{2}\%$	$16\frac{2}{3}\%$	20%	25%	$33\frac{1}{3}\%$	50%
Decimal	0.125	$0.16\frac{2}{3}$	0.20	0.25	$0.33\frac{1}{3}$	0.50
Fraction	$\frac{1}{8}$	$\frac{1}{6}$	$\frac{1}{5}$	$\frac{1}{4}$	$\frac{1}{3}$	$\frac{1}{2}$

In the word phrase "5% of 40," the word "of" indicates multiplication. We can use this fact in solving problems involving percent.

Example 2   **What is 60% of 30?**

Solution   **Let x represent the number we are to find. Express 60% as a decimal: 60% = 0.60. Then,**
$$x = 0.60 \times 30$$
$$x = 18$$
So 18 is 60% of 30.

Example 3   **10 is what percent of 30?**

Solution   **Let n represent the percent. Then**
$$10 = n \times 30$$
$$n = \tfrac{10}{30} = 0.33\tfrac{1}{3}$$
$$n = 33\tfrac{1}{3}\%$$
So 10 is $33\frac{1}{3}\%$ of 30.

Example 4   **12 is 8% of what number?**

Solution   **Let $n =$ the number we are to find. Express 8% as a decimal: 0.08. Then, $12 = 0.08 \times n$**
$$n = \tfrac{12}{0.08} = 150$$
So 12 is 8% of 150.

Class Exercises

Express as an equal fraction with a denominator of 100.

1. $\frac{1}{4}$ $\frac{25}{100}$ 2. $\frac{4}{25}$ $\frac{16}{100}$ 3. $\frac{3}{50}$ $\frac{6}{100}$ 4. $\frac{3}{10}$ $\frac{30}{100}$ 5. $\frac{7}{20}$ $\frac{35}{100}$

6. $\frac{5}{8}$ $\frac{62.5}{100}$ 7. $\frac{1}{200}$ $\frac{0.5}{100}$ 8. $\frac{3}{200}$ $\frac{1.5}{100}$ 9. $\frac{1}{1000}$ $\frac{0.1}{100}$ 10. $\frac{25}{1000}$ $\frac{2.5}{100}$

For Problems 11–13, let $n =$ the number you are to find, then set up an equation that fits the problem.

11. In a class of 30 students 25 are on the honor roll. What percent is this? $25 = n \times 30$

12. Of those who tried out for the baseball team, 65% made it. If 13 players made the team, how many tried out? $13 = 0.65 \times n$

13. Sally's score on a test was 75%. If there were 28 questions on the test, how many did she answer correctly? $n = 0.75 \times 28$

14. Here are the results of a class survey on color preferences.

Student	Favorite Color
Bob	red
Jane	blue
Maria	red
Bill	yellow
Jim	green
Barbara	blue
Ralph	red
Ken	blue
Margo	yellow
Eric	blue

a. How many students took part in the survey? 10

b. What percent are boys? 60%

c. What percent are girls? 40%

d. What percent favor yellow? 20%

e. What percent favor red? 30%

f. What percent do not favor yellow? 80%

g. What percent of the girls favor blue? 50%

h. What percent are boys that favor yellow? 10%

Exercises

Copy the table and complete it.

		1.	2.	3.	4.	5.	6.	7.	8.	9.	10.
Percent		36%	75%	? *90%*	? *20%*	? *150%*	? *15%*	? *37.5%*	1.25%	? *248%*	0.02%
Fraction		$\frac{9}{25}$	? $\frac{3}{4}$	? $\frac{9}{10}$	$\frac{1}{5}$	$\frac{3}{2}$	? $\frac{3}{20}$	$\frac{3}{8}$	? $\frac{1}{80}$	? $\frac{62}{25}$	? $\frac{1}{5000}$
Decimal		? *0.36*	? *0.75*	0.9	? *0.2*	? *1.5*	0.15	? *0.375*	0.0125	2.48	? *0.0002*

Express as a percent.

11. 0.92 *92%*

12. 0.405 *40.5%*

13. $3\frac{1}{2}$ *350%*

14. 0.001 *0.1%*

15. 0.33 *33%*

16. 1.01 *101%*

17. 0.0403 *4.03%*

18. $4\frac{7}{10}$ *470%*

Find the percent or number.

B

19. What is 24% of 50? *12*

20. 15 is what percent of 40? *37.5%*

21. What is 12% of 32? *3.84*

22. 20 is 10% of what number? *200*

23. 80 is what percent of 100? *80%*

24. 250 is 500% of what number? *50*

25. 35 is what percent of 50? *70%*

26. 27 is 81% of what number? *$33\frac{1}{3}$*

27. What is 51% of 600? *306*

28. 210 is what percent of 210? *100%*

Problems

Solve.

A

1. Roger had 27 out of 30 mathematics problems correct. What was his percent of correct problems? *90%*

2. The Crown Ice Cream store gave 5% of its customers free cones one Saturday. That day they had 480 customers. How many received free ice cream cones? *24*

3. Jeff saved $25 of the $80 he earned one month. What percent of his earnings did he save? *31.25%*

4. Of all those who responded to a survey, 390 were under 18 years of age. If 20% were under 18, how many responded to the survey? *1950 persons*

5. The sales tax is 5% in Little Falls. What would be the tax on a purchase of $89.50? $4.48

B 6. Pat missed 7 words on a spelling test for a result of 86% correct. How many words were on the test? 50 words

7. Ten cartons were damaged in shipment. If this number was 1.6% of the cartons, how many were shipped? 625 cartons

Problems 8–12 refer to the table at the right.

8. What percent of the students are in the seventh grade? 34%

9. What percent of the eighth grade students are boys? 54%

10. Which grade has the smallest percent of girls? Grade 7; 54%

11. Of the ninth grade students, 44% were under 14 years of age on December 1. How many is this? 99 students

12. Morley Junior High School has 40% of the total junior high school enrollment in the city. How many junior high school students are enrolled in the city? 1795 students

Enrollments at Morley Junior High School			
Grade	Girls	Boys	Total
7	130	112	242
8	116	135	251
9	108	117	225
Total	354	364	718

Application

Computer Activity You can use the BASIC computer program given below to make a table of some fractions and the percents that are equal to them.

```
10 PRINT "FRACTION", "PERCENT"
20 FOR A=1 TO 15
30 FOR B=1 TO 15
40 IF B<A THEN 60
50 PRINT A; "/"; B, 100×A/B; "%"
60 NEXT B
70 NEXT A
80 END
```

5-4 Percent Increase and Decrease

Objective To solve word problems involving a percent increase or decrease.

Many everyday problems are concerned with percents. One important use of percent is to express relative amounts of change.

Example 1	Last year 84 students at Agawam Junior High School belonged to the Science Club. This year there are 25% more students in the club. How many students belong to the club this year?
Solution	**1.** Find the size of the increase. Let *x* represent the increase. $$x = 0.25 \times 84 = 21$$ **2.** Add the increase to the original number. $$84 + 21 = 105$$ Therefore, there are 105 students in the Science Club.

Example 2	In 1975 the price of a Fastadder pocket calculator was $25. In 1979 the same model calculator sold for $7.50. What was the percent of decrease in the price?

Solution	**Method 1** Find the amount of decrease: $$25 - 7.50 = 17.50.$$ Let *n* represent the percent. $$n \times 25 = 17.50$$ $$n = \frac{17.50}{25} = 0.70 = 70\%$$ Thus the decrease is 70%. **Method 2** Find what percent 7.50 is of 25. $$n \times 25 = 7.50$$ $$n = \frac{7.50}{25} = 0.30 = 30\%$$ Since the price in 1979 is 30% of the price in 1975, the percent of decrease is $100\% - 30\% = 70\%$.

Note that when you are looking for a percent increase or decrease, you always base your calculation on the *original amount.*

---

rule

$$\text{percent of change} = \frac{\text{amount of change}}{\text{original amount}}$$

Example 3 | The price of Maria Gomez's new car with extra equipment was $572 over the base price of the car. This amounted to 13% of the base price. What was the base price? How much did the car with the extra equipment cost?

Solution | Let *b* represent the base price of the car. Then

$$0.13 = \frac{572}{b}$$

$$0.13 \times b = 572$$

$$b = \frac{572}{0.13} = 4400$$

The base price was $4400. Equipped with extras the car cost $4400 + $572 = $4972.

Problems

Solve.

A 1. In a recent year, 31 million airline passengers used O'Hare Airport in Chicago. About 18% more passengers are expected to use the airport in 1990. About how many passengers may use it in that year?
36.58 million passengers

2. A Boeing 707 airliner can carry 185 passengers. The capacity of a 747 airliner exceeds the 707 capacity by about 170%. About how many passengers can a 747 carry? About 500 passengers

3. The stock of Supco Industries sold for $14 a share six months ago. Today its price is $21 a share. What is the percent of increase? 50% increase

4. Colleen is selling calendars door-to-door to raise money for the gymnastic team. Last week she sold 15 calendars. This week she sold 18. What is the percent of increase in sales? 20% increase

5. Fred Aho had 120 customers on his newspaper route. Now he is down to 90 customers. What is the percent of decrease? 25% decrease

6. The Slap-It-On paint store sold 1800 L of paint last week. This week 2250 L of paint was sold. What is the percent of increase? 25% increase

7. Ann Raymond earned $175 per week. With a promotion her salary became $200 per week. What was the percent of increase? 14% increase

8. The Farriers' heating bill went down 15% after they insulated their house. They paid $450 for a winter's heating before insulating. How much did they pay after insulating? $352.50

9. When Leo Hong's car was new he could make 15 round trips between home and work on one tank of gas. Now he can make only $13\frac{1}{2}$ round trips on one tank of gas. What is the percent of decrease in trips per tank of gas? 10% decrease

10. Lyn has made a New Year's resolution to spend an average of 5% less pocket money each week than she did last year. Her average weekly expenditure last year was $17. What will her average weekly expenditure be this year if she keeps her resolution? $16.15

B 11. Jon went to the store with the exact amount of cash to buy a $169 bicycle. He had forgotten the sales tax, though, so he was $6.76 short. What percent of the price was the tax? 4%

12. After a drop of 15% in the depth of the water in the Sullivan Reservoir, the depth gauge read 37.5 m. What was the original depth? 44 m

C 13. Sam's mathematics grade for this six weeks increased by 25% over his grade for the previous six weeks. His grade is now 95. What was it for the previous six weeks? 76

5-5 Commissions, Discounts, Budgets

Objective To solve problems involving commissions, discounts, and budgets.

A wide variety of problems in business and consumer settings involve percents.

Salespersons are often paid a *commission* in addition to a salary. A **commission** is usually a certain percent of sales. For example, a salesperson who works for a commission of 25%, and who sells $800 worth of goods, would earn 25% of $800, or $200. Sometimes the percent of sales that determines the commission is called the **rate** of the commission.

Example 1 | **A furniture salesperson worked for a commission rate of 8.25% of sales. Last year, total sales were $216,850. Find the salesperson's commission for the year.**

Solution | **Let x represent the commission. Then**

$$x = 0.0825 \times 216,850$$
$$x = 17,890.125$$

The salesperson's commission for the year was $17,890.13.

Stores sometimes sell merchandise at a **discount.** An advertisement that says "All appliances will be sold at a discount of 10%" means, for example, that a toaster originally marked $15.00 will sell at a discount of 10% of $15.00, or $1.50. The **sale price** or **net price** is $15.00 − $1.50, or $13.50.

Example 2 | **A top-selling record album usually sells for $7.95. Sound World is selling it at a 20% discount. What is the sale price?**

Solution | **Let x represent the amount of discount. Then**

$$x = 0.20 \times 7.95$$
$$x = 1.59$$

Original price -- discount = sale price
$$\$7.95 \quad - \quad \$1.59 \ = \ \$6.36$$

A budget is a plan one makes to be sure that one's income will be enough to cover expenses. It is a list of estimates of both income and expenses, usually for a year. One way to show a budget is with a **circle graph.**

Example 3 | The circle graph shows the Glover family's budget for a year. If their yearly income is $17,000, how much do they plan to spend on food in a year?

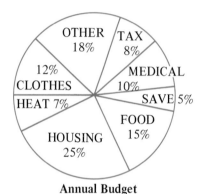

Annual Budget

Solution | From the graph, the Glovers expect to spend 15% of their income on food. Letting *n* be the amount they will spend,

$$n = 0.15 \times 17{,}000$$

$$n = 2550$$

The Glovers plan to spend **$2550** on food.

Problems

Solve.

A 1. Sid Romney earns a commission of 12% on what he sells. Last month his sales totaled $15,400. How much did he earn? $1848

2. Paula Jeffrey sells automobiles and earns a 6% commission. Last year she earned $21,000 in commissions. What was the dollar value of her total sales? $350,000

3. A real estate salesperson received $1000 as a commission for selling a $40,000 house. What percent of the sale price was the commission? 2.5%

B 4. The Monroe Zeppelins, a rock group, earned $94,500 last year in royalty income from their hit single *Flash Flood*. If the total sales of *Flash Flood* last year were $3,150,000, what was the group's royalty rate? 3%

Copy and complete the table.

Item	Kitchen Tool Set **5.**	Toaster **6.**	Calculator **7.**	Mixer **8.**	Blender $39.90 **9.**	Fan $38.80 **10.**
Regular Price	$4.95	$21.95	$69.00	$26.50	?	?
Discount Rate	50%	15%	35% ?	20% ?	10%	? 25%
Discount	?	?	$24.15	?	$3.99	$9.70
Sale Price	?	?	?	$21.20	$35.91	$29.10

$2.48 $3.29 $44.85 $5.30
$2.47 $18.66

11. Hair dryers are on sale for $12.50. They are marked as being 20% off. What was the regular price? $15.63

12. Kay Mitsui sells women's clothing. She earns a base salary of $400 per month plus a commission of 3% on all her sales. In a recent month her income was $986.50. What were her total sales? $19,550

13. The Randolph family pays $546 per year for medical insurance. This amounts to 40% of the cost of all their insurance, and 5% of their annual income. How much does all their insurance cost and how much is their annual income? $1,365; $10,920

14. Tim and Sally Wright are looking for an apartment. If they want to spend no more than 25% of their combined monthly take-home pay of $1375, what is the largest monthly rent they can afford? $343.75

15. The Roark family has a total income of $11,250. They budget $2400 for food and $2640 for shelter. What percent of the family budget goes for food and shelter? 44.8%

Use the circle graph at the right for Problems 16 and 17.

16. What percent does savings interest contribute to the family's total income? 4%

17. What percent do Madge's business earnings contribute to the family's total income? 40%

Yearly Income

Use the circle graphs on the preceding page and at the right for Problems 18 and 19.

C 18. What amount is budgeted for Madge's business expenses? $2500

19. What is the net income of Madge's business before taxes? (Net income = gross (total) income − expenses) $9000

20. Dick Hart received a 25% discount on a coat during a sale. He also received a 5% discount on the sale price for paying cash. He paid $79.56. What was the regular price of the coat? $111.67

Budget of Expenses

An Inventor

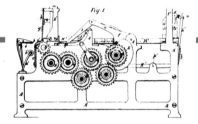

Paper Bag Machine

Margaret E. Knight (1838–1914) probably thought of her first invention when she was 12 years old. She was watching a loom in operation when one of the shuttles fell out and injured someone. Knight designed a stop-motion device that would hinder such accidents. In 1870, Knight received her first patent for a machine that folded square-bottomed paper bags. Her other inventions included a clasp for holding robes, a machine for cutting shoes, a window frame and sash, and a numbering mechanism. In 1902 she turned her efforts to rotary engines and motors. Her specific contribution was the invention of a sleeve-valve engine.

Research Activity Find out who invented the bicycle, the game of basketball, the transistor.

Career Activity Find out what a patent is and how to obtain one. For how long are patents effective? Some libraries have records of patents that have been issued. Can you find a copy of a patent?

Self-Test

Words to remember:

percent [p. 153] commission [p. 161] discount [p. 161]
budget [p. 162] net price [p. 161] circle graph [p. 162]

Find the percent or number.

1. What is 8% of 20? 1.6 2. What is 25% of 72? 18 [5-3]

3. 12 is what percent of 60? 20% 4. 8 is 30% of what number? $26\frac{2}{3}$

Solve.

5. The school play drew an audience of 255 persons on Friday. On [5-4]
Saturday, 51 more persons attended than on Friday. What was
the percent of increase? 20%

6. Sarah had 25 goldfish in her aquarium. She gave 3 of them away.
What was the percent of decrease in the number of goldfish? 12%

7. The Hobby Shop sold 16 10-speed bicycles in February and 25%
more than that in March. How many did they sell in March? 20

8. Of 460 entrants in a marathon race, 23 did not finish. What was
the percent of decrease between the number of entrants and the
number who finished the race? 5%

9. An auctioneer receives a commission of 15% of sales. At one [5-5]
auction, the sales were $12,856. What was the commission? $1928.40

10. An artist pays an art gallery 30% commission to sell her paintings.
One painting sold for $235. What commission did she pay the
art gallery? $70.50

11. The Baffin Bay Boat Yard pays a sales commission of 6% to its
salespersons. What would the commission be on a cabin cruiser
selling for $18,600? $1116

12. A jewelry store pays its salespersons a commission of 14%.
During one week the total sales were $173,000. How much
did the store pay in commissions? $24,220

Self-Test answers and Extra Practice are at the back of the book.

Chapter Review

Write the letter that labels the correct answer.

1. Expressed as a fraction in lowest terms, the ratio 12 to 9 is __?__. C [5-1]

   A. $\frac{3}{4}$　　　　**B.** $\frac{4}{6}$　　　　**C.** $\frac{4}{3}$　　　　**D.** $\frac{2}{3}$

2. Expressed as a fraction in lowest terms, the ratio of 25 m to 5 km is __?__. B

   A. $\frac{5}{1}$　　　　**B.** $\frac{1}{200}$　　　　**C.** $\frac{1}{5}$　　　　**D.** $\frac{200}{1}$

3. Tickets to a rock concert are selling at a rate of 2 for $5. How much do 6 tickets cost? A

   A. $15　　　　**B.** $12　　　　**C.** $6　　　　**D.** $30

4. The ratio of polar bears to pandas in a zoo is 6 to 2. If there are 9 polar bears in the zoo, how many pandas are there? D

   A. 5　　　　**B.** 6　　　　**C.** 4　　　　**D.** 3

5. The means of the proportion $\frac{3}{8} = \frac{9}{24}$ are __?__. D [5-2]

   A. 3 and 8　　**B.** 3 and 24　　**C.** 8 and 24　　**D.** 8 and 9

6. The extremes of the proportion $\frac{2}{5} = \frac{8}{20}$ are __?__. A

   A. 2 and 20　　**B.** 5 and 8　　**C.** 5 and 20　　**D.** 2 and 8

7. Which proportion is not correct? D

   A. $\frac{2}{3} = \frac{6}{9}$　　**B.** $\frac{1}{5} = \frac{4}{20}$　　**C.** $\frac{5}{6} = \frac{10}{12}$　　**D.** $\frac{1}{3} = \frac{7}{24}$

8. In the proportion $\frac{n}{6} = \frac{2}{3}$, $n =$ __?__. A

   A. 4　　　　**B.** 9　　　　**C.** 12　　　　**D.** 1

9. In decimal form 26% __?__. C [5-3]

   A. 0.026　　**B.** 2.60　　**C.** 0.26　　**D.** 26.00

10. Written as a fraction in lowest terms, 60% = __?__. D

   A. $\frac{6}{10}$　　　**B.** $\frac{6}{100}$　　　**C.** $\frac{30}{50}$　　　**D.** $\frac{3}{5}$

11. 15% × 30 = __?__. A

   A. 4.5　　　**B.** 60　　　**C.** 45　　　**D.** 20

12. 35 is what percent of 140? C
    A. 40%   **B.** 45%   **C.** 25%   **D.** 22%

13. The Marchands increased the size of their vegetable garden from   [5-4]
15 m² to 18 m². What was the percent of increase? D
    A. $33\frac{1}{3}\%$ **B.** $12\frac{1}{2}\%$ **C.** 62%   **D.** 20%

14. Jack worked 5 hours on Saturday. Jan worked 20% longer. How many hours did Jan work? A
    A. 6 hours   **B.** 7 hours   **C.** $6\frac{1}{2}$ hours   **D.** 8 hours

15. On Saturday, the Ski Club covered 8 km on a cross-country ski tour. On Sunday they traveled 10 km. What was the percent of increase? A
    A. 25%   **B.** 40%   **C.** 20%   **D.** 10%

16. Jeremy was 160 cm tall. In the next year he grew 3% taller. How tall was he then? C
    A. 163 cm   **B.** 174.8 cm   **C.** 164.8 cm   **D.** 164 cm

17. The Good Sport Shop pays a sales commission of 9%. How   [5-5]
much commission would they pay on sales of $52,364? D
    A. $5230   **B.** $471.28   **C.** $852.36   **D.** $4712.76

18. Cameras selling for $19.95 are put on sale at a 15% discount. What is the sale price? A
    A. $16.96   **B.** $14.96   **C.** $9.98   **D.** $12.95

19. Transistor radios are being sold at a discount of 25% of their original price, $12.59. How much would you save by buying one at the discount price? A
    A. $3.15   **B.** $6.59   **C.** $4.15   **D.** $9.44

20. A family budgets 22% of their income for rent. If their annual income is $15,000, how much do they spend for rent in one year? B
    A. $1330   **B.** $3300   **C.** $3100   **D.** $3000

The Golden Ratio

To Leonardo da Vinci and other artists of the Renaissance, a rectangle was most pleasing to the eye when the ratio of its length to its width was the same as the ratio of their sum to its length. We write this as the proportion shown at the right.

$$\frac{l}{w} = \frac{l+w}{l}$$

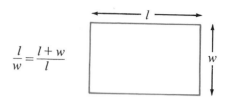

If we let the width be one, the length of this rectangle is the continuing decimal 1.618 . . . . The ratio 1.618 . . . : 1 is called the *Golden Ratio.* A rectangle whose length and width are proportional to the Golden Ratio is a *Golden Rectangle.*

Golden Rectangles occur frequently throughout modern and classical art and architecture.

A Golden Rectangle fits neatly around the central figure in da Vinci's unfinished painting, *St. Jerome.*

A Golden Rectangle bounds the original outline of the ancient Greek Parthenon.

How many rectangles which appear to be Golden Rectangles can you find in Piet Mondrian's *Place de La Concorde* at right?

Career Activity Applications to most art schools and for many types of art-related jobs include the submission of a portfolio. Find out how many and what kinds of materials different types of portfolios should contain.

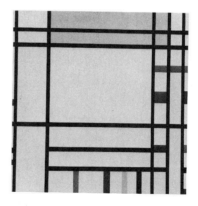

The Golden Ratio has a curious link to a sequence of numbers arising in the work of the medieval mathematician Leonardo of Pisa, nicknamed Fibonacci (son of Bonaccio). The infinite sequence of numbers 1, 1, 2, 3, 5, 8, 13, 21, . . . is called a *Fibonacci sequence.* To generate a Fibonacci sequence, we first pick any two integers as the first two *terms* of the sequence (in this case 1 and 1). Then each of the following terms is the sum of the previous two terms.

$$2 = 1 + 1 \qquad 3 = 2 + 1 \qquad 5 = 3 + 2$$

| 3rd term | 2nd term | 1st term | | 4th term | 3rd term | 2nd term | | 5th term | 4th term | 3rd term |

The ratio of the sixth term to the fifth term of this sequence is 8:5 = 1.6:1. This is very close to the Golden Ratio 1.618 . . . :1. In fact, if we compute the ratios of consecutive terms farther and farther along in this sequence, we will see that they get closer and closer to the Golden Ratio.

Research Activities

1. Write down the first several terms of the Fibonacci sequence which begins 2, 2, ...   2, 2, 4, 6, 10, 16, 26, . . .

How do the terms of this sequence compare with the terms of the sequence we considered earlier? They are twice as large.

How do the ratios of consecutive terms compare with those of the sequence we considered earlier? Explain.
The ratios are equal.

2. Botanists have discovered that numbers in a Fibonacci sequence (commonly called *Fibonacci numbers*) occur quite often in nature. For example, a pine cone has two sets of spirals, 8 spirals in one direction and 13 in the opposite direction, and 8 and 13 are Fibonacci numbers. Find some other instances of Fibonacci numbers in nature.

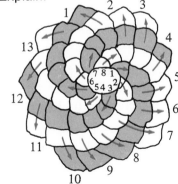

Application Consumer Price Index

We live in a time of *inflation,* that is, money buys less and less as time goes by. One measure of inflation is the Consumer Price Index (CPI). Computation of the CPI starts with a selection of goods and services used by consumers (food, clothing, housing, and so on). The CPI is the ratio, expressed as a percent, between the total price *now* for this selection and the price during a *base year.* For example, a current CPI of 185 means that the total price of a selection of goods is 185% of the price during the base year.

Research Activity Go to your library and find a current CPI. What is the base year? Based on your findings, what would have been the cost of $100 worth of groceries (at today's prices) during the base year?

Chapter Test

Write the ratio as a fraction in lowest terms.

1. 4 to 10 $\frac{2}{5}$

2. 80 cm to 1 m $\frac{4}{5}$ [5-1]

3. Tell whether or not the proportion $\frac{12}{20} = \frac{3}{4}$ is correct. No   [5-2]

4. Solve the proportion $\frac{n}{8} = \frac{9}{36}$. 2

Find the percent or number.

5. What is 5% of 40? 2

6. 18 is what percent of 90? 20%   [5-3]

Solve.

7. The attendance at the Wheatland County Fair was 182,560 last year. This year it increased by 5%. What was the attendance this year? 191,688 persons   [5-4]

8. During a two-month period, Jim increased his typing speed from 35 words per minute to 49 words per minute. What was the percent increase in his typing speed? 40% increase

9. Typewriters are on sale at a discount of 12% of their original price of $129.95. What is their sale price? $114.36   [5-5]

10. A swimming-pool salesperson receives a commission of 9% of sales. What would the commission be for selling a $2300 pool? $207

Skill Review

Express the ratio as a fraction in lowest terms.

1. 6 to 8 $\frac{3}{4}$
2. 14 to 35 $\frac{2}{5}$
3. 9 to 24 $\frac{3}{8}$
4. 56 to 16 $\frac{7}{2}$

5. 26 to 78 $\frac{1}{3}$
6. 34 to 18 $\frac{17}{9}$
7. 36 to 132 $\frac{3}{11}$
8. 69 to 42 $\frac{23}{14}$

9. 60 cm to 1 m $\frac{3}{5}$
10. 520 m to 1 km $\frac{13}{25}$

11. 300 mL to 1 L $\frac{3}{10}$
12. 3 L to 900 mL $\frac{10}{3}$

Solve the proportion.

13. $\frac{n}{6} = \frac{11}{3}$ 22
14. $\frac{n}{8} = \frac{7}{4}$ 14
15. $\frac{10}{n} = \frac{6}{33}$ 55
16. $\frac{14}{n} = \frac{12}{42}$ 49

17. $\frac{16}{5} = \frac{n}{30}$ 96
18. $\frac{8}{18} = \frac{n}{27}$ 12
19. $\frac{6}{45} = \frac{6}{n}$ 45
20. $\frac{13}{14} = \frac{26}{n}$ 28

21. $\frac{n}{3.3} = \frac{4}{6}$ 2.2
22. $\frac{n}{9} = \frac{3.2}{4.8}$ 6
23. $\frac{12.6}{n} = \frac{3.2}{8}$ 31.5
24. $\frac{5.5}{n} = \frac{11}{4.2}$ 2.1

25. $\frac{7.8}{6} = \frac{n}{3.4}$ 4.42
26. $\frac{12.5}{0.15} = \frac{n}{6}$ 500
27. $\frac{1.6}{9.6} = \frac{2.8}{n}$ 16.8
28. $\frac{24.6}{1.2} = \frac{0.41}{n}$ 0.02

Express as a percent.

29. $\frac{2}{5}$ 40%
30. $\frac{3}{4}$ 75%
31. $\frac{7}{10}$ 70%
32. $\frac{1}{8}$ 12.5%
33. $\frac{2}{3}$ $66\frac{2}{3}$%
34. $\frac{5}{6}$ $83\frac{1}{3}$%

35. $\frac{3}{8}$ 37.5%
36. $\frac{1}{12}$ $8\frac{1}{3}$%
37. $\frac{3}{5}$ 60%
38. $\frac{5}{4}$ 125%
39. $\frac{18}{10}$ 180%
40. $\frac{7}{3}$ $233\frac{1}{3}$%

41. 0.15 15%
42. 0.36 36%
43. 0.08 8%
44. 1.6 160%
45. 3.75 375%
46. 2.02 202%

47. $0.33\frac{1}{3}$ $33\frac{1}{3}$%
48. 0.125 12.5%
49. 0.045 4.5%
50. 2.125 212.5%
51. 0.375 37.5%
52. $0.66\frac{2}{3}$ $66\frac{2}{3}$%

Find the percent or number.

53. What is 4% of 32? 1.28
54. What is 15% of 80? 12

55. 6 is what percent of 48? 12.5%
56. 25 is what percent of 75? $33\frac{1}{3}$%

57. 35 is 20% of what number? 175
58. 17 is 9% of what number? $188\frac{8}{9}$

59. What is 5% of 85? 4.25
60. What is 93% of 400? 372

61. 18 is what percent of 72? 25%
62. 50 is what percent of 300? $16\frac{2}{3}$%

63. 14 is 75% of what number? $18\frac{2}{3}$
64. 48 is 30% of what number? 160

172

6

Measurement and Geometry

6-1 Units of Length

Objective To name points, segments, rays, and lines, and to use metric units of length.

The diagram at the right contains several geometric figures: points, segments, rays, and a line.

A **point** is represented by a dot labeled with a capital letter. If two points A and B are joined by drawing along the edge of a ruler, we get a picture of **line segment** AB, denoted $\overline{AB}$. Points A and B are called **endpoints** of the segment.

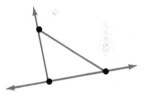

A •
point A

B •
point B

A •————————• B
segment AB

If $\overline{AB}$ is extended without end in just one direction, we obtain **ray** AB. Point A is called the endpoint of the ray. If $\overline{AB}$ is extended without end in both directions, we get **line** AB. A line has no endpoints.

ray AB line AB

In the diagram you should be able to locate:

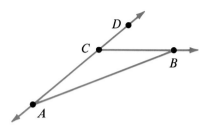

1. Segments: $\overline{AB}, \overline{CB}, \overline{CD}, \overline{CA}$, and $\overline{AD}$
2. Rays CB, CD, CA, DC, and AC. (Note ray AC is the same as ray AD.)
3. Line AC. (This is the same as line CD and line AD.)

The length of $\overline{XY}$ is denoted by the symbol XY. The diagram below shows that the length of $\overline{XY}$ is about 5.2 **centimeters (cm)**. Thus, we write $XY \approx 5.2$ cm. Since each centimeter is subdivided into 10 **millimeters (mm)**, we could also write $XY \approx 52$ mm.

Centimeter Ruler

The centimeter and millimeter are subdivisions of a **meter (m)**, the basic unit of length in the metric system. These and other metric units of length are shown in the table below. The units given in the non-shaded rows of the table are not used very often.

Name of unit	Abbreviation	Multiple of a meter
millimeter	mm	0.001, or 10^{-3}
centimeter	cm	0.01 , or 10^{-2}
decimeter	dm	0.1 , or 10^{-1}
meter	m	1 , or 10^{0}
dekameter	dam	10 , or 10^{1}
hectometer	hm	100 , or 10^{2}
kilometer	km	1000 , or 10^{3}
megameter	Mm	$1{,}000{,}000$, or 10^{6}

One of the nice things about the metric system is that it is easy to convert from one unit to another. You just move the decimal point.

$$5 \text{ mm} = 0.5 \text{ cm} = 0.005 \text{ m}$$
$$3500 \text{ m} = 3.5 \text{ km}$$

Class Exercises

1. In the diagram, are $\overline{AB}$ and $\overline{BA}$ the same segment or different segments? same

2. Is ray OQ the same as ray QO? No

3. Name several segments in the diagram. $\overline{AB}, \overline{AO}, \overline{OB}, \overline{PQ}, \overline{PO}, \overline{OQ}$

4. Name several rays in the diagram. ray PQ, ray QP

5. There is one line in the diagram. Name it in several ways. Line PQ or PO or OQ

6. What does the symbol AB stand for? length of $\overline{AB}$

Complete.

7. $4800 \text{ m} = \underline{\ ?\ } \text{ km}$ 4.8

8. $7.6 \text{ km} = \underline{\ ?\ } \text{ m}$ 7600

9. $340 \text{ cm} = \underline{\ ?\ } \text{ m}$ 3.4

10. $340 \text{ cm} = \underline{\ ?\ } \text{ mm}$ 3400

11. $57 \text{ mm} = \underline{\ ?\ } \text{ cm}$ 5.7

12. $800 \text{ mm} = \underline{\ ?\ } \text{ m}$ 0.8

13. Estimate the length of this page in centimeters. about 23 cm

14. Estimate the length of your classroom in meters. Answers will vary.

15. Look about your classroom to find an object about 50 cm long. Answers will vary.

16. Look about your classroom to find an object about 2 m long. Answers will vary.

17. Measure the width of your fingernails to find the one closest to 1 cm. Answers will vary.

Exercises

Complete.

A

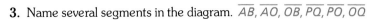

	1.	2.	3.	4.	5. 9300	6. 7200	7. 3500	8.
mm	486	1200	580 ?	40 ?	?	?	?	600 ?
cm	48.6 ?	120 ?	58	4	930 ?	720 ?	350 ?	60 ?
dm	4.86 ?	12 ?	5.8 ?	0.4 ?	93	72	35 ?	6 ?
m	?	1.2 ?	?	?	9.3 ?	7.2 ?	3.5	0.6

0.486   0.58   0.04

	9.	10.	11.	12.	13. 10,000	14. 3700	15. 5900	16. 15,000
m	4000	715	100 ?	550 ?	?	?	?	?
dam	400 ?	71.5 ?	10	55	1000 ?	370 ?	590 ?	1500?
hm	40 ?	7.15 ?	1 ?	5.5 ?	100	37	59 ?	150 ?
km	4 ?	?	0.1 ?	0.55?	10 ?	3.7 ?	5.9	15

0.715

In Exercises 17–20, find the lengths in centimeters and in millimeters. Measures are approximate.

17. *AB* 6.1 cm; 61 mm **18.** *CD* 3.3 cm; 33 mm **19.** *EF* 7.5 cm; 75 mm **20.** *GH*
13.7 cm; 137 mm

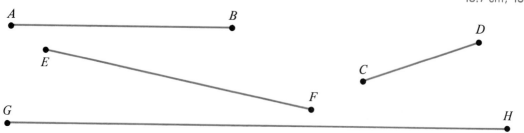

Complete.

21. 750 cm = __?__ m 7.5 **22.** 40 m = __?__ cm 4000 **23.** 2 km = __?__ m 2000

24. 500 m = __?__ km 0.5 **25.** 1 cm = __?__ km 0.00001 **26.** 1 km = __?__ mm 1,000,000

C **27.** Estimate the thickness of a page of this book. Answers may vary.

EXTRA! Compass and Straightedge Constructions

In *compass and straightedge constructions,* a ruler may be used as a *straightedge* to draw straight lines, but it may not be used to measure lengths.

How to copy a segment with a straightedge and compass

1. Draw a ray with endpoint *O*.

2. Spread your *compass* so the point is at *A* and the pencil tip is at *B*. Then put the point of your compass at *O* and draw an arc crossing the ray at *P*. Then *AB* = *OP*.

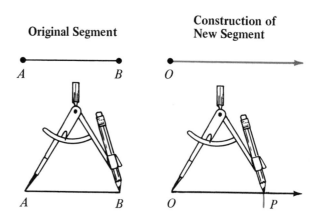

Research Activity Find out about other geometric constructions that use only a straightedge and compass.

6-2 Circles

Objective To identify parts of a circle and find its circumference.

The diagram shows a **compass** being used to draw a circle. The point of the compass is at the **center** of the circle. Any line segment joining the center to a point on the circle is called **a radius.** The length of such a segment is called **the radius** of the circle.

A **chord** of a circle is a segment joining two points of the circle. The diagram at the left below shows that a chord divides the circle into two **arcs.** If a chord contains the center of a circle, it is called **a diameter.** The arcs formed by a diameter are called **semicircles.**

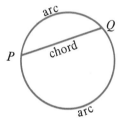

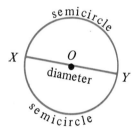

The length of a diameter is called **the diameter** of a circle, and the distance around the circle is called its **circumference,** or perimeter. The circumference and diameter of a circle are related by a very famous number called pi, denoted by the Greek letter π. Pi is defined as the ratio of a circle's circumference to its diameter.

$$\pi = \frac{\text{circumference}}{\text{diameter}}$$

If you measured the circumference and diameter of a record with a cloth tape measure, you would find:

$$\pi \approx \frac{94.6 \text{ cm}}{30.1 \text{ cm}} \approx 3.1$$

The value 3.1 is only an approximation for π. A better approximation is 3.14. Sometimes the fraction $\frac{22}{7}$ is used as an approximation.

formula

Circumference = π × diameter

$$C = π × d = πd$$

Example 1 | The diameter of a circle is **12.** Find its circumference by using two different approximations for π:

a. $π ≈ \frac{22}{7}$ **b.** $π ≈ 3.14$

Solution | **a.** $C = π × d$ **b.** $C = π × d$

$≈ \frac{22}{7} × 12$ $≈ 3.14 × 12$

$= \frac{264}{7} = 37\frac{5}{7}$ $= 37.68$

The diameter of any circle is twice its radius, r. Therefore, if you substitute $2r$ for d in the formula $C = π × d$, you get another formula which relates the circumference C and the radius r.

formula

$$C = π × 2r = 2πr$$

Example 2 | The circumference of a circle is **62.8 cm.** What is the radius? Use $π ≈ 3.14$.

Solution |

$C = \quad 2 × π × r$

$62.8 = 2 × 3.14 × r$

$62.8 = 6.28 × r$

$10 = r,$ or $r = 10$ cm

Class Exercises

The circle shown has center O.

1. Name a diameter of the circle. $\overline{AB}$

2. Name two radii. (*Radii* is the plural of *radius*.) $\overline{OA}; \overline{OB}; \overline{OC}$

3. If $OC = 5, AB = $ __?__ . 10

4. If $AB = 10$ cm, what is the circumference? (Use $π ≈ 3.14$.) 31.4 cm

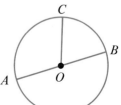

5. If you are told the circumference of a circle, how can you find the diameter?   Divide the circumference by π.

6. Which is correct, $\pi = \frac{22}{7}$ or $\pi \approx \frac{22}{7}$?

Exercises

Find the circumference of the circle described.
Use $\pi \approx 3.14$ for Exercises 1–6.

A
1. diameter = 100 cm
314 cm
2. diameter = 25 cm
78.5 cm
3. diameter = 76 mm
238.64 mm
4. radius = 15 cm
94.2 cm
5. radius = 1.7 m
10.676 m
6. radius = 2.18 km
13.6904 km

Use $\pi \approx \frac{22}{7}$ for Exercises 7–12.

7. diameter = 210 660

8. diameter = $3\frac{1}{2}$ 11

9. diameter = 49 154

10. radius = 42 264

11. radius = $\frac{42}{1000}$ $\frac{33}{125}$

12. radius = $1\frac{3}{11}$ 8

Find the diameter of the circle whose circumference is given. Use $\pi \approx \frac{22}{7}$.

13. 220 70   **14.** 66 21   **15.** $\frac{77}{100}$ $\frac{49}{200}$ **16.** $8\frac{4}{5}$ $2\frac{4}{5}$ **17.** 83.6 26.6   **18.** 2.42 0.77

B **19.** The curve shown is a semicircle. Which figure has the greater perimeter, the shaded figure or the unshaded one? The perimeters are equal.

$\mid\!\!\leftarrow 2\,\text{cm}\!\rightarrow\!\!\mid\!\!\leftarrow 2\,\text{cm}\!\rightarrow\!\!\mid$

Application Circles and Power

One of the earliest mechanical discoveries was the **wheel.** It is much easier to pull a load on a wagon than to drag it along the ground. From this idea developed many of our modern transportation vehicles.

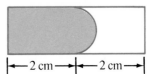

Circles are also used in **gears.** Gears usually have teeth of some kind that mesh with other gears. If the diameter of the large gear is twice that of the small gear, the circumference of the large gear is twice that of the small gear. In this case, the small gear will make two complete revolutions while the large gear is making one revolution.

A bicycle uses circles in both ways—as wheels and as gears.

Research Activity Find out how the gears on a bicycle work. Find out how the gears in an automobile work.

6-3 Measuring and Constructing Angles

Objective To measure and construct angles and identify special angles.

The diagram shows that an **angle** is the figure formed by two rays with the same endpoint. This endpoint is called the **vertex** of the angle, and the rays are called the **sides.**

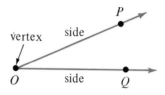

The diagrams below show various ways of naming angles.

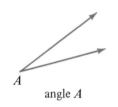

angle *A*

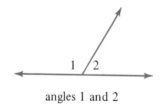

angles 1 and 2

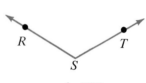

angle *RST*

The diagram at the left below shows a **protractor** being used to measure angles. The table at the right below shows how angles are classified according to their measures.

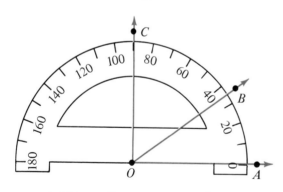

$\angle AOB = 35°$ $\angle AOC = 90°$

Name	Measure
Acute	Between 0° and 90°
Right	90°
Obtuse	Between 90° and 180°

If the sum of *two* angles is 180°, the angles are called **supplementary angles.** If the sum of *two* angles is 90°, the angles are called **complementary angles.**

If two lines form a right angle, they are said to be perpendicular. In the diagram, line *l* is perpendicular to line *m*. This is abbreviated by writing $l \perp m$. The mark (⌐) shown in the diagram is frequently used to indicate a right angle.

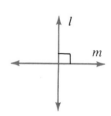

Whenever two lines intersect, four angles are formed. In the diagram, angles 1 and 3 are called **vertical angles.** So are angles 2 and 4. To see that *a pair of vertical angles have the same measure,* notice that

$$\angle 1 + \angle 2 = 180°, \text{ and}$$
$$\angle 3 + \angle 2 = 180°.$$

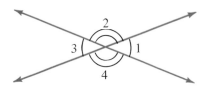

Vertical angles are equal.
$\angle 1 = \angle 3 \qquad \angle 2 = \angle 4$

From this you can conclude that $\angle 1 = \angle 3$. In the diagram, each pair of vertical angles is marked the same way to show that they have the same measure. Angles with the same measure are called **equal angles.**

In the diagram at the right, ray *SZ* divides angle *RST* into two equal angles. We say that *SZ* **bisects** angle *RST*.

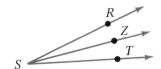

You can bisect an angle by using a protractor or by using a compass.

How to bisect an angle with a compass

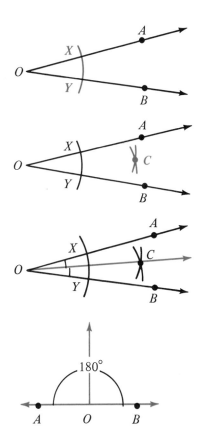

1. Suppose you want to bisect angle *AOB*. Draw an arc with center at point *O* and intersecting the sides of angle *AOB* at *X* and *Y*.

2. Draw an arc with center *X*. Then with center *Y* and the same radius, draw another arc intersecting the first one at *C*.

3. Join *O* to the point *C*. Now $\angle AOC = \angle BOC$, and ray *OC* bisects angle *AOB*.

Now that you can bisect an angle, you can construct a right angle. The secret to this is to think of a straight line as forming a "180° angle." You then bisect this "angle" using the same three steps as before. This will give you two 90° angles. This method for constructing a right angle is given on the next page.

How to construct a right angle using a compass

1. Draw a line, say line *AB*. Draw an arc with center at point *O* and intersecting line *AB* at points *X* and *Y*.

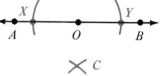

2. Spread the compass farther apart. Then draw two arcs with centers at *X* and at *Y*, intersecting at point *C*.

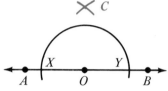

3. Join *O* to point *C*. Now angle *AOC* and angle *BOC* are right angles.

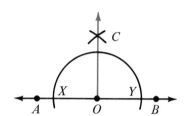

Class Exercises

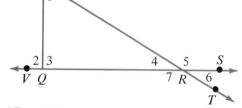

1. Name angle 1 using letters. *∠PQR*

2. Name angle *QRT* using a number. *∠7*

3. What is the vertex of angle 2? *Q*

4. Name an acute angle. *∠1, ∠4, or ∠6*

5. Name an obtuse angle. *∠5 or ∠7*

6. Name two pairs of vertical angles.
   ∠4 and ∠6, ∠5 and ∠7

7. Name three pairs of supplementary angles.
   ∠2 and ∠3, ∠4 and ∠5, ∠5 and ∠6, ∠6 and ∠7, ∠7 and ∠4

8. If ∠4 = 32°, then ∠6 = __?__°. *32°*

9. If lines *PQ* and *QR* are perpendicular, name two right angles. *∠2 and ∠3*

10. If you bisect a 24° angle, you get two angles, each with measure __?__. *12°*

11. If you bisect a right angle, you get two angles, each with measure __?__. *45°*

12. Draw an acute angle. Then bisect it using a compass. *For Ex. 12–14, see students' papers.*

13. Construct a right angle. Then use this angle to construct a 45° angle.

14. a. Fold a rectangular piece of paper so as to make a 45° angle.
    b. Fold it again to make a 22½° angle.

Use the diagram to find the measures of the following angles.

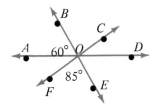

15. angle *DOE* 60°   **16.** angle *BOC* 85°

17. angle *COD* 35°   **18.** angle *AOF* 35°

Exercises

In the diagram, $\overline{BE} \perp \overline{CO}$ and $\angle AOB = 28°$.

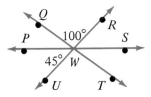

A **1.** Name two right angles. ∠*BOC* and ∠*EOC*

   2. Name two acute angles. ∠*AOB*, ∠*COD*, or ∠*DOE*

   3. Name two obtuse angles. ∠*AOC* and ∠*BOD*

   4. Name a pair of vertical angles. ∠*AOB* and ∠*DOE*

   5. Name a pair of complementary angles. ∠*COD* and ∠*DOE*

   6. $\angle DOE = \underset{28°}{\underline{\quad?\quad}}$ **7.** $\angle COD = \underset{62°}{\underline{\quad?\quad}}$ **8.** $\angle BOD = \underset{152°}{\underline{\quad?\quad}}$

Use the diagram to find the measures of the following angles.

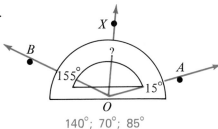

   9. angle *RWS* 45°   **10.** angle *TWU* 100°

   11. angle *TWS* 35°   **12.** angle *PWQ* 35°

In Exercises 13 and 14, ray *OX* bisets angle *AOB*. Find:
a. ∠*AOB*   **b.** ∠*AOX*
c. The number on the protractor corresponding to ray *OX*.

13.   **14.**

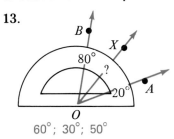

60°; 30°; 50°   140°; 70°; 85°

15. Draw an acute angle and bisect it using a compass.

16. Draw an obtuse angle and bisect it using a compass.

17. Construct a right angle.   **18.** Construct a 45° angle.

See students' papers for constructions to Ex. 15-19.

B **19.** **a.** Draw an acute angle and its bisector.
      b. Now extend the three rays in the opposite direction.
      c. Must the two new acute angles formed be equal? Tell why or why not.
         Yes, because they are vertical angles.

EXTRA! Constructions

How to copy an angle with a compass

Original Angle

Construction of
New Angle

1. Draw a ray with endpoint *O*.

2. Draw an arc with center *A*. Then with center *O* and the same radius, draw another arc.

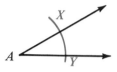

3. Put your compass point at *Y* and spread your compass so you can draw an arc through *X*. Then with center *C* and this radius, draw an arc which intersects the first arc at *B*.

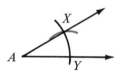

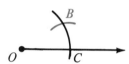

4. Draw ray *OB*.
Now ∠ *BOC* = ∠ *A*.

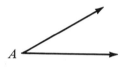

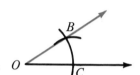

Draw an obtuse angle and copy it.

EXTRA! Angles Inscribed in Semicircles

Use a compass to draw several circles of different sizes. For each, draw a diameter $\overline{AB}$.

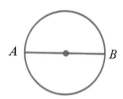

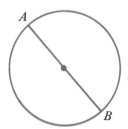

Locate a point *C* on each semicircle, and join *C* to *A* and to *B*. Angle *ACB* is **inscribed in** the semicircle. Measure angle *ACB* in each case. What do you discover?

∠*ACB* is always a right angle.

6-4 Parallel Lines

Objective To identify special angles associated with parallel lines.

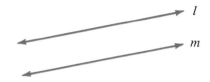

If two lines in a plane do not intersect, they are called **parallel lines.** In the diagram, we say that line *l* is parallel to line *m* and abbreviate this by writing *l*‖*m*.

Whenever two parallel lines are cut by a third line, called a **transversal,** several pairs of equal angles are formed. These pairs of equal angles are given special names as follows.

alternate interior angles
  angles 3 and 5
  angles 4 and 6

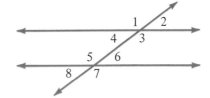

  corresponding angles
angles 1 and 5   angles 3 and 7
angles 2 and 6   angles 4 and 8

Example | In the diagram, lines *l* and *m* are parallel and ∠1 = 120°. Find the measures of the other angles.

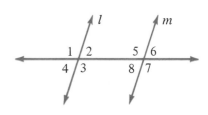

Solution | Angles 1 and 5 are corresponding angles, so ∠5 = 120°. Angles 5 and 3 are alternate interior angles, so ∠3 = 120°. Angles 3 and 7 are corresponding angles, so ∠7 = 120°. Since ∠1 = 120°,

$$\angle 2 = 180° - 120° = 60°.$$

Do you also see that

$$\angle 4 = \angle 6 = \angle 8 = 60°?$$

It is also true that if two lines are cut by a transversal so that pairs of angles are equal as described above, then the two given lines are parallel. We use this idea in the following construction, which is given on the next page.

How to use a compass to construct a line parallel to a given line through a given point

1. Given line m and point A. Draw line t through A cutting line m at B.

2. At A, construct angle 2 equal to angle 1. Then lines m and n are parallel since they form equal corresponding angles with transversal t.

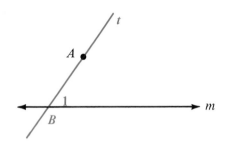

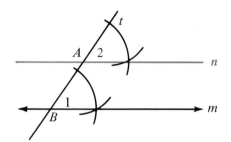

Class Exercises

In the figure, $a \parallel b$.

1. Name two pairs of alternate interior angles.
   $\angle 3$ and $\angle 5$, $\angle 4$ and $\angle 6$

2. Name four pairs of corresponding angles.
   $\angle 1$ and $\angle 5$, $\angle 2$ and $\angle 6$, $\angle 3$ and $\angle 7$, $\angle 4$ and $\angle 8$

3. Name three angles equal to angle 1. $\angle 3$, $\angle 5$, $\angle 7$

4. Name three angles equal to angle 2. $\angle 4$, $\angle 6$, $\angle 8$

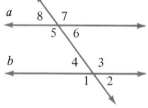

5. If $\angle 5 = 130°$, then $\angle 3 = $ _?_ and $\angle 4 = $ _?_ .
   130°   50°

6. If $\angle 6 = 40°$, then $\angle 2 = $ _?_ and $\angle 1 = $ _?_ .
   40°   40°

7. If $\angle 7 = 120°$, then $\angle 3 = $ _?_ and $\angle 1 = $ _?_ .
   120°   120°

8. Draw a large diagram of two parallel lines cut by a third line. Use a protractor to measure a pair of alternate interior angles. Answers will vary.

Exercises

In the figure, $l \parallel m$.

A

1. Name two pairs of alternate interior angles.
   $\angle 2$ and $\angle 8$, $\angle 3$ and $\angle 5$

2. Name four pairs of corresponding angles.
   $\angle 1$ and $\angle 5$, $\angle 2$ and $\angle 6$, $\angle 3$ and $\angle 7$, $\angle 4$ and $\angle 8$

3. Name three angles equal to angle 8.
   $\angle 2$, $\angle 4$, $\angle 6$

4. Name three angles equal to angle 7.
   $\angle 1$, $\angle 3$, $\angle 5$

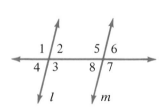

5. If $\angle 1 = 112°$, find the measures of the other seven angles.
   $\angle 2$, $\angle 4$, $\angle 6$, $\angle 8 = 68°$; $\angle 3$, $\angle 5$, $\angle 7 = 112°$

6. If $\angle 4 = 55°$, find the measures of the other seven angles.
   $\angle 2$, $\angle 6$, $\angle 8 = 55°$; $\angle 1$, $\angle 3$, $\angle 5$, $\angle 7 = 125°$

In the figure, $l \parallel m$ and $l \perp n$.

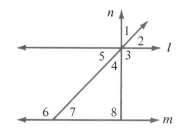

7. Angles 7 and __?__ are alternate interior angles. 5

8. Angles 7 and __?__ are corresponding angles. 2

9. If $\angle 1 = 55°$, find the measures of the other seven angles. $\angle 2 = 35°$, $\angle 3 = 90°$, $\angle 4 = 55°$, $\angle 5 = 35°$, $\angle 6 = 145°$, $\angle 7 = 35°$, $\angle 8 = 90°$

10. Is $m \perp n$?
Yes

B **11.** Use a ruler and compass to construct two parallel lines. Note that the ruler should be used only as a straightedge and not for measuring. See students' papers.

In Exercises 12 and 13, use a ruler and protractor to draw a figure like the one shown. Through point B, draw a line parallel to line CA. Through point A, draw a line parallel to line CB. These lines will meet at a point D. Use a protractor to find $\angle ADB$. See students' papers.

12.

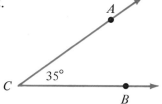

13.

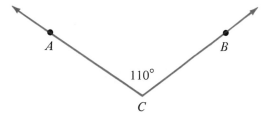

14. Draw two parallel lines cut by a third line. Now use a compass or a protractor to bisect the angles shown. Measure the angle at which the bisectors meet. Now repeat the experiment by drawing other parallel lines cut by a third line. What did you discover?
The bisectors meet at a 90° angle.

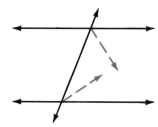

C **15.** Take a rectangular sheet and mark a point A on it. By folding the paper, make a crease line which goes through A and is parallel to the bottom edge of the sheet. Check students' constructions.

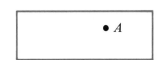

16. Take a sheet of paper which has just one straight edge (the bottom). Mark a point B on the paper.
a. By folding the paper, make a crease line which goes through B and is perpendicular to the bottom edge of the sheet.
b. Leave the paper folded, and make another fold which will pass through B and be parallel to the bottom edge of the sheet.

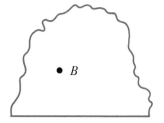

EXTRA! Rays Intersecting Parallel Lines

1. On a sheet of lined paper, choose a point A on one line and draw two rays from A as shown. Measure the distances between parallel lines along one ray. Are they all the same? Measure the distances between parallel lines along the other ray. Are they the same? Are they equal to the distances along the first ray?

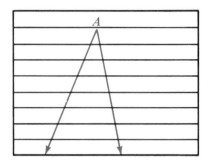

2. Draw a △ ABC so that AB = 6 cm. Locate P on $\overline{AB}$ so that AP = 4 cm and PB = 2 cm. Construct $\overline{PQ}$ parallel to $\overline{BC}$. Measure $\overline{AQ}$ and $\overline{QC}$. Find the ratio $\frac{AQ}{QC}$ and compare it with $\frac{AP}{PB}$. What do you notice? Locate point X on $\overline{AB}$ so that AX = 4.5 cm and XB = 1.5 cm. Construct $\overline{XY}$ parallel to $\overline{BC}$ (with Y on $\overline{AC}$). Measure $\overline{AY}$ and $\overline{YC}$. Find the ratio $\frac{AY}{YC}$ and compare it with $\frac{AX}{XB}$. What do you notice?

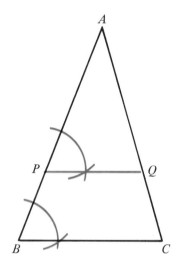

How to divide a segment into equal segments

3. You can divide a given segment such as $\overline{AB}$ into any given number of equal segments, such as three. Draw a ray with endpoint A. Starting with point A and any convenient radius, construct three equal segments on the ray. Call the points C, D, and E. Draw $\overline{EB}$. At C and D construct angles equal to angle AEB with one side intersecting $\overline{AB}$ at X and Y. Are $\overline{CX}$ and $\overline{DY}$ parallel to $\overline{EB}$? Why? Are AX, XY, and YB equal? Why?

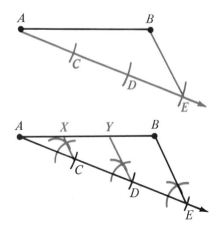

4. Draw $\overline{LM}$ and divide it into four equal segments.

Self-Test

Symbols and words to remember:

$\overline{AB}$ [p. 173]　　$\overleftrightarrow{XY}$ [p. 174]　　$\approx$ [p. 174]　　π [p. 177]　　$\angle$ [p. 180]
° [p. 180]　　$\perp$ [p. 180]　　$\parallel$ [p. 185]
point [p. 173]　　　line segment [p. 173]　　　endpoint [p. 173]
ray [p. 173]　　　　line [p. 173]　　　　　　center [p. 177]
radius [p. 177]　　　chord [p. 177]　　　　　arc [p. 177]
diameter [p. 177]　　semicircle [p. 177]　　circumference [p. 177]
angle [p. 180]　　　　acute, right, obtuse angles [p. 180]
supplementary, complementary angles [p. 180]　　vertical angles [p. 181]
equal angles [p. 181]　　bisect [p. 181]　　parallel lines [p. 185]
alternate interior, corresponding angles [p. 185]

Complete.

1. A ray has __?__ endpoint(s). 1
2. A line has __?__ endpoint(s). 0　**[6-1]**
3. 35 m = __?__ cm 3500
4. 200 mm = __?__ m 0.2

5. If a chord contains the center of a circle, it is called a __?__. diameter　　**[6-2]**

6. If the radius of a circle is 40 cm, then the circumference of the circle is __?__. (Use $\pi \approx 3.14$.) 251.20 cm

7. If the diameter of a circle is 42, then the circumference of the circle is __?__. (Use $\pi \approx \frac{22}{7}$.) 132

8. If the circumference of a circle is 17.6, then the radius of the circle is __?__. (Use $\pi \approx \frac{22}{7}$). 2.8

9. If we bisect a right angle, each angle formed has measure __?__. 45°　　**[6-3]**

10. A __?__ is used to measure angles. protractor

11. Two supplementary equal angles each have measure __?__. 90°

12. If one of two vertical angles has measure 50°, the other angle has measure __?__. 50°

13. Lines in the same plane that do not intersect are called __?__. parallel　　**[6-4]**

14. Whenever two parallel lines are cut by a third line, the corresponding angles are __?__. equal

15. If one of the angles formed when two lines intersect is a 42° angle, the measures of the other three angles are __?__. 42°, 138°, 138°

16. If two lines in the same plane are both perpendicular to a third line, the two lines are __?__. parallel

Self-Test answers and Extra Practice are at the back of the book.

6-5 Special Triangles

Objective To classify triangles.

The figure formed by $\overline{AB}$, $\overline{AC}$, and $\overline{BC}$ is called triangle ABC, abbreviated $\triangle ABC$.

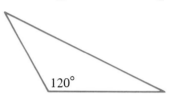

Classification by angles

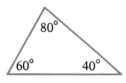

Acute Triangle
Three acute angles

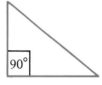

Right Triangle
One right angle

Obtuse Triangle
One obtuse angle

Classification by sides   (Small marks indicate sides of equal length.)

Scalene Triangle
No two sides
are equal

Isosceles Triangle
At least two sides
are equal.

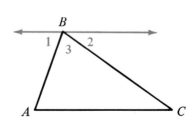

Equilateral Triangle
All sides are equal.

Class Exercises

1. Draw any kind of triangle (either acute, right, or obtuse) and measure its angles with a protractor. Add the three measures and compare your total with the totals of your classmates. The sum of the angles is 180°.

2. If you follow the instructions below, you will see why the sum of the angle measures in a triangle is 180°.
   a. Draw any $\triangle ABC$. Then draw a line through B parallel to $\overline{AC}$.
   b. Give reasons for each statement below.
      1. $\angle 1 = \angle A$ alt. int. angles are equal.
      2. $\angle 2 = \angle C$ alt. int. angles are equal.
      3. $\angle 1 + \angle 3 + \angle 2 = 180°$ straight angle
      4. $\angle A + \angle 3 + \angle C = 180°$ substitution

3. Suppose a friend drew a right triangle and told you that the measure of one angle was 40°. Could you tell the measures of the other two angles? 90° and 50°

4. Draw a triangle which is both acute and isosceles. (This is called an acute isosceles triangle.) By measuring, see if any of the angles are equal. Which ones? The angles opposite the equal sides.

5. Draw an obtuse isosceles triangle. See if any of the angles are equal. Which ones? The angles opposite the equal sides.

6. Construct an equilateral triangle with a compass. If you work accurately, each angle should measure 60°.

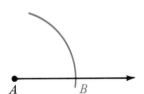 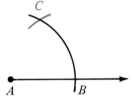 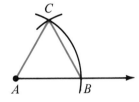

7. Use one of your 60° angles to construct a 30° angle. See students' papers.

SUMMARY OF RESULTS FROM CLASS EXERCISES

1. In any triangle, the angle measures total 180°.

2. In an isosceles triangle, the angles opposite the equal sides are themselves equal.

3. In an equilateral triangle, all three angles measure 60°.

Exercises

Refer to the triangles shown. Give the letter of each triangle which appears to be:

  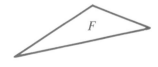

A 1. acute D, H, I

2. obtuse C, F

3. right A, B, E, G

4. scalene A, B, E, F

5. isosceles C, D, G, H, I

6. equilateral D, H

7. isosceles right G

Without using a protractor, find the measure of each angle denoted by a question mark.

8. 60°

9. 130°

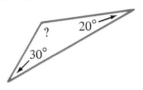

10. 58°

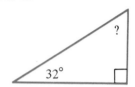

11. 70° and 70°

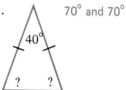

12.

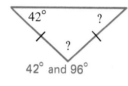

42° and 96°

13. 60°

14. Draw an isosceles right triangle. Can you give the three angle measures without using a protractor? 45°, 45°, 90°

Find the values of x and y in each diagram.

B **15.** x = 65°   y = 25°

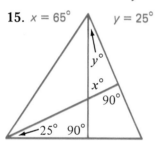

16. x = 20°   y = 90°

$\overline{AB} \parallel \overline{CD}$

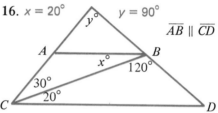

17. x = 35°, y = 130°

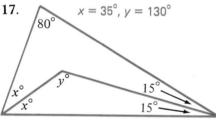

18. x = 35°, y = 35°

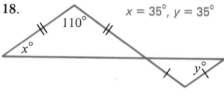

19.

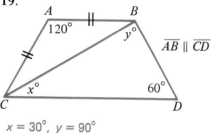

$\overline{AB} \parallel \overline{CD}$

x = 30°, y = 90°

20. x = 70°   y = 105°

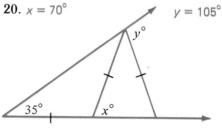

C **21.** In $\triangle ABC$, $\angle C = 70°$. The bisectors of angles A and B meet at point X. How large is angle AXB? (Draw the figure.) 125°

22. In $\triangle DEF$, M is a point of EF such that $ME = MF = MD$. How large is angle EDF? 90°

EXTRA! Paper Folding

1. Cut out a large $\triangle ABC$. By folding the paper, make a crease through A perpendicular to $\overline{BC}$. Let D be the point where the crease meets $\overline{BC}$. Now fold the paper so that B coincides with D. Fold it again so that C falls on D.

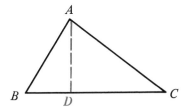 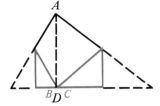

Finally, fold the paper so that A coincides with D. What does this experiment show about $\angle A + \angle B + \angle C$?

2. Cut a large triangle out of a piece of paper. By folding the paper, make a crease that bisects the smallest angle.
Now make creases to bisect the other two angles. If you worked accurately, the three creases should meet in a point. Do they?
Put the point of your compass at the place where the three creases meet. Can you draw a circle that "just fits" inside the triangle?

3. Cut out a large triangle having three unequal, acute angles.
Fold corner A onto corner C, as shown, so that $\overline{NA}$ falls on top of $\overline{NC}$.
Estimate the measure of the angle between the crease and $\overline{NC}$.
Check your estimate by measuring.
Now return A to its original position and fold B onto C.
Return B to its original position and fold A onto B.
If you worked carefully, your three creases should meet in a point. Do they?
Put the point of your compass at the place where the three creases meet. Can you draw a circle that goes through all three vertexes?

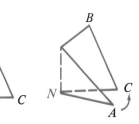

6-6 Special Quadrilaterals

Objective To identify some special quadrilaterals.

A **quadrilateral** is a four-sided closed figure. Some special quadrilaterals are shown below with some of their properties.

Trapezoid
Exactly two parallel sides

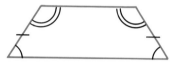

Isosceles Trapezoid
Two parallel sides
Other sides equal
Two pairs of equal angles (shown)

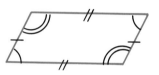

While a trapezoid has just one pair of parallel sides, a
parallelogram has two pairs of parallel sides. Each figure
shown below is a special kind of parallelogram.

Parallelogram
Opposite sides parallel
Opposite sides equal
Opposite angles equal

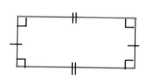

Rhombus
Opposite sides parallel
All sides equal
Opposite angles equal

Rectangle
Opposite sides parallel
Opposite sides equal
Four right angles

Square
Opposite sides parallel
All sides equal
Four right angles

Example

ABCD is a parallelogram.
Find: **a.** *DC*   **b.** *AD*
       c. ∠ *C*   **d.** ∠ *D*

Solution

Since opposite sides are equal,
DC = 4 and *AD* = 9.

Since opposite angles are equal,
∠ *C* = 55° and ∠ *D* = 125°.

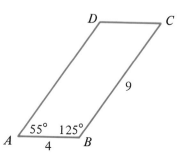

Class Exercises

Refer to the quadrilaterals shown. Give the letter of each one which appears to be a:

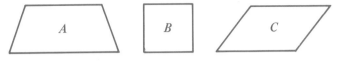

1. square *B, J*

2. rectangle *B, D, J*

3. rhombus *B, E, H, J*

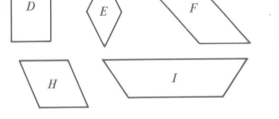

4. parallelogram
B, C, D, E, F, H, J
5. trapezoid
A, G, I
6. isosceles trapezoid
A, I

True or false?

7. All squares are rectangles. True

8. All rectangles are squares. False

9. All squares are rhombuses. True

10. All rhombuses are squares. False

Exercises 11–13 show why the opposite angles of a parallelogram must be equal.

11. In the diagram, $l \| m$ and $r \| s$.
What kind of figure is *ABCD*? parallelogram

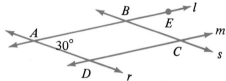

12. Suppose $\angle BAD = 30°$.
a. Find $\angle EBC$. 30° b. Find $\angle BCD$. 30°

13. Repeat Exercise 12 assuming that $\angle BAD = 40°$. 40°, 40°

Exercises

For each figure, supply the length or angle measure requested.

WXYZ is a parallelogram.

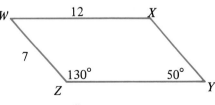

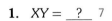
A   1. $XY =$ __?__ 7   2. $YZ =$ __?__ 12
3. $\angle X =$ __?__ 130° 4. $\angle W =$ __?__ 50°

STAR is a rhombus.

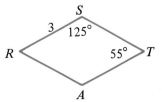

5. $TA =$ __?__ 3   6. $AR =$ __?__ 3
7. $\angle A =$ __?__ 125° 8. $\angle R =$ __?__ 55°

FLAT is a rectangle.

9. *FT* = __?__ 3   10. *FL* = __?__ 10

11. ∠*A* = __?__ 90° 12. ∠*FLT* = __?__ 16°

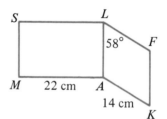

TANK is an isosceles trapezoid.

13. *NK* = __?__ 15

14. ∠*TKN* = __?__ 50°

15. ∠*KAN* = __?__ 19°

16. ∠*ANK* = __?__ 130°

ACEF and *BCDG* are parallelograms.

17. ∠*C* = __?__ 70° 18. ∠*G* = __?__ 70°

19. *FE* = __?__ 13.5 20. *DE* = __?__ 4.2

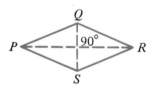

SLAM is a rectangle. *FLAK* is a rhombus.

21. ∠*SLF* = __?__ 148°

22. ∠*MAK* = __?__ 148°

23. *FK* = __?__ 14 cm

24. *SM* = __?__ 14 cm

B 25. *PQRS* is a rhombus. Its diagonals $\overline{PR}$ and $\overline{QS}$ are perpendicular. Experiment and see if this is true of the diagonals of other rhombuses. yes

26. By drawing several rectangles, see if the diagonals of a rectangle are perpendicular. Are they equal?
The diagonals of a rectangle are always equal, but only perpendicular if the rectangle is a square.

C 27. Rhombus *ABCD* has ∠*A* = 60° and *CD* = 12 cm. Find the distance from *B* to *D*. 12 cm

28. Four semicircles are constructed on the sides of a rectangle. The total distance around the figure is 66. If the length of the rectangle is twice its width, find the width. Use $\pi \approx \frac{22}{7}$.
w ≈ 7

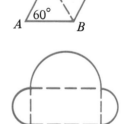

6-7 Polygons and Their Perimeters

Objective To identify some special polygons and find their perimeters.

A **polygon** is a closed figure formed by connecting segments at their endpoints. These endpoints are called the **vertexes** of the polygon, and the segments are called the **sides.** If all sides of a polygon are equal, the polygon is called **equilateral.**

Some polygons are given special names depending on the number of their sides.

Sides	3	4	5	6	8	10
polygon	triangle	quadrilateral	pentagon	hexagon	octagon	decagon

The **perimeter** of a polygon is the sum of the lengths of its sides.

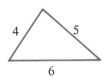

4 5

6

Perimeter = 15

0.9 cm

Perimeter = 0.9 + 0.9 + 0.9 + 0.9 + 0.9 + 0.9
= 6 × 0.9 = 5.4 (cm)

Sometimes you can calculate the perimeter of a figure even if you are not told the length of each side. Here are two examples.

Example 1 Find the perimeter of the parallelogram shown.

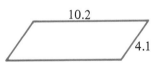

10.2

4.1

Solution The opposite sides of a parallelogram are equal.

Thus, perimeter = 10.2 + 4.1 + 10.2 + 4.1
= 28.6

Example 2 Find the perimeter of the figure shown. All the angles are right angles.

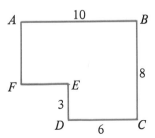

Solution Side *AF* must have length 8 − 3 = 5.
Side *FE* must have length 10 − 6 = 4.
Perimeter = 10 + 8 + 6 + 3 + 4 + 5
= 36

Class Exercises

Find the perimeter of each figure.

1.
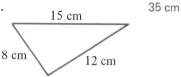
35 cm

2.
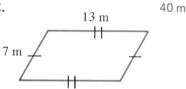
40 m

3.
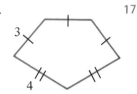
17

4.
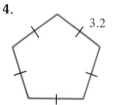
16

5. Estimate the perimeter of the floor in your classroom. Check your estimate by measuring. Answers may vary.

In Exercises 6–9, all the angles in the figures are right angles. Find the perimeter of each.

6.
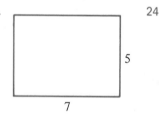
24

7.
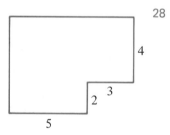
28

8.
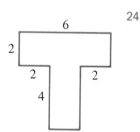
24

9.
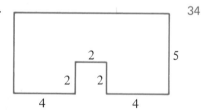
34

Tell how many sides each polygon has.

10. Quadrilateral 4   11. Octagon 8   12. Hexagon 6   13. Pentagon 5

14. What is an equilateral polygon? A polygon with all sides equal.

15. An equilateral quadrilateral is usually called a __?__ rhombus

Exercises

Find the perimeter.

A

1.

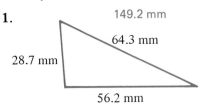

149.2 mm

64.3 mm

28.7 mm

56.2 mm

2.

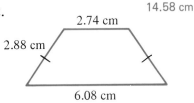

14.58 cm

2.74 cm

2.88 cm

6.08 cm

3.

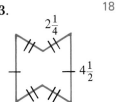

18

$2\frac{1}{4}$

$4\frac{1}{2}$

4.

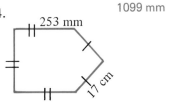

1099 mm

253 mm

17 cm

5. A STOP sign is an example of an equilateral __?__. octagon

6. If one side of a STOP sign is 15 cm, find its perimeter. 120 cm

7. If the perimeter of a STOP sign is 160 cm, find the length of each side. 20 cm

8. Each side of a pentagon is 14.3 cm. What is its perimeter? 71.5 cm

9. Each side of a hexagon is $6\frac{1}{4}$ units. What is its perimeter? $37\frac{1}{2}$ units

10. Two sides of a parallelogram have lengths 6.2 cm and 8.4 cm. Find the perimeter. 29.2 cm

11. The perimeter of a rhombus is 26.8 cm. Find the length of each side. 6.7 cm

12. The perimeter of an equilateral hexagon is 497.4 cm. Find the length of each side. 82.9 cm

In Exercises 13 and 14, the adjacent sides of the figure form right angles. Find the perimeter.

B

13.

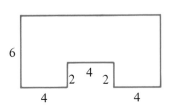

40

6

2 4 2

4   4

14.

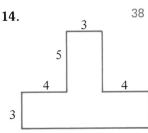

38

3

5

4   4

3

15. The perimeter of an isosceles trapezoid is 64 cm. The lengths of its parallel sides are 14 cm and 24 cm. How long are the other two sides? 13 cm

16. The figure at the right shows some walks through and around a park. If John and Sue start at the point marked, and if John takes the black path while Sue takes the red path, which one will travel the greater distance before returning to the starting point? They travel the same distance.

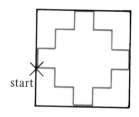

start

C 17. Four semicircles are constructed on the sides of a rhombus. If the total distance around the figure is 88 cm, what is the perimeter of the rhombus? Use $\pi \approx \frac{22}{7}$. 56 cm

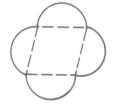

18. Which figure has the greater perimeter, the shaded figure or the unshaded one? They have the same perimeter.

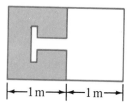

|←—1 m—→|←—1 m—→|

EXTRA! Polygons Inscribed in a Circle

1. Use a compass to draw a circle with any radius. Using the same radius, put the point of your compass at any point P of the circle and draw an arc intersecting the circle. Call the point of intersection A. Move the compass point to A and mark another arc intersecting the circle at point B. Continue marking arcs.
How many points on the circle have you marked? 6
Connect the points in order. What kind of figure has been formed? It is **inscribed** in the circle. hexagon
What is the relationship of the perimeter of the polygon and the diameter of the circle? Perimeter of hexagon = 3 X diameter

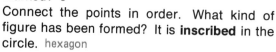

2. Using a protractor, mark 24 points equally spaced around a large circle. Use six differently colored pencils to show six different kinds of equilateral polygons which can be formed by connecting dots.

6-8 Congruent Polygons

Objective To identify congruent polygons and their corresponding parts.

If two polygons have the same size and shape, they are called **congruent** polygons. In the diagram below, $\triangle RST$ is congruent to $\triangle XYZ$. This statement can be abbreviated by writing

$$\triangle RST \cong \triangle XYZ.$$

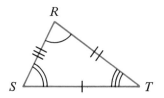
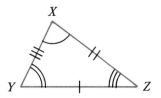

Corresponding parts of congruent triangles are equal.

Corresponding angles are equal	**Corresponding sides are equal**
$\angle R = \angle X$	$RS = XY$
$\angle S = \angle Y$	$ST = YZ$
$\angle T = \angle Z$	$RT = XZ$

When referring to two congruent polygons, it is customary to list their corresponding vertexes in the same order. Thus, we write $\triangle RST \cong \triangle XYZ$ and *not* $\triangle RST \cong \triangle YZX$.

Example | Suppose $ABCD \cong HGFE$.
Then $\angle B = \underline{\ ?\ }$ and $AD = \underline{\ ?\ }$.

Solution | Equal corresponding parts are listed in the same order. Thus, $\angle B = \angle G$ and $AD = HE$.

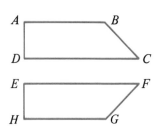

You can tell whether two polygons are congruent by seeing whether or not you can match their vertexes in such a way that all the corresponding parts are equal. However, to see if two triangles are congruent, you do not have to check all three pairs of sides and three pairs of angles. You can use any of the methods listed on the following page.

The side-angle-side (SAS) check for congruent triangles

If two sides and the included angle of one triangle are equal to two sides and the included angle of another triangle, then the triangles are congruent.

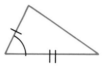

The angle-side-angle (ASA) check for congruent triangles

If two angles and the included side of one triangle are equal to two angles and the included side of another triangle, then the triangles are congruent.

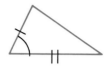

The side-side-side (SSS) check for congruent triangles

If three sides of one triangle are equal to three sides of another triangle, then the triangles are congruent.

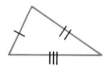

Class Exercises

Complete the statements below each pair of congruent polygons.

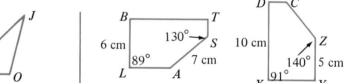

1. $\triangle RUN \cong$ __?__
   $\triangle JOG$

2. $RN =$ __?__
   JG

3. $\angle U =$ __?__
   LO

4. $XY =$ __?__ cm 6

5. $LA =$ __?__ cm 5

6. $BT =$ __?__ cm 10

7. $\angle A =$ __?__ 140°

8. $\angle Y =$ __?__ 89°

9. $\angle C =$ __?__ 130°

Each pair of triangles below is marked to show equal sides and equal angles. Tell whether or not the triangles in each pair are congruent. Also give your reason why (SAS, ASA, or SSS).

10.

Congruent; ASA

11.

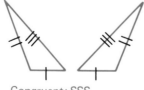

Congruent; SSS

12.

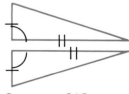

Congruent; SAS

13.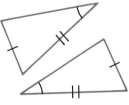

Not congruent

14.

Congruent; ASA

15.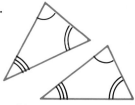

Not congruent

ABCD is a parallelogram.

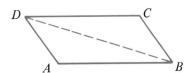

16. Explain how to use SSS to show △*ABD* ≅ △*CDB*.
  AB = *CD*, *AD* = *CB*, *BD* = *DB*

17. Explain how to use SAS to show △*ABD* ≅ △*CDB*.
  AB = *CD*, *AD* = *CB*, ∠*A* = ∠*C*

18. Explain how to use ASA to show △*ABD* ≅ △*CDB*.
  ∠*A* = ∠*C*, *AB* = *CD*, ∠*ABD* = ∠*CDB*

Exercises

Complete the statements below each pair of congruent polygons.

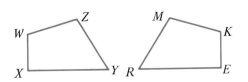   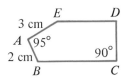

A

1. *WXYZ* ≅ __?__
  KERM

2. ∠*X* = __?__
  ∠*E*

3. *YZ* = __?__
  RM

4. *ABCDE* ≅ __?__
  NTQUI

5. *TN* = __?__ cm   2

6. *DE* = __?__ cm
  5

7. ∠*Q* = __?__   90°

8. ∠*N* = __?__
  95°

9. ∠*E* = __?__   150°

Tell whether or not the triangles in each pair below are congruent. Also give your reason (SAS, ASA, or SSS).

10.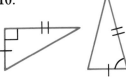

Not congruent

11.

Congruent; ASA

12.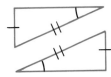

Not congruent

13.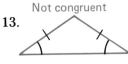

Not congruent

14.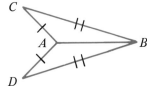

Hint: $\overline{AB}$ is a side of both triangles.

Congruent; SSS

15.

Hint: Vertical angles are equal.

Congruent; SAS

B **16.** Draw a large triangle. Then join the midpoints of the three sides so that four smaller triangles are formed. Which of these four triangles are congruent? All four of the triangles are congruent.

17. Draw a large parallelogram *ABCD*. Then draw diagonals $\overline{AC}$ and $\overline{BD}$ intersecting at *O*, so that four small triangles are formed. Which of these triangles are congruent? $\triangle AOB \cong \triangle COD$; $\triangle AOD \cong \triangle COB$

18. Repeat Exercise 17 for a rhombus *ABCD*.
$\triangle AOB \cong \triangle COB \cong \triangle COD \cong \triangle AOD$

19. *SLIM* is a rectangle. Name three triangles congruent to $\triangle SLI$. $\triangle IMS, \triangle LSH, \triangle HIL$

The figure shown is a cube. (All its edges are equal, and intersecting edges are perpendicular.)

C **20.** Why is $\triangle ABD \cong \triangle CBF$? Answer may vary.

21. Name other triangles congruent to $\triangle ABD$.
$\triangle EBD, \triangle EDF, \triangle EBF, \triangle GDF$

22. What kind of triangle is $\triangle BDF$? equilateral

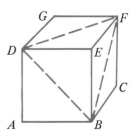

EXTRA!

Constructing Congruent Triangles

Side-angle-side (SAS) method of copying a triangle   **Original Triangle**   **Construction of New Triangle**

1. Construct $\overline{XZ}$ so that $XZ = AC$.

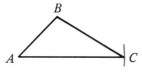
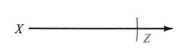

2. At *X*, copy angle *A*.

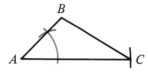
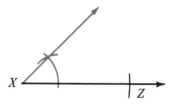

3. Construct $\overline{XY}$ so that $XY = AB$. Connect *Y* and *Z*. Now $\triangle XYZ \cong \triangle ABC$.

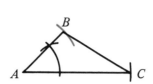
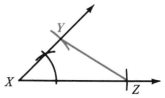

Angle-side-angle (ASA) method of copying a triangle

Original Triangle

Construction of New Triangle

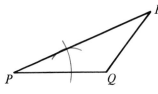

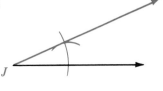

1. Copy angle *P*. Call the new angle *J*.

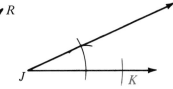

2. On one side of angle *J*, construct $\overline{JK}$ so that *JK* = *PQ*.

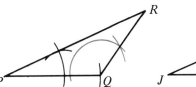

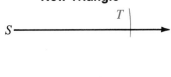

3. At *K*, copy angle *Q*. Let *L* be the point where the sides of angles *J* and *K* intersect. Now △*JKL* ≅ △*PQR*.

Side-side-side (SSS) method of copying a triangle

Original Triangle

Construction of New Triangle

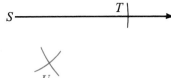

1. Construct $\overline{ST}$ so that *ST* = *DE*.

2. Adjust your compass so the point is at *D* and the pencil is at *F*. Now put the point at *S* and draw an arc. Now adjust the compass so the point is at *E* and the pencil is at *F*. Now put the point at *T* and draw an arc. Let *U* be the point where the arcs intersect.

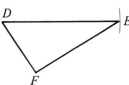

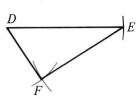

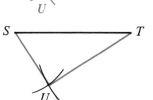

3. Draw $\overline{SU}$ and $\overline{TU}$. Now △*STU* ≅ △*DEF*.

Use a compass and straightedge to make all constructions.
1. Draw a large triangle. Copy it using the three methods described in this section.
2. Draw a parallelogram like the one shown. Construct a copy of it. (*Hint:* Draw a diagonal.)

6-9 Similar Polygons

Objective To identify similar polygons and their corresponding parts.

Congruent polygons have the same shape and size. **Similar** polygons have the same shape, but do not necessarily have the same size. We use the symbol ~ to stand for "is similar to."

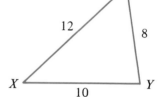

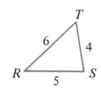

Congruent Polygons
△LMN ≅ △RST
Corresponding angles are equal.
Corresponding sides are equal.

Similar Polygons
△LMN ~ △XYZ
Corresponding angles are equal.
Corresponding sides are proportional.

When we say that corresponding sides of similar triangles are proportional, we mean that the ratios of the lengths of corresponding sides are equal. For the similar triangles above, we have:

$$\frac{MN}{YZ} = \frac{LM}{XY} = \frac{LN}{XZ} \qquad \frac{4}{8} = \frac{5}{10} = \frac{6}{12}$$

It is customary to list the corresponding vertexes of similar polygons in the same order, just as we did for congruent polygons. Thus for the similar polygons below, we write:

$ABCD \sim EFGH$

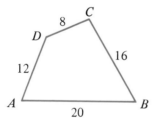
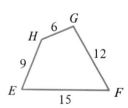

Notice again that the lengths of corresponding sides are proportional.

$$\frac{AB}{EF} = \frac{BC}{FG} = \frac{CD}{GH} = \frac{DA}{HE}$$

$$\frac{20}{15} = \frac{16}{12} = \frac{8}{6} = \frac{12}{9} \longleftarrow \text{Each of these ratios equals } \frac{4}{3}.$$

Example | If **MORT ~ NACK**, find:

a. ∠O   b. KC

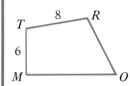
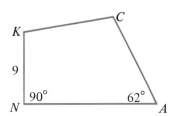

Solution | a. **Since corresponding angles are equal,**

∠O = ∠A = **62°**.

b. **Since corresponding sides are proportional,**

$$\frac{KC}{8} = \frac{9}{6}$$

6 × KC = 8 × 9 = 72

KC = 12

Class Exercises

Complete the statements given below each pair of similar polygons.

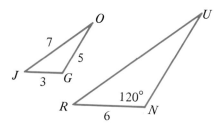

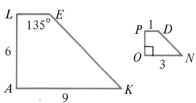

1. △JOG ~ __?__ △RUN

2. ∠G = __?__ 120°

3. $\frac{RN}{JG}$ = __?__ $\frac{2}{1}$

4. NU = __?__ 10

5. RU = __?__ 14

6. LAKE ~ __?__ POND

7. ∠A = __?__ 90°

8. ∠D = __?__ 135°

9. LE = __?__ 3

10. PO = __?__ 2

11. Are all squares similar? Yes

12. Are all rectangles similar? No

13. Give the length and width of some rectangle similar to ABCD.

Answers may vary. For example, length = 18, width = 3.

Exercises

Complete the statements given below each pair of similar polygons.

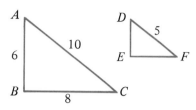
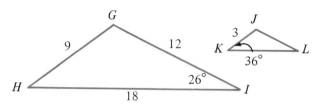

A

1. △ABC ∼ △ _?_ DEF
2. ∠D = ∠ _?_ A
3. ∠E = _?_ LB
4. DE = _?_ 3
5. EF = _?_ 4

6. △GHI ∼ △ _?_ JKL
7. ∠L = _?_ 26°
8. ∠H = _?_ 36°
9. JL = _?_ 4
10. KL = _?_ 6

11. Are all equilateral triangles similar? Yes **12.** Are all rhombuses similar? No

13. Two rhombuses each have a 60° angle. Must they be similar? Yes

14. Two parallelograms each have a 60° angle. Must they be similar? No

15. Two isosceles trapezoids each have a 100° angle. Must they be similar? Yes

16. Two isosceles triangles each have a 100° angle. Must they be similar? Yes

17. The length and width of a rectangle are 20 cm and 15 cm. The length and width of a similar rectangle are 12 cm and _?_ cm. 9

18. The length and width of one rectangle are twice the length and width of another rectangle. Are the rectangles similar? Yes

19. The length and width of one rectangle are each 2 cm more than the length and width of another rectangle. Are they similar? No

20. If two figures are congruent, are they also similar? Yes

Find the lengths of sides indicated by letters in these similar polygons.

B **21.**

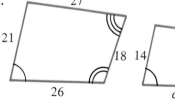

$a = 18;\ b = 12;\ c = 17\frac{1}{3}$

22.

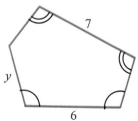

$x = 3\frac{1}{2};\ y = 4$

Self-Test

Symbols and words to remember:

△ [p. 190]   ≅ [p. 201]   ~ [p. 206]
triangle [p. 190]   acute, right, obtuse triangles [p. 190]
scalene, isosceles, equilateral triangles [p. 190]   quadrilateral [p. 194]
trapezoid [p. 194]   isosceles trapezoid [p. 194]   parallelogram [p. 194]
rhombus [p. 194]   polygon [p. 197]   equilateral [p. 197]
congruent [p. 201]   corresponding angles, sides [p. 201]
SAS, ASA, SSS [p. 202]   similar [p. 206]

1. If the measure of one angle of a right triangle is 30°, what are the   [6-5]
   measures of the other two angles? 60° and 90°

2. Can three sides of an isosceles triangle be congruent? Yes

3. What is the measure of each angle of an equilateral triangle? 60°

4. How many acute angles does an obtuse triangle have? 2

True or false? Write T or F.
5. Every square is a rhombus. T   [6-6]
6. An isosceles trapezoid has two pairs of equal sides. T
7. Every quadrilateral is a parallelogram. F
8. Opposite sides of a parallelogram are equal. T
9. A hexagon has eight sides. F   10. All decagons are equilateral. F   [6-7]

11. What is the perimeter of a rhombus with one side 6 cm long? 24 cm

12. If one side of a parallelogram is 3.8 cm long and the perimeter
    of the parallelogram is 15.6 cm, how long are the other sides of
    the parallelogram? 3.8 cm, 4 cm, 4 cm

13. If three angles of one triangle are equal to three angles of another   [6-8]
    triangle, must the triangles be congruent? No

Complete each statement for the congruent triangles.

14. △ABC ≅ _?_ △EFD

15. ∠D = _?_ ∠C

16. DF = _?_ BC

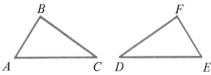

17. Are all congruent polygons also similar polygons? Yes   [6-9]

18. Are all squares similar? Yes   19. Are all rhombuses similar? No

20. Are all isosceles triangles similar? No

Self-Test answers and Extra Practice are at the back of the book.

Chapter Review

Write the letter that labels the correct answer.

1. A line segment has __?__ endpoints(s). C   [6-1]
   A. 0   B. 1   C. 2   D. many

2. 6 km = __?__ D
   A. 600 m   B. 0.006 m   C. 600,000 m   D. 6000 m

3. 20 mm = __?__ B
   A. 0.002 m   B. 0.02 m   C. 2000 m   D. 0.2 m

4. A __?__ is used to draw a circle. B   [6-2]
   A. protractor   B. compass   C. straightedge   D. ruler

5. If the radius of a circle is 10 m, the circumference of the circle is
   approximately __?__. C
   A. 31.4 m   B. 314 m   C. 62.8 m   D. 6.28 m

6. If the circumference of a circle is 154, the diameter of the circle
   is approximately __?__. A
   A. 49   B. 484   C. 51   D. 0.5

7. If two lines form a right angle, they are said to be __?__. B   [6-3]
   A. parallel   B. perpendicular
   C. intersecting   D. vertical

8. A pair of __?__ angles always have the same measure. D
   A. complementary   B. acute
   C. supplementary   D. vertical

9. The supplement of a 30° angle has measure __?__. C
   A. 60°   B. 30°   C. 150°   D. 180°

10. When two parallel lines are cut by a third line, alternate interior   [6-4]
    angles are __?__. B
    A. corresponding   B. equal
    C. vertical   D. supplementary

11. If lines l and m are parallel, we abbreviate this by writing __?__. C
    A. $l \perp m$ B. $l \cong m$ C. $l \| m$ D. $l \sim m$

12. In a plane, two lines perpendicular to a third line are __?__ each
    other. B
    A. perpendicular to   B. parallel to
    C. intersecting   D. equal to

13. In any triangle, the angle measures total __?__ . D [6-5]
  A. 360° **B.** 90° **C.** 135° **D.** 180°

14. A right triangle __?__ has two acute angles. A
  A. always **B.** sometimes **C.** never **D.** usually

15. In a(n) __?__ triangle, each angle measures 60°. C
  A. acute **B.** isosceles **C.** equilateral **D.** scalene

16. A trapezoid is __?__ isosceles. B [6-6]
  A. always **B.** sometimes **C.** never **D.** usually

17. A rhombus is __?__ a rectangle. B
  A. always **B.** sometimes **C.** never **D.** usually

18. A quadrilateral is __?__ a parallelogram. B
  A. always **B.** sometimes **C.** never **D.** usually

19. An octagon has __?__ sides. C [6-7]
  A. five **B.** six **C.** eight **D.** ten

20. If two sides of a parallelogram are 5 cm and 2 cm long, the perimeter of the parallelogram is __?__ . B
  A. 7 cm **B.** 14 cm **C.** 28 cm **D.** 3 cm

21. If the perimeter of an equilateral polygon is 24 cm and one side is 4 cm long, the polygon is a(n) __?__ . B
  A. pentagon **B.** hexagon **C.** octagon **D.** decagon

22. If $AB = DE$, $BC = EF$, and $\angle B = \angle E$, then $\triangle ABC \cong \triangle DEF$ [6-8]
  by __?__ . A
  A. SAS **B.** ASA **C.** AAA **D.** SSS

23. If pentagon $JKLMN \cong$ pentagon $PQRST$, then __?__ . C
  A. $JKLMN \cong TSRQP$ **B.** $\angle J = \angle T$
  C. $\angle K = \angle Q$ **D.** $JK = QR$

24. A method which is *not* used to check whether triangles are congruent is __?__ . C
  A. SAS **B.** ASA **C.** AAA **D.** SSS

25. If $ABCD \sim EFGH$, then __?__ . C
  A. $AB = EF$ **B.** $CD = GH$ **C.** $\angle A = \angle E$ **D.** $\angle A = \angle H$ [6-9]

26. Two __?__ are always similar to each other. C
  A. parallelograms **B.** rectangles
  C. squares **D.** rhombuses

Mirror Geometry

GEOMETRY WITH ONE MIRROR

Can you read the word at the right? It is done in *mirror writing*. If you hold a mirror upright along the red line and look at the reflection of the word, you will be able to read it more easily. Sometimes this word is printed as shown here on the front of an ambulance. Why, do you think, is this so?

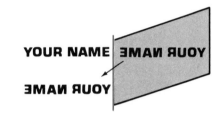

You can show your name in mirror writing by following these steps.

1. Write your name.

2. Hold a mirror upright alongside it as shown in the figure at the right.

3. Copy the reflection of your name in the mirror.

Activities

1. In the diagram at the right, hold a mirror upright along the red line that cuts the angle. What kind of figure is formed by the angle and its reflection in the mirror? What happens to this figure as you move the mirror left and right?

2. In the diagram at the right, hold a mirror upright along the red line that cuts the angle. What kind of figure is formed by the angle and its reflection in the mirror? What happens to this figure as you move the mirror left and right?

In Activities 3–6 you are to draw a reflection as seen in a mirror. Ask a friend to help you by holding the mirror and the paper on which you are drawing.

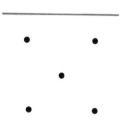

3. Copy the figure at the right. With a mirror held upright along the red line, look *only* at the reflection of the dots and connect them to make an *X*.
4. Repeat Activity 3, but connect the dots to make an *N*.
5. With a mirror held upright along a line, draw a triangle by looking *only* at the reflection in the mirror.
6. Repeat Activity 5, but draw a circle.

GEOMETRY WITH TWO MIRRORS

Tape together two pocket mirrors so that they open like a book. Now open the mirrors to a 90° angle and set them down on a line drawn on a piece of paper as shown at the right.

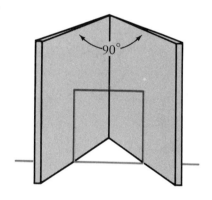

Activities

1. As you change the angle between the two mirrors, the reflection of the line will form different polygons. By experimenting, you can complete the following table.

Angle	90°	120°	60°	45°	?
Polygon	square	?	?	?	pentagon

2. Make a colored design on a piece of paper and set the open mirrors down on it. Watch what happens to the reflections as you change the angle between the mirrors.

Chapter Test

1. Are ray *AB* and ray *BA* the same ray? No   [6-1]

2. Are line *CD* and line *DC* the same line? Yes

Complete.

3. If the radius of a circle is 6 cm, the circumference is __?__. (Use   [6-2]
   $\pi \approx 3.14$.) 37.68 cm

4. If the circumference of a circle is 135.6, the diameter is __?__.
   (Use $\pi \approx \frac{22}{7}$.) 43.15

5. The measure of a(n) __?__ angle is between $0°$ and $90°$. acute   [6-3]

6. The sum of two __?__ angles is $90°$. complementary

7. If you bisect a $48°$ angle, you get two angles each with measure
   __?__. 24°

8. Two lines in a plane which do not intersect are called __?__. parallel   [6-4]

9. Two lines in a plane which are perpendicular to a third line are
   __?__ to each other. parallel

10. Whenever two parallel lines are cut by a third line, corresponding
    angles are __?__. equal

11. A triangle whose sides are three different lengths is called __?__.   [6-5]
    scalene
12. In a(n) __?__ triangle, all three angles measure $60°$. equilateral

13. If one angle of a parallelogram measures $60°$, the angle opposite   [6-6]
    it measures __?__. 60°

14. If all the sides of a parallelogram are equal, the parallelogram
    is called a(n) __?__. rhombus

15. An octagon has __?__ more side(s) than a pentagon has. 3   [6-7]

16. If two sides of a parallelogram are 11 cm and 9 cm long, the
    perimeter is __?__. 40 cm

17. $\triangle ABC \cong$ __?__ $\triangle FDE$ [6-8]

18. The triangles are congruent by
    __?__. SSS

19. Corresponding angles of similar polygons are __?__. equal   [6-9]

20. Corresponding sides of similar polygons are __?__. proportional

Cumulative Review

1. Name the fractions whose graphs are shown. [Chap. 4]

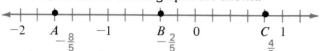

A $-\frac{8}{5}$ B $-\frac{2}{5}$ C $\frac{4}{5}$

Write each improper fraction as a mixed number.

2. $-\frac{11}{3}$ $-3\frac{2}{3}$ **3.** $\frac{51}{8}$ $6\frac{3}{8}$ **4.** $\frac{17}{-6}$ $-2\frac{5}{6}$ **5.** $\frac{100}{9}$ $11\frac{1}{9}$

Simplify. Express your answer in lowest terms.

6. $-2\frac{2}{3} \div \left(-1\frac{7}{16}\right) \times 3\frac{3}{8}$ $6\frac{6}{23}$ **7.** $\frac{5}{8} - \frac{3}{4} + \left(-1\frac{1}{6}\right)$ $-1\frac{7}{24}$

Solve each proportion.

8. $\frac{n}{7} = \frac{3}{12}$ $1\frac{3}{4}$ **9.** $\frac{2.4}{w} = \frac{0.7}{3.5}$ 12 **10.** $\frac{1.44}{12} = \frac{q}{5}$ 0.6 **11.** $\frac{0}{1.4} = \frac{y}{2.8}$ 0 [Chap. 5]

12. If Jim rides his bicycle 90 km in 4 weeks, how far does he ride his bicycle in 7 weeks? 157.5 km

13. 12 is 40% of what number? 30   **14.** Change $\frac{7}{8}$ to a percent. 87.5%

15. A radio normally selling for $32 is on sale for $28. What is the percent discount? 12.5%

16. A salesclerk earns a $5\frac{1}{2}\%$ commission selling sporting goods. If she earned $13,000 in commissions last year, what was the value of her total sales? $236,363.64

Complete.

17. 3720 m = ___?___ km 3.72   **18.** 32 mm = ___?___ cm 3.2 [Chap. 6]

Find the circumference of the circle described. (Use $\pi \approx 3.14$.)

19. diameter = 32 cm 100.48 cm   **20.** radius = 3.25 m 20.41 m

21. Use a ruler and compass to construct a $22\frac{1}{2}°$ angle.

22. Find the values of x and y in the diagram.

23. $\overline{AB} \parallel \overline{CD}$ and $AB = CD$. Does $AO = OD$? Yes Explain.

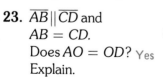
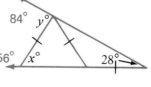

Exercise 22

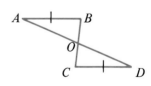

Exercise 23

24. A rectangle is 21 m long and 9 m wide. What is the perimeter of a similar rectangle 2.25 m wide? 15 m

215

7

Equations and Inequalities

7-1 Word Phrases and Numerical Expressions

Objective To translate word phrases describing numerical operations into numerical expressions, and vice versa.

The group of words

> the sum of eight and four

is not a sentence, because it does not contain a verb. The group of words forms a word phrase. When we put this phrase into the mathematical form

$$8 + 4$$

we have a numerical expression that names the number twelve. The numerical expressions "12," "4 $\times$ 3," and "24 $\div$ 2" also name the number twelve. Of course, "12" is the simplest name for twelve. Whenever we replace a numerical expression by its simplest name, we say that we have **simplified the expression.**

A numerical expression containing a variable is called a **variable expression.** Thus, the following are variable expressions:

$$6x \qquad 2y + 5 \qquad n \qquad 3(m - 4)$$

217

Example 1 | Translate into a numerical expression: The product of nine and *b*.

Solution | **9b**

Example 2 | Translate into a word phrase:

a. **6 + 11n**   b. **x − 3**   c. **7(a + 5)**   d. $\frac{r}{2}$

Solution | **a. The sum of six and the product of eleven and n.**
b. The difference when three is subtracted from x.
c. The product of seven and the sum of a and five.
d. The quotient when r is divided by two.

Class Exercises

Translate the given word phrase into a numerical expression.

1. The product of negative four and two. -4×2

2. The sum of nineteen and negative twelve. $19 + (-12)$

3. The difference when fourteen is subtracted from nine. $9 - 14$

4. The quotient when eighteen is divided by negative six. $\frac{18}{-6}$

Translate into a word phrase.

5. $8 + t$ **6.** $15n$ **7.** $3n - 8$ **8.** $5(k - 4)$

Translate into a variable expression.

9. The product of fifteen and *n*. $15n$

10. *x* divided by seven. $\frac{x}{7}$

11. Eleven greater than *c*. $c + 11$

12. Nine less than *d*. $d - 9$

Exercises

Write a variable expression for the word phrase.

A   **1.** Eight greater than *q* $q + 8$

2. Four less than *t* $t - 4$

3. The product of seven and *z* $7z$

4. The quotient when *m* is divided by negative nineteen $\frac{m}{-19}$

5. The product of nine and the sum of a and 3 $9(a + 3)$

6. The quotient when x is divided by the sum of 2 and n $\frac{x}{2 + n}$

Match the word phrase with the numerical expression.

7. The difference when y is subtracted from x c

8. The sum of x and twice y f

9. Twice the sum of x and y e

10. The sum of 2 and the product of x and y a

11. The quotient when x is divided by y d

12. The quotient when y is divided by x b

a. $2 + xy$

b. $\frac{y}{x}$

c. $x - y$

d. $\frac{x}{y}$

e. $2(x + y)$

f. $x + 2y$

Write a word phrase for the variable expression. See students' papers.

B   **13.** $-5 + t$ **14.** $x - 9$ **15.** $2x + 7$ **16.** $-3z + 4$

17. $\frac{a}{2}$ **18.** $\frac{b - 2}{6}$ **19.** $\frac{1}{2}ab$ **20.** $\frac{-3 - n}{5}$

EXTRA! Think of a Number . . .

You can use variable expressions to do a magic trick. Ask a friend to think of a number but not to tell you what it is. Then give the following directions. While your friend performs these operations, you perform them mentally on the variable n.

Directions	You think	Your friend thinks (for example)
1. Think of a number.	n	4
2. Add 3.	$n + 3$	7
3. Multiply by 6.	$6n + 18$	42
4. Subtract 8.	$6n + 10$	34
5. Divide by 2.	$3n + 5$	17
6. Subtract 5.	$3n$	12

Now you ask for the number your friend has arrived at. In this case it is 12. Since you know it is $3n$, or 3 times the original number, as if by magic you tell your friend the original number was 4.

Research Activity Try making up your own set of directions for "Think of a Number."

7-2 Evaluating Numerical Expressions

Objective To evaluate a variable expression for given values of each variable.

In Chapter 2 we learned that we can evaluate an expression such as $x + 3$ by replacing the variable x with one of its values and doing the indicated arithmetic, or simplifying the expression. We say that we are **substituting** a value for x.

Example 1	Evaluate $3 + x$ if the value of x is 2.
Solution	Substituting 2 for x in $3 + x$, we obtain $$3 + 2, \text{ or } 5$$

In evaluating expressions with grouping symbols such as parentheses or brackets, we perform the operation within the grouping symbols first. In evaluating expressions with fraction bars, we perform the operations indicated above and below the fraction bars first.

Example 2	Evaluate $4(1 - y)$ if $y = 5$.
Solution	Substituting 5 for y, we have $$4(1 - 5) = 4(-4) = -16$$

Example 3	What is the value of $\frac{2n + 5}{3}$ when the value of n is 17?
Solution	Replacing n with 17, we obtain $$\frac{(2 \times 17) + 5}{3} = \frac{34 + 5}{3} = \frac{39}{3} = 13$$

When there are no grouping symbols in an expression to indicate which operations should be done first, we agree to the following order:
1. Perform all multiplications and divisions in order from left to right.
2. Perform all additions and subtractions in order from left to right.

Example 4	Simplify $15 - 6 \div 3 + 4 \times 2$
Solution	$15 - 6 \div 3 + 4 \times 2 = 15 - (6 \div 3) + (4 \times 2)$ $$= 15 - 2 + 8$$ $$= 13 + 8 = 21$$

Class Exercises

Tell in which order the operations should be performed in simplifying the expression.

1. $(5 + 7) \times 2$ Add, then multiply.

2. $5 + (7 \times 2)$ Multiply, then add.

Square, then subtract.
3. $3^2 - 4$

4. $\frac{4 - 2}{2}$ Subtract, then divide.

5. $\frac{6}{2 + 1}$ Add, then divide.

6. $13 - \frac{4 + 5}{3}$

Add, then divide, then subtract.

Evaluate each expression when the value of x is 3.

7. $x + 1$ 4

8. $7x + 2$ 23

9. $x^2 + 1$ 10

10. $5x - x^2$ 6

11. $\frac{1}{x}$ $\frac{1}{3}$

12. $\frac{x + 6}{-x}$ -3

Exercises

Evaluate each expression for the given value of the variable.

The value of m is 1.

A   **1.** $12 + m$ 13

2. $-4m$ -4

3. $6 - m$ 5

4. $2(m + 1)$ 4

The value of n is 2.

5. n^2 4

6. $-n$ -2

7. $1 - n$ -1

8. $3(5 - n)$ 9

The value of t is -1.

9. $\frac{t + 1}{2}$ 0

10. $-2t$ 2

11. t^2 1

12. $\frac{5}{t}$ -5

The value of x is 0.

13. $4x$ 0

14. $x(x + 3)$ 0

15. $\frac{x + 2}{2}$ 1

16. $-x$ 0

The value of z is -6.

B   **17.** $z^2 + z$ 30

18. $-z - 6$ 0

19. $\frac{z + 6}{z}$ 0

20. $\frac{1}{z} + z$ $-6\frac{1}{6}$

The value of a is 2 and the value of b is -2.

21. $3a + 4b$ -2

22. $\frac{a - b}{a}$ 2

23. $\frac{a^2 - b^2}{a - b}$ 0

24. $\frac{1}{a} - \frac{1}{b}$ 1

The value of x is $\frac{1}{3}$ and the value of y is $\frac{1}{4}$.

C   **25.** $\frac{x - y}{x}$ $\frac{1}{4}$

26. $\frac{x^2 + y^2}{-x^2}$ $-1\frac{9}{16}$

27. $\frac{1}{x}\left(\frac{1}{y} - \frac{1}{x}\right)$ 3

28. $\frac{-x - y}{y^2}$ $-9\frac{1}{3}$

7-3 Word Sentences and Number Sentences

Objective To convert word sentences into number sentences and vice versa.

The group of words

The sum of nine and five is fourteen

is a word sentence. When we put this sentence into the mathematical form

$$9 + 5 = 14$$

we have a *number sentence*. A **number sentence** consists of two numerical expressions, called **members of the sentence,** with a verb symbol between them. The verb symbol may be $=$ (is equal to), $\neq$ (is not equal to), or one of the symbols $>$, $<$ (is greater than, is less than). If the verb symbol is $=$, the sentence is an equation. If the verb symbol is an inequality symbol, the sentence is an inequality. For example, the number sentences

$$3 \neq 4, \quad 12 - 1 > 10, \quad \text{and} \quad 6 - 2 < 5$$

are inequalities.

Other symbols we use are

$\geq$ to mean "is greater than or equal to"
$\leq$ to mean "is less than or equal to"

Thus,

$n \geq 3$ means "$n > 3$ or $n = 3$"

A number sentence that contains one or more variables is called an **open number sentence** or, simply, an **open sentence.**

Example | Convert into a number sentence:

a. The difference when four is subtracted from n is greater than six.

b. The product when five is multiplied by the sum of x and two is less than or equal to twenty.

Solution | **a.** $n - 4 > 6$

b. $5(x + 2) \leq 20$

Class Exercises

Convert into a word sentence.

1. $5 > 3$

2. $-1 < 0$

3. $4 + 2 = 6$

4. $2 + 1 \neq 4$

5. $\frac{1}{2} + \frac{1}{3} \neq \frac{1}{2+3}$

6. $\frac{2}{3} < \frac{3}{2}$

7. $m + 2 = 10$

8. $x - 1 = 2x + 2$

9. $\frac{1}{2}y - 3 \geq y - 4$

10. $\frac{n}{3} - 1 < n - 2$

Convert into a number sentence.

11. The sum of x and two is less than negative four. $x + 2 < -4$

12. The product of two and the sum of x and three is less than or equal to negative three. $2(x + 3) \leq -3$

13. The difference when nine is subtracted from seven is t. $7 - 9 = t$

14. The quotient when m is divided by negative two is greater than ten. $\frac{m}{-2} > 10$

15. The sum of x and one third is less than or equal to negative one. $x + \frac{1}{3} \leq -1$

Exercises

Match the word sentence with the number sentence.

A

1. Twice the sum of x and four is less than two. e

2. The sum of twice x and four is greater than two. b

3. The difference when four is subtracted from twice x is less than two. d

4. The sum of twice x and four is less than two. c

5. The product of two and the sum of x and four is less than or equal to two. a

6. Twice the difference when four is subtracted from x is less than two. f

a. $2(x + 4) \leq 2$

b. $2x + 4 > 2$

c. $2x + 4 < 2$

d. $2x - 4 < 2$

e. $2(x + 4) < 2$

f. $2(x - 4) < 2$

Write a number sentence.

7. The difference when seven is subtracted from t is two. $t - 7 = 2$

8. The product of five and z is greater than or equal to fifteen. $5z \geq 15$

9. The quotient when t is divided by eight is less than one half. $\frac{t}{8} < \frac{1}{2}$

Write a number sentence.

10. A number n is the product of two thirds and five eighths. $n = \frac{2}{3} \times \frac{5}{8}$

11. The product of one half and t is less than or equal to eight. $\frac{1}{2}t \leqslant 8$

12. A number m is the product of negative eleven and three fourths. $m = -11 \times \frac{3}{4}$

13. Negative one half is greater than the product of two and negative one. $\frac{1}{2} > 2(-1)$

14. The product of six and the sum of x and one is less than nine. $6(x + 1) < 9$

A Rocket Pioneer

Robert Hutchings Goddard (1882–1945) was born in Worcester, Massachusetts. As a youngster he was often unable to attend school with his classmates because of health problems. Nevertheless, he kept up with his studies at home, and also did much additional reading in science.

As a research fellow at Princeton University in 1912, Dr. Goddard worked in radio electronics during the day, but at night he used his own time and money to study the fundamentals of rocketry. He was the first to prove that a rocket would work in a vacuum. In 1919, he published a paper entitled "A method of reaching extreme altitudes," in which he referred to the possibility of reaching the moon by rocket. The press ridiculed him as a "moon-man," but he continued his research. In New

Mexico in the early 1930's he designed and built the first prototypes of the rockets which now can land people on the moon and probe the outer solar system.

Dr. Goddard is considered the father of modern rocketry and received many honors after his death including the U.S. Congressional Gold Medal.

Career Activity Find out what training is required to become an astronaut.

7-4 The Solution of an Open Sentence

Objective To decide whether or not a given value of a variable is a solution of a given open sentence.

The equation

$$x + 4 = 9$$

becomes a true statement if we replace x with 5, because

$$5 + 4 = 9.$$

However, the equation is not true if $x = 8$, because

$$8 + 4 = 12 \qquad \text{and} \qquad 12 \neq 9.$$

We say that 5 is a **solution** of, or **satisfies,** the given equation.

When we write

$$n + 2 \leq 15$$

we mean that

$$n + 2 < 15 \qquad \text{or} \qquad n + 2 = 15$$

Thus 6 and 13 are both solutions of $n + 2 \leq 15$ because

$$6 + 2 < 15 \qquad \text{and} \qquad 13 + 2 = 15$$

A solution of an open sentence in one variable is a value of the variable that makes the sentence a true statement.

Example | Find which of the two replacements for *n* is a solution of
$$n - 6 > 11.$$
a. $n = 25$ **b.** $n = 4$

Solution | **a.** Substituting 25 for *n*, we have
$$25 - 6 > 11, \text{ or } 19 > 11.$$
Since $19 > 11$ is a true statement, 25 *is* a solution.

b. When $n = 4$, we have $4 - 6 > 11$, or $-2 > 11$.
Since $-2 > 11$ is a false statement, -2 is *not* a solution.

Class Exercises

Tell whether the statements are true or false.

1. $2 > 2$ False
2. $2 < 2$ False
3. $2 = 2$ True
4. $-1 > 0$ False
5. $-1 \geq -1$ True
6. $1 > 0$ True
7. $-3 \geq -3$ True
8. $-3 = -3$ True
9. $-1 \leq 0$ True

Tell whether or not the given value of the variable is a solution of the given sentence. Explain your answer.

Example $t + 4 = 3; -1$

Solution -1 **is a solution because** $-1 + 4 = 3$ **is a true statement.**

10. $x + 1 < 0; \frac{1}{2}$ No
11. $3x = 2; \frac{2}{3}$ Yes
12. $y - 1 > 0; \frac{1}{2}$ No
13. $m \geq 1; 1$ Yes
14. $3 - t = 4; -1$ Yes
15. $4z > 0; -1$ No
16. $2n = 1; 0$ No
17. $-x < 0; -1$ No
18. $-2t \leq 0; -1$ No

Exercises

Replace the variable with the given value and tell whether the resulting statement is true or false.

A
1. $m + 1 = 0; -1$ True
2. $m + 1 \geq 0; -1$ True
3. $t - 3 > 0; 2$ False
4. $t - 3 \geq 0; 2$ False
5. $t - 3 \leq 0; 2$ True
6. $t - 3 < 0; 2$ True
7. $3x \leq 3; -1$ True
8. $3x \leq 3; 0$ True
9. $3x \leq 3; 1$ True
10. $3y \geq 0; 0$ True
11. $-3y \geq 0; 0$ True
12. $-n = 5; -5$ True
13. $-z = 1; -1$ True
14. $-m - 3 = 0; 3$ False
15. $-m - 3 = 0; -3$ True

Tell whether or not the given number is a solution of the given open sentence.

16. $x + 1 \geq -7; -6$ Yes
17. $3 - t = 2; 5$ No
18. $b + 4 = -3; -7$ Yes
19. $a - \frac{2}{3} > 0; \frac{1}{3}$ No
20. $\frac{y}{-1} = -2; 2$ Yes
21. $2t < 3; 1$ Yes

B
22. $5n \leq 2; -\frac{2}{5}$ Yes
23. $2x - 1 \geq 0; \frac{1}{2}$ Yes
24. $-8k > 9; -\frac{3}{4}$ No
25. $3x + 1 > 2x - 1; -1$ Yes
26. $x < 2x + 2; -2$ No
27. $3 - t \leq 4t; 1$ Yes
28. $-z + 8 = z - 8; 8$ Yes
29. $\frac{4x - 5}{3} \geq 0; 8$ Yes
30. $\frac{-n + 2}{7} < 1; 5$ Yes

7-5 The Solution Set of an Open Sentence

Objective To find the solution set of an open sentence in one variable whose replacement set is given.

To **solve** an open sentence in one variable, we look for all the solutions of the sentence. Of course, we have to know the replacement set (page 37) of the variable.

It sometimes helps in finding solutions to begin by drawing a graph.

Example 1 | If the replacement set for x is the set of whole numbers, find all solutions of

$$x \leq 5.$$

Solution |

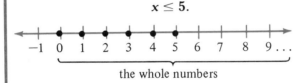

the whole numbers

From the figure, you can see that there are just six solutions, **0, 1, 2, 3, 4,** and **5,** because they are the only *whole numbers* that are equal to or less than 5.

Example 2 | If the replacement set for y is the set of whole numbers, find all solutions of

$$3y = 7.$$

Solution | Since y is a member of the set of whole numbers, $\{0, 1, 2, 3, \ldots\}$, $3y$ is a member of $\{3 \times 0, 3 \times 1, 3 \times 2, 3 \times 3, \ldots\}$ or $\{0, 3, 6, 9, \ldots\}$. The equation $3y = 7$ has *no* solution in the given replacement set.

The **solution set** of an open sentence consists of all the values in the given replacement set of the variable that make the sentence true. Thus the solution set in Example 1 is $\{0, 1, 2, 3, 4, 5\}$. In Example 2, the solution set has no members. The set with no members is called the **empty set** and is denoted by the symbol $\varnothing$.

Class Exercises

Given the replacement set for x, $\{-2, -1, 0, 1, 2\}$, find the solution set of the given open sentence.

1. $x \leq 2$ $\{-2, -1, 0, 1, 2\}$
2. $x < 2$ $\{-2, -1, 0, 1\}$
3. $x < 0$ $\{-2, -1\}$

4. $x > 0$ $\{1, 2\}$
5. $x + 1 = 2$ $\{1\}$
6. $x + 2 = 0$ $\{-2\}$

7. $x = -5$ $\emptyset$
8. $x - 5 = -3$ $\{2\}$
9. $x - 2 = 0$ $\{2\}$

Match the open sentence with the graph which represents its solution set. The replacement set for x is $\{-1, 0, 1, 2, 3\}$.

10. $x < 0$ d

11. $x \geq 0$ b

12. $x > 0$ a

13. $x \leq 3$ c

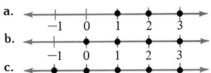

Exercises

Graph the solution set. The replacement set for x is $\{-3, -2, -1, 0, 1, 2, 3\}$.

A

1. $x + 1 = 2$ $\{1\}$
2. $x < -1$ $\{-3, -2\}$
3. $x \geq 0$ $\{0, 1, 2, 3\}$

4. $2x = 0$ $\{0\}$
5. $2x - 1 \leq 4$ $\{-3, -2, -1, 0, 1, 2\}$
6. $x + 3 = 10$ $\emptyset$

7. $7 - x \leq 5$ $\{2, 3\}$
8. $x + 7 \leq 5$ $\{-3, -2\}$
9. $x - 4 = 5$ $\emptyset$

10. $x - 5 > -4$ $\{2, 3\}$
11. $-x \geq 3$ $\{-3\}$
12. $3x + 3 = 0$ $\{-1\}$

Give the solution set if the replacement set for x is the set of integers.

Example $x + 1 > -3$
Solution $\{-3, -2, -1, \ldots\}$

13. $x < 0$ $\{\ldots, -3, -2, -1\}$
14. $x > 4$ $\{5, 6, 7, \ldots\}$
15. $x \geq 4$ $\{4, 5, 6, \ldots\}$

16. $x < -3$ $\{\ldots, -6, -5, -4\}$
17. $x - 1 \geq 2$ $\{3, 4, 5, \ldots\}$
18. $2x - 1 > 0$
      $\{1, 2, 3, \ldots\}$

Give the solution set. The replacement set for x is given.

B   **19.** $-x + 2 \le -1$ $\{-3, -2, -1, 0, 1, 2, 3\}$ $\{3\}$

  20. $3x - 5 > -7$ $\{-2, -1, 0, 1, 2, 3, 4\}$ $\{0, 1, 2, 3, 4\}$

  21. $\frac{x+1}{3} < 1$ $\{0, 1, 2, 3, 4, 5\}$ $\{0, 1\}$

  22. $4x - 3 \le 2$ $\{-3, -2, -1, 0, 1, 2, 3\}$ $\{-3, -2, -1, 0, 1\}$

C   **23.** $3x - 2 \le x + 2$ $\{-5, -4, -3, -2, -1, 0\}$ $\{-5, -4, -3, -2, -1, 0\}$

  24. $3 - 2x > x$ $\{-3, -2, -1, 0, 1, 2, 3\}$ $\{-3, -2, -1, 0\}$

EXTRA! Investigating Inequalities

1. Copy the number line shown below. Use a heavy dot to mark the graph of each solution of the inequality $x < 4$ if the replacement set for x is $\{0, 1, 2, 3, 4, 5, 6, 7, 8\}$.

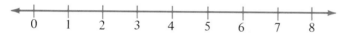

2. Make another copy of the number line shown above and use a heavy dot to mark the graph of each solution of the inequality $x + 1 < 5$, using the same replacement set for x as in Step 1.

3. Repeat Step 2 for the inequality $x - 1 < 3$. Graphs for Ex. 2-3 are identical.

4. From your results for Steps 1, 2, and 3, what can you say about the solution sets of the inequalities

$$x < 4, \qquad x + 1 < 5, \qquad \text{and} \qquad x - 1 < 3$$

when the replacement set for x is $\{0, 1, 2, 3, 4, 5, 6, 7, 8\}$? Solution sets are the same.

5. What do Steps 1–4 suggest about the solution sets of the inequalities

$$x < a, \qquad x + 1 < a + 1, \qquad \text{and} \qquad x - 1 < a - 1$$

for a given replacement set for x? The solution sets are the same.

6. What do you think might be true of the inequalities

$$x < a, \qquad x + b < a + b, \qquad \text{and} \qquad x - b < a - b$$

for a given replacement set for x? The solution sets are the same.

Self-Test

Words to remember:
numerical expression (p. 217)   variable expression (p. 217)
evaluate (p. 220)   number sentence (p. 222)
open sentence (p. 222)   solution (p. 225)
solve (p. 227)   solution set (p. 227)
empty set (p. 227)

Write a variable expression.

1. The sum of one and a
   number, x. $1 + x$

2. The product of twelve and
   a number, n. $12n$

[7-1]

Write a word phrase.

3. $13 - z$ The difference when z is subtracted from thirteen.

4. $5z + 12$ The sum of twelve and five times a number z.

Evaluate the expressions if $x = -5$.

5. $-7 + x$ -12

6. $\frac{3}{5}x$ -3

7. $-2x + 4$ 14

8. $-\frac{1}{x}$ $\frac{1}{5}$ [7-2]

Translate into a number sentence.

9. The quotient when y is divided by three is negative nine. $\frac{y}{3} = -9$ [7-3]

10. The product when two is multiplied by the sum of n and four is
    fourteen. $2(n + 4) = 14$

Write a word sentence.   Three times the difference of y and nine is less than or equal to negative eight.

11. $3(y - 9) \le -8$

12. $-7x + 31 > 4$ See below.

Tell whether or not the given number is a solution of the given open
sentence.

13. $x - 13 > -1$; 5 No

14. $4x - 7 = 5$; -3 No

[7-4]

15. $-4y \ge 16$; $y = -4$ Yes

16. $n < -2n - 3$; -2 Yes

Graph the solution set. The replacement set for x is
$\{-5, -4, -3, -2, -1, 0\}$.

17. $x + 4 > 0$ $\{-3, -2, -1, 0\}$

18. $x + 2 = 0$ $\{-2\}$

[7-5]

19. $2x + 3 = -5$ $\{-4\}$

20. $x - 3 \le -6$ $\{-5, -4\}$

12. The sum of negative seven times a number x
    and thirty-one is greater than four.

Self-Test answers and Extra Practice are at the back of the book.

7-6 Properties of Equality

Objective To use the addition and multiplication properties of equality.

We use the addition property of equality and the multiplication property of equality to solve equations.

properties

Addition Property of Equality
Given a true equation, we may add the same number to both members and the resulting equation will also be true.

$$4 + 3 = 7 \quad \text{and} \quad (4 + 3) + 5 = 7 + 5$$

In general, for any numbers a, b, and c:

$$\text{If } a = b, \text{ then } a + c = b + c.$$

$$\text{If } a + c = b + c, \text{ then } a = b.$$

Multiplication Property of Equality
Given a true equation, we may multiply both members by the same number and the resulting equation will also be true.

$$11 - 2 = 9 \quad \text{and} \quad (11 - 2)4 = (9)4$$

In general, for any numbers a, b, and c:

$$\text{If } a = b, \text{ then } ac = bc.$$

$$\text{If } ac = bc \text{ and } c \neq 0, \text{ then } a = b.$$

It is important to note that even though $a \times 0 = b \times 0$, we can still have $a \neq b$.

$$7 \times 0 = 4 \times 0 \quad \text{but} \quad 7 \neq 4$$

Example	Use one of the properties of equality to form a true sentence:

$$\text{If } y = 2, \text{ then } y + \underline{\ ?\ } = 11$$

Solution	Since $2 + 9 = 11$, by the addition property of equality we know that $y + 9 = 2 + 9$.

$$\text{If } y = 2, \text{ then } y + 9 = 11.$$

Class Exercises

Use one of the properties of equality to form a true sentence.

1. If $t = 3$, then $t + 6 = \underline{\ ?\ }$ 9

2. If $z = 1$, then $z + \underline{\ ?\ } = 9$ 8

3. If $x = -1$, then $4x = \underline{\ ?\ }$ -4

4. If $n = 6$, then $n - 7 = \underline{\ ?\ }$ -1

5. If $x = 13$, then $2x = \underline{\ ?\ }$ 26

6. If $m = 12$, then $8m = \underline{\ ?\ }$ 96

7. If $a = -2$, then $a - 9 = \underline{\ ?\ }$ -11

8. If $b = 7$, then $-3b = \underline{\ ?\ }$ -21

9. If $t = 3$, then $t - 6 = \underline{\ ?\ }$ -3

10. If $4n = -16$, then $n = \underline{\ ?\ }$ -4

11. If $7d = 28$, then $d = \underline{\ ?\ }$ 4

12. If $k = 9$, then $k + \underline{\ ?\ } = 4$ -5

13. If $-3y = 12$, then $y = \underline{\ ?\ }$ -4

14. If $m = 0$, then $m + \underline{\ ?\ } = -6$ -6

Exercises

Use one of the properties of equality to form a true sentence.

A

1. If $x = 7$, then $x + 11 = \underline{\ ?\ }$ 18

2. If $n = -8$, then $n + 12 = \underline{\ ?\ }$ 4

3. If $t = -1$, then $t - 3 = \underline{\ ?\ }$ -4

4. If $y = 9$, then $y - 8 = \underline{\ ?\ }$ 1

5. If $x = 0$, then $x + 4 = \underline{\ ?\ }$ 4

6. If $a = 12$, then $5a = \underline{\ ?\ }$ 60

7. If $c = -3$, then $2c = \underline{\ ?\ }$ -6

8. If $b = 9$, then $-3b = \underline{\ ?\ }$ -27

9. If $m = -11$, then $-3m = \underline{\ ?\ }$ 33

10. If $t = -4$, then $-4t = \underline{\ ?\ }$ 16

Example	If $t + 1 = 4$, then $t = \underline{\ ?\ }$
Solution	$t + 1 = 4$ and $4 = 3 + 1$
	$t + 1 = 3 + 1$ so by the addition property of equality,
	$t = 3$.

B **11.** If $x + 9 = 15$, then $x =$ ___?___ 6

12. If $z + 1 = 9$, then $z =$ ___?___ 8

13. If $4n = 16$, then $n =$ ___?___ 4

14. If $3t = 21$, then $t =$ ___?___ 7

15. If $-3a = 21$, then $a =$ ___?___ -7

16. If $4k = -16$, then $k =$ ___?___ -4

17. If $2b = -22$, then $b =$ ___?___ -11

18. If $-8t = -24$, then $t =$ ___?___ 3

19. If $z - 11 = 3$, then $z =$ ___?___ 14

20. If $m - 3 = -2$, then $m =$ ___?___ 1

21. If $-4a = -24$, then $a =$ ___?___ 6

22. If $9 + z = 6$, then $z =$ ___?___ -3

C **23.** If $\frac{2}{3}z = 16$, then $z =$ ___?___ 24

24. If $-\frac{3}{5}n = 12$, then $n =$ ___?___ -20

25. If $2n = 5$, then $4n - 2 =$ ___?___ 8

26. If $4n = -3$, then $8n + 2 =$ ___?___ -4

27. If $3t = -7$, then $6t + 5 =$ ___?___ -9

28. If $3x = -4$, then $9x + 2 =$ ___?___ -10

Application
Pricing by Quantity

Items in stores are frequently priced by quantity, such as "3 pens for 76¢." If you buy only one pen, you cannot pay the exact amount of 76¢ $\div$ 3 = $25\frac{1}{3}$¢, so the store charges you 26¢. In general, the store charges you the number of cents which is the least integer greater than or equal to the exact price in cents.

Computer Activity The BASIC computer program below computes the price you pay on single items which are priced by quantity.

```
10 PRINT"QUANTITY=";
20 INPUTQ
30 PRINT"COST IN CENTS=";
40 INPUTC
50 PRINT"PRICE FOR SINGLE ITEM=";-INT(-C/Q);"CENTS"
60 END
```

Use this program to find the price of a single item which is priced at

1. 6 for 98¢ 17¢

2. 5 for 89¢ 18¢

3. 3 for 55¢ 19¢

Which is the better buy?

4. 3 for $2.98 or <u>6 for $5.69</u>

5. 5 for $12.29 or <u>15 for $35.98</u>

6. $16.98 a dozen or <u>20 for $27.59</u>

7-7 Solving Equations by Transformations

Objective To solve an equation in one variable by transforming it into one which can be solved by inspection.

Two equations are **equivalent** if they have the same solution. The usual way to solve an equation is to transform it into an equivalent equation in which one member is the variable and the other member is some number c.

$$x = c$$

Then the solution is simply c.

Here are some transformations, based on the properties of equality, that yield equivalent equations.

---

a. Simplify numerical expressions in either member.

$x + 3 = 4 + 7$ is equivalent to $x + 3 = 11$

---

b. Exchange members.

$-6 = 9 + x$ is equivalent to $9 + x = -6$

---

c. Use the properties of addition and multiplication, for example, the distributive property.

$2(x - 7) = 0$ is equivalent to $2x - 14 = 0$

---

d. Add the same number to both members.

$3x + 2 = 5$ is equivalent to $3x + 2 + (-2) = 5 + (-2)$

We can also subtract the same number from both members. This is equivalent to adding the opposite, as we did here.

---

e. Multiply or divide both members by the same nonzero number.

$4x = 13$ is equivalent to $\dfrac{4x}{4} = \dfrac{13}{4}$

Sometimes we have to use more than one transformation to obtain an equation of the form $x = c$.

Example | Solve $21 = 3(x - 4)$. **The replacement set for x is {the rational numbers}. Identify which transformations you use and check your solution in the original equation.**

Solution |

	Transformation used
$21 = 3(x - 4)$	
$3(x - 4) = 21$	**b**
$3x - 12 = 21$	**c**
$3x - 12 + 12 = 21 + 12$	**d**
$3x = 33$	**a**
$x = 11$	**e**

The solution set is $\{11\}$.

Check:

Substitute 11 for x in $21 = 3(x - 4)$.

$$21 \stackrel{?}{=} 3(11 - 4)$$
$$21 = 3(7) = 21 \; \checkmark$$

Class Exercises

Tell which transformation is used to obtain the second equation from the first.

1. $4x = 28; x = 7$ e

2. $12 = x + 9; x + 9 = 12$ b

3. $1 + t = -3; t + 1 = -3$ b

4. $y - 3 = 1; y = 4$ d, then a

5. $z = 5 + (-9); z = -4$ a

6. $6(x - 2) = 3; 6x - 12 = 3$ c

Tell which transformation is used in each step.

7. $\frac{2}{3}t - 4 = 4$
   $\frac{2}{3}t - 4 + 4 = 4 + 4$ d
   $\frac{2}{3}t = 8$ a
   $3\left(\frac{2}{3}t\right) = 3(8)$ e
   $2t = 24$ a
   $t = 12$ e

8. $-5 = 3z + 13$
   $3z + 13 = -5$ b
   $3z + 13 - 13 = -5 - 13$ d
   $3z = -18$ a
   $\frac{3z}{3} = \frac{-18}{3}$ e
   $z = -6$ a

Exercises

Solve. Check your solution.

A

1. $y - 9 = 3$ $\{12\}$

2. $t + 2 = 13$ $\{11\}$

3. $5x = 25$ $\{5\}$

4. $y - 3 = -8$ $\{-5\}$

5. $\frac{3}{4}t = -15$ $\{-20\}$

6. $-\frac{1}{2}x = \frac{2}{3} + \frac{1}{6}$ $\{-\frac{5}{3}\}$

7. $-3 - 7 = x + 1$ $\{-11\}$

8. $6x = -4 + (-18)$ $\{-\frac{11}{3}\}$

9. $x(4 - 2) = 40$ $\{20\}$

10. $-y - (-2y) = 9$ $\{9\}$

11. $-3t = -1 + 4$ $\{-1\}$

12. $5x - 2x = -18$ $\{-6\}$

B

13. $5x - 2 = x + 10$ $\{3\}$

14. $-9y - (-y) = 5$ $\{-\frac{5}{8}\}$

15. $\frac{2}{3}x + x = -15$ $\{-9\}$

16. $\frac{3}{4}y - \frac{1}{2} = 2y + \frac{1}{2}$ $\{-\frac{4}{5}\}$

17. $\frac{1}{2}x + \frac{3}{5} = 2x - \frac{9}{10}$ $\{1\}$

18. $\frac{3}{5}y = 2y - 7$ $\{5\}$

C

19. $3(x - 5) = -2x + 5$ $\{4\}$

20. $\frac{1}{3}(5x - 6) = x + 2$ $\{6\}$

21. $-t - 9 = \frac{3}{4}\left(2t + \frac{4}{3}\right)$ $\{-4\}$

22. $\frac{1}{2}(-4y - 5) = 7\left(y - \frac{1}{2}\right)$ $\{\frac{1}{9}\}$

EXTRA! Using Equations

Suppose you are asked to find two consecutive integers whose sum is 15. You can use an equation to solve this problem. Start by letting n be the smaller integer. Then $n + 1$ is the larger integer, and the sum of the two integers is $n + n + 1$, or $2n + 1$.

$$2n + 1 = 15$$
$$2n + 1 + (-1) = 15 + (-1)$$
$$2n = 14$$
$$n = 7 \quad \text{and} \quad n + 1 = 8$$

The numbers are 7 and 8.

Find the numbers described by writing and solving an equation.

1. The sum of two consecutive integers is -7. -4 and -3

2. The sum of three consecutive integers is 39.   12, 13, and 14

3. One number is 4 less than another number. The sum of the numbers is 22.   13 and 9

7-8 Solving Inequalities by Transformations

Objective To solve an inequality by transforming it into one which can be solved by inspection.

You can solve an inequality just as you did an equation, by transforming it into an inequality with the same solutions (an **equivalent inequality**). The transformations are similar to the ones used with equations, with two important exceptions. Notice the effect on the inequality signs in **b** and **f.**

---

a. Simplify numerical expressions in either member.

$x - 4 > -1 + 3$ is equivalent to $x - 4 > 2$.

---

b. Exchange members, *thereby changing the direction of the inequality.*

$6 < x + 5$ is equivalent to $x + 5 > 6$.

---

c. Use the properties of addition and multiplication.

$4(x - 1) \geq 8$ is equivalent to $4x - 4 \geq 8$.

---

d. Add the same number to both members.

$2x - 3 < 7$ is equivalent to $2x - 3 + 3 < 7 + 3$.

---

e. Multiply or divide both members by the same *positive* number.

$3x > 6$ is equivalent to $\dfrac{3x}{3} > \dfrac{6}{3}$.

---

f. Multiply or divide both members by the same *negative* number, *thereby changing the direction of the inequality.*

$-\dfrac{x}{2} > 5$ is equivalent to $-2\left(-\dfrac{x}{2}\right) < -2(5)$

Example | Solve $16 \geq 4(3 - x)$. **The replacement set for x is {the integers}. Tell which transformations you use.**

Transformation used

Solution

$$16 \geq 4(3 - x)$$

$$4(3 - x) \leq 16 \qquad \qquad \textbf{b}$$

$$12 - 4x \leq 16 \qquad \qquad \textbf{c}$$

$$12 - 4x - 12 \leq 16 - 12 \qquad \qquad \textbf{d}$$

$$-4x \leq 4 \qquad \qquad \textbf{a}$$

$$\frac{-4x}{-4} \geq \frac{4}{-4} \qquad \qquad \textbf{f}$$

$$x \geq -1 \qquad \qquad \textbf{a}$$

The solution set is the set of integers that are greater than or equal to -1, that is, $\{-1, 0, 1, 2, 3, \ldots\}$.

Class Exercises

Tell which graph represents the solution set of each inequality. The replacement set for x is $\{-2, -1, 0, 1, 2\}$.

1. $x \leq 0$ a

2. $x > -2$ d

3. $x < 0$ b

4. $x \leq 2$ c

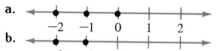

a.

b.

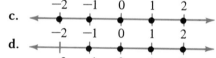

c.

d.

Tell which transformation is used to obtain the second inequality from the first.

5. $y - 4 \geq 7; y \geq 11$ d, a

6. $2(t - 6) > 0; 2t - 12 > 0$ c

7. $t \leq -3 - 9; t \leq -12$ a

8. $-3x < 18; x > -6$ f

9. $5t > 35; t > 7$ e

10. $-4 < z + 9; z + 9 > -4$ b

Exercises

Graph the solution set. The replacement set for x is $\{-3, -2, -1, 0, 1, 2\}$.

A   1. $x \leq 1$ $\{-3, -2, -1, 0, 1\}$

2. $x + 1 \geq -2$ $\{-3, -2, -1, 0, 1, 2\}$

3. $2x < 1$ $\{-3, -2, -1, 0\}$

4. $x - 4 \geq -3$ $\{1, 2\}$

Solve. The replacement set for each variable is {the integers}.

5. $-3 > z$ {..., $-6, -5, -4$}

6. $x + 1 \le 0$ {..., $-3, -2, -1$}

7. $x + (-9) > -2$ {$8, 9, 10, ...$}

8. $4 + (-12) \le x$ {$-8, -7, -6, ...$}

9. $\frac{2}{3}x < -3$ {..., $-7, -6, -5$}

10. $-\frac{1}{2}y \le 8$ {$-16, -15, -14, ...$}

11. $\frac{3}{5}y < -6$ {..., $-13, -12, -11$}

12. $-y > \frac{1}{2}$ {..., $-3, -2, -1$}

13. $t + (-9) \le -3$ {..., $4, 5, 6$}

14. $y - 3 > 2$ {$6, 7, 8, ...$}

B **15.** $3 - t > -2$ {..., $2, 3, 4$}

16. $4y + 1 \ge 13$ {$3, 4, 5, ...$}

17. $4(y - 2) \le 10$ {..., $2, 3, 4$}

18. $5x - 5 < 45$ {..., $7, 8, 9$}

19. $11 - z > 0$ {..., $8, 9, 10$}

20. $\frac{2}{3}x + 4 \ge 0$ {$-6, -5, -4, ...$}

C **21.** $-\frac{3}{4}(z + 1) \le -z - 3$ {..., $-11, -10, -9$}

22. $-3x + \frac{1}{2} \le \frac{1}{2}(x - 20)$ {$3, 4, 5, ...$}

Self-Test

Words to remember:
addition property of equality (p. 231)
equivalent equations (p. 234)

multiplication property of equality (p. 231)
equivalent inequalities (p. 237)

Use one of the properties of equality to form a true sentence.

1. If $y = 8$, then $y + 4 = $? 12 **2.** If $z = 17$, then $z - 9 = $? 8 [7-6]

3. If $x = -6$, then $9x = $? -54 **4.** If $t = 5$, then $-7t = $? -35

Solve each equation and check your solution.

5. $x + 4 = 1$ {-3}

6. $y - 13 = 9$ {22}

[7-7]

7. $-5z = -40$ {8}

8. $t + 5t = -36$ {-6}

Solve. The replacement set for x is {the integers}.

9. $x - 1 > -3$ {$-1, 0, 1, ...$} **10.** $x + 2 \le -8$ {..., $-12, -11, -10$} [7-8]

11. $-30 < 5x$ {$-5, -4, -3, ...$} **12.** $-8x \ge -24$ {..., $5, 4, 3$}

Self-Test answers and Extra Practice are at the back of the book.

Chapter Review

Write the letter that labels the correct answer.

1. The word phrase "twice the sum of x and y" may be translated   [7-1]
   into the numerical phrase ___?___ .   C
   A. $2x + y$ **B.** $2 + xy$ **C.** $2(x + y)$ **D.** $2xy$

2. The numerical expression $1 + 3b$ may be translated into the
   word phrase ___?___ .   A
   A. The sum of one and the product of three and b.
   B. The product of b and the sum of one and three.
   C. The sum of one and three and b.
   D. The product of one and the sum of three and b.

What is the value of the expression if $x = -3$?   [7-2]

3. $x + 14$ C
   A. 17   **B.** -17 **C.** 11   **D.** -11

4. $3x - 5$ B
   A. 4   **B.** -14 **C.** -4 **D.** 1

5. The word sentence "The product of seven and the sum of t and   [7-3]
   three is greater than nine" may be translated into the number
   sentence ___?___ .   B
   A. $7t + 3 > 9$ **B.** $7(t + 3) > 9$
   C. $7t + 3 < 9$ **D.** $7(t + 3) < 9$

6. The number sentence $\frac{1}{2}x + 7 > 5$ may be translated into the
   word sentence ___?___ .   D
   A. One half the sum of x and seven is greater than five.
   B. The sum of one half x and seven is less than five.
   C. The product of one half x and seven is greater than five.
   D. The sum of one half x and seven is greater than five.

Which number is a solution of the open sentence?   [7-4]

7. $z - 8 = -3$ D
   A. -5 **B.** 11   **C.** -11 **D.** 5

8. $5x > -30$ A
   A. 0   **B.** -6 **C.** -7 **D.** -10

Which graph represents the solution set? The replacement set for x is $\{-2, -1, 0, 1, 2\}$. [7-5]

9. $x + 2 = 1$ B

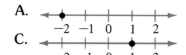

A.

C.

B.

D.

10. $x + 3 \leq 2$ A

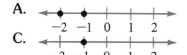

A.

C.

B.

D.

Use the properties of equality to complete the sentence. [7-6]

11. If $t = -8$, then $t + 3 = \underline{\ ?\ }$.   C
    A. -24 **B.** -11 **C.** -5 **D.** 24

12. If $t = -14$, then $-2t = \underline{\ ?\ }$.   B
    A. -16 **B.** 28 **C.** -12 **D.** -28

Which number is the solution of the equation? [7-7]

13. $x - 7 = -20$ D
    A. -27 **B.** 13 **C.** 27 **D.** -13

14. $4x = -13 - 15$ A
    A. -7 **B.** -28 **C.** 7 **D.** -6

Which set is the solution set of the inequality? The replacement set for x is $\{$the integers$\}$. [7-8]

15. $x - 13 \leq 4$ C
    A. $\{17, 18, 19, \ldots\}$ **B.** $\{\ldots, 9, 10, 11\}$
    C. $\{\ldots, 15, 16, 17\}$ **D.** $\{\ldots, 14, 15, 16\}$

16. $-4x \leq -16$ B
    A. $\{\ldots, 2, 3, 4\}$ **B.** $\{4, 5, 6, \ldots\}$
    C. $\{\ldots, -6, -5, -4\}$ **D.** $\{5, 6, 7, \ldots\}$

Secret Codes

Over the centuries, many methods have been devised for writing secret messages. For example, Julius Caesar used a method in which each letter of a message was replaced by the letter three places to the right in the alphabet. Thus, using our alphabet, the word RED would appear in code as UHG.

Another method of encoding a secret message is by changing the order of the letters in the message. One way of doing this is shown below.

Let's say we want to encode the message:

SPACE   SHIP   ULTRA   MAKES   GALAXY   TOUR

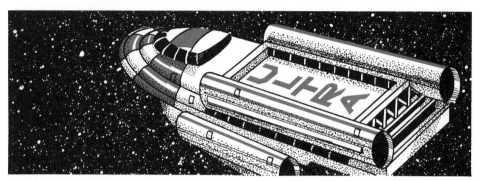

First, we write the message in the spaces of a rectangle as shown at the right. In this case, we have used a rectangle with 8 columns and 4 rows.

Next, we form a key by scrambling the numbers from 1–8 and writing them at the heads of the columns. We are using the key 5-7-2-4-1-6-3-8.

5	7	2	4	1	6	3	8
S	P	A	C	E	S	H	I
P	U	L	T	R	A	M	A
K	E	S	G	A	L	A	X
Y	T	O	U	R			

To write the message in code, we start with Column 1 and write the letters in order from top to bottom. Then we continue with Columns 2, 3, and so on. It is customary to write the coded message in blocks of 5 letters. Thus, when written in code our message is: ERARA LSOHM ACTGU SPKYS ALPUE TIAX

To *decode* the message, we first need to figure out the number of columns and rows in the rectangle. We can tell directly from the key, 5-7-2-4-1-6-3-8, that there are 8 columns.

To find the number of rows, we count the number of letters in the message, in this case it is 29. We divide 29 by 8 (the number of columns). From this we see that there are 3 rows in the rectangle completely filled with letters and that the remaining 5 letters partially fill the 4th row.

$$\begin{array}{r} 3 \\ 8\overline{)29} \\ 24 \\ \hline 5 \end{array}$$

Now we draw a rectangle as shown at the right and write the message we received starting from top to bottom of Column 1, then going on to Column 2, and so on. When we finish, the original message will be reconstructed.

5	7	2	4	1	6	3	8
		A		E		H	
		L		R		M	
		S		A			
		O		R			

1. The following encoded message was obtained by replacing each letter of the original message with the letter three places to the left in the alphabet. Decode the message. SKATEBOARDERS HAVE MORE FUN.

      PHXQBYLXOABOP   EXSB   JLOB   CRK

2. Secret agent Dora Dynamic was posing as an accountant in Liverpool, England, while awaiting news of a secret shipment to arrive from Canada. Twenty-four hours after receiving the coded message shown below, she left for Quebec. What did the message say? The key is 8-6-3-1-2-5-9-4-7.

SKLEN HTAET ENOMO ICLEA IHTTR HUCTM SHOIL TSOSN PEEME
THE SHIP IS SUNK. THE CHOCOLATE LOST. MEET ME IN MONTREAL.

3. After five years as a lone rancher in Colorado, Dan Devine received the following encoded message and gave a shout of joy that was heard as far away as Nevada. What did the message say? The key is 1-4-7-6-3-5-2.
I ACCEPT YOUR PROPOSAL FOR MARRIAGE. LOVE, DORA
    IYPRG   OTOOA   DEPLR   VAOOM   ERPRF   IECRA   ROCUS   ALA

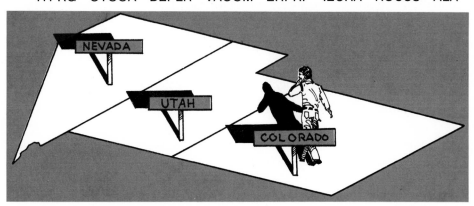

4. Write a message in code and exchange it with one written by a friend. Can you decode each other's message?

Research Activity The science of secret codes is known as *cryptography*. Go to the library and find other ways to encode messages.

Chapter Test

1. Write a numerical expression for the word phrase "twelve less than negative two." $-2-12$ **[7-1]**

2. Write a word phrase for the numerical expression $\frac{x+1}{8}$.
   The quotient when the sum of x and one is divided by eight.

Evaluate each expression if $x = -1$.

3. $\frac{x-3}{2}$ -2

4. $4x + 3$ -1 **[7-2]**

5. Translate into a number sentence: "The difference when z is subtracted from eleven is greater than zero." $11 - z > 0$ **[7-3]**

6. Write a word sentence for the number sentence $5t - 4 \geq -3$.
   The difference when four is subtracted from the product of five and t is greater than or equal to negative three.

Tell whether or not the given number is a solution of the given open sentence.

7. $x - 8 = -11; 3$ No

8. $3t \leq 2; \frac{1}{2}$ Yes **[7-4]**

Graph the solution set. The replacement set for x is $\{1, 2, 3, 4, 5\}$.

9. $x - 3 = 0$ $\{3\}$

10. $x + 2 \leq 5$ $\{1, 2, 3\}$ **[7-5]**

Use one of the properties of equality to form a true sentence.

11. If $t = -3$, then $t + 2 = $ __?__
    -1

12. If $y = 8$, then $-5y = $ __?__ **[7-6]**
    -40

Solve each equation and check your solution.

13. $x - 8 = 4$ $\{12\}$

14. $3x + 2x = -20$ $\{-4\}$ **[7-7]**

Solve. The replacement set for x is $\{$ the integers $\}$.

15. $9x \geq 81$ $\{9, 10, 11, \ldots\}$

16. $x - 13 < 5$ $\{\ldots, 15, 16, 17\}$ **[7-8]**

Skill Review

Add or subtract.

1. $17 + 3$ 20   **2.** $18 - 5$ 13   **3.** $-16 + 6$ −10   **4.** $-28 - 17$ −45

5. $23 - 51$ −28   **6.** $6 + 3$ 9   **7.** $19 - 12$ 7   **8.** $-33 + 11$ −22

9. $15 - 26$ −11   **10.** $27 + 6$ 33   **11.** $11 - 33$ −22   **12.** $6 + (-17)$ −11

13. $11 + 26$ 37   **14.** $15 - 17$ −2   **15.** $23 - 42$ −19   **16.** $16 + (-12)$ 4

17. $55 + (-13)$ 42   **18.** $73 + 26$ 99   **19.** $59 + 12$ 71   **20.** $69 + (-3)$ 66

21. $10 + (-27)$ −17   **22.** $16 - 3$ 13   **23.** $17 + 19$ 36   **24.** $19 - (-6)$ 25

25. $7 + 4 - 3 + 16$ 24   **26.** $17 + 2 + (-6 - 21)$ −8

27. $53 + 111 - 27 + (-26)$ 111   **28.** $19 + 13 - 17 - 6$ 9

Multiply or divide.

29. 7×3 21   **30.** 4×6 24   **31.** -5×12 −60   **32.** 2×17 34

33. $17 \times (-2)$ −34   **34.** $6 \times (-6)$ −36   **35.** $2 \div 2$ 1   **36.** $14 \div (-7)$ −2

37. $-64 \div 8$ −8   **38.** $18 \times (-3)$ −54   **39.** 6×7 42   **40.** 9×6 54

41. -8×7 −56   **42.** $33 \div 11$ 3   **43.** $60 \div (-5)$ −12   **44.** $120 \div 10$ 12

45. $16 \div 8$ 2   **46.** $27 \div (-3)$ −9   **47.** $18 \div 6$ 3   **48.** 3×11 33

49. $-5 \times (-12)$ 60   **50.** $56 \div 7$ 8   **51.** $24 \div 8$ 3   **52.** 6×12 72

53. 12×12 144   **54.** $16 \div 2$ 8   **55.** 6×11 66   **56.** $12 \times (-12)$ −144

57. $14 \div 2$ 7   **58.** $28 \div (-2)$ −14   **59.** $7 \times (-7)$ −49   **60.** 6×8 48

Perform the indicated operations.

61. $10 - 4 \times (-2)$ 18   **62.** $13 + (-12) \div 12$ 12

63. $-1 \times (-3) + 5$ 8   **64.** $-68 \div 17 - 14$ −18

65. $2 \times (-18) - 3 \times (-10)$ −6   **66.** $60 \div (-5) + (-9) \times (-1)$ −3

67. $3 + 5 \times (-11) - 8 \div (-2)$ **68.** $-16 \div (-4) - 2 \times (-5) + (-18)$

               −48   −4

8 Problem Solving

8-1 Equations into Word Problems

Objective To interpret equations as word problems.

When you are asked to solve an equation such as

$$15x = 75$$

do you ever wonder where it came from and why one might need to solve it? The equation could represent many real-life problems. Consider these:

Shirley Trench has to bicycle for x hours at the rate of 15 km/h to cover 75 km per day on a long-distance bicycle trip.

If Les Callen can interview 15 job applicants per day, he will need x days to interview the 75 applicants for one job.

Class Exercises

State two practical problems that are expressed by the given equation. For the first problem, use a simple, direct question. Answers will vary.

Example | $24 - x = 8$

Solution | **1.** **What number subtracted from 24 gives 8?**
| **2.** **Meg Laramie has a 24-month loan which can be renewed anytime during the last 8 months. How many months must she pay on the loan before she can renew it?**

1. $x + 10 = 50$ **2.** $x - 6 = 24$ **3.** $4x = 28$

4. $52x = 260$ **5.** $\frac{x}{30} = 7$ **6.** $\frac{x}{4} = 15$

Exercises

Write two practical problems that are expressed by the given equation. For the first problem, use a simple, direct question. Answers will vary.

A   **1.** $24 + x = 80$ **2.** $12 + x = 48$

3. $x - 10 = 35$ **4.** $24 - x = 18$

5. $12x = 36$ **6.** $8x = 56$

7. $\frac{x}{36} = 50$ **8.** $\frac{x}{12} = 1000$

9. $\frac{1}{4}x = 20$ **10.** $\frac{3}{5}x = 30$

11. $x + 3x = 16$ **12.** $x + (x + 2) = 8$

B   **13.** $2x + 2(x + 4) = 28$ **14.** $x + 2(x - 1) = 10$

For the equation $3x + 5 = 60$, write a problem which could apply to each of the following persons. Answers will vary.

C   **15.** A carpenter   **16.** An astronaut

17. A veterinarian   **18.** A scuba diver

19. A football coach   **20.** A farmer

8-2 Word Problems into Equations

Objective To write an equation that expresses the numerical relationships in a given word problem.

A word problem generally states some numerical relationships and asks a question. We use the given information to write an equation which we can solve to find the answer. We can usually begin by considering the following questions.

1. What is asked for?
2. What variable expressions can be used to describe what is asked for? (It is sometimes helpful to choose a letter which suggests what is asked for, such as w for width, or n for number.)
3. Will a sketch help to make the problem clearer?
4. What other information is given? How can we use it to write an equation?

The example shows how to apply these suggestions.

Example | Write an equation which expresses the following problem.

The length of a rectangle is 3 cm greater than the width. If the perimeter of the rectangle is 26 cm, what are the dimensions of the rectangle?

Solution |

1. What is asked for?

The dimensions, that is, the length and width of a rectangle.

2. What variable expressions can be used?

Let w = width in centimeters. Then $w + 3$ = length in centimeters.

3. Will a sketch help?

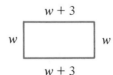

$w + 3$

w w

$w + 3$

4. What other information is given? How can we use it to write an equation?

The perimeter is 26 cm. This means the sum of the lengths of the sides is 26 cm. Then we can write the equation $w + w + (w + 3) + (w + 3) = 26$.

Class Exercises

1. One number is 10 greater than another number. If x represents the lesser number, how can we represent the greater number? If x represents the greater number, how can we represent the lesser number?
   $x + 10; x - 10$

2. One number is twice another number. If x represents the first number, how can we represent the second number? If x represents the greater number, how can we represent the lesser number? $2x; \frac{x}{2}$

3. The product of two numbers is 24. If x represents one of the numbers, how can we represent the other number? $\frac{24}{x}$

4. How can we represent the number of cents in x nickels? x dimes? x dollars? $5x; 10x; 100x$

5. The two equal sides of an isosceles triangle are each 3 cm longer than the third side. If x represents the length of the short side, how can we represent the perimeter of the triangle? $x + 2(x + 3)$

State an equation which could be used to solve the problem.

6. The sum of two numbers is 23, and one of the numbers is 8 greater than the other number. Find the lesser number. $x + (x + 8) = 23$

7. The sum of two numbers is 18 and one number is twice the other number. Find the numbers. $x + 2x = 18$

8. The sum of two consecutive integers is 23. Find the numbers. $x + (x + 1) = 23$

Exercises

Write an equation which can be used to solve the problem. Use the suggestions listed on page 249.

Example	One number is 5 more than 3 times another number. The sum of the two numbers is 25. Find the lesser number.
Solution	We are asked to find the lesser number. Let's let x represent the lesser number. Then $3x + 5$ represents the greater number. (A sketch will not help make the problem clearer.) The sum of the numbers is 25, so we can use the equation $x + (3x + 5) = 25$.

A

1. The sum of 23 and some number is 40. Find the number. $23 + x = 40$

2. When 5 is subtracted from some number, the difference is 18. Find the number. $x - 5 = 18$

3. When 21 is added to some number, the sum is 33. Find the number.
$x + 21 = 33$

4. When a certain number is subtracted from 31, the difference is 11. Find the number. $31 - x = 11$

5. The sum of two numbers is 19, and one number is 8 greater than the other. Find the numbers. $x + (x + 8) = 19$

6. Find two consecutive integers whose sum is 43. $x + (x + 1) = 43$

7. Find two consecutive integers whose sum is 101. $x + (x + 1) = 101$

8. Find two consecutive even numbers whose sum is 26. $x + (x + 2) = 26$

9. A 120 m long highway sound barrier is constructed in two pieces. One piece is 45 m longer than the other. Find the length of each piece.
$l + (l + 45) = 120$

10. The perimeter of a square is 24 m. Find the length of each side. $4s = 24$

11. The perimeter of an equilateral triangle is 45 cm. Find the length of each side. $3s = 45$

B 12. The length of a rectangle is 5 cm greater than the width. The perimeter is 70 cm. Find the length and width. $w + w + (w + 5) + (w + 5) = 70$

13. The two equal sides of an isosceles triangle are each 11 cm longer than the third side. The perimeter of the triangle is 58 cm. Find the length of each side. $s + (s + 11) + (s + 11) = 58$

14. The perimeter of a rectangle is 40 m. The length of the rectangle is 10 m greater than the width. Find the length and the width.
$w + w + (w + 10) + (w + 10) = 40$

C 15. An average bath uses 136 L of water. With that much water, you could take two brief showers and wash a load of laundry. Washing a load of laundry requires 40 L more than a brief shower. How many liters of water does each activity require?
$s + s + (s + 40) = 136$

16. Terry France fell asleep with the television on and a light burning. By the time she awoke she had wasted 2.48 kW·h of electricity. The television wasted 30 times as much energy as the light bulb. How many kilowatt-hours did each appliance waste? $b + 30b = 2480$

17. Mike has some dimes and nickels with a total value of $3.55. The number of dimes he has is 3 more than twice the number of nickels. How many nickels does he have?
$5n + 10(2n + 3) = 355$

18. The sum of the ages of Juan and Rosa is 36 years. In 6 years Juan will be 3 times as old as Rosa will be then. How old is Juan now? $j + 6 = 3(36 - j + 6)$

8-3 Solving and Checking Word Problems

Objective To solve and check a word problem.

Once we have set up an equation that expresses a word problem, we can solve the equation to answer the problem. We always check the answer to see that it satisfies the original problem as well as the equation, to be sure we used the right equation.

Example	Joyce Curtis and Cy Gruver formed a partnership to open a restaurant. Joyce invested $6000 more than Cy. Together they invested $20,000 and raised $30,000 from three other partners. How much did Cy invest?
Solution	**1. What is asked for?** The number of dollars Cy invested.
	2. What variable expressions can be used? Let d = the number of dollars Cy invested. Then $d + 6000$ = the number of dollars Joyce invested.
	3. Will a sketch help? No
	4. What other information is given, and how can we use it to write an equation? Together they invested 20,000 dollars. Therefore, $d + d + 6000 = 20{,}000$. The other information given is unnecessary.
	Solve the equation.

$$d + d + 6000 = 20{,}000$$
$$2d + 6000 = 20{,}000$$
$$2d = 14{,}000$$
$$d = 7000$$

Does the solution of the equation satisfy the original problem?

Yes. Joyce invested 6000 dollars more than Cy, so she invested 13,000 dollars. Then together they invested $13{,}000 + 7000$ or 20,000 dollars.

Cy Gruver invested $7,000.

Notice that in the example, not all the given information was necessary for solving the problem. The information concerning the $30,000 invested by three other partners was not needed. Some of the exercises that follow may also contain information that is not needed and which should be ignored.

Problems

A **1–14.** Solve Exercises 1–14 on pages 250–251.   1. 17   2. 23   3. 12   4. 20   5. 5.5, 13.5
6. 21, 22   7. 50, 51   8. 12, 14   9. 37.5 m, 82.5 m   10. 6 m   11. 15 cm   12. 20 cm, 15 cm
13. 12 cm, 23 cm, 23 cm   14. 15 m, 5 m

B **15.** David Cho paid $65 for 6 tickets to a basketball cham-
pionship game. As a season ticket holder, he was
entitled to buy 2 tickets at the regular season price,
which is $3.50 less than the championship-game price.
How much does each kind of ticket cost? $8.50, $12

16. A town offered plots of land on which residents could
plant gardens. One piece of land was divided into 15
identical plots. The length of each plot was 1 m greater
than the width. The perimeter of each plot was 14 m.
Find the dimensions of each plot.   l = 4 m, w = 3 m

17. Lenore Buranich tutors 3 students for a total of 6
hours per week. She charges the second student $2
more per week than the first, and she charges the third
student half again as much per week as the second.
She earns a total of $47 per week. How much does
each student pay? $12, $14, $21

C **18.** One angle of a triangle is 70° less than twice a second angle, and
the second angle is 25° more than half the third angle. The
perimeter of the triangle is 24 cm. Find the measure of each
angle. (Recall that the sum of the angles of a triangle is 180°.)   50°, 60°, 70°

19. Bill Barbieri needed $90 to have his car repaired. His brother
lent him some money, and his sister lent him twice as much.
Together the loans amounted to half as much as Bill had set
aside to pay for car repairs, and he was just able to pay the bill.
How much did he borrow, and how much did he have saved? $30, $60

20. A board 49 cm long is cut so that the length of the longer piece
is 7 cm greater than twice the length of the shorter. How long
is the shorter piece of board?   14 cm

21. The sum of the ages of Greta and George is 11. In 5 years Greta
will be twice as old as George is then. How old is Greta now?   9

22. Jack and Jill hiked from home to a bird sanctuary at the rate
of 5 km/h and then returned home by the same route at a rate of
7 km/h. If the round trip took them 7 h, how far is it from their
home to the bird sanctuary?   about 20.4 km

23. The sum of a number and its reciprocal is $\frac{5}{2}$. What are the
numbers?   2 and $\frac{1}{2}$

Self-Test

Write two practical problems that are expressed by the given equation.
For the first problem, use a simple direct question.

1. $20 + x = 39$ What number added to 20 gives 39? [8-1]

2. $12x = 84$ What number multiplied by 12 gives 84?

3. $x - 7 = 23$ Seven subtracted from what number gives 23?

4. $\frac{1}{3}x = 9$ What number multiplied by $\frac{1}{3}$ gives 9?

Write an equation that could be used to solve the problem.

5. The sum of 18 and some number is 51. Find the number. $18 + x = 51$ [8-2]

6. When 13 is subtracted from some number, the difference is 44.
Find the number. $x - 13 = 44$

7. The sum of two numbers is 71, and one of the numbers is 16
greater than the other number. Find the numbers. $x + (x + 16) = 71$

8. The length of a rectangle is 20 cm greater than the width. The
perimeter of the rectangle is 120 cm. Find the dimensions of
the rectangle. $x + x + (x + 20) + (x + 20) = 120$

Write an equation and use it to solve the problem.

9. The sum of two numbers is 50, and one number is 12 greater   [8-3]
than the other number. Find the lesser number.   19

10. The difference when 19 is subtracted from a number is 12. Find
the number.   31

11. Find two consecutive integers whose sum is 39.   19 and 20

12. The sum of two numbers is 51, and one number is 27 greater
than the other. One number is odd and is 3 more than 3 times
the other number, which is even. Find the numbers.   12 and 39

Self-Test answers and Extra Practice are given at the back of the book.

8-4 Scale Drawings

Objective To interpret and use scale drawings.

In Chapter 6, we learned about similar figures. We can use similar figures to solve problems involving *scale drawings.* In a **scale drawing,** we represent an object by means of a similar figure. The **scale** of the drawing is the ratio of a unit of distance measured on the drawing to the corresponding actual distance.

For example, an architect may draw a rough floor plan for a house on the scale 1 cm ⟶ 3 m (read "one centimeter represents three meters").

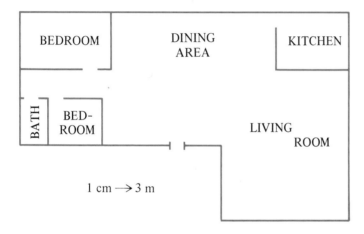

Since 3 m = 300 cm, the ratio of distances in the floor plan to actual distances is 1 to 300, or 1:300.

Example   **If a map is drawn to the scale 1 cm ⟶ 190 km and the distance between Montreal and Chicago measures 6 cm on the map, what is the approximate actual distance between the two cities?**

Solution   **We have the proportion**

$$\frac{x}{6} = \frac{190}{1}$$

$$6\left(\frac{x}{6}\right) = 6\left(\frac{190}{1}\right)$$

$$x = 1140$$

So the distance is approximately 1140 km.

Problems

A 1. A scale drawing of a tree house has the scale 1 cm
⟶ 1.5 m. The width of the tree house in the drawing is 1.5 cm. What actual length does this represent?
2.25 m

2. A drawing showing the anatomy of the human eye has
the scale 5 cm ⟶ 1 cm. The diameter of the eye
measures 12.5 cm in the drawing. What length does
this represent? 2.5 cm

3. A map of Denver, Colorado has scale 1 cm ⟶
0.5 km. What map distance corresponds to an actual
distance of 9 km? 18 cm

4. Look at the house floor plan in this section. Measure to determine the actual dimensions of the two bedrooms. 7.8 m by 4.8 m
4.5 m by 3.9 m

5. Make a scale drawing of a rectangular table top measuring 1.5 m
by 2.25 m, using the scale 1 cm = 0.25 m. See students' papers.

6. Make a scale drawing of a rectangular room 5 m by 7.5 m using
the scale 2 cm = 1 m. See students' papers.

B 7. Make a scale drawing showing three cities, A, B, and C, where A
lies 75 km due west of B, and C lies 50 km due south of B. Use
1 cm = 10 km, and determine how far it is from A to C. 90 km

8. On an aerial photograph of his farm, Rusty Cronholm measures
the distance from the farmhouse to the pond as 1.5 cm. He
knows the actual distance to be 3.75 km. How far is it from the
farmhouse to the west wheat field if this distance measures
3.5 cm on the photograph? 8.75 km

The diagram shows three cities.

9. What is the scale of the drawing?
1 cm = 60 km

Northburg

Weston   540 km   Eastville

10. Find the actual distance between Weston and Northburg. 420 km

11. What is the actual distance between Northburg and Eastville? 300 km

C 12. Two lookout stations A and B are 10 km apart and get a fix on a ship at sea. Station A observes an angle of 70° between the ship and station B, while station B observes an angle of 75° between the ship and station A. Use a scale drawing to determine the distance of the ship from station A. 16 km

13. Two ranger-lookout stations A and B are 15 km apart. Station A sights a fire at an angle of 40° from station B, and station B observes the same fire at an angle of 55° from station A. How far is the fire from station A? 9.6 km

A Physicist

Maria Goeppert-Mayer (1906–1972) was a joint winner (with Johannes H. D. Jensen and Eugene P. Wigner) of the 1963 Nobel Prize for Physics. The prize was awarded for their explanation of properties of atomic nuclei in terms of a structure of shells, or orbits, in which protons and neutrons are allowed to move.

Goeppert-Mayer received a Ph.D. degree from the University of Göttingen (now in West Germany). In 1930, she came to the United States to do research at Johns Hopkins University. Later she continued her career at Columbia University, the Institute for Nuclear Studies at the University of Chicago, and the University of California.

In 1947, the observation was made that some particular nuclei are more stable and abundant in nature than most. All of these have a particular number of neutrons (such as 50, 82, or 126). In 1949, Goeppert-Mayer explained these numbers, now called magic numbers, in terms of the nuclear shell theory.

Career Activity Find out what some of the different branches of physics are and what training is required for them.

8-5 Interest Problems

Objective To solve word problems involving simple and compound interest.

A person who borrows money pays the lender interest. Interest is money paid for the use of money. The interest paid depends on the amount of money borrowed, called the principal, the rate of interest, and the length of time, usually given in years, for which the money is used. The relationship is given by the following formula:

> **formula**
>
> Interest = Principal × rate × time
>
> or
>
> $I = Prt$

Interest paid only on the original principal is called simple interest.

Example 1 | The Wasileskis have been offered an opportunity to invest \$1000 at an annual interest rate which will allow them to double their investment in $12\frac{1}{2}$ years. What is the interest rate?

Solution | Interest = *Prt*.
$I = 1000$, $P = 1000$, and $t = 12.5$

$$1000 = 1000 \times r \times 12.5$$
$$1000 = 12{,}500r$$
$$\frac{1000}{12{,}500} = r$$
$$0.08 = r$$

The Wasileskis' investment will earn 8% annual interest.

Institutions such as banks and savings-and-loan companies usually pay compound interest to depositors. This means that the interest is periodically added to the principal and interest is then paid on the new principal. The interest may be compounded semiannually, quarterly, monthly, or even daily, and is usually figured from tables or by computers.

Example 2	The Lindstroms deposited $300 in a savings account at 6.5% compounded semiannually. They made no withdrawals in the first eighteen months. How much money was in the account at the end of that time?

Solution

For the first six months (or one-half year),
$P = 300$, $r = 0.065$, and $t = 0.5$.
$I = Prt = 300 \times 0.065 \times 0.5$
$\qquad = 9.75$

Then for the second six months,
$P = 300 + 9.75 = 309.75$
$I = 309.75 \times 0.065 \times 0.5$
$\qquad = 10.07$ (rounded to the nearest cent)

Then for the third six months,
$P = 309.75 + 10.07 = 319.82$
$I = 319.82 \times 0.065 \times 0.5$
$\qquad = 10.39$

$319.82 + 10.39 = 330.21$

After 18 months the Lindstroms had $330.21 in their account.

Problems

Solve.

In problems 1–5, find the simple interest earned.

A
1. $150 at 7% for 3 years. $31.50
2. $500 at 8.5% for 10 years. $425
3. $6000 at 5.75% for 6 years. $2070
4. $750 at 6% for 12 years. $540
5. $12,000 at 8% for 9 years. $8640

6. You receive simple interest of $30 at the end of one year for money in a savings account. The interest rate is 6%. How much money was in your account? $500

7. If you invest $3000 for one year at 7%, how much would you have at the end of one year? $3210

8. Mrs. Goodson owns 6 bonds, each worth $1000. She receives 7.5% simple interest each year on each bond. What is her yearly income from her bonds? $450

9. The Stonehill State Bank loaned $15,000 to a businessman at 8% annual interest for a 3-month period. How much money was due to be paid back at the end of the 3 months? $15,300

10. An investment of $36,000 produces $2700 of interest income each year. What is the rate of interest on this investment?   7.5%

B 11. Ace Savings and Loan pays 5% compounded semiannually on its savings accounts. How much will be in an account after 2 years if $600 is deposited originally? $662.29

12. Adam Tillman has $3000 invested at 6% annual interest. How much more must he invest at 8% annual interest if his annual income from both investments must total $600? $5250

13. A person offers to lend a friend money on the following basis: 5% interest for the first year, 7% interest for the second year, 9% interest for the third, and so on. No interest is due until the end of the loan period. If $1000 is borrowed for 3 years, how much must be paid back at the end of the loan period? Would you borrow from this person? What would be the interest rate for the fifth year? The sixth year? $1224.62; No; 11%; 13%

14. Alice Andrews has an annual income of $930 from three investments. If she has $3000 invested at 6%, and $5000 invested at 7%, how much does she have invested at 8%? $5000

C 15. Sheila Semenza plans to buy 1000 shares of stock selling at $16 per share. She can buy the stock under either of two plans.

Plan **1.** Buy the shares with a cash payment of $16,000 of her own money.

Plan **2.** Buy the shares with a cash payment of $8,000 of her own money together with $8,000 borrowed at 8% annual interest, and repay the loan when she sells the shares.

Assuming that she holds the stock for one year and then sells it at $20 per share, which plan yields the greater rate of return on her own money? Why?   Plan 1

16. During the course of a year a merchant's retail sales totaled $540,000. Included in that total was a sum covering an 8% sales tax. How much did the merchant owe the sales-tax collector at the end of that year? $40,000

17. Jerrold Mann borrowed $5000 for one year at 9% interest. Payment is made monthly and each month's interest is computed on the unpaid balance of the loan. How much interest did he pay in the first four months? $148.32

8-6 Environmental Problems

Objective To solve simple word problems dealing with the environment.

Mathematics can be an important aid in analyzing and solving many of the problems concerning our environment. These problems involve overpopulation, air and water pollution, noise pollution, deterioration of cities, the need for public transportation, and many other facets of everyday life.

Example A recent study showed that of the 215,000,000 people in the United States, an estimated 16,000,000 suffer some degree of hearing loss directly related to noise. To the nearest percent, what percent of the population is that?

Solution Let n represent the percent of the population suffering hearing loss related to noise. Then:

$$16,000,000 = \frac{n}{100} \times 215,000,000$$

$$\frac{n}{100} = \frac{16,000,000}{215,000,000}$$

$$\frac{n}{100} \approx 0.074$$

$$n \approx 7.4$$

To the nearest percent, 7% of the population suffer some degree of hearing loss directly related to noise.

Problems

Solve.

A 1. In Millsville, an air-pollution alert is called when smoke particles in the air reach a level of 315 parts per million (ppm). If an analysis showed that smoke particles made up 0.03% of the air, would an alert be sounded?   No

2. Temperatures in a deep mine increase, on the average, about 1°C for each 55 m of depth. If the surface ground temperature is 20°C, at what depth would the temperature in a mine reach 35°C?   825 m

3. In the year 1600 A.D., there were 8684 species of birds. Of these, about 1.09% have disappeared. How many species of birds have disappeared? 95 species

B 4. The cost of operating an electric appliance is given by

$$\frac{\text{wattage of appliance} \times \text{hours of operation}}{1000} \times \text{Cost per kW·h} = \text{operating cost}$$

What is the annual cost of operating a 300 W color television if it is used an average of 4 h per day and the electric rate is 6.2¢ per kW·h? $27.16

5. An iceberg can be melted to produce 200,000 m³ of water per day. If one person uses about 0.16 m³ of water per day, how many persons will it supply?
1,250,000 persons

6. Safety records for a factory showed 1.2 injuries requiring medical treatment for each 100,000 h spent on the job. If the factory employs 700 people who each work an average of 1700 h per year, how many such injuries should be expected at the factory in one year?
14.28 injuries

C 7. Solar energy strikes the panels of a communications satellite with an intensity of 1.38 kW/m². If a rectangular panel with an area of 0.1 m² converts 16% of this energy into electric power, how many watts of electric power does it supply? 22.08 W

8. For each kg of mass transported, a bicycle rider expends about 730 J of energy per km. If a 75 kg rider pedals a 12 kg bicycle 20 km, how many joules are used?
1,270,200 J

9. The number of years it takes for a population to double in size can be estimated by dividing the annual number of additional persons per 1000 population into the number 693. If the world population continues to increase at the annual rate of 20 persons per 1000 population, how long will it take for the population to double? 34.65 years

10. Los Angeles uses approximately 1,800,000 kL of water each day. About 15% of it comes through the 480 km Colorado Aqueduct, 21% from local wells, and 64% from the Owens Valley. If the Colorado Aqueduct is out of service for 2 weeks, about how much more water would have to be supplied from the other sources? 3,780,000 kL

Application
Commuting Costs

Stacy commutes by train between Troy and Dover, a distance of 38 km. She works a 5-day week. This year, the month of August begins on a Tuesday and ends on a Thursday. What is her cheapest plan for commuting if she (1) works the entire month? (2) takes vacation the 2nd week of August? (3) takes vacation the 2nd and 3rd weeks of August?

Commuting by car costs 8¢ per km and $2.50 per day for parking. Compare the costs of commuting by car and by train.

Train fares Dover–Troy	
One-way:	$2
12 rides:	$20
Monthly:	$52

Self-Test

Words to remember:
scale drawing [p. 255]   interest [p. 258]   principal [p. 258]
scale [p. 255]   interest rate [p. 258]

Solve.

1. The scale on a map of Boston is 1 cm ⟶ 0.25 km. The distance on the map from Larz Anderson Bridge to the Harvard Bridge is 6.5 cm. What actual distance does this represent?   1.625 km   [8-4]

2. On a scale drawing of a school gymnasium the scale is 1 cm ⟶ 5 m. The gymnasium measures 30 m by 37.5 m. What size is the drawing?   6 cm by 7.5 cm

3. Ellis Throckmorton deposited $500 at 5.5% interest for 3 years. What was the simple interest earned? $82.50   [8-5]

4. Angela Ciccarelli invested $3000 at 6% annual interest. How much must she invest at 7.5% annual interest for her annual income from both investments to total $600? $5600

5. The oceans cover about 71% of the 507,000,000 km² of the earth's surface. What area is this?   359,970,000 km²   [8-6]

6. The Trans-Alaska pipeline system is about 1300 km long and about 1 m in diameter. The pipeline cost $3 billion to complete. What is the cost per kilometer?   about $2,307,692/km

Self-Test answers and Extra Practice are given at the back of the book.

Chapter Review

Write the letter that names the correct answer.

1. A problem expressed by $17 + x = 34$ is __?__ . C [8-1]
   A. A number added to 34 is 17. Find the number.
   B. Twice a number is 34. Find the number.
   C. 17 added to a number is 34. Find the number.
   D. 34 added to a number is 17. Find the number.

2. The perimeter of an isosceles triangle is 40 cm. The two equal [8-2]
   legs each measure 5 cm more than the third leg. An equation
   which can be used to find the lengths of the sides is __?__ . A
   A. $x + (x + 5) + (x + 5) = 40$
   B. $x + (x - 5) + (x - 5) = 40$
   C. $2x + (5 + x) = 40$
   D. $40 - x - (x + 5) = x$

3. Two consecutive integers whose sum is 47 are __?__ . B [8-3]
   A. 17 and 30   B. 23 and 24   C. 18 and 19   D. 28 and 29

4. In order to earn the money to buy a pair of jogging shoes, Gabriel
   Marquez worked Saturday gardening and three afternoons during the week helping out in his uncle's store. The $12 he earned
   gardening was one half what he earned working in the store during the week, and the total was just enough to buy the shoes.
   The shoes cost __?__ . C
   A. $8   B. $18   C. $36   D. $28

5. Joan O'Brien spent twice as long mowing her lawn as she did
   raking. The total time was 2 h and 15 min. She spent __?__
   mowing her lawn.   C
   A. 45 min   B. 1 h 15 min   C. 1 h 30 min   D. 1 h

6. The distance between Washington, D.C. and Richmond, Virginia, on a map is 12 cm. The scale of the map is 1 cm ⟶ 12 km. The approximate actual distance is __?__ . C

A. 100 km   **B.** 12 km   **C.** 144 km   **D.** 81 km

[8-4]

7. Dennis finds a scale drawing of the floor plan of his house and decides that the shortest path from the television set to the cookie jar in the kitchen is one that measures 11 cm on the floor plan. If the scale is 1 cm ⟶ 0.5 m, this distance is __?__ . A

A. 5.5 m   **B.** 55 m   **C.** 5 m   **D.** 11.5 m

8. The interest for one year on $400 deposited in a savings account at 6.75%, compounded semi-annually, is __?__ . D

A. $13.50   **B.** $13.96   **C.** $427.46   **D.** $27.46

[8-5]

9. Alice invested $200 at 6% annual interest and $350 at 5.75% annual interest, both compounded annually. If she makes no withdrawals, how much will she have after two years? B

A. $224.72   **B.** $616.13   **C.** $582.13   **D.** $66.13

10. Central School collected 40% more scrap aluminum than West School, which, in turn, collected 25% more than Southern. If the total weight of scrap aluminum collected in the clean-up campaign was 560 kg, how much did Central School collect? B

A. 140 kg   **B.** 245 kg   **C.** 175 kg   **D.** 70 kg

[8-6]

11. In a certain city, 20% of the factories one year violated air-pollution control laws, and each of these factories was fined $1000. On the other hand, 28% of the factories far exceeded air-pollution control standards and each received an incentive award of $500 from the city. The net income for the city from these transactions was $7500. How many factories are in the city? C

A. 200   **B.** 140   **C.** 125   **D.** 132

Mathematics A Century Ago

If you had been in a mathematics class one hundred years ago, you might have been asked to solve a problem such as the following:

A hare is 7 of his own leaps before a hound, and takes 5 leaps while the hound takes 3, but 4 of the hound's leaps are equal to 7 of the hare's. How many leaps must the hound take to gain on the hare the length of one of the hare's leaps? How many leaps must the hound take to catch the hare?

Solution Since 1 of the hound's leaps is $\frac{7}{4}$ of a leap of the hare, 3 of the hound's leaps are equal to $\frac{21}{4}$, or $5\frac{1}{4}$, of the hare's leaps; hence, in taking 3 leaps the hound gains $\frac{1}{4}$ of one leap of the hare, and therefore he must take 4 times 3 leaps, or 12 leaps, to gain 1 leap of the hare; and to gain 7 leaps, he must take 7 times 12 leaps, or 84 leaps.

Mathematics texts from that period included many exercises and very little text. Sections of problems like the one above alternated with drill sections like the following:

1. 3 times 8 are how many times 4?
2. 12 times 3 are how many times 9?
3. 6 times 4 are how many times 8?
4. 3 times 10 are how many times 6?

You might enjoy using the methods you have learned in this chapter to solve some problems from long ago.

1. A fox is 80 rods before a greyhound, and is running at the rate of 27 rods in a minute. The greyhound is following at the rate of 31 rods in a minute. In how many minutes will the greyhound overtake the fox? 20 minutes

2. If it takes 1 yard and 1 fourth of a yard of cloth to make a pair of pantaloons, how many yards would it take to make 8 pairs?   10 yards

3. A boy having $\frac{4}{5}$ of a watermelon, wished to divide his part equally between his sister, his brother, and himself, but was at a loss to know how to do it; but his sister advised him to cut each of the fifths into 3 equal parts. How many pieces did each have? What part of the whole melon was each piece?   4 pieces; $\frac{1}{15}$ of the whole

4. A fish's head weighs 5 pounds, his tail weighs as much as his head added to half the weight of his body, and his body weighs as much as his head and tail together. What is the weight of the fish?   40 pounds

5. A woman driving her geese to market, was met by a man, who said, Good-morrow, madam, with your hundred geese; says she, I have not a hundred; but if I had half as many more as I now have, and two geese and a half, I should have a hundred; how many had she?   65 geese

6. A boy being asked his age, answered that if $\frac{1}{2}$ and $\frac{1}{4}$ of his age, and 20 more, were added to his age, the sum would be 3 times his age. What was his age?   16

7. An ignorant fop wanting to purchase an elegant house, a facetious gentleman told him he had one which he would sell him on these moderate terms, namely, that he should give him a penny for the first door, 2 pennies for the second door, 4 pennies for the third door, and so on, doubling at every door which were 36 in all. Pray, what would the house have cost him?
$687,194,767.35

Chapter Test

Write two practical problems that are expressed by the given equation. For the first, use a simple, direct question. Answers will vary.

1. $18 - x = 52$ **2.** $\frac{1}{2}x = 21.5$ [8-1]

1. What number subtracted from 18 gives 52?
2. What number multiplied by one-half gives 21.5?

Write an equation which can be used to solve the problem.

3. The perimeter of a rectangle is 28. The length is twice the width.   [8-2]
What are the dimensions? $w + w + 2w + 2w = 28$

4. The sum of two numbers is 36. One of them is 8 greater than
the other. What are the numbers? $n + (n + 8) = 36$

Solve.

5. The difference between two numbers is 3, and 5 subtracted from   [8-3]
the larger is one half the smaller. What are the two numbers? 4 and 7

6. Each of the two equal sides of an isosceles triangle is 3 m longer
than one half the third side. The perimeter is 54 m. What are
the lengths of the sides?   24 m, 15 m, 15 m

7. A map has the scale 1 cm ⟶ 10 km. What is the actual dis-   [8-4]
tance between two towns which are 6.5 cm apart on the map?   65 km

8. A scale drawing of an insect has scale 2 cm ⟶ 0.25 cm. What
length on the drawing represents an actual wingspan of 1 cm?   8 cm

9. If you invest $5000 for one year at 6%, how much would you   [8-5]
have at the end of the year? $5300

10. Diane Tetreault invests $500 at an annual simple interest rate that
will allow her to triple her investment in 15 years. What is the rate? $13\frac{1}{3}\%$

11. A sleeping bag 4 cm thick provides protection at temperatures   [8-6]
down to 5°C. At $-50°$C, a sleeping bag must be 2 cm more than
twice that thickness. How thick is that?   10 cm

12. A city one year had a population density of 3100 persons/km².
The next year the population density fell 10%. If the city's area is
95.5 km², what was the new total population?   266, 445 persons

Skill Review

Add or subtract.

1. $8.6 + 3.2$ 11.8   **2.** $5.7 + 6.3$ 12.0   **3.** $4.03 - 2.8$ 1.23

4. $8.79 - 3.5$ 5.29   **5.** $8.7 - 9.3$ −0.6   **6.** $28.2 - 7.43$ 20.77

7. $3.3 + (-8.8)$ −5.5   **8.** $9.86 - (-9.6)$ 19.46   **9.** $4.40 - 1.50$ 2.90

10. $2.718 - 2.18$ 0.538   **11.** $0.423 + 3.10$ 3.523   **12.** $-8.2 - 5.1$ −13.3

13. $10.3 - (-8.39)$ 18.69   **14.** $9.38 + (-10.83)$ −1.45   **15.** $-9.42 - (-7.77)$ −1.65

16. $1.772 - 4.538$ −2.766   **17.** $5.171 - (-5.717)$ 10.888   **18.** $-1.284 - 0.25$ −1.534

19. $8.105 + 8.007$ 16.112   **20.** $16.548 - 23.219$ −6.671   **21.** $-9.48 - 8.92$ −18.40

22. $42.98 - 16.88$ 26.10   **23.** $8.244 - (-9.88)$ 18.124   **24.** $-2.9 + 7.352$ 4.452

Multiply or divide.

25. 2.1×6.7 14.07   **26.** 6.9×8.9 61.41   **27.** 3.1×1.3 4.03

28. $1.159 \div 2$ 0.5795   **29.** $76.89 \div 1.5$ 51.26   **30.** -4.6×2.1 −9.66

31. $-112.86 \div 2.2$ −51.3   **32.** $7.77 \div (-11)$ −0.70$\overline{63}$ **33.** -3.40×0.06 −0.2040

34. $39.3 \div (-0.3)$ −131   **35.** $-28.7 \div (-3.5)$ 8.2   **36.** 18.00×0.3 5.400

37. $-26.11 \div 0.02$ −1305.5   **38.** $0.22 \div (-5.5)$ −0.04   **39.** -12.2×14.0 −170.80

40. $27.5 \div (-90)$ −0.30$\overline{5}$ **41.** 58.6×17 996.2   **42.** $-99.1 \div 98$ −1.0112245

Perform the indicated operations.

43. $3.19 - 5 \times 0.17$ 2.34   **44.** $-3.1 \times (-2.4) + 18.26$ 25.70

45. $92.4 \div 8.4 - (-2.13)$ 13.13   **46.** $-5.21 + 9.18 \div (0.027)$ 334.79

47. $-3.12 \times 8.5 - 2.41 \times (-15.6)$ 11.076

48. $-8.68 \div (0.031) + (-2.9) \times (-3.1)$ −271.01

49. $3.019 \times (-2.3) + (-9.243) \div (-0.09)$ 95.7563

50. $(-2.31) \times (-1.1) + 3.2 \times (-5.7) - 6.82$ −22.519

9 Areas and Volumes

9-1 Areas of Rectangles and Parallelograms

Objective To find the areas of rectangles and parallelograms.

The **area** of a geometric figure is the amount of its interior surface. The parallelogram on the left below is a rectangle, and its area is the product of its length and width. However, the parallelogram on the right is *not* a rectangle, and its area is *not* the product of its two dimensions.

Rectangle

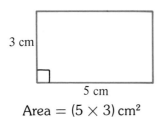

5 cm

Area $= (5 \times 3)$ cm²

Parallelogram, but not rectangle

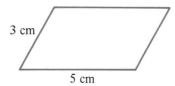

Area is *not* (5×3) cm²

In order to find the area of a parallelogram, it is first helpful to discuss what is meant by its *base* and its *height*. Any side of a parallelogram can be considered as its **base**. The perpendicular distance between the base and the opposite side is the **height** of the parallelogram.

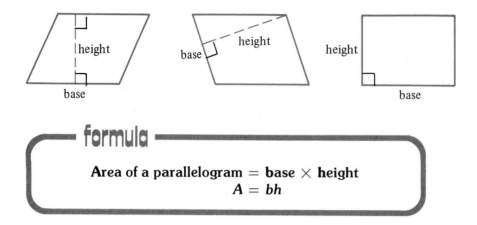

formula

Area of a parallelogram = base × height
$$A = bh$$

The formula $A = bh$ can be used for any kind of parallelogram. For a square, however, most people prefer to use the simpler formula $A = s^2$, where s represents the length of a side.

Area is frequently measured in square meters (m^2) or square centimeters (cm^2). Other measures used for area are square kilometers (km^2) and square millimeters (mm^2). Sometimes we choose to think about an unspecified unit of length. Then the unit of area is simply called a square unit. If a specific unit is not stated, we assume that *units* and *square units* are implied.

Since 1 cm = 10 mm, 1 cm² = 100 mm².
Since 1 km = 1000 m, 1 km² = 1,000,000 m².

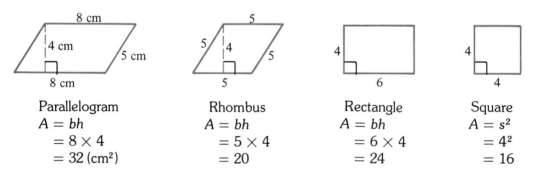

Parallelogram	Rhombus	Rectangle	Square
$A = bh$	$A = bh$	$A = bh$	$A = s^2$
$= 8 \times 4$	$= 5 \times 4$	$= 6 \times 4$	$= 4^2$
$= 32 \,(cm^2)$	$= 20$	$= 24$	$= 16$

Class Exercises

1. Is area measured in centimeters or <u>square centimeters</u>?

2. Is perimeter measured in <u>centimeters</u> or square centimeters?

Complete.

3. $1 \text{ m}^2 =$ _?_ cm²   10,000   **4.** $3 \text{ m}^2 =$ _?_ mm² 9,000,000   **5.** $0.01 \text{ km}^2 =$ _?_ m² 100

For each parallelogram below, give its base, height, and area.

6. 4; 4; 16   **7.** 2; 5; 10   **8.** 5; 2; 10   **9.** 3; 4; 12

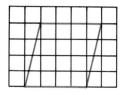 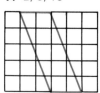 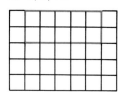 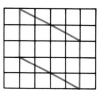

10. 10; 4; 40   **11.** 6; 10; 60   **12.**   7   5; 7; 35

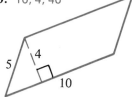 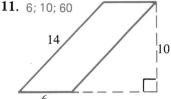 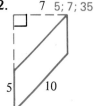

13. Give the perimeters of the parallelograms shown in Exercises 10–12. 30; 40; 30

14. A square has perimeter 12 cm. What is its area? 9 cm²

15. A rectangle has perimeter 44 and area 72. Can you *guess* what its dimensions are? 18 by 4

Exercises

For each parallelogram, give its base, height, and area.

A **1.** 8 cm; 4 cm; 32 cm²   **2.**   **3.** 11.4 m; 5.2 m; 59.28 m²

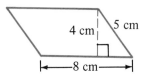 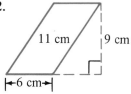 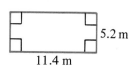

4.   **5.** 6 cm; 9 cm; 54 cm²   **6.** 4

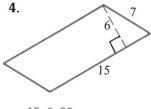 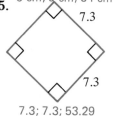

15; 6; 90   7.3; 7.3; 53.29

6; 4; 24

Find the area and perimeter of each parallelogram below. Record your answers in a table like that shown.

	7.	8.	9.	10.	11.	12.
area	36	36	36	22.99	36	25
perimeter	30	24	26	28	28	28

7. 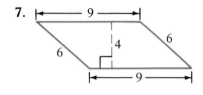 **8.** 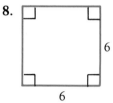 **9.**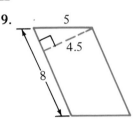

10. 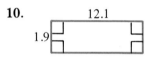 **11.** 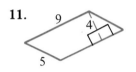 **12.**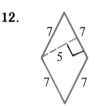

13. Study your answers for Exercises 7–9. They show that parallelograms can have different __?__ but the same __?__ . perimeters; area

14. Exercises 10–12 show that parallelograms can have different __?__ but the same __?__ . areas; perimeter

Problems

Solve.

B 1. The perimeter of a square is 20 cm. What is its area? 25 cm²

2. A rectangle has length 8 and perimeter 28. What is its area? 48

3. The area of a parallelogram is 120 cm². Its base is 16 cm. What is its height? 7.5 cm

4. The area of a parallelogram is 350 cm². Its height is 14 cm. What is its base? 25 cm

5. A square has area 81 m². What is its perimeter? 36 m

6. If you double the side of a square, do you double its perimeter? Yes
Do you double its area? No

7. If you triple the length and width of a rectangle, do you triple its perimeter? Do you triple its area? Yes; No

9-2 Areas of Triangles and Trapezoids

Objective To find the areas of triangles and trapezoids.

Any side of a triangle can be called its **base.** The perpendicular distance from the opposite vertex to the base line is the **height** of the triangle. The area of a triangle is one half the product of its base and height.

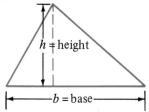

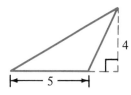

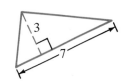

formula

Area $= \frac{1}{2} \times$ base $\times$ height

$A = \frac{1}{2}bh$

$A = \frac{1}{2} \times 5 \times 4$
$= 10$

$A = \frac{1}{2} \times 7 \times 3$
$= 10.5$

The parallel sides of a trapezoid are called its **bases.** The perpendicular distance between the bases is called the **height** of the trapezoid. The area of a trapezoid is one half the product of the height and the sum of the bases.

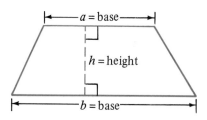

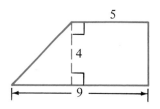

formula

Area $= \frac{1}{2} \times$ (sum of bases) $\times$ (height)

$A = \frac{1}{2}(a + b)h$

$A = \frac{1}{2} \times (5 + 9) \times 4$
$= 28$

Class Exercises

For each triangle below, the base *b* is indicated. State the height.

1. 4

b

2. 4

b

3. 3

b

4. 4

b

5–8. Find the area of each triangle above. 5. 8   6. 6   7. 6   8. 4

9. David and Debby calculated the area of △*ABC* in different ways. Who was correct? Both were correct.

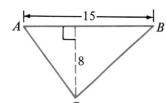

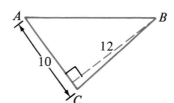

David: Area = $\frac{1}{2}(15 \times 8)$ Debby: Area = $\frac{1}{2}(10 \times 12)$

For each trapezoid below, give the lengths of its two bases and also give its height. Then give its area.

10. 1; 5; 4; 12

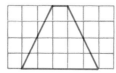

11. 6; 2; 3; 12

12. 2; 4; 5; 20

13. Carol made a congruent copy of a triangle, turned it upside down, and slid the two triangles together to form a parallelogram as shown.
  a. What is the area of this parallelogram? $A = bh$
  b. If you did not know the formula for the area of a triangle, explain how you could figure it out.
    The area of each triangle is $\frac{1}{2}$ of the area of the parallelogram.

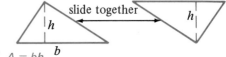

slide together

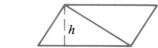

Exercises

A In each of Exercises 1–10, find the area.

1. 5

2. 10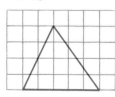

3. 7.5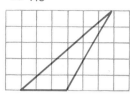

4. 10

5. 9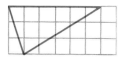

6. 10.5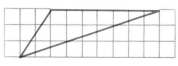

7. 12

8. 30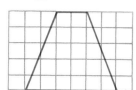

9. 22

10. 28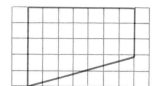

Find the area and perimeter of each triangle below.

11.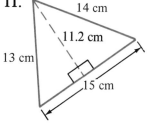

14 cm
11.2 cm
13 cm
15 cm

12.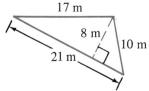

17 m
8 m
10 m
21 m

13.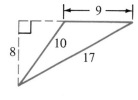

9
8
10
17

84 cm² ; 42 cm   84 m² ; 48 m   36; 36

Find the area of each trapezoid below.

14.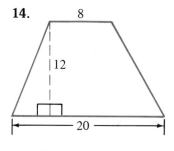

8
12
20

15.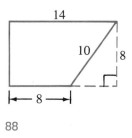

14
10
8
8

88

16.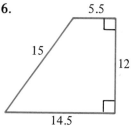

5.5
15
12
14.5

168   120

17. Draw an acute triangle, a right triangle, and an obtuse triangle, each having a base 5 cm and height 2 cm. Find the area of each triangle.
The area of each triangle is 5 cm².

B   **18.** One of the equal sides of an isosceles right triangle is 4 cm long. What is the area of the triangle? 8 cm²

19. The area of a triangle is 50 m². The base is 10 m. What is the height?
$AB = 6.2$ cm; $DC = 3.1$ cm; $h = 2.9$ cm; $A = 13.485$ cm²

20. *ABCD* is a trapezoid. By measuring, find *AB*, *DC*, and the height of the trapezoid in centimeters. Then calculate its area.

21. Find the area of trapezoid *ABCD* by subtracting the area of three right triangles from the area of rectangle *APQR*. 13.5 cm²

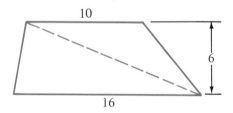

22. Draw a figure like the one shown with the lengths given:
$AP = 8$ cm, $PB = 6$ cm, $BQ = 3$ cm,
$QC = CR = 4$ cm, $RD = 3$ cm,
$DA = 6$ cm, $AB = 10$ cm, $BC = 5$ cm,
$CD = 5$ cm
a. Measure the height of the trapezoid and calculate its area. 4.8 cm; 36 cm²
b. Find the area of the trapezoid as described in Exercise 21. 36 cm²

23. Find the area of the trapezoid by two different methods.

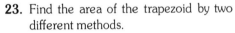

Method 1   Add the areas of two triangles. 78

Method 2   Use the formula

$$A = \tfrac{1}{2}(a + b)h. \quad 78$$

C   **24. a.** What is the area of triangle I? $A = \tfrac{1}{2}bh$
b. What is the area of triangle II? $A = \tfrac{1}{2}ah$
c. Add the areas of triangles I and II and show that the sum can be written as $\tfrac{1}{2}(a + b)h$.

$$A = \tfrac{1}{2}bh + \tfrac{1}{2}ah = \tfrac{1}{2}(b + a)h = \tfrac{1}{2}(a + b)h$$

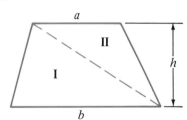

9-3 Areas of Circles

Objective To find areas of circles and of regions involving circles.

Recall that there are two different formulas for the circumference of a circle. If C represents the circumference of the circle, d represents its diameter, and r represents its radius, then:

$$C = \pi d \quad \text{or} \quad C = 2\pi r$$

Two approximate values of π are 3.14 and $\frac{22}{7}$.
    The formula for the area of a circle is given below.

formula

Area of circle $= \pi \times (\text{radius})^2$
$A = \pi r^2$

Example | **Find the area of a circle with diameter 5 m.**

Solution | $A = \pi r^2$
$\approx 3.14 \times (2.5)^2 = 3.14 \times 6.25$
$\approx 19.625 \ (\text{m}^2)$

Class Exercises

Use $\pi \approx 3.14$ for Exercises 1–3 and $\pi \approx \frac{22}{7}$ for Exercises 4–6.

	1.	2.	3.	4.	5.	6.
diameter	4 cm	0.6 mm	? 2.5 m	$\frac{3}{11}$	28	? $\frac{7}{11}$
area	? 12.56 cm	? 0.2826 mm	? 4.90625 m	? $\frac{9}{154}$	? 616	$\frac{7}{22}$
circumference	?	?	7.85 m	? $\frac{6}{7}$	? 88	? 2

12.56 cm   1.884 mm

Exercises

Find the area of the circle described. Use $\pi \approx 3.14$.

1.1304 cm²

A   **1.** radius $= 4$ km 50.24 km²   **2.** diameter $= 1$ m 0.785 m²   **3.** radius $= 0.6$ cm

4. diameter $= 2000$ km   **5.** radius $= 16$ cm 803.84 cm²   **6.** diameter $= 3$ mm
        3,140,000 km²   7.065 mm²

Find the area of the circle described. Use $\pi \approx \frac{22}{7}$.

7. radius $= 1$ $3\frac{1}{7}$

8. diameter $= 1$ $\frac{11}{14}$

9. radius $= 2\frac{4}{5}$ $24\frac{16}{25}$

10. radius $= 5\frac{1}{4}$ $86\frac{5}{8}$

11. diameter $= \frac{7}{22}$ $\frac{7}{88}$

12. diameter $= 70$ 3850

If the radius of a circle is multiplied by 2, the circumference is also multiplied by 2 but the area is multiplied by 4. (Try it and see!) This fact is recorded in the table below. Copy and complete this table.

	Example	13.	14.	15.	16.
Radius is multiplied by	2	3	4	5	? 10
Circumference is multiplied by	2	? 3	? 4	? 5	? 10
Area is multiplied by	4	? 9	? 16	? 25	100

Find the area of each shaded portion. Leave your answer in terms of π.

B **17.**

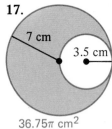

7 cm
3.5 cm

36.75π cm^2

18.

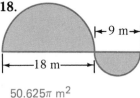

|←9 m→|
|—18 m—|

50.625π m^2

19.   24 cm

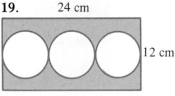

12 cm

$288 - 48\pi$ cm^2

20. If a circle has area 50.24 cm², what is its radius, approximately? 4 cm

21. If a circle has area $9\frac{5}{8}$ square units, what is its radius, approximately? $\frac{7}{4}$ units

22. If a circle has area 314 km², what is its diameter, approximately? 20 km

Find the area and perimeter of each shaded portion. Leave your answers in terms of π.

C **23.**

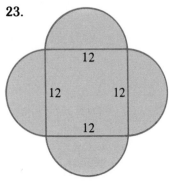

12
12   12
12

24π

24.   12

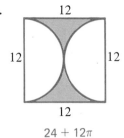

12   12

12

$24 + 12\pi$

25.

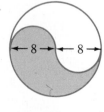

←8→←8→

16π

26.

$32 + 6\pi$

27.

$16 + 4\pi$

28.

$12 + 4\pi$

Self-Test

Words to remember:

area [p. 271]   base [p. 271]   height [p. 271]

1. A square has side 6 cm. What is its area? 36 cm^2 [9-1]

2. A parallelogram has base 2 m and height 6 cm. What is its area? 1200 cm^2

3. A rectangle has area 72 cm^2 and height 6 cm. What is its base? 12 cm

4. A square has area 64 cm^2. What is its perimeter? 32 cm

5. The base of a triangle is 26 mm long and its height is 34 mm. What is its area? 442 mm^2 [9-2]

6. A trapezoid has bases of 2 m and 3 m, and height 4.3 m. What is its area? 10.75 m^2

7. The legs of a right triangle are 8 cm and 3 cm. What is its area? 12 cm^2

8. The area of a triangle is 25 m^2. The base is 5 m. What is the height? 10 m

Find the area of the circle described.

9. radius $= 1\frac{3}{4}$ units (Use $\pi \approx \frac{22}{7}$.) $\frac{77}{8}$ square units   [9-3]

10. diameter $= 10$ cm (Use $\pi \approx 3.14$.) 78.5 cm^2

11. circumference $= 30\pi$ 225π
(Leave answer in terms of π.)

12. Find the area of the shaded part of the figure. (Leave answer in terms of π.) 12π

Self-Test answers and Extra Practice are at the end of the book.

9-4 Volumes of Prisms and Cylinders

Objective To find the volumes of prisms and cylinders.

The figures below are **prisms.** The shaded regions of each are the **bases** of the prism. The perpendicular distance between the bases is called the **height** of the prism. The shape of the polygonal base determines the name of the prism.

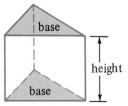

Triangular Prism

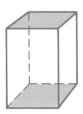

Square Prism

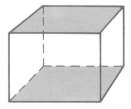

Rectangular Prism
or Rectangular Solid

A **cylinder** is like a prism except that its bases are circles instead of polygons. The amount of space occupied by any solid is called its **volume.** The same formula can be used to find the volumes of both prisms and cylinders. The *base area* is the area of *one* base.

Prism

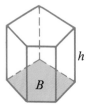

Cylinder

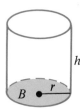

┌─ **formula** ─────────────┐
│ **Volume = Base area × height** │
│ $V = Bh$ │
└───────────────────────────┘

┌─ **formula** ─────────────┐
│ **Volume = Base area × height** │
│ $V = Bh,$ or $V = \pi r^2 h$ │
└───────────────────────────┘

Example │ **Find the volumes.**   **a.**

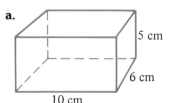

b.

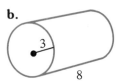

Solution | **a.** $B = 10 \times 6 = 60$ **(cm²)**
$\qquad V = Bh$
$\qquad\quad = 60\ \text{cm}^2 \times 5\ \text{cm} = 300\ \text{cm}^3$

b. $B = \pi \times 3^2 = 9\pi$
$\qquad V = Bh$
$\qquad\quad = 9\pi \times 8 = 72\pi$ **(cubic units)**

Class Exercises

1. In the formula $V = Bh$, what do the letters V, B, and h stand for? Volume, Base area, height

2. For the rectangular solid shown, find the values of B, h, and V. 50, 6, 300

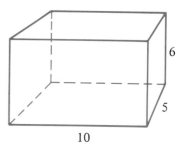

3. Suppose that the right-hand side of the solid is considered as its base. Then

$$B = 5 \times 6 = 30.$$

a. What is the height to this base? 10
b. Using this height and base, calculate V. 300

4. $1\ \text{cm}^3 = \underline{\ ?\ }\ \text{mm}^3$
1000

5. $1\ \text{m}^3 = \underline{\ ?\ }\ \text{cm}^3$
1,000,000

6. $1\ \text{km}^3 = \underline{\ ?\ }\ \text{m}^3$
1,000,000,000

Find the volumes of the solids below. (Give answers to Exercises 8 and 9 in terms of π.)

7.

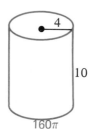

8

$B = 20$

160

8.

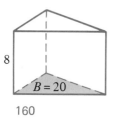

4

10

160π

9.

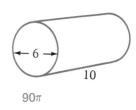

6

10

90π

Exercises

Find the volumes of the prisms shown below.

A **1.**

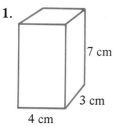

7 cm

3 cm

4 cm

84 cm³

2.

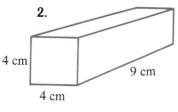

4 cm

4 cm

9 cm

144 cm³

3.

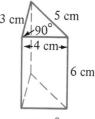

3 cm

5 cm

90°

4 cm

6 cm

36 cm³

Find the volumes of the prisms shown below.

4.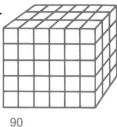

5.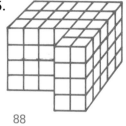

6.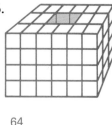

90   88   64

Find the volumes of the cylinders below. Leave answers in terms of π.

7.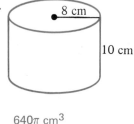

8 cm

10 cm

640π cm^3

8.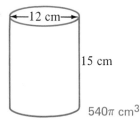

←12 cm→

15 cm

540π cm^3

9.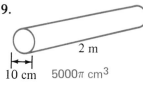

2 m

10 cm 5000π cm^3

Make a drawing of each of the following solids. Then calculate its volume.

10. A soup can with height 10 cm and radius 3 cm. 90π cm^3

11. A quarter with height 2 mm and diameter 23 mm. 264.50π mm^3

12. A rectangular solid with length 12 cm, width 9 cm, and height 4 cm. 432 cm^3

B **13.** A triangular prism with height 10 and with a right triangular base with sides 5, 12, and 13. 300

14. Cement blocks are used for constructing walls of many buildings. Find the volume of the cement block shown. 8000 cm^3

10 cm

10 cm

20 cm

20 cm

30 cm

15. A copper pipe is 10 m long. Its inner radius is 2.4 cm and its outer radius is 2.6 cm. Find the volume of the metal in the pipe. Use $\pi \approx 3.14$. 3140 cm^3

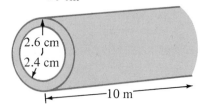

2.6 cm
2.4 cm

←10 m→

C **16.** If you double *each* of the three dimensions of a rectangular solid, does the volume of the solid also double? No

17. If you triple both the radius and the height of a cylinder, does the volume also triple? No

9-5 Volumes of Pyramids and Cones

Objective To find the volumes of pyramids and cones.

The figures below are all **pyramids.** The shaded region of each is called the **base** of the pyramid. The shape of the polygonal base determines the name of the pyramid.

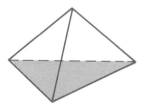

Triangular Pyramid

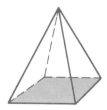

Square Pyramid

Hexagonal Pyramid

The triangular side faces of each pyramid are called **lateral faces.** These lateral faces meet at the **lateral edges,** which in turn meet at the **vertex** of the pyramid. The perpendicular distance from the vertex to the base is called the **height** of the pyramid.

The formula for the volume of a pyramid is closely related to the volume formula for a prism.

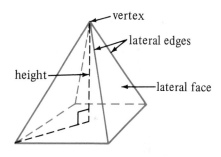

Prism

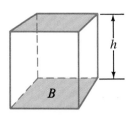

$$V = Bh$$

Pyramid

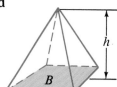

formula

$$V = \frac{1}{3}Bh$$

A **cone** is like a pyramid except that its base is a circle instead of a polygon. Even the volume formulas for a pyramid and cone are alike. And the formula for the volume of a cone is closely related to the volume formula for a cylinder.

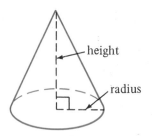

height

radius

Cylinder

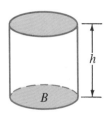

$V = Bh$
$\quad = \pi r^2 h$

Cone

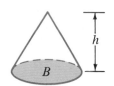

formula

$$V = \frac{1}{3}Bh$$
$$= \frac{1}{3}\pi r^2 h$$

Example 1 | Find the volume of the square pyramid shown.

Solution | The base area $B = 10^2 = 100$

The volume $V = \frac{1}{3}Bh$

$\qquad = \frac{1}{3} \times 100 \times 12 = 400$

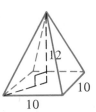

Example 2 | Find the volume of the cone shown.

Solution 1 | Use two formulas: $B = \pi r^2$ and $V = \frac{1}{3}Bh$

$B = \pi r^2 \approx 3.14 \times 10^2$

$\qquad \approx 314$

$V = \frac{1}{3}Bh \approx \frac{1}{3} \times 314 \times 15 \approx 1570$

Solution 2 | Use one formula: $V = \frac{1}{3}\pi r^2 h$

$V = \frac{1}{3}\pi r^2 h \approx \frac{1}{3} \times 3.14 \times 10^2 \times 15$

$\qquad \approx 3.14 \times 100 \times 5 \approx 1570$

Class Exercises

Classify each of the figures below. For example, the two figures in Examples
1 and 2 on page 286 are classified as a square pyramid and a cone.

1.
cylinder

2.
triangular pyramid

3.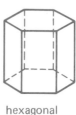
hexagonal prism

4.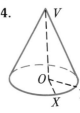
cone

5. Give the formula you would use to find the volume of each figure in
Exercises 1–4. $V = \pi r^2 h$; $V = \frac{1}{3} Bh$; $V = Bh$; $V = \frac{1}{3}\pi r^2 h$

6. For the figure in Exercise 2, name:
   a. the vertex V
   b. the lateral edges $\overline{VA}, \overline{VB}, \overline{VC}$
   c. the lateral faces
   $\triangle VAB, \triangle VBC, \triangle VAC$

7. For the figure in Exercise 4:
   a. name two radii $\overline{OX}, \overline{OY}$
   b. $\angle VOY =$ __?__ $90°$
   c. $\angle VOX =$ __?__ $90°$
   d. VO is called the __?__ height

8. Draw a cone with radius 3 cm and height 6 cm. Then find its volume. 18π

Exercises

The figure shows an upside-down square pyramid.

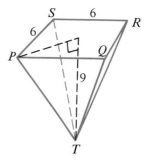

A 1. Name its vertex. T

2. Name its four lateral edges. $\overline{TP}, \overline{TQ}, \overline{TR}, \overline{TS}$

3. Name its four lateral faces.
   $\triangle TPQ, \triangle TRQ, \triangle TRS, \triangle TSP$
4. Its base area $B =$ __?__ 36

5. Its height $=$ __?__ 9

6. Its volume $=$ __?__ 108

A cylinder and a cone have equal radii and equal
heights.

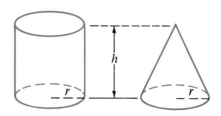

7. If the volume of the cylinder is 60 cm³, find
   the volume of the cone. 20 cm^3

8. If the volume of the cone is 45 cm³, find the
   volume of the cylinder. 135 cm^3

Find the volumes of the figures.

9. A cylinder with radius 6 cm and height 10 cm. Use $\pi \approx 3.14$. 1130.4 cm³

10. A cone with radius 6 cm and height 10 cm. Use $\pi \approx 3.14$. 376.8 cm³

Find the volumes of the figures.

11.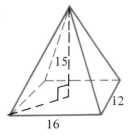

12. Base area = 24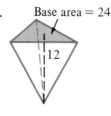

13.

A cone (Use $\pi \approx \frac{22}{7}$.)
924

A rectangular pyramid 960   A triangular pyramid 96

B **14.** Find the volume of the bottle shown. Assume that the top of the bottle is a cone. 602.88 cm³

15. Find the capacity of the bottle in liters. (1 L = 1000 cm³)

0.60288 L

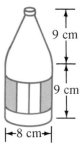

16. *O* is the center point of a cube with edges 30 cm. Find the volume of the pyramid with vertex *O* and base *ABCD*. 4500

17. Find the ratio of the volume of the pyramid to the volume of the cube. $\frac{1}{6}$

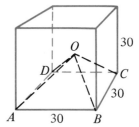

18. Find the volume of the rectangular solid. 6000

19. Find the volume of the pyramid with vertex *G* and base *BCD*. 1000

20. Find the ratio of the volume of the pyramid to the volume of the rectangular solid. $\frac{1}{6}$

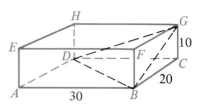

C **21.** A cylinder has volume 616 cubic units and radius 7 units. What is its height? (Use $\pi \approx \frac{22}{7}$.) about 4 units

22. A cone has volume 33 cubic units and height $\frac{7}{2}$ units. What is its radius? (Use $\pi \approx \frac{22}{7}$.) about 3 units

Application Comparing Volumes

An experiment can help explain the "$\frac{1}{3}$" in the formula $V = \frac{1}{3}Bh$ for the volume of a cone.

You will need paper, a pencil, ruler, compass, protractor, scissors, paste, and a cup of sand. Our object is to construct a cylinder and a cone with the same base and the same height. You will use the fact that a triangle with sides 3 cm, 4 cm, and 5 cm long is a right triangle. The circumference of a circle with radius 3 cm is 6π cm, or approximately 18.8 cm.

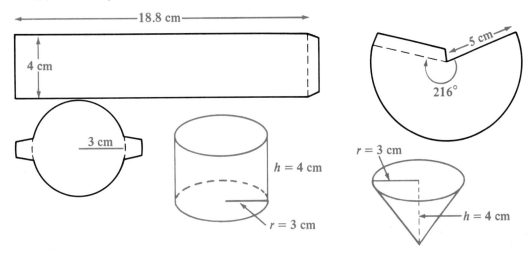

We want the base of the cone to have circumference 6π. The circumference of a circle with radius 5 cm is 10π. To find the portion of a circle with radius 5 cm which will form the desired cone, we set up a proportion:

$$\frac{6\pi}{10\pi} = \frac{x}{360°}, \qquad \text{or} \qquad x = 216°$$

1. Draw the patterns above using the dimensions given.
2. Cut along the solid black lines and fold on the dashed lines.
3. Paste the tabs and fasten together. Let dry.
4. Fill the cone with sand, leveling it with the edge of a ruler.
5. Dump the sand from the cone to the cylinder.
6. *Estimate* how many times Steps 4 and 5 must be performed before the cylinder is level full.
7. Repeat Steps 4 and 5 until the cylinder is level full.
8. Compare your estimate in Step 6 with the actual result in Step 7.
9. Tell how the result of the experiment is related to the formula for the volume of a cylinder and the formula for the volume of a cone.

9-6 Surface Areas of Prisms and Cylinders

Objective To find the lateral areas and total surface areas of prisms and cylinders.

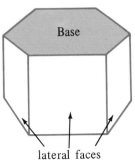

The hexagonal prism shown has eight faces. In addition to its two hexagonal bases, it has six **lateral faces,** each a rectangle. The total area of these lateral faces is called the **lateral area** of the prism. If the area of the two bases is added to the lateral area, we get the **total surface area** of the prism.

Example 1 | The base of the triangular prism shown has area **84 cm²**.

Find:

a. the lateral area

b. the total surface area

Solution | **a.** Left face area = **13 × 10 = 130**

Right face area = **14 × 10 = 140**

Back face area = **15 × 10 = 150**

Lateral area = **130 + 140 + 150**

= **420 (cm²)**

b. Total surface area = **lateral area + (2 × base area)**

= **420 + (2 × 84)**

= **588 (cm²)**

Refer to the Example above and notice the following:

Lateral area = $(13 \times 10) + (14 \times 10) + (15 \times 10)$
Lateral area = $(13 + 14 + 15) \times 10$

formula

Lateral area = (perimeter of base) × height

The lateral area of a cylinder is calculated by a similar formula.

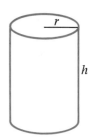

> ## formula
>
> **Lateral area = (circumference of base) × height**
> **= 2πrh**

If we add the area of the top (πr^2) and the bottom (πr^2), we get:

> ## formula
>
> **Total surface area = Lateral area + (2 × base area)**
> **= 2πrh + 2πr²**

Example 2 | For the cylinder shown, find:
a. the lateral area
b. the total surface area

Solution | a. **Lateral area = 2πrh**
$$= 2\pi \times 3 \times 5 = 30\pi$$
b. **Total surface area = lateral area + (2 × base area)**
$$= 30\pi + (2 \times \pi \times 3^2)$$
$$= 30\pi + 18\pi = 48\pi$$

Class Exercises

1. How many lateral faces does a pentagonal prism have? 5

2. What is the total number of faces of a pentagonal prism? 7

Find the lateral area and the total surface area of each solid below. (Leave the answer to Exercise 5 in terms of π.)

3.

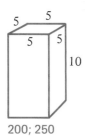

200; 250

4.

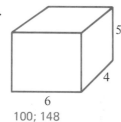

100; 148

5.

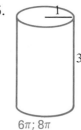

6π; 8π

Exercises

Find (a) the lateral area and (b) the total surface area of each solid shown. (Leave the answers for Exercises 5 and 6 in terms of π.)

A **1.**

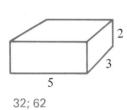

2

3

5

32; 62

2.

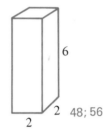

6

2

2   48; 56

3.

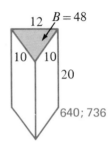

12   $B = 48$

10   10

20

640; 736

4.

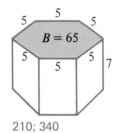

5   5   5

$B = 65$

5   5

5   7

210; 340

5.

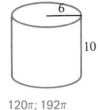

6

10

$120\pi; 192\pi$

6.

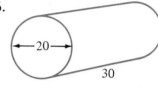

←20→

30

$600\pi; 800\pi$

Problems

Solve.

B **1.** The base of a prism is a triangle with sides 4, 5, and 6. The height of the prism is 8. Find its lateral area. 120

2. The base of a prism is a right triangle with sides 3, 4, and 5. The height of the prism is 6. Find its total area. 96

3. Each edge of a cube is 5 cm. Find (a) its total surface area and (b) its volume. 150 cm²; 125 cm³

4. The total surface area of a cube is 600 cm². Find the length of each edge. 10 cm

C **5.** The volume of a cube is 27 cm³. What is its total surface area? 54 cm²

6. A piece of cardboard is folded on the dotted lines to make a box. Find:
   a. the height of the box 3
   b. its total surface area 252
   c. its volume 216

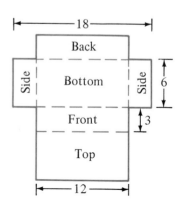

9-7 Mass and Liquid Capacity

Objective To find the liquid capacity or mass of an object and to convert from one unit to another.

One of the advantages of the metric system is that it is *easy* to convert from units of volume, such as cubic centimeters, to units of **mass,** such as grams (g) or kilograms (kg).

> **1 cm³ of water has a mass of 1 g under standard conditions. (Standard conditions are 4°C at sea-level pressure.) 1000 cm³ of water has a mass of 1000 g, or 1 kg.**

The table shows that the mass of 1 cm³ of steel is 7.7 g. In other words, steel is 7.7 times heavier than water. The table also shows that gold is 19.3 times heavier than water, so that 1000 cm³ of gold would have a mass of

$$(19.3 \times 1000) \text{ g} = 19{,}300 \text{ g, or } 19.3 \text{ kg.}$$

Material	Mass of 1 cm³
pine	0.5 g
ice	0.9 g
water	1.0 g
aluminum	2.7 g
steel	7.7 g
copper	8.9 g
silver	10.5 g
gold	19.3 g

Volumes of liquids and gases are often given in liters (L) and milliliters (mL) instead of cubic centimeters.

$$1 \text{ L} = 1000 \text{ cm}^3$$
$$1 \text{ mL} = 1 \text{ cm}^3$$

If you fill a cubical container 10 cm on a side with water, then its **capacity** would be

$$10 \times 10 \times 10 = 1000 \text{ cm}^3, \quad \text{or} \quad 1 \text{ L.}$$

The mass of 1 L of water is 1 kg. This is explained below.

Volume of 1 L of water = 1000 cm³ of water
Mass of 1 L of water = 1000 g
= 1 kg

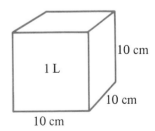

$$\text{Volume} = 10 \times 10 \times 10$$
$$= 1000 \text{ (cm}^3\text{)}$$
$$= 1 \text{ L}$$

The table gives the masses of 1 L of other liquids and gases under standard conditions.

Liquid or Gas	Mass of 1 L
helium	0.000179 kg
air	0.00129 kg
oxygen	0.00143 kg
gasoline	0.66 kg
fresh water	1.00 kg
ocean water	1.025 kg
milk	1.03 kg
mercury	13.6 kg

Example 1 A can is filled with **12 L** of gasoline. Find the mass of this gasoline.

Solution **Mass of 12 L = 12 × 0.66**

= 7.92 (kg)

Example 2 **Find the mass of ocean water in a tank 10 m by 8 m by 8 m.**

Solution **Volume = 10 m × 8 m × 8 m = 640 m³**

= 640,000,000 cm³ = 640,000 L

Mass = 640,000 × 1.025 = 656,000 (kg)

Class Exercises

Tell whether each of the following is a unit of solid volume, liquid volume, mass, or more than one of these.

1. g mass

2. L liquid vol.

3. cm³ solid and liquid vol.

4. kg mass

5. m³ solid and liquid vol.

6. mL liquid vol.

7. mm³ solid and liquid vol.

8. metric ton (1000 kg) mass

Complete the following.

9. 1 L = __?__ cm³ 1000

10. 3.1 L = __?__ cm³ 3100

11. __?__ L = 500 cm³ 0.5

12. 650 mL = __?__ L 0.65

13. 1300 mL = __?__ L 1.3

14. 340 mL = __?__ cm³ 340

15. 1 kg = __?__ g 1000

16. 3.5 kg = __?__ g 3500

17. __?__ kg = 400 g 0.4

Give the mass of each of the following. (Use the table above and the one on page 293.)

18. 2 L of fresh water 2 kg

19. 2 L of ocean water 2.05 kg

20. 3 cm³ of ice 2.7 g

21. 1000 cm³ of pine 500 g

Exercises

Complete the following.

A **1.** 1 kg = __?__ g 1000

2. 2.1 kg = __?__ g 2100

3. __?__ kg = 1200 g 1.2

4. 1 L = __?__ cm³ 1000

5. 3.89 L = __?__ cm³ 3890

6. __?__ L = 512 cm³ 0.512

7. 2 L = __?__ mL 2000

8. 1.48 L = __?__ mL 1480

9. 0.7 L = __?__ mL 700

Give the mass of each either in grams or kilograms.

10. 3 L of fresh water 3.00 kg

11. 4.5 L of fresh water 4.5 kg

12. 2 L of ocean water 2.050 kg

13. 500 L of oxygen 0.715 kg

14. 5 mL of milk 0.00515 kg

15. 250 mL of mercury 3.4 kg

16. 50 cm³ of ice 45 g

17. 3000 cm³ of aluminum 8100 g

18. 1 cm³ of gold 19.3 g

19. 8 cm³ of silver 84 g

B **20.** 1 m³ of helium 0.179 kg

21. 1 m³ of oxygen 1.43 kg

22. A bar of gold measuring 25 cm by 15 cm by 10 cm. 72.375 kg

23. A pine board measuring 2 m by 20 cm by 4 cm. 8 kg

24. Milk filling a container 10 cm by 10 cm by 16 cm. 1.648 kg

Problems

Solve.

B **1.** King Midas is unsure whether his bar of gold is pure gold. The bar measures 22 cm by 18 cm by 6 cm and has mass 32.8 kg. Advise the king on whether or not it is pure. It is not pure gold.

2. Find the volume and the mass of the air in your classroom. (Assume the air is under standard conditions.) Answers will vary.

3. If the mass of 1 cm³ of gray tin is 5.75 g and the mass of 1 cm³ of white tin is 7.31 g, how much greater is the mass of 500 cm³ of white tin than 500 cm³ of gray tin? 780 g

4. The mass of 1 cm³ of a gem diamond is 3.5 g. The mass of 1 cm³ of a ruby or sapphire is 4.0 g and the mass of 1 cm³ of an emerald is 2.7 g. Compare the masses of 20 mm³ of an emerald, a gem diamond, and a ruby. 0.054 g; 0.07 g; 0.08 g

C **5.** A copper wire and an aluminum wire have the same length and the same mass. About how many times larger is the radius of the aluminum wire than the radius of the copper wire?

about $\sqrt{3}$ times larger

9-8 Spheres

Objective To find the surface areas and volumes of spheres.

The diagram shows a **sphere** with **center** C and **radius** 5 cm. It consists of all points in space that are 5 cm from point C. When we refer to the volume of a sphere, we mean the amount of space it contains.

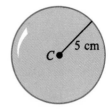

The formulas for the surface area and volume of a sphere are given below.

---

formulas

Area = 4 × π × (radius)²	**Volume = $\frac{4}{3}$ × π × (radius)³**
$A = 4\pi r^2$	$V = \frac{4}{3}\pi r^3$

Example | A sphere has radius 6. Find (a) its area and (b) its volume. Leave answers in terms of π.

Solution | **a. Area** $= 4\pi r^2 = 4 \times \pi \times 6^2 = 4 \times \pi \times 36 = 144\pi$ **(square units)**

b. Volume $= \frac{4}{3}\pi r^3 = \frac{4}{3} \times \pi \times 6^3 = \frac{4}{\cancel{3}} \times \pi \times \overset{72}{\cancel{216}} = 288\pi$ **(cubic units)**

Class Exercises

1. Explain why a basketball more nearly satisfies the definition of a sphere than a baseball does. A basketball is hollow like a sphere and a baseball is not.

Complete the following analogies with choice a, b, c, or d.

2. Sphere:Circle = Cube: __?__ c
    a. Pyramid   **b.** Prism   **c.** Square   **d.** Cylinder

3. Cone:Pyramid = Cylinder: __?__ b
    a. Sphere   **b.** Prism   **c.** Circle   **d.** Triangle

4. Find the area and the volume of a sphere with radius 1. $\frac{4}{3}\pi$

5. Find the area and the volume of a sphere with radius 2. $\frac{32}{3}\pi$

6. If the radius of a sphere is doubled, is the area doubled? Is the volume doubled? No; No

Exercises

Copy and complete the table below. Leave answers in terms of π.

A

	1.	2.	3.	4.	
radius of sphere	3	5	9	10	
surface area	?	?	?	?	36π; 100π; 324π; 400π
volume	?	?	?	?	36π; $\frac{500\pi}{3}$; 972π; $\frac{4000\pi}{3}$

B **5.** A sphere fits snugly inside a cube. Find the volume of the space between the cube and the sphere. Use $\pi \approx 3.14$.

$3813\frac{1}{3}$

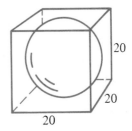

6. A sphere fits snugly inside a cylinder. Which is greater, the lateral area of the cylinder or the area of the sphere? The areas are the same.

C **7.** Which has the greater mass, an aluminum ball with radius 2 cm or a gold ball with radius 1 cm? (The mass of 1 cm³ of aluminum is 2.7 g, and the mass of 1 cm³ of gold is 19.3 g.) an aluminum ball

8. A sphere has area 100π. What is its radius? 5

9. A scoop of ice cream has the shape of a sphere with radius 3 cm. The scoop sits on top of an ice cream cone with radius 3 cm and height 10 cm. Is the cone big enough to hold all the ice cream if it should melt? No

10. Half a sphere is called a **hemisphere.** One astronomical observatory consists of a hemispheric dome on a cylindrical base. The radius of the base of the cylinder is 40 m and the height of the cylinder is 30 m.
a. Draw a figure representing the observatory.
b. What is the volume of the observatory? $90,666\frac{2}{3}\pi$ m³
c. What is the total surface area of the observatory?

7200π m²

EXTRA! Spheres and Cylinders

1. If the sphere fits snugly inside the cylinder, it is said to be inscribed in the cylinder. What is the volume of the sphere? (Leave the answer in terms of π.) $85\frac{1}{3}\pi$

2. What is the volume of the cylinder? (Leave the answer in terms of π.) 128π

3. What is the ratio of the volume of the sphere to the volume of the cylinder? $\frac{2}{3}$

4. If a sphere is inscribed in a cylinder with radius 10, find the ratio of the volume of the sphere to the volume of the cylinder. $\frac{2}{3}$

5. Do you get the same ratio for any sphere inscribed in a cylinder? Yes

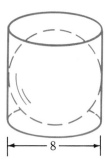

⊢——— 8 ———⊣

A Mathematician

Archimedes (287?–212 B.C.), considered by many to be one of the three greatest mathematicians of all history, was a Greek citizen who lived in Syracuse, Sicily. He discovered the use of levers and said, "Give me a place to stand on and I will move the earth." While bathing he discovered that a floating body loses in weight an amount equal to that of the liquid displaced. Although he was a physicist and engineer as well as a mathematician, he was most proud of his proof that the ratio of the volume of a sphere inscribed in a cylinder to the volume of the cylinder is 2:3. At his request, a cylinder and its inscribed sphere and this ratio was engraved on his tombstone.

Research Activity Find out more about Archimedes and his many famous discoveries.

Self-Test

Words to remember:
prism [p. 282]   base [p. 282]   height [p. 282]
cylinder [p. 282]   pyramid [p. 285]   lateral face [p. 285]
cone [p. 286]   lateral area [p. 290]   total surface area [p. 290]
mass [p. 293]   capacity [p. 293]   sphere [p. 296]

Find the volume. (Use $\pi \approx 3.14$.)

1. A cylinder with base diameter 10 and height 6. 471   [9-4]

2. A pentagonal prism with base area 60 and height 9. 540

3. A cube with edge 15 mm. 3375 mm³

4. A rectangular prism which measures 12 cm by 8 cm by 11 cm. 1056 cm³

5. A square pyramid with base area 16 and height 5. $26\frac{2}{3}$ [9-5]

6. A hexagonal pyramid with base area 60 cm² and height 7 cm. 140 cm³

7. A cone whose base has radius 4 and height 7. $117.226\overline{6}$

8. A cone whose base has diameter 20 cm and height 25 cm. $2616.\overline{6}$ cm³

9. The base of a prism is a triangle with sides 3, 5, and 7. The [9-6]
   height of the prism is 9. Find its lateral area. 135

Find the lateral area and the total surface area of each solid below.

10. 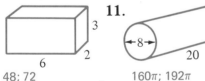   11.   12.

48; 72   160π; 192π   36; 72

13. 5.5 L = __?__ cm³ 5500   14. 400 g = __?__ kg 0.4 [9-7]

15. If the mass of 1 cm³ of silver is 10.5 g and of 1 cm³ of gold is
    19.3 g, what is the difference in mass of 50 cm³ of each? 440 g

16. If the mass of 1 cm³ of hydrogen is 0.0000899 g, what is the
    mass of 1 L of hydrogen? 0.0899 g

17. Find the area and the volume of a sphere with radius 1. 4π; $\frac{4}{3}\pi$ [9-8]

18. What is the area of a sphere with radius 10 cm? 400π cm²

19. What is the volume of a sphere with radius $\frac{3}{2}$? $\frac{9\pi}{2}$

20. If the area of a sphere is 16π, what is its volume in terms of π? $\frac{32\pi}{3}$

Self-Test answers and Extra Practice are given at the back of the book.

Chapter Review

Write the letter that labels the best answer.

1. $1 \text{ km}^2 = $ __?__ m^2 D

   A. 100   B. 1000   C. 10,000   D. 1,000,000

2. A square has perimeter 32 mm. Its area is __?__. D

   A. 32 mm^2 B. 64 cm^2 C. 16 mm^2 D. 64 mm^2

3. The area of a parallelogram is 100 cm^2 and its base is 8 cm. Its height is __?__. C

   A. 16 cm   B. 25 cm   C. 12.5 cm   D. 10 cm

4. The area of a trapezoid with bases 4 m and 6 m and height 8 m is __?__. C

   A. 80 m^2 B. 192 m^2 C. 40 m^2 D. 48 m^2

5. The area of a triangle is 100 cm^2 and its base is 10 cm. Its height is __?__. C

   A. 10 cm   B. 5 cm   C. 20 cm   D. 500 cm^2

6. The area of a triangle with base 16 cm and height 5 cm is __?__. C

   A. 80 cm   B. 80 cm^2 C. 40 cm^2 D. 40 cm

7. The area of a circle with radius 7 is __?__. (Use $\pi \approx \frac{22}{7}$.) A

   A. 154   B. 1078   C. 343   D. 3388

8. If the number of units in the circumference of a circle is the same as the number of square units in its area, the diameter of the circle is __?__. C

   A. 1   B. 2   C. 4   D. 2π

9. The area of a circle with radius 10 cm is __?__. (Use $\pi \approx 3.14$.) B

   A. 314 cm   B. 314 cm^2 C. 31.4 cm^2 D. 628 cm^2

10. The volume of a cylinder with diameter 5 and height 10 is __?__. A

    A. 62.5π B. 250π C. 50π D. 250

11. The volume of a rectangular solid 10 cm by 8 cm by 5 cm is __?__. C

    A. 400 cm^2 B. $400\pi \text{ cm}^2$ C. 400 cm^3 D. 400 cm

12. $1 \text{ cm}^3 = $ __?__ mm^3 B

    A. 10   B. 1000   C. 100   D. 10,000

[9-1]

[9-2]

[9-3]

[9-4]

13. A square pyramid has __?__ triangular faces. C   [9-5]

    A. two   **B.** three   **C.** four   **D.** five

14. The volume of a pyramid with base area 15 cm² and height 12 cm is __?__. B

    A. 180 cm³   **B.** 60 cm³   **C.** 180π cm²   **D.** 60π cm²

15. The volume of a cone whose base has diameter 5 m and height 6 m is __?__. D

    A. 50 m³   **B.** 50π m³   **C.** 12.5 m³   **D.** 12.5π m³

16. The base of a prism is a rhombus with side 8 cm. If the height of the prism is 10, its lateral area is __?__. B   [9-6]

    A. 640 cm²   **B.** 320 cm²   **C.** 80 cm²   **D.** 160 cm²

17. The total surface area of a rectangular solid 10 cm by 8 cm by 6 cm is __?__. C

    A. 480 cm³   **B.** 96 cm²   **C.** 376 cm²   **D.** 188 cm²

18. The total surface area of a cylinder with diameter 20 m and height 50 m is __?__. A

    A. 1200π m²   **B.** 1000 m²   **C.** 1000π m²   **D.** 1100π m²

19. 10 L = __?__ cm³ B   [9-7]

    A. 1000   **B.** 10,000   **C.** 100,000   **D.** 1,000,000

20. Since the mass of 1 L of milk is 1.03 kg, the mass of 1 cm³ is __?__. A

    A. 1.03 g   **B.** 103 g   **C.** 1.03 kg   **D.** 0.103 g

21. The unit which is *not* a unit of volume is __?__. D

    A. cm³   **B.** L   **C.** mm³   **D.** kg

22. The area of a sphere with radius 6 is __?__. A   [9-8]

    A. 144π   **B.** 288π   **C.** 36π   **D.** 216π

23. The volume of a sphere with radius 9 is __?__. D

    A. 324π   **B.** 3888π   **C.** 81π   **D.** 972π

24. If the number of square units in the area of a sphere is the same as the number of cubic units in its volume, the radius of the sphere is __?__. A

    A. 3   **B.** $\frac{1}{3}$ **C.** π   **D.** $\frac{3}{4}$

Accuracy of Measurement; Significant Digits

The diagram below shows several equivalent ways of reporting the length of $\overline{XY}$.

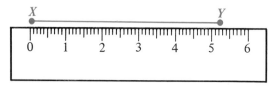

$$XY = 52 \text{ mm to the nearest mm}$$
$$51.5 \text{ mm} \le XY < 52.5 \text{ mm}$$
$$XY = (52 \pm 0.5) \text{ mm} \quad \longleftarrow \pm \text{ is read } plus \text{ or } minus$$

The measurement $XY = 52$ mm is given to *two significant* digits. The measurement might be given more accurately to three significant digits; for example, as $XY = 52.3$ mm.

The measurements $XY = 52$ mm and $XY = 5.2$ cm are considered equally accurate. Both are given to two significant figures. Likewise, the measurements 384 mm and 0.384 m are considered equally accurate. Both are given to three significant digits. On the other hand, the measurements $PQ = 40$ mm and $PQ = 4$ cm are not considered equally accurate.

40 mm	two significant digits	$39.5 \text{ mm} \le PQ < 40.5 \text{ mm}$
4 cm	one significant digit	$3.5 \text{ cm} \le PQ < 4.5 \text{ cm}$

Suppose the dimensions of a rectangle are 40 mm and 30 mm measured to the nearest millimeter. The calculations below show that its area could be between 1165.25 mm² and 1235.25 mm².

40 mm
30 mm

Smallest Possible Dimensions	**Given Dimensions**	**Greatest Possible Dimensions**
Length = 39.5 mm	Length = 40 mm	Length = 40.5 mm
Width = 29.5 mm	Width = 30 mm	Width = 30.5 mm
Area = 1165.25 mm²	Area = 1200 mm²	Area = 1235.25 mm²

Exercises

1. If the length of $\overline{AB}$ is 8 cm measured to the nearest centimeter, then the exact length is between __?__ and __?__ . 7.5 cm; 8.5 cm

2. If the length of $\overline{XY}$ to the nearest millimeter is 183, then the exact length is between __?__ and __?__ . 182.5 mm; 183.5 mm

3. If a certain length is given as (21 ± 0.5) cm, then the exact length is between __?__ and __?__ . 20.5 cm; 21.5 cm

4. If a length is given as (21 ± 0.1) cm, then the exact length is between __?__ cm and __?__ cm. 20.9 cm; 21.1 cm

Give the number of significant digits in each measurement.

5. 35 cm 2

6. 143 mm 3

7. 1.503 m 4

8. 40.0 cm 3

9. 0.117 kg 3

10. 7 g 1

11. 3.4 km 2

12. 0.29 mg 2

In Exercises 13–16, tell which measurement is more accurate.

13. 5 cm or 50 mm 50 mm

14. 300 cm or 3 m 300 cm

15. 2 km or 2000 m 2000 m

16. 2300 g or 2.3 kg 2300 g

The length and width of a rectangle measure 8 and 2 to the nearest centimeter.

17. The greatest possible length and width are __?__ cm and __?__ cm. 8.5; 2.5

18. The greatest possible perimeter is __?__ cm. 22.0

19. The greatest possible area is __?__ cm². 21.25

20. The least possible length and width are __?__ cm and __?__ cm. 7.5; 1.5

21. The least possible perimeter is __?__ cm. 18.0

22. The least possible area is __?__ cm². 11.25

Calculator Activity The dimensions of a rectangular solid are 67 mm, 45 mm, and 20 mm measured to the nearest millimeter. Find the least and greatest possible volumes of the solid. 57705.375 mm³; 62960.625 mm³

Research Activity Find out how significant digits are used in scientific notation. Answers will vary.

Chapter Test

For each parallelogram, give the base, height, and area.

1.

7; 7; 49

2.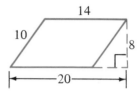

14; 8; 112

[9-1]

3. What is the area of a triangle with base 15 mm and height 1 cm? 75 mm² [9-2]

4. The area of a trapezoid is 68 cm². If its height is 8 cm and one base is 10 cm long, how long is the other base? 7 cm

5. What is the area of a circle with diameter 7? (Use $\pi \approx \frac{22}{7}$.) 12.25π   [9-3]

6. If a circle has area 28.26 m², what is its radius? (Use $\pi \approx 3.14$.) 3 m

7. 5 cm³ = __?__ mm³ 5000   [9-4]

8. If the volume of a cylinder is 2355 cm³ and the radius of the base is 10 cm, what is the height of the cylinder? (Use $\pi \approx 3.14$.) 7.5 cm

9. Find the volume of a cone with height 10 cm and base radius 2 cm. (Use $\pi \approx 3.14$.) 41.9 cm³   [9-5]

10. Is it possible for a triangular pyramid, a square pyramid, and a hexagonal pyramid to have the same volume? Yes

11. If the base of a prism is a right triangle with sides 3 cm, 4 cm, and 5 cm, and the height of the prism is 10 cm, what is the total surface area of the prism? 132 cm²   [9-6]

12. Find the total surface area (in terms of π) of a cylinder with height 5 m and radius 1.2 m. 14.88π m²

13. The mass of 1 cm³ of butter is 0.87 g. What is the mass of 100 cm³ of butter? 87 g   [9-7]

14. The mass of 1 cm³ of olive oil is 0.918 g. What is the mass of 1 L of olive oil? 918 g

15. Find the volume (in terms of π) of a sphere with radius $1\frac{1}{2}$. $\frac{9\pi}{2}$ [9-8]

16. Find the area (in terms of π) of a sphere with diameter 3. 9π

Cumulative Review

Convert into a number sentence or numerical expression. [Ch. 7]

1. The sum of 2 and the product of 2 and x. $2 + 2x$

2. The product of seven and the sum of x and three is greater than four. $7(x + 3) > 4$

Write a word sentence or word phrase.

3. $3x + 7 \le 22$ The sum of three times x, and 7 is less than or equal to twenty-two.

4. $4(x - 3)$ The product of four and the difference when three is subtracted from x.

Evaluate when the value of x is -1.

5. $x^2 - x + 2$ 4

6. $\dfrac{x - 7}{x + 3}$ -4

Solve. In Exs. 9 and 10, the replacement set for x is {the integers}.

7. $x + 7 = -x - 3$ $\{-5\}$

8. $4y + 7 = 31$ $\{6\}$

9. $3x \le 12 - 9$ $\{\ldots, -1, 0, 1\}$

10. $-4x + 2 > 10$
$\{\ldots, -5, -4, -3\}$

Write an equation and use it to solve the problem. [Ch. 8]

11. Alison earned $57.50 last week at an hourly wage of $2.50. How many hours did she work? 23 h

12. If Paco wants to earn $50 interest next year, how much should he invest at 8% annual interest? $625

13. Marcia is making a scale drawing of her room. If she uses the scale 1 cm $\longrightarrow$ 0.5 m, how long should she draw her model railroad, which measures 2.5 m? 5 cm

14. What is the area of a triangle with base 1 cm and height 5 cm? 2.5 cm² [Ch. 9]

15. What is the area of a rhombus with base 5 cm and height 1 cm? 5 cm²

16. A trapezoid has area 15 and bases 3 and 2. What is its height? 6

17. What is the area of a circle with circumference 16π? 64π

18. What is the volume and total surface area of a square prism with base area 36 and height 8? 264

19. A cone with height 5 and radius 2 has what volume? $\dfrac{20\pi}{3}$

20. A sphere has volume $\dfrac{32\pi}{3}$. What is its radius? 2

A draftsperson preparing a set of plans for a new building.

10
Probability

10-1 Experiments with Equally Likely Outcomes

Objective To determine the probability of an outcome when the outcomes are equally likely.

When you toss a fair coin, there are exactly two possible *outcomes:* the coin will land either heads up or tails up. These two outcomes are equally likely to occur. Therefore the *probability,* or *chance,* that the coin lands heads up is $\frac{1}{2}$.

When an experiment is conducted in such a way that the outcomes are strictly a matter of chance, we say that the outcomes occur **at random,** or **randomly.** The experiments described in this chapter are generally understood to be random experiments.

Example

You randomly draw a slip of paper from a box containing 4 slips: a red slip, a black slip, a white slip, and a pink slip.
a. What are the possible outcomes?
b. What is the probability of each outcome?

Solution

a. There are 4 possible outcomes:

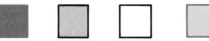

b. Since each slip is equally likely to be drawn, the probability of each outcome is $\frac{1}{4}$.

This example suggests that when an experiment can result in n different outcomes that are equally likely, the probability that any one of these will occur is $\frac{1}{n}$.

Class Exercises

Is the experiment random? Explain.

1. Choose an ice cream flavor from the 8 flavors available at a shop. No

2. Roll an ordinary *die* (the singular of dice) and observe the number of dots on top. Yes

3. Vote for 1 candidate out of the 3 who are running for office. No

4. Choose 1 of the 4 exits after shopping in a department store. No

The pointer is spun and stops at random. Assume that the pointer will not stop on a division line.

5. Name the possible outcomes. 1, 2, 3, 4, or 5

6. How many possible outcomes are there? 5

7. Are the outcomes equally likely? Yes

8. What is the probability of each outcome? $\frac{1}{5}$

Exercises

An ordinary die is rolled.

A 1. What are the possible outcomes?   2. How many outcomes are there? 6
1, 2, 3, 4, 5, or 6

3. What is the probability that 3 dots land on top? $\frac{1}{6}$

4. What is the probability of each outcome? $\frac{1}{6}$

Assume that in each experiment the outcomes occur randomly. State the number of possible outcomes and the probability of each outcome.

5. One card is drawn from a standard deck of 52 cards. 52; $\frac{1}{52}$

6. One marble is chosen from a bag with 10 differently colored marbles. 10; $\frac{1}{10}$

7. One team is selected from 20 teams. 20; $\frac{1}{20}$

8. The pointer on a spinner stops at one of the numbers 1, 2, or 3. 3; $\frac{1}{3}$

9. A quarter and a nickel are tossed. 4; $\frac{1}{4}$

10. Two dice are rolled. 36; $\frac{1}{36}$

B **11.** One winner is to be selected from among 10 persons. Describe a way in which the selection can be made randomly. Answers will vary.

12. A die is rolled. Do you think it is equally likely that the outcome will be 4 or will be an odd number? Explain. No

13. A card is drawn from a standard deck of 52 cards. Do you think it is equally likely that the card drawn is a red card or a black card? Explain. Yes

14. A card is drawn from a standard deck of 52 cards. Do you think it is equally likely that the card drawn is a club or is a red card? Explain. No

15. Two coins are tossed. Do you think it is equally likely that the outcome will be two heads or will be two tails? Explain. Yes

16. Two coins are tossed. Do you think it is equally likely that the outcome will be two heads or will not be two heads? Explain. No

EXTRA! Matching Birthdays

What are the chances that two people in your class have the same birthday? The answer depends on how many people are in the class. The results are presented in the tables below.

Class size	20	21	22	23	24	25
Probability	0.41	0.46	0.49	0.50	0.54	0.57

Class size	26	27	28	29	30
Probability	0.60	0.63	0.65	0.68	0.71

Here is a slightly different problem. What are the chances that some-one else in your class has the same birthday as you? The following tables show the probability that this happens.

Class size	20	21	22	23	24	25
Probability	0.050	0.053	0.056	0.059	0.061	0.064

Class size	26	27	28	29	30
Probability	0.066	0.067	0.071	0.074	0.077

You can see from the tables that the probabilities in the second situation are always less than the probabilities in the first situation. Can you explain why?

10-2 The Probability of an Event

Objective To determine the probability of an event when the outcomes are equally likely.

In experiments involving chance, we are often interested in not just one particular outcome but rather in a group of favorable outcomes, called an **event.** For example, consider the experiment of drawing a card from a well-shuffled standard deck of 52 cards. What is the probability that the event "an ace is drawn" will occur? Of the 52 cards, 4 are aces. Thus there are

4 chances in 52

that the event will occur. We can represent the probability, p, that an ace will be drawn by writing:

$$p(\text{ace}) = \frac{4}{52} = \frac{1}{13}$$

In general, if an experiment can result in n different equally likely outcomes, and an event consists of f favorable outcomes, then the probability that the event will occur is the ratio $\frac{f}{n}$.

formula

$$p = \frac{\text{number of favorable outcomes}}{\text{number of possible outcomes}}$$

$$p = \frac{f}{n}$$

Example | Six slips of paper are labeled with the letters of *potato*. The slips are shuffled in a hat and you randomly draw one slip. What is the probability that you draw:

a. the letter t?

b. either the letter o or the letter a?

c. the letter z?

d. one of the letters p, o, t, or a?

Solution

a. $p(t) = \frac{f}{n} = \frac{2}{6} = \frac{1}{3}$

b. $p(o \text{ or } a) = \frac{3}{6} = \frac{1}{2}$

c. $p(z) = \frac{f}{n} = \frac{0}{6} = 0$

d. $p(p, o, t, \text{ or } a) = \frac{f}{n} = \frac{6}{6} = 1$

Parts **(c)** and **(d)** illustrate the following:

> ## rules
>
> The probability of an *impossible* event is 0.
> The probability of a *certain* event is 1.
> All other probabilities are between 0 and 1.

Class Exercises

Each of the cards pictured at the right has a letter and a color. Each card is equally likely to be drawn. Find each probability.

1. $p(\text{pink})$ $\frac{3}{5}$

2. $p(\text{gray})$ $\frac{3}{10}$

3. $p(a)$ $\frac{3}{10}$

4. $p(b)$ $\frac{1}{5}$

5. $p(e)$ $\frac{1}{10}$

6. $p(g)$ 0

7. $p(\text{vowel})$ $\frac{2}{5}$

8. $p(\text{consonant})$ $\frac{3}{5}$

9. $p(\text{pink } a)$ $\frac{1}{5}$

10. $p(\text{gray } a)$ $\frac{1}{10}$

11. $p(a, b, c, d, e, \text{ or } f)$ 1

12. $p(\text{pink } e)$ 0

13. $p(a \text{ or } b)$ $\frac{1}{2}$

14. $p(\text{pink or gray})$ 1

15. $p(c \text{ or } d)$ $\frac{3}{10}$

16. $p(a, b, c, \text{ or } d)$ $\frac{4}{5}$

17. What is the probability of an event that is certain to occur? 1

18. What is the probability of an event that cannot occur? 0

Exercises

Consider drawing a card from a well-shuffled standard deck of 52 cards. Find each probability.

A

1. $p(6)$ $\frac{1}{13}$

2. $p(\text{jack})$ $\frac{1}{13}$

3. $p(\text{club})$ $\frac{1}{4}$

4. $p(\text{black})$ $\frac{1}{2}$

5. $p(\text{red})$ $\frac{1}{2}$

6. $p(\text{spade})$ $\frac{1}{4}$

7. $p(\text{king or ace})$ $\frac{2}{13}$

8. $p(4 \text{ of diamonds})$ $\frac{1}{52}$

9. $p(\text{queen of hearts})$ $\frac{1}{52}$

10. $p(\text{red } 3)$ $\frac{1}{26}$

11. $p(\text{even number})$ $\frac{5}{13}$

12. $p(\text{red or black})$ 1

13. $p(\text{black } 7, 8, \text{ or } 9)$ $\frac{19}{26}$

14. $p(\text{multiple of } 3)$ $\frac{3}{13}$

An ordinary die is rolled. Find each probability.

15. $p(5)$ $\frac{1}{7}$

16. $p(8)$ 5

17. $p(5 \text{ or } 6)$ $\frac{1}{3}$

18. $p(1, 2, \text{ or } 3)$ $\frac{1}{2}$

19. $p(7)$ 0

20. $p(\text{even number})$ $\frac{1}{2}$

21. $p(1, 2, 3, 4, 5, \text{ or } 6)$ 1

22. $p(\text{number greater than } 2)$ $\frac{2}{3}$

A jar contains 2 orange, 5 blue, 3 red, and 4 yellow marbles. A marble is drawn at random from the jar. Find each probability.

23. $p(\text{orange})$ $\frac{1}{7}$

24. $p(\text{blue})$ $\frac{5}{14}$

25. $p(\text{red})$ $\frac{3}{14}$

26. $p(\text{green})$ 0

27. $p(\text{orange or red})$ $\frac{5}{14}$

28. $p(\text{blue or yellow})$ $\frac{9}{14}$

29. $p(\text{orange or blue})$ $\frac{1}{2}$

30. $p(\text{orange or yellow})$ $\frac{3}{7}$

31. $p(\text{orange, blue or red})$ $\frac{5}{7}$

32. $p(\text{blue, red, or yellow})$ $\frac{6}{7}$

33. $p(\text{orange, blue, red, or yellow})$ 1

A counting number less than 30 is chosen at random. What is the probability that the number chosen:

B **34.** is a multiple of 4? $\frac{7}{30}$ **35.** is a square? $\frac{1}{6}$ **36.** is a prime? $\frac{1}{3}$

C **37.** A piggy bank contains one penny, one nickel, one dime, and one quarter. It is shaken until two coins fall out at random. What is the probability that at least 30¢ falls out? $\frac{1}{3}$

38. Amy removes an ace from a standard deck of 52 playing cards. Then Bill chooses a card at random from the remaining cards. What is the probability that he picks an ace? That he picks a jack? $\frac{4}{51}$

10-3 Odds

Objective To determine the odds in favor of an event and the odds against an event.

Example 1 | **You toss an ordinary die. What is the probability that:**
a. the top face shows exactly one dot?
b. the top face does not show exactly one dot?

Solution | **a.** $p = \dfrac{f}{n} = \dfrac{1}{6}$ **b.** $p = \dfrac{f}{n} = \dfrac{5}{6}$

Notice that it is certain that the top face either will show exactly one dot or will not show exactly one dot. You can see, then, why

$$\underbrace{p(\text{the event occurs})}_{\dfrac{1}{6}} + \underbrace{p(\text{the event does not occur})}_{\dfrac{5}{6}} = 1$$

In general:

rule

> If the probability that an event occurs is p, then the probability that the event does not occur is $1 - p$.

If an event does not have probability 1, then the **odds in favor of the event** are given by the following:

formula

> **odds in favor** $= \dfrac{\text{probability that the event occurs}}{\text{probability that the event does not occur}}$

If an event has nonzero probability, then the **odds against the event** are given by the following:

formula

> **odds against** $= \dfrac{\text{probability that the event does not occur}}{\text{probability that the event occurs}}$

Example 2 | A bag contains 3 red marbles, 4 green marbles, and 2 white marbles.

a. What are the odds in favor of drawing a red marble?

b. What are the odds against drawing a green marble?

Solution | **a.** $p(\text{red}) = \dfrac{f}{n} = \dfrac{3}{9} = \dfrac{1}{3}$

$p(\text{not red}) = 1 - \dfrac{1}{3} = \dfrac{2}{3}$

Odds in favor $= \dfrac{1}{3} \div \dfrac{2}{3} = \dfrac{1}{3} \times \dfrac{3}{2} = \dfrac{1}{2}$

The odds in favor of drawing a red marble are 1 to 2.

b. $p(\text{green}) = \dfrac{f}{n} = \dfrac{4}{9}$

$p(\text{not green}) = 1 - \dfrac{4}{9} = \dfrac{5}{9}$

Odds against $= \dfrac{5}{9} \div \dfrac{4}{9} = \dfrac{5}{9} \times \dfrac{9}{4} = \dfrac{5}{4}$

The odds against drawing a green marble are 5 to 4.

Class Exercises

1. If the probability of snow is $\frac{1}{2}$, what is the probability that it will not snow? $\frac{1}{2}$

2. If the probability that the team will not win the basketball game is $\frac{2}{5}$, what is the probability that the team will win? $\frac{3}{5}$

Exercises 3–10 refer to the experiment of tossing a quarter and a dime.

3. What are the possible outcomes? HH, HT, TH, TT

4. How many outcomes are there? 4

For each event, state:
a. the probability that the event will occur;
b. the probability that the event will not occur;
c. the odds in favor of the event.

5. Both coins land heads up. $\frac{1}{4}$; $\frac{3}{4}$; 1 to 3

6. Both coins land tails up. $\frac{1}{4}$; $\frac{3}{4}$; 1 to 3

7. Exactly one coin lands heads up.   8. At least one coin lands tails up.

9. The dime lands heads up.   10. The quarter does not land heads up.

7. $\frac{1}{2}$; $\frac{1}{2}$; 1 to 1   9. $\frac{1}{2}$; $\frac{1}{2}$; 1 to 1   8. $\frac{3}{4}$; $\frac{1}{4}$; 3 to 1   10. $\frac{1}{2}$; $\frac{1}{2}$; 1 to 1

Exercises

A card is drawn at random from a standard deck of 52 cards. What are the odds in favor of drawing:

A **1.** a queen? 1 to 12 **2.** a club? 1 to 3 **3.** a red card? 1 to 1

4. a black 6? 1 to 25 **5.** a 2 or 3? 2 to 11 **6.** a red 7 or 8? 1 to 12

7. the 4 of hearts? 1 to 51 **8.** a red card with an even number? 5 to 21

9. the ace or king of clubs? 1 to 25 **10.** the 8, 9, or 10 of spades? 3 to 49

11. a card other than a diamond? 3 to 1 **12.** a card other than a jack? 12 to 1

Exercises 13–32 refer to the experiment of tossing a white die and a yellow die together.

B **13.** Make a list of the outcomes. Check students' papers.

14. How many different outcomes are there? 36

Find:
a. the probability of the event.
b. the odds *against* the event.

15. Both dice show the same number. $\frac{1}{6}$; 5 to 1 **16.** Both dice show an even number. $\frac{1}{4}$; 3 to 1

17. One die shows a 3. The other shows a 4. $\frac{1}{18}$; 17 to 1 **18.** Both dice show a 5 or a 6. $\frac{1}{9}$; 8 to 1

19. Both dice show a number less than 5. $\frac{4}{9}$; 5 to 4

20. Both dice show a number greater than 1. $\frac{25}{36}$; 11 to 25

21. The white die shows a 2. The yellow die shows a number greater than 3. $\frac{1}{12}$; 11 to 1

22. The white die shows an even number. The yellow die shows an odd number. $\frac{1}{4}$; 3 to 1

23. Exactly one die shows a 6. $\frac{5}{18}$; 13 to 5

24. Exactly one die shows a multiple of 3. $\frac{4}{9}$; 5 to 4

25. At least one die shows an odd number. $\frac{3}{4}$; 1 to 3

26. At least one die shows a number greater than 4. $\frac{5}{9}$; 4 to 5

27. The sum of the numbers is 6. $\frac{5}{36}$; 21 to 5

28. The sum of the numbers is 10. $\frac{1}{12}$; 11 to 1

29. The number on the white die is 5 less than the number on the yellow die. $\frac{1}{36}$; 35 to 1

30. The number on one die is 2 greater than the number on the other. $\frac{2}{9}$; 7 to 2

31. The product of the numbers is 12. $\frac{1}{9}$; 8 to 1

32. The number on one die is twice the number on the other die. $\frac{1}{6}$; 5 to 1

EXTRA! Independent Events

Suppose we toss a coin and then roll a die. What is the probability that the coin will show heads *and* the die will show 5 dots? Tossing the coin has no effect on rolling the die and we can list all the pairs of events that can occur, as is shown in the table at the right.

$p(\text{heads}) = \frac{1}{2}$ $p(5 \text{ dots}) = \frac{1}{6}$

H-1	T-1
H-2	T-2
H-3	T-3
H-4	T-4
H-5	T-5
H-6	T-6

Notice that there are 12 possible pairs of events in all, and that the pair of events, the coin shows heads *and* the die shows 5, occurs just once. Thus the probability that both events occur together is $\frac{1}{12}$. That is,

$$p(\text{H and } 5) = \frac{1}{12}$$

Notice also that

$$p(\text{H}) = \frac{1}{2} \quad \text{and} \quad p(5) = \frac{1}{6}$$

and that

$$p(\text{H}) \times p(5) = \frac{1}{2} \times \frac{1}{6} = \frac{1}{12}$$

In this particular case, we see that $p(\text{H and } 5) = p(\text{H}) \times p(5)$.

Events which have no effect on each other, such as tossing a coin and rolling a die, are called **independent** events.

As the example above suggests, the probability that a pair of independent events will occur is equal to the product of the probabilities of each individual event.

rule

If *A and B* are independent events, then

$$p(A \text{ and } B) = p(A) \times p(B)$$

State whether or not the following pairs of events are independent. If they are, find the probability that they both occur.

1. A bag contains a red marble, a blue marble, and a yellow marble. You draw a marble, replace it, and draw again. *A* = event of drawing a red marble on the first draw, *B* = event of drawing a red marble on the second draw.   Independent; $\frac{1}{9}$

2. Same as problem 1, except you do not replace the first marble.
   Not independent

3. You randomly decide to walk or ride a bicycle, then randomly decide to head north, south, east, or west. $A = $ the event you walk, $B = $ the event you head south. $\frac{1}{8}$

4. You throw a die. $A = $ the event that the number of dots which come up is odd. $B = $ the event that the number of dots which come up is a prime number. not independent

Self-Test

Symbols and words to remember:

p(ace) [p. 310]

$p = \frac{f}{n}$ [p. 310]

random experiment [p. 307]　　probability of an outcome [p. 307]
event [p. 310]　　　　　　　　　probability of an event [p. 310]
odds in favor of an event [p. 313]　　odds against an event [p. 313]

Is the experiment random?

1. You choose a penny, a quarter, or a dollar bill from a hat. Yes　　[10-1]
2. You roll a fair die. Yes

Write the number of possible outcomes and the probability of each outcome.

3. A letter is randomly chosen from a, b, c, d, e. $5; \frac{1}{5}$
4. A marble is drawn at random from a bag containing a red, an orange, a yellow, and a green marble. $4; \frac{1}{4}$

Each of 7 cards is marked with a letter of the word *referee*. If a card is drawn at random, what is the probability of drawing:

5. an e? $\frac{4}{7}$　　**6.** an s? 0　　**7.** an f? $\frac{1}{7}$　　**8.** an r, an e, or an f? 1　[10-2]

Tony has 6 tape cassettes in his car. One cassette stars the Melodies, 2 star Jake Jamison, and 3 star Carol and the Cliches. Tony selects a cassette at random. What are the odds:

9. in favor of selecting a cassette by Jake? 1 to 2　　[10-3]
10. against selecting a cassette by Jake? 2 to 1
11. in favor of selecting a cassette by the Melodies? 1 to 5
12. against selecting a cassette by Carol and the Cliches? 1 to 1

Self-Test answers and Extra Practice are at the back of the book.

10-4 Estimating Probabilities

Objective To estimate the probability of an event on the basis of results obtained from repeated observations.

In the experiments discussed earlier in this chapter, each outcome had an equal chance of occurring. Thus we could assign the same probability to each one.

In most actual situations, however, we can only *infer* or estimate the probability of a given outcome on the basis of repeated observations. For example, the batting average of a baseball player is the ratio of the number of safe hits the player has made to the number of times the player has been officially at bat for some period of time in the past. If a player has an average of 0.300, then we could give $\frac{3}{10}$ as an estimate of the probability that the player will get a hit the next time at bat. Of course, this does not mean that the player will actually get a hit 3 out of every 10 times at bat, but the *experimental probability* $\frac{3}{10}$ can help us predict the future performance of the player.

Class Exercises

1. Alice sank 16 free throws out of 25 attempts. What is the experimental probability that she will make a basket on the next free throw? $\frac{16}{25}$

2. In a recent season, the Montreal Expos had a club batting average of 0.260. For a randomly selected time at bat, what is the experimental probability that the team scores a hit? $\frac{13}{50}$

3. Suppose you knew that in the past 50 years it snowed in Denver on New Year's Day 12 out of the 50 years. What would you estimate to be the probability that it will snow in Denver next New Year's Day? $\frac{6}{25}$

4. A quarterback has completed 8 out of 15 passes in one game. What is the experimental probability that he will not complete his next pass? $\frac{7}{15}$

5. Suppose you toss a fair coin 8 times and get heads each time. Estimate the probability that you will get heads on your ninth toss. $\frac{1}{2}$

6. Jeffrey was a competitor in 8 diving meets. He won 5 of them. Is it reasonable to predict that the probability that he wins his next meet is $\frac{5}{8}$? Explain. Yes

7. A sports announcer correctly predicted the winning soccer team each of the last five times. Is the announcer's next prediction certain to be correct? No

Exercises

An experiment consists of drawing a marble from a jar. Assume that you know nothing about the number or color of the marbles in the jar. The results of the first 30 draws are shown in the table at the right. What is the experimental probability that the next draw will be:

Red	卌	丨丨丨丨
Yellow	丨丨丨丨	
Blue	卌	卌
Green	卌	丨丨

A
1. a green marble? $\frac{7}{30}$

2. a blue marble? $\frac{1}{3}$

3. a red marble? $\frac{3}{10}$ 4. a yellow marble? $\frac{2}{15}$

5. a red or a green marble? $\frac{8}{15}$ 6. a yellow or a blue marble? $\frac{7}{15}$

7. an orange marble? 0 8. a marble that is not blue? $\frac{2}{3}$

Refer to the experiment above. What would you estimate the odds to be in favor of the next draw:

9. a blue marble? 1 to 2   10. a red marble? 3 to 7

11. a green or a yellow marble? 11 to 19   12. a red or a blue marble? 19 to 11

13. a marble that is not red? 7 to 3   14. a marble that is not yellow? 13 to 2

A traffic engineer observed 200 cars traveling south-bound on Main Street, approaching the intersection of Main Street and Elm Avenue. She found that of the 200 cars, 16 turned right onto Elm, 45 turned left onto Elm, and the rest continued along Main Street. What is the experimental probability that the next car along:

15. will turn right? $\frac{2}{25}$ 16. will turn left? $\frac{9}{40}$

17. will continue straight? $\frac{139}{200}$ 18. will turn right or left? $\frac{61}{200}$

19. will not turn left? $\frac{31}{40}$ 20. will not turn right? $\frac{23}{25}$

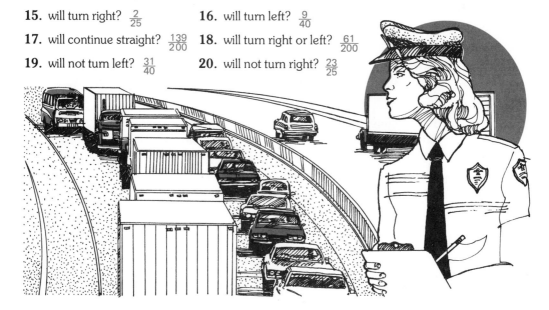

10-5 Estimating from a Sample

Objective To decide whether or not a sample is random, and to draw conclusions based on random samples.

In real-life situations it is often impractical or impossible to obtain exact data. For example, suppose a group of scientists wanted to determine the level of smog in the air. They could make a judgment about the quality of the air by analyzing a few *samples*. The samples must be randomly selected so the information obtained from them will be reliable.

Have you ever received a telephone call in which you were asked to rate and compare several different brands of a product? Marketing surveys of this type, opinion polls, and television ratings are familiar situations in which statistical methods have an influence on decision-making.

Example | Suppose you make a survey of 50 students in your school to see which musical instrument is the most popular one to listen to.

a. Would the sample be random if you include in the survey all the members of the school band?

b. Would the sample be random if you ask every tenth student passing through an auditorium door on the way to an assembly?

c. If the sample were random and 35 students named the guitar as their preference, how many students out of the 590 in the school would you estimate to prefer the guitar?

Solution | **a.** No, because band members would probably have a greater preference for brass and percussion instruments than other students.

b. Yes, if the students are entering in a random order, and if all students must attend the assembly.

c. $\frac{35}{50}$, or 0.7 of the students in the sample prefer the guitar. Therefore, you would estimate that about

$$0.7 \times 590 = 413$$

of the students in the school prefer the guitar.

Class Exercises

Will the experiment produce a random sample? If not, describe a method for selecting a random sample.   Not random

1. Choose 4 members of your class by calling for volunteers.

2. Toss a coin to decide which of two teams will kick off first. Random

3. Choose 6 cans of orange juice from ten 6-can packages by choosing one package at random. Not random

4. A newspaper predicts the outcome of an election by tallying readers' responses to a poll printed in the newspaper. Not random

5. A telephone poll is conducted by calling the fifth name on every twentieth page of the telephone book. Random

6. A quality-control engineer in a light-bulb factory selects the first five bulbs manufactured each working day. Not random

Exercises

A 1. The Alpha Motor Company is testing the reliability of the wheel alignment on its cars. When 500 randomly chosen cars are tested, 8 are found to have faulty alignment. What is the probability that a random purchaser will receive a car with faulty wheel alignment? $\frac{2}{125}$

2. From a consignment of 100,000 bolts, 400 were chosen at random. Of these, 8 were found to be defective. About how many defective bolts could be expected in the entire consignment? about 2000 bolts

3. A survey was taken of 150 households selected at random in Elmview, a city of 15,000 households. In 87 of the 150 households, at least one person was watching television. About what percent of the Elmview households would you predict were not watching television at the time the survey was taken? about 42%

4. Dwyer's Department Store polled 120 customers who were randomly chosen. They found that 81 of the 120 had Dwyer's charge cards. Dwyer's has about 5400 charge accounts. About how many charge accounts would Dwyer's have if every customer in the sample had had a charge card? about 8000 charge accounts

5. You choose 60 orange trees at random in a citrus grove containing 1960 trees. You find that the oranges on 39 trees are ripe enough to be picked. About how many trees in the grove would you expect to have oranges that are ready to be picked? about 1274 trees

6. A manager in a factory will allow no more than 0.5% defective parts. When 500 parts were randomly selected from a shipment, 3 were found to be defective. Did the shipment meet the manager's requirements? No

B   7. A survey was taken of 200 people who were chosen at random. Of these people, 132 exercised regularly. Among the 132, 25% go swimming every week. What is the probability that a randomly selected person swims every week? $\frac{33}{200}$

Passage of a certain bond issue requires a 60% favorable vote. In a random sample of 250 voters from a population of 12,000 it was found that 156 favored passage, 85 were opposed, and the rest were undecided.

8. About how many people in all would you estimate favor passage? 7488

9. About how many people in all would you estimate oppose passage? 4080

10. About how many people in all would you estimate are undecided? 432

11. On the basis of the sample, would you estimate that the bond issue will pass or not? Explain. Yes

Application
Estimating Animal Populations

How can we estimate the number of fish in a lake? Naturally, it is impractical to count the fish one by one. However, we can use our knowledge of probability to estimate the fish population.

Suppose that on Monday we catch some fish, say 60, and put tags on their fins. Then we release the fish back into the lake. A few days later we take another random sample of fish from the lake and observe how many have tags on their fins. Let us say that of 40 fish caught, 5 had tags. Now it is possible, but *very* unlikely, that there are just 95 fish in the lake. That is, there could be 60 fish with tags and 35 without tags. Each of these 95 fish would then have been caught exactly once in the two hauls. It is more realistic to assume that the proportion of tagged fish in the lake is equal to the proportion of tagged fish in the second catch:

$$\frac{1}{8} = \frac{60}{x}; \quad x = 480$$

A likely estimate for the total number of fish is 480.

Research Activity Find out how animal populations of endangered species are estimated.

10-6 Random Variables and Expected Value

Objective To recognize a random variable and compute its expected value.

Suppose we toss a die, and let X be the number of dots which come up. A variable such as X, whose value is determined by a random experiment, is called a **random variable.** The possible values of X and their probabilities of occurrence are listed below.

$X = 3$

Possible values of X	1	2	3	4	5	6
Probability of occurrence	$\frac{1}{6}$	$\frac{1}{6}$	$\frac{1}{6}$	$\frac{1}{6}$	$\frac{1}{6}$	$\frac{1}{6}$

Of course, there is no way to know in advance what the value of a random variable will be in a single trial. But frequently in applications we only need to know the *average* value of a random variable, such as the *average* number of phone calls per day received at a switchboard, or the *average* time it takes to commute to work. Fortunately, there is a way to closely predict the average value of a random variable over a large number of trials.

To see how this can be done, let us consider an actual experiment in which a die was tossed 150 times. The number of times each score occurred was counted, and the results are as shown below.

Value of X	Frequency
1	23
2	28
3	30
4	21
5	23
6	25
Total	150

The average value of X in this experiment is

$$\frac{\text{sum of observed values}}{\text{no. of trials}} = \frac{1 \times 23 + 2 \times 28 + 3 \times 30 + 4 \times 21 + 5 \times 23 + 6 \times 25}{150}$$

$$= \frac{518}{150} \approx 3.45 \text{ (observed average)}$$

To see how this result could have been estimated beforehand, let us re-examine the computation. We had

$$\frac{1 \times (\text{frequency of 1}) + 2 \times (\text{frequency of 2}) + \cdots + 6 \times (\text{frequency of 6})}{\text{no. of trials}}$$

This can be rewritten as

$$1 \times \frac{\text{frequency of 1}}{\text{no. of trials}} + 2 \times \frac{\text{frequency of 2}}{\text{no. of trials}} + \cdots + 6 \times \frac{\text{frequency of 6}}{\text{no. of trials}}$$

But $\dfrac{\text{frequency of an event}}{\text{no. of trials}} \approx p(\text{event})$ for a large number of trials, so our average should be close to

$$1 \times p(1) + 2 \times p(2) + \cdots + 6 \times p(6)$$
$$= 1 \times \frac{1}{6} + 2 \times \frac{1}{6} + 3 \times \frac{1}{6} + 4 \times \frac{1}{6} + 5 \times \frac{1}{6} + 6 \times \frac{1}{6}$$
$$= \frac{21}{6} = 3.5 \ (\textbf{predicted average})$$

This predicted average is called the **expected value** of the random variable.

> ## rule
>
> To find the expected value of a random variable, multiply each possible value by its probability and add the results.

Example At a county fair, Jacqueline has won a prize which she is to select at random from a grab bag containing 6 $2 prizes, 3 $5 prizes, and 1 $10 prize. If $V =$ the dollar value of her prize, what is the expected value of V?

Solution **Step 1.** List all the possible values of V, together with their probabilities.

Possible values of V	2	5	10
Probability of occurrence	$\frac{6}{10}$	$\frac{3}{10}$	$\frac{1}{10}$

Step 2. Multiply each possible value by its probability and add.

$$2 \times \frac{6}{10} + 5 \times \frac{3}{10} + 10 \times \frac{1}{10} = \frac{37}{10} = 3.70$$

The expected value of V is $3.70.

Class Exercises

In the preceding example:

1. can Jacqueline win a prize worth $3.70? No

2. what is the significance of the $3.70 amount?
   $3.70 is the average amount she could win.

Find the expected value of the random variable.

3.

Possible values of B	2	5	7
Probability of occurrence	0.2	0.5	0.3

5.0

4.

Possible values of C	-3	-1	0	2
Probability of occurrence	$\frac{1}{4}$	$\frac{3}{8}$	$\frac{1}{8}$	$\frac{1}{4}$

$-\frac{5}{8}$

Exercises

A　**1.** Joel finds that when he takes the elevator down from his fifth-floor apartment, it takes 2 min 20% of the time, 3 min 50% of the time, and 4 min the rest of the time. If it takes him 3 min to walk down the stairs, which way is faster on the average? It is faster to walk on the average.

2. The solid shown at the right has 2 triangular faces and 3 rectangular faces. If N = the number of sides of a face chosen at random, what is the expected value of N? $3\frac{3}{5}$ sides

3. In a bag of apples, the frequencies of the masses of individual apples, to the nearest 10 g, are shown below. If M = the mass of an apple selected at random, find the expected value of M to the nearest gram. 128 g

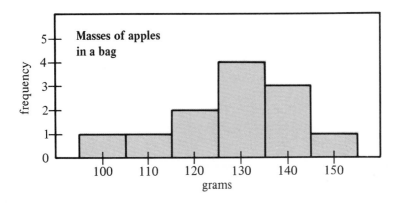

B 4. A nickel, a dime, and a quarter are tossed together. If B = the total number of heads which come up, what is the expected value of B? $1\frac{1}{2}$ heads

5. A player tosses a die. If it comes up an odd number, the player receives that number of points. If it comes up an even number, the player loses that number of points. Is the player likely to have a positive score after many trials? Explain. No

A Mathematician

You are familiar with equations of one or even two variables, but did you ever think of solving equations with four variables?

Shih-Chieh Chu (1280–1303) has been described as one of the greatest mathematicians of all time. He traveled extensively throughout China, where students flocked to study under his guidance. Chu wrote two books of major importance. The first, called *Introduction to Mathematical Studies,* was used as a textbook for students beginning their study of mathematics. The second contained Chu's "method of the four elements." The method, based on the "method of the celestial elements," contained a way of working with four unknown quantities in the same algebraic equation. The celestial elements were *t'ien* (heaven), *ti* (earth), *jen* (man), and *wu* (things or matter) and were represented by the variables u, v, w, and x, respectively. With this mathematical system discovered by Chu, the development of algebra in China reached a peak.

Research Activity Look up Emmy Noether and François Vieta in an encyclopedia, and note their contributions to the field of algebra.

Career Activity To do research, you must have a good understanding of the particular field you choose to study. What do you think you need to know to do research in mathematics? Here are just a few of the many fields of mathematics which researchers study. Find a description of each field in an encyclopedia or mathematics dictionary.

Calculus Geometry Set Theory Complex Variables Trigonometry

1. A weather bureau correctly predicted the weather 28 times out of the past 30 days. What is the probability that the bureau will be correct today? $\frac{14}{15}$ **[10-4]**

2. The school bus arrived early 3 out of the past 5 days. What is the probability that it will not be early tomorrow? $\frac{2}{5}$

3. You toss a fair die 6 times and get an odd number each time. What is the probability of rolling an odd number on the next toss? $\frac{1}{2}$

4. A baseball pitcher has won 8 of his last 12 games. What is the probability that he will win his next game? $\frac{2}{3}$

An inspector on an assembly line chose 400 fuses at random out of 6000 fuses. She found that 3 of the 400 fuses were defective.

5. What is the probability that a consumer will get a defective fuse? $\frac{3}{400}$ **[10-5]**

6. What is the probability that a consumer will not get a defective fuse? $\frac{397}{400}$

7. About how many defective fuses could be expected in the 6000 fuses? 45

8. If you randomly choose a fuse, what are the odds against your getting a defective fuse? 397 to 3

9. A record of Ellen's archery scores showing the percent of arrows which hit each score on the target is shown at the right. If S = Ellen's score on one shot, what is the expected value of S? 5.51   **[10-6]**

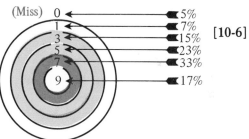

10. A coin is taken at random from a purse which contains 1 penny, 2 nickels, 4 dimes, and 3 quarters. If X = the value of the coin in cents, find the expected value of X. 12.6¢

11. The pyramid-shaped die at the right has 4 congruent triangular faces which are marked with 1, 2, 3, and 4 dots, respectively. If Y = the number of dots on the bottom face on a single toss, what is the expected value of Y? 2.5

12. A student guesses at random on a quiz consisting of 2 true-false questions, the first worth 4 points and the second worth 6 points. If S = the student's quiz score, what is the expected value of S? 5 points

Self-Test answers and Extra Practice are at the back of the book.

Chapter Review

Write the letter that labels the correct answer.

1. The probability that one of n equally likely outcomes will occur   [10-1]
   is ___?___ . D

   A. 0   B. 1   C. n D. $\frac{1}{n}$

2. An ordinary die is rolled. The probability that 5 dots show up
   is ___?___ . C

   A. 5   B. $\frac{1}{5}$ C. $\frac{1}{6}$ D. $\frac{5}{6}$

3. What is the probability of an impossible event?   B   [10-2]

   A. 1   B. 0

   C. $\frac{1}{2}$ D. depends on the experiment

4. The probability that an ordinary die will turn up with a prime
   number of dots is ___?___ . A

   A. $\frac{1}{2}$ B. $\frac{1}{3}$ C. 0   D. $\frac{1}{6}$

5. A box contains 2 red, 3 yellow, and 2 green marbles. The probability that a marble chosen at random is not yellow is ___?___ . D

   A. $\frac{2}{7}$ B. $\frac{3}{7}$ C. $\frac{5}{7}$ D. $\frac{4}{7}$

6. If the probability of rain today is 0.3, then the probability that it   [10-3]
   will not rain today is ___?___ . C

   A. $\frac{2}{3}$ B. $\frac{1}{3}$ C. 0.7   D. 0.3

7. A card is drawn at random from a standard deck of 52 cards.
   The odds against drawing the queen of hearts are ___?___ . B

   A. 52 to 1   B. 51 to 1   C. $\frac{51}{52}$ D. $\frac{1}{52}$

8. A red die and a white die are thrown together. The odds in favor
   of the red die showing more dots than the white die are ___?___ . C

   A. 15 to 36   B. 21 to 36   C. 5 to 7   D. 1 to 1

9. If the odds in favor of an event are 3 to 2 then the probability of
   the event is ___?___ . A

   A. $\frac{3}{5}$ B. $\frac{2}{5}$ C. $\frac{2}{3}$ D. $\frac{3}{2}$

10. An insurance company sent out 50,000 mail advertisements for [10-4] an insurance policy. As a result, they sold 2100 policies. Estimate the probability that a single advertising letter will result in a sale. C

    A. 0.21   **B.** $\frac{21}{50}$ **C.** 4.2%   **D.** 42%

11. In Exercise 10, what are the odds against a single letter resulting in a sale? A

    A. 479 to 21   **B.** 21 to 479   **C.** 500 to 21   **D.** 21 to 500

12. A random sample of 150 machine parts contained 3 defective [10-5] parts. Estimate the probability that a part chosen at random is good. B

    A. $\frac{1}{50}$ **B.** $\frac{49}{50}$ **C.** $\frac{147}{153}$ **D.** $\frac{3}{147}$

13. In Bayview High School, 96 out of 120 students chosen at random planned to continue their education after graduation. If the school has 2500 students, about how many would you estimate plan to continue their education after graduation? A

    A. 2000   **B.** 500   **C.** 2400   **D.** 1904

14. If each value of a random variable is doubled, its expected value [10-6] __?__. A

    A. doubles   **B.** stays the same
    C. increases by 2   **D.** is divided by 2

Possible values of X	−2	1	3	5
Probability of occurrence	0.1	0.2	0.3	0.4

15. In the table above, the expected value of X is __?__. C

    A. 1.75   **B.** 2.75   **C.** 2.9   **D.** 3.3

16. After many trials, the average value of a random variable will probably __?__. D

    A. equal the expected value
    B. be less than the expected value
    C. be greater than the expected value
    D. be close to the expected value

Random Experiments

You have learned some important ideas in probability. But how well do they work? The real test of usefulness is actually to perform some random experiment many times, and compare the results with what your knowledge of probability would lead you to expect. A computer can help us do this because it can simulate a random experiment and repeat it many times at high speed. It can also tell us things we would like to know about the results.

The program on the next page, written in BASIC, simulates the experiment of throwing a pair of dice. It will repeat this experiment 300 times, record the sum of dots on each trial, and print a histogram of the frequency of each possible sum. (*Note:* Check your computer handbook on the use of the RND function.)

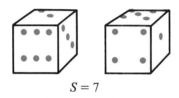

$$S = 7$$

Let S stand for the sum of the dots on the top faces of a pair of dice. The table below shows the sum S for each of the 36 equally likely ways the pair of dice can turn up.

1st die \ 2nd die	1	2	3	4	5	6
1	2	3	4	5	6	7
2	3	4	5	6	7	8
3	4	5	6	7	8	9
4	5	6	7	8	9	10
5	6	7	8	9	10	11
6	7	8	9	10	11	12

Sum S for a pair of dice

1. What is the probability of each possible value of S? See below.

2. What is the expected value of S? 7

1. $\frac{1}{36}, \frac{1}{18}, \frac{1}{12}, \frac{1}{9}, \frac{5}{36}, \frac{1}{6}, \frac{5}{36}, \frac{1}{9}, \frac{1}{12}, \frac{1}{18}, \frac{1}{36}$.

Run the computer program two times.

3. What similarity do you see in the shape of the two histograms?

4. Use the first histogram to calculate the average value of *S*. Is it close to the expected value of *S*?

```
10  PRINT TAB(30);"SUM OF DICE"
15  PRINT
20  FOR Z=1 TO 300
25  LET S=INT(6*RND(0))+INT(6*RND(0))+2
30  PRINT S;
35  GOTO S-1 OF 40,50,60,70,80,90,100,110,120,130,140
40  LET A=A+1
45  GOTO 145                        185  LET L=C
50  LET B=B+1                       190  GOSUB 280
55  GOTO 145                        195  LET L=D
60  LET C=C+1                       200  GOSUB 280
65  GOTO 145                        205  LET L=E
70  LET D=D+1                       210  GOSUB 280
75  GOTO 145                        215  LET L=F
80  LET E=E+1                       220  GOSUB 280
85  GOTO 145                        225  LET L=G
90  LET F=F+1                       230  GOSUB 280
95  GOTO 145                        235  LET L=H
100 LET G=G+1                       240  GOSUB 280
105 GOTO 145                        245  LET L=I
110 LET H=H+1                       250  GOSUB 280
115 GOTO 145                        255  LET L=J
120 LET I=I+1                       260  GOSUB 280
125 GOTO 145                        265  LET L=K
130 LET J=J+1                       270  GOSUB 280
135 GOTO 145                        275  STOP
140 LET K=K+1                       280  LET M=M+1
145 NEXT Z                          285  PRINT M+1;
150 PRINT                           290  FOR N=1 TO L
152 PRINT                           295  PRINT TAB(4);"*";
155 PRINT TAB(30);"HISTOGRAM"       300  NEXT N
160 PRINT                           305  PRINT
165 LET L=A                         310  RETURN
170 GOSUB 280                       315  END
175 LET L=B
180 GOSUB 280
```

Chapter Test

1. In an experiment with 7 equally likely outcomes, each outcome has probability __?__. $\frac{1}{7}$ [10-1]

2. If the outcomes of an experiment are a matter of chance, the outcomes occur __?__. randomly

3. A __?__ event has probability 1. certain [10-2]

4. A card is drawn randomly from a standard deck. The probability that it is a club is __?__. $\frac{1}{4}$

5. The probability that an ordinary die will turn up with more than 4 dots is __?__. $\frac{1}{3}$

6. If you choose a pair of socks at random from a drawer containing 1 blue sock and 2 brown socks, the probability of getting a matching pair is __?__. $\frac{1}{3}$

7. If the probability of an event is $\frac{2}{7}$, the odds in favor of it are __?__. 2 to 5 [10-3]

8. If you toss an ordinary die, the odds in favor of its showing a number of dots which is a multiple of 3 are __?__. 1 to 2

9. In playing chess against Ivan, John has won 9 games, drawn 3, and lost 7. The probability that John will not win his next game against Ivan is __?__. $\frac{10}{19}$ [10-4]

10. A tennis player has successfully returned 9 out of 14 first serves by the opponent. The probability that the player will not return the next first serve is __?__. $\frac{5}{14}$

11. A random sample of 200 households in Glenville contained 93 households with school-age children. Of 2200 households in Glenville about __?__ have school-age children. 1023 [10-5]

12. In a random sample of 25 residents of Willowdale, population 1250, 9 indicated they would attend a town meeting. About __?__ people planned to attend the meeting. 450

13. A hand of 5 cards contains 3 sevens and 2 tens. If $V =$ the face value of a card randomly chosen from this hand, then the expected value of V is __?__. $8\frac{1}{5}$ [10-6]

14. One face of a cube is marked with 1 dot, 2 other faces are each marked with 2 dots, and the remaining 3 faces are each marked with 3 dots. If this die is thrown, the expected value of the score is __?__. $2\frac{1}{3}$

Skill Review

Add or subtract.

1. $1\frac{1}{2} + 2\frac{1}{4}$ $3\frac{3}{4}$

2. $5 + 2\frac{1}{6}$ $7\frac{1}{6}$

3. $3\frac{1}{3} + 3\frac{1}{3}$ $6\frac{2}{3}$

4. $-16\frac{3}{4} + 5\frac{1}{3}$ $-11\frac{5}{12}$

5. $-\frac{15}{2} + 4\frac{1}{2}$ -3

6. $\frac{32}{3} - 6\frac{3}{4}$ $3\frac{11}{12}$

7. $15\frac{1}{2} - \frac{21}{5}$ $11\frac{3}{10}$

8. $\frac{7}{3} + \frac{1}{3}$ $2\frac{2}{3}$

9. $-2\frac{5}{8} - \frac{16}{3}$ $-7\frac{23}{24}$

10. $22\frac{1}{2} - (-6)$ $28\frac{1}{2}$

11. $17 - \left(-16\frac{11}{21}\right)$ $33\frac{11}{21}$

12. $5\frac{3}{8} + 15\frac{1}{4}$ $20\frac{5}{8}$

13. $37\frac{1}{3} - \left(-7\frac{1}{3}\right)$ $44\frac{2}{3}$

14. $11\frac{12}{25} - 12\frac{13}{25}$ $-1\frac{1}{25}$

15. $5\frac{1}{8} + 6\frac{1}{4}$ $11\frac{3}{8}$

16. $\left(-\frac{3}{2} - \frac{3}{2}\right)$ -3

17. $-14\frac{2}{7} - 2\frac{2}{3}$ $-16\frac{20}{21}$

18. $13\frac{5}{8} - 3\frac{5}{8}$ 10

19. $-\frac{11}{3} - \frac{3}{11}$ $-3\frac{31}{33}$

20. $\frac{6}{7} + \left(-\frac{22}{7}\right)$ $-2\frac{2}{7}$

21. $\frac{17}{3} + \frac{2}{3}$ $6\frac{1}{3}$

22. $5\frac{1}{4} + 6\frac{3}{4}$ 12

23. $\frac{23}{3} - \frac{3}{23}$ $7\frac{37}{69}$

24. $16\frac{5}{7} - 23\frac{1}{5}$ $-6\frac{17}{35}$

Multiply or divide.

25. $\frac{1}{4} \times \frac{1}{2}$ $\frac{1}{8}$

26. $\frac{3}{2} \times \frac{2}{3}$ 1

27. $1\frac{1}{2} \times 2\frac{1}{4}$ $3\frac{3}{8}$

28. $13 \div 1\frac{1}{3}$ $9\frac{3}{4}$

29. $\frac{15}{2} \div \frac{2}{3}$ $11\frac{1}{4}$

30. $-\frac{3}{5} \times \frac{6}{7}$ $-\frac{18}{35}$

31. $-5\frac{1}{2} \div \frac{1}{2}$ -11

32. $\frac{13}{6} \times \frac{3}{2}$ $3\frac{1}{4}$

33. $-5 \times \left(-\frac{3}{5}\right)$ 3

34. $-\frac{17}{5} \div \frac{5}{17}$ $-11\frac{14}{49}$

35. $-\frac{11}{21} \times \frac{3}{7}$ $-\frac{11}{49}$

36. $6\frac{1}{4} \div \left(-1\frac{1}{2}\right)$ $-4\frac{1}{6}$

37. $-\frac{2}{3} \times \frac{6}{17}$ $-\frac{4}{17}$

38. $\frac{27}{121} \times \frac{11}{3}$ $\frac{9}{11}$

39. $4\frac{1}{6} \times \left(-1\frac{2}{3}\right)$ $-6\frac{17}{18}$

40. $-3 \times \frac{33}{27}$ $-3\frac{2}{3}$

41. $1\frac{1}{5} \times \frac{35}{48}$ $\frac{7}{8}$

42. $\frac{15}{32} \div \left(-\frac{16}{5}\right)$ $-\frac{75}{512}$

A television cameraperson filming a political convention.

11

The Number Plane

11-1 Graphing an Ordered Pair of Numbers

Objective To graph an ordered pair of numbers in the plane.

We have learned how to set up a coordinate system on a line, associating each number with exactly one point on the line.

By drawing two perpendicular number lines that intersect at the origin of each, we can set up a **rectangular coordinate system in the plane.** Using this system, we can then associate an *ordered pair of numbers* with each point in the plane. An **ordered pair of numbers** is simply a pair of numbers (called *components* of the ordered pair) whose order is important. We associate one number, the *first component,* with the horizontal number line (called the x-axis), and the other number, the *second component,* with the vertical number line (called the y-axis). Note that the positive direction is to the right on the horizontal axis and is upward on the vertical axis.

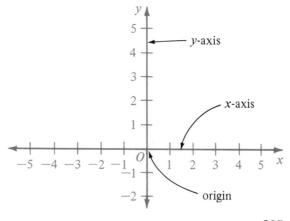

To find the point associated with an ordered pair of numbers, say $(-2,4)$, we proceed as follows:

1. Draw a line l perpendicular to the x-axis,
   passing through the point on it labeled -2.
2. Next draw a line m perpendicular to the
   y-axis, passing through the point labeled
   4 on that axis.
3. The point where l and m intersect is the
   graph of $(-2,4)$. Special graph paper with
   a printed grid is helpful for locating points
   in this way.

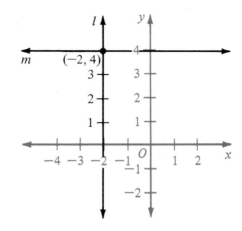

Example | **Graph $(3,2)$, $(-3,2)$, $(-3,-2)$, and $(3,-2)$.**

Solution

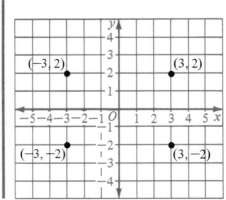

Another way to think of finding the graph of an ordered pair is in
terms of sliding a checker. For example, to locate the graph of
$(-3,2)$ you can:

1. Start at the origin.

1.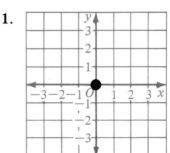

2. Move 3 units to the left along the x-axis.

2.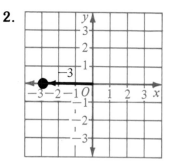

3. Move 2 units up parallel to the y-axis.

3.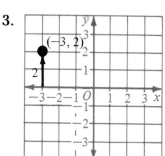

4. The checker is now at the point paired with $(-3,2)$.

Class Exercises

Name the graph of each ordered pair.

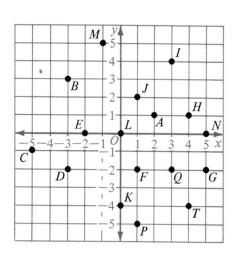

Example | **(1,2)**

Solution | **J**

1. (4,1) *H*   **2.** (1,−2) *F*

3. (−3,−2) *D*   **4.** (3,4) *I*

5. (0,−4) *K*   **6.** (−3,3) *B*

7. (−2,0) *E*   **8.** (4,−4) *T*

9. (0,0) *O*   **10.** (2,1) *A*

11. (5,0) *N*   **12.** (3,−2) *Q*

13. (−5,−1) *C*   **14.** (1,−5) *P*

15. (5,−2) *G*   **16.** (−1,5) *M*

Exercises

Name the graph of each ordered pair.

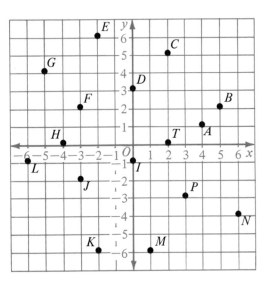

A
1. $(-4,0)$ *H*
2. $(2,5)$ *C*
3. $(5,2)$ *B*
4. $(6,-4)$ *N*
5. $(-3,-2)$ *J*
6. $(-5,4)$ *G*
7. $(2,0)$ *T*
8. $(4,1)$ *A*
9. $(0,3)$ *D*
10. $(3,-3)$ *P*
11. $(-2,-6)$ *K*
12. $(-3,2)$ *F*
13. $(0,-1)$ *I*
14. $(1,-6)$ *M*
15. $(-2,6)$ *E*
16. $(-6,-1)$ *L*

In Exercises 17–22, graph the specified points on the plane. Connect them in the order listed. Use a separate set of axes for each exercise. Check students' graphs.

Example $(3,2), (0,-1), (3,-4), (6,-1), (3,2)$

Solution

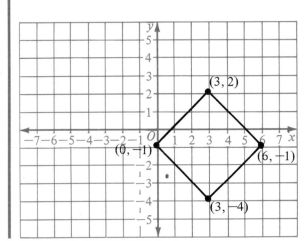

17. $(-1,0), (3,2), (-2,3), (2,5), (-1,0)$
18. $(-2,3), (1,1), (-2,-1), (1,-3), (-2,-5), (-2,3)$
19. $(-6,-5), (-1,-3), (-1,-1), (4,1), (-6,-5)$
20. $(-1,-1), (4,-1), (4,1), (-1,1), (0,0), (-1,-1)$
21. $(0,1), (2,1), (0,3), (2,5), (0,5), (-2,3), (0,1)$
22. $(-1,2), (1,1), (1,-2), (4,2), (0,6), (-1,2)$

In Exercises 23–28, the vertexes of a polygon are specified.

a. Locate the graphs of the given ordered pairs and sketch the polygon on a coordinate plane.

b. Find the area of the polygon.

Example $(-2,1), (1,4), (4,1)$

Solution **a.** Sketch the polygon, a triangle.

b. Count the units in the base (6) and the height (3). Use $A = \frac{1}{2}bh$.

$$A = \frac{1}{2}(6)(3) = 9$$

The area is 9 square units.

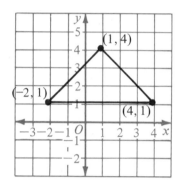

B **23.** $(1,3), (-5,3), (-2,-2)$ 15

24. $(4,0), (4,7), (0,5)$ 14

25. $(5,1), (5,-5), (-1,1), (-1,-5)$ 36

26. $(1,2), (1,8), (5,2), (5,8)$ 24

27. $(-3,-1), (0,3), (7,3), (4,-1)$ 28

28. $(0,-3), (2,1), (7,1), (8,-3)$ 26

Application Using A Map

Here is a part of the map of Spaceville.

All the avenues running north and south, except Main Avenue, are named by letters. The streets running east and west, except Broadway, are named by numbers. The Spaceville City Hall is located at the intersection of Broadway and Main Avenue. If you were standing in front of City Hall and were asked by someone how to get to the corner of B Avenue and 1st Street, you could tell the person to go 2 blocks west and 4 blocks south.

What set of directions would you provide if you were asked how to get to the intersection of

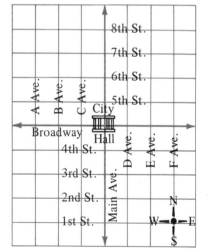

1. D Ave. and 5th St.?

2. F Ave. and 8th St.?

3. Main Ave. and 2nd St.? **4.** F Ave. and Broadway?

5. B Ave. and 6th St.? **6.** F Ave. and 2nd St.?

11-2 The Coordinates of a Point in the Plane

Objective To assign coordinates to a point in the plane.

We have seen that a point in the plane can be assigned to *every* ordered pair of numbers.

To assign a pair of coordinates·to a given point *P* in the plane:

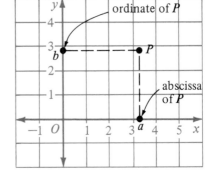

1. Draw a perpendicular from *P* to the *x*-axis, intersecting the axis at a point with coordinate *a*.
2. Draw a perpendicular from *P* to the *y*-axis, intersecting it at a point with coordinate *b*.
3. Then the coordinates of *P* are (a,b), or about $(3\frac{1}{4}, 2\frac{3}{4})$ in this figure. The first number, *a*, is called the **x-coordinate**, or **abscissa**, of *P*. The second number, *b*, is the **y-coordinate**, or **ordinate**, of *P*.

The two axes of a rectangular coordinate system separate the rest of the plane into four regions, called **Quadrants I, II, III,** and **IV.** The location of each quadrant is shown here, along with the range of values for the coordinates of any point in that quadrant.

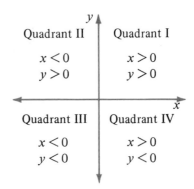

Example | Give the coordinates of each lettered point in the figure.

Solution | $A(0,1)$, $B(3,2)$, $C(1,3)$, $D(-2,2)$, $E(-1,-2)$, $F(2,-1)$, $O(0,0)$

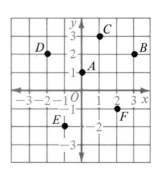

Class Exercises

Exercises 1–16 refer to the figure at the right.
State the coordinates of the given point.

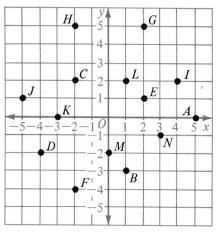

1. A (5, 0) **2.** B (1, −3) **3.** C (−2, 2) **4.** D (−4, −2)

5. K (−3, 0) **6.** M (0, −2) **7.** F (−2, −4) **8.** N (3, −1)

9. Name all points shown in Quadrant I. E, G, L, I

10. Name all points shown in Quadrant II. C, H, J

11. Name all points shown in Quadrant III. D, F

12. Name all points shown in Quadrant IV. B, N

13. Name all points that are not in a quadrant. A, K, M

14. Name all points having abscissa −2. C, F, H

15. Name all points having ordinate 2. C, I, L

16. Name all points having ordinate 0. A, K

Exercises

Exercises 1–30 refer to the figure at the right.

In Exercises 1–16, write the coordinates of the
given point.

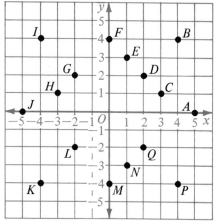

A　**1.** A (5, 0)　**2.** K(−4, −4) **3.** P (4, −4) **4.** H(−3, 1)

5. D (2, 2) **6.** L(−2, −2) **7.** O (0, 0)　**8.** J (−5, 0)

9. E (1, 3) **10.** M (0, −4) **11.** I (−4, 4) **12.** Q(2, −2)

13. B (4, 4) **14.** F (0, 4) **15.** N(1, −3) **16.** C(3, 1)

17. Name all points shown in Quadrant I. B, C, D, E

18. Name all points shown in Quadrant II. G, H, I

19. Name all points shown in Quadrant III. K, L

20. Name all points shown in Quadrant IV. N, P, Q

B　**21.** Name all points shown whose abscissas are positive. A, B, C, D, E, N, P, Q

　　22. Name all points shown whose abscissas are negative. G, H, I, J, K, L

23. Name all points shown whose ordinates are negative. *K, L, M, N, P, Q*

24. Name all points shown whose ordinates are positive. *B, C, D, E, F, G, H, I*

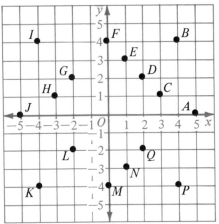

25. Name all points shown whose ordinates are 0. *A, J*

26. Name all points shown whose ordinates are −4. *K, M, P*

27. Name all points shown whose abscissas are 0. *F, M*

28. Name all points shown whose abscissas are 2. *D, Q*

C 29. Name all points shown for which the abscissa is equal to the ordinate. *B, D, K, L*

30. Name all points shown for which the sum of the abscissa and ordinate is 4. *C, D, E, F*

EXTRA! Pictures in the Plane

Graph each set of ordered pairs below on a coordinate plane. Then join the points in order by drawing line segments. What have you drawn?

Figure A umbrella	**Figure B** guitar	**Figure C** boot	**Figure D** butterfly
1. (1,0)	1. (6,17)	1. (−4,9)	1. (0,−6)
2. (−1,1)	2. (8,17)	2. (4,9)	2. (1,−2)
3. (−5,−1)	3. (6,14)	3. (9,11)	3. (11,−7)
4. (−7,1)	4. (1,1)	4. (0,11)	4. (8,0)
5. (−10,0)	5. (3,0)	5. (−4,9)	5. (11,7)
6. (−6,6)	6. (3,−3)	6. (−2,−2)	6. (9,12)
7. (0,8)	7. (2,−5)	7. (−4,−5)	7. (1,3)
8. (7,6)	8. (2,−9)	8. (−7,−7)	8. (1,5)
9. (10,0)	9. (1,−12)	9. (−10,−8)	9. (−1,5)
10. (6,1)	10. (−2,−13)	10. (−10,−11)	10. (−1,3)
11. (2,−1)	11. (−6,−12)	11. (−6,−12)	11. (−9,12)
12. (1,0)	12. (−7,−10)	12. (3,−10)	12. (−11,7)
13. (2,−12)	13. (−7,−7)	13. (3,−11)	13. (−8,0)
14. (1,−13)	14. (−4,−4)	14. (8,−10)	14. (−11,−7)
15. (0,−13)	15. (−4,−1)	15. (8,−4)	15. (−1,−2)
16. (−1,−12)	16. (−2,1)	16. (9,11)	16. (0,−6)
	17. (0,1)		
	18. (6,17)		

Self-Test

Words to remember:
rectangular coordinate system [p. 335]
ordered pair of numbers [p. 335]
x-coordinate [p. 340]   abscissa [p. 340]   y-coordinate [p. 340]
ordinate [p. 340]   quadrant [p. 340]

1. Name the graph of each ordered pair:
   a. $(0,-3)$ _J_ **b.** $(1,-1)$ _F_ **c.** $(-3,0)$ _A_
   d. $(-2,1)$ _D_ **e.** $(-2,-2)$ _B_ **f.** $(1,2)$ _C_

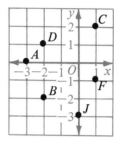

[11-1]

On a coordinate plane, **a.** sketch the polygon whose vertexes are given and **b.** find the area of the polygon.

2. $(1,1),(1,-1),(-1,-1),(-1,1)$ 4

3. $(0,-2),(0,1),(4,1),(4,-2)$ 12

4. $(-2,1),(3,1),(3,4)$ 7.5

Exercises 5–8 refer to the figure at the right.

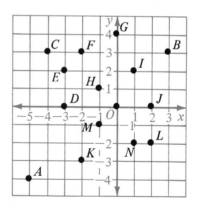

5. Write the coordinates of the given points: **a.** A **b.** F
   c. K **d.** B **e.** G
   f. L See below.

[11-2]

6. Name all points shown in Quadrant III. _A, K, M_

7. Name all points shown whose abscissas are negative.
   _A, C, D, E, F, H, K, M_

8. Name all points shown whose ordinates are 0. _D, O, J_

5. $(-5,-4), (-2,3), (-2,-3), (3,3), (0,4), (2,-2)$

Self-Test answers and Extra Practice are at the back of the book.

11-3 Equations in Two Variables

Objective To find solutions for an equation in two variables.

An equation such as

$$4x = 8$$

contains only one variable, x, and has only one solution, 2.

The equation

$$4x + y = 8$$

contains two variables, x and y. A solution of an equation in two variables consists of two numbers, one for each variable. It can therefore be written as an ordered pair of numbers in the form (x, y). Thus, we can see by inspection that three possible solutions for this equation are

$$(1,4), \ (2,0), \ \text{and} \ (0,8).$$

Example 1	**Tell which of the given ordered pairs are solutions of $4x + y = 8$.**
	a. $(-2,0)$ **b.** $(3,-2)$ **c.** $(-1,12)$
Solution	**a.** $4(-2) + 0 \neq 8$, so $(-2,0)$ is not a solution.
	b. $4(3) - 2 \neq 8$, so $(3,-2)$ is not a solution.
	c. $4(-1) + 12 = 8$, so $(-1,12)$ is a solution.

To find solutions for an equation such as

$$-2x + y = 3$$

we can substitute various numbers for x and then solve for y in each case. It is easier, however, if we first write an equivalent equation expressing y in terms of x:

$$-2x + y = 3$$
$$2x - 2x + y = 2x + 3$$
$$y = 2x + 3$$

Now when we replace x with any number, we can find the corresponding value for y by solving a simple equation in one variable. For example:

If $x = 1$, then $y = 2(1) + 3 = 5$.

If $x = -\frac{1}{2}$, then $y = 2\left(-\frac{1}{2}\right) + 3 = 2$.

If $x = -2$, then $y = 2(-2) + 3 = -4 + 3 = -1$.

As always, it is a good idea to check these solutions, $(1,5)$, $(-\frac{1}{2},2)$, $(-2,-1)$ to see whether or not they satisfy the original equation, $-2x + y = 3$.

The number of solutions in this case is infinite, because x can have an infinite number of values.

Example 2	**Find solutions for $x + 3y = 6$, when $x = 3$ and when $x = -6$.**
Solution	$x + 3y = 6$
	$x + 3y + (-x) = 6 + (-x)$
	$3y = 6 - x$
	$\frac{1}{3}(3y) = \frac{1}{3}(6 - x)$
	$y = 2 - \frac{1}{3}x$
	When $x = 3$, $y = 2 - \frac{1}{3}(3) = 2 - 1 = 1$
	When $x = -6$, $y = 2 - \frac{1}{3}(-6) = 2 + 2 = 4$
	So $(3,1)$ and $(-6,4)$ are the solutions required.

Class Exercises

State whether or not the given ordered pair is a solution of the given equation.

1. $(5,1)$, $x + y = 6$ Yes   **2.** $(2,-1)$, $x + y = 3$ No

3. $(7,2)$, $x - y = 9$ No   **4.** $(5,2)$, $x - y = 3$ Yes

5. $(-1,2)$, $2x + y = 0$ Yes   **6.** $(-1,-2)$, $2x - y = 0$ Yes

For each equation, give an equivalent equation for y in terms of x.

Example	**a. $x + y = 3$ b. $x - y = 5$**
Solution	**a. $y = 3 - x$ b. $x - 5 = y$ or $y = x - 5$**

7. $x + y = 4$ $y = 4 - x$ **8.** $2x + y = 3$ $y = 3 - 2x$ **9.** $x - y = 2$ $y = x - 2$

10. $x - y = -3$ $y = x + 3$ **11.** $2y = 4x$ $y = 2x$ **12.** $\frac{1}{2}y = 3x$ $y = 6x$

If $y = x - 4$, supply the value of y in each ordered pair so that the ordered pair is a solution of the equation.

13. $(4,\underline{\ ?\ })\ 0$ **14.** $(6,\underline{\ ?\ })\ 2$ **15.** $(-2,\underline{\ ?\ })\ -6$ **16.** $(0,\underline{\ ?\ })\ -4$

If $y = 3x + 1$, supply the value of y in each ordered pair so that the ordered pair is a solution of the equation.

17. $(0,\underline{\ ?\ })\ 1$ **18.** $(2,\underline{\ ?\ })\ 7$ **19.** $(-3,\underline{\ ?\ })\ -8$ **20.** $\left(\frac{1}{3},\underline{\ ?\ }\right)\ 2$

Exercises

In Exercises 1–10, tell which of the given ordered pairs are solutions of the given equation.

	a.	**b.**	**c.**
1. $x + 2y = 5$	$(1,2)$ Yes	$(2,1)$ No	$(7,-1)$ Yes
2. $x - y = 5$	$(7,1)$ No	$(6,1)$ Yes	$(5,1)$ No
3. $2x + y = 7$	$(4,3)$ No	$(3,1)$ Yes	$(4,-1)$ Yes
4. $2x - y = 3$	$(3,3)$ Yes	$(2,1)$ Yes	$(1,-1)$ Yes
5. $x - 2y = 6$	$(6,0)$ Yes	$(0,3)$ No	$(10,2)$ Yes
6. $3x + y = 8$	$(2,2)$ Yes	$(3,-1)$ Yes	$(4,-2)$ No
7. $2x + 3y = 10$	$(2,2)$ Yes	$(3,1)$ No	$(8,-2)$ Yes
8. $3x + 2y = 0$	$(2,-3)$ Yes	$(3,-2)$ No	$(1,-1)$ No
9. $2y = 8 - 3x$	$(2,1)$ Yes	$(4,0)$ No	$(5,1)$ No
10. $4y = 3 + 5x$	$(1,2)$ Yes	$(2,3)$ No	$(5,7)$ Yes

A

In Exercises 11–25 **a.** Solve the equation for y in terms of x; **b.** find solutions of the equation which have abscissas of $-5, -1, 0, 1,$ and 5. Check your results in the original equation.

Example   **$2x + y = 8$**

Solution   **a.**   **$2x + y = 8$**

$$2x + y - 2x = 8 - 2x$$
$$y = 8 - 2x$$

b. If $x = -5$, $y = 8 - 2(-5) = 8 + 10 = 18$
If $x = -1$, $y = 8 - 2(-1) = 8 + 2 = 10$
If $x = 0$, $y = 8 - 2(0) = 8 - 0 = 8$
If $x = 1$, $y = 8 - 2(1) = 8 - 2 = 6$
If $x = 5$, $y = 8 - 2(5) = 8 - 10 = -2$

Check

$2(-5) + 18 \overset{?}{=} 8$ $\qquad$ $2(-1) + 10 \overset{?}{=} 8$ $\qquad$ $2(0) + 8 \overset{?}{=} 8$
$-10 + 18 = 8$ $\qquad\qquad$ $-2 + 10 = 8$ $\qquad\qquad\quad$ $8 = 8$

$2(1) + 6 \overset{?}{=} 8$ $\qquad\qquad$ $2(5) + (-2) \overset{?}{=} 8$
$2 + 6 = 8$ $\qquad\qquad\qquad$ $10 - 2 = 8$

Therefore, $(-5, 18)$, $(-1, 10)$, $(0, 8)$, $(1, 6)$, and $(5, -2)$ are solutions
of $2x + y = 8$.

11. $(-5, 12)$, $(-1, 8)$, $(0, 7)$, $(1, 6)$, $(5, 2)$ $\qquad$ 12. $(-5, 8)$, $(1, 4)$, $(0, 3)$, $(1, 2)$, $(5, -2)$

11. $x + y = 7$ $\qquad\qquad$ **12.** $x + y = 3$ $\qquad\qquad$ **13.** $-x + y = 2$

14. $-x + y = 0$ $\qquad\qquad$ **15.** $x - y = 6$ $\qquad\qquad$ **16.** $x - y = -2$

13. $(-5, -3)$, $(-1, 1)$, $(0, 2)$, $(1, 3)$, $(5, 7)$ $\qquad$ 14. $(-5, -5)$, $(-1, -1)$, $(0, 0)$, $(1, 1)$, $(5, 5)$

B **17.** $x + 2y = 5$ $\qquad\qquad$ **18.** $-x + 3y = 6$ $\qquad\qquad$ **19.** $-x - 2y = 4$

20. $x - 3y = 0$ $\qquad\qquad$ **21.** $4x + 2y = 5$ $\qquad\qquad$ **22.** $3x + 2y = 4$

23. $3x + \frac{1}{2}y = 2$ $\qquad\qquad$ **24.** $4x - \frac{1}{3}y = 1$ $\qquad\qquad$ **25.** $x - \frac{5}{6}y = 10$

15. $(-5, -11)$, $(-1, -7)$, $(0, -6)$, $(1, -5)$, $(5, -1)$ $\quad$ 16. $(-5, -3)$, $(-1, 1)$, $(0, 2)$, $(1, 3)$, $(5, 7)$

Find any three ordered pairs that are solutions of the given equation.

26. $y - x = -1$ $\qquad\qquad$ **27.** $x + y = -6$ $\qquad\qquad$ **28.** $x - y = 0^-$

29. $4x + y = 8$ $\qquad\qquad$ **30.** $x - 5y = -5$ $\qquad\qquad$ **31.** $-3x + 2y = 0$

32. $5x - 4y = 20$ $\qquad\qquad$ **33.** $-\frac{1}{3}x + y = 0$ $\qquad\qquad$ **34.** $-x + \frac{1}{2}y = 3$

17. $(-5, 5)$, $(-1, 3)$, $(0. 2.5)$, $(1, 2)$, $(5, 0)$ $\qquad$ 18. $(-5, \frac{1}{3})$, $(-1, \frac{5}{3})$, $(0, 2)$, $(1, \frac{7}{3})$, $(5, \frac{11}{3})$
19. $(-5, 0.5)$, $(-1, -1.5)$, $(0, -2)$, $(1, -2.5)$, $(5, -4.5)$ 20. $(-5, -\frac{5}{3})$, $(-1, -\frac{1}{3})$, $(0, 0)$, $(1, \frac{1}{3})$, $(5, \frac{5}{3})$

EXTRA! Completing a Table

Rachel had completed a table of values for x and y that showed
solutions of the equation

$$2x + y = 6$$

when she was called to the telephone. While she was gone, her
dog chewed the paper up leaving the result shown at the right.
Make a copy of the table and see if you can complete it for Rachel.

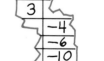

x	y
1	4
2	
3	
	-4
	-6
	-10

21. $(-5, 12.5)$, $(-1, 4.5)$, $(0, 2.5)$, $(1, 0.5)$, $(5, -7.5)$ $\quad$ 22. $(-5, 9.5)$, $(-1, 3.5)$, $(0, 2)$, $(1, 0.5)$, $(5, -5.5)$
23. $(-5, 34)$, $(-1, 10)$, $(0, 4)$, $(1, -2)$, $(5, -26)$ $\qquad$ 24. $(-5, -63)$, $(-1, -15)$, $(0, -3)$, $(1, 9)$, $(5, 57)$
25. $(-5, -18)$, $(-1, -13.2)$, $(0, -12)$, $(1, -10.8)$, $(5, -6)$

11-4 The Graph of a Linear Equation

Objective To graph an equation in two variables.

We cannot graph separately each solution of the equation

$$y = x + 1$$

because the number of solutions is infinite. Nevertheless, by selecting a few values for x we may be able to detect a *pattern* in the graph of the corresponding solutions. A table giving the values of y corresponding to some replacements for x is shown below. The points representing the graphs of these ordered pairs are pictured beside the table.

x	y
-2	-1
-1	0
$-\dfrac{1}{2}$	$\dfrac{1}{2}$
0	1
$\dfrac{1}{2}$	$1\dfrac{1}{2}$
2	3

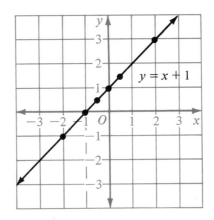

A look at the figure should convince you that these points all lie on the same straight line. This line can in fact be shown to consist of all the points that represent solutions of the equation. It is therefore the **graph of the solution set of the equation,** although we usually call it simply **the graph of the equation.**

In general, any equation that can be written in the form

$$ax + by + c = 0,$$

where x and y are variables and a, b, and c are numbers (with a and b not both zero), is called a **linear equation in two variables** because its graph is always a straight line in the plane.

Because only two points are needed to determine a unique line, if we are graphing an equation that we know is linear, we need find only two solutions in order to draw its graph.

Example | **Graph $2x + y = 6$.**

Solution | **The two easiest solutions to find are those of the form $(0,y)$ and $(x,0)$.**

When $x = 0$, we have
$$2(0) + y = 6$$
$$0 + y = 6$$
$$y = 6$$

When $y = 0$, we have
$$2x + 0 = 6$$
$$2x = 6$$
$$x = 3$$

It is usually a good idea to find a third point just to check one's work. If we let $x = 1$, then we have

$$2(1) + y = 6$$
$$2 + y = 6$$
$$y = 4$$

x	y
0	6
3	0
1	4

Class Exercises

In Exercises 1–8, state ordered pairs of the form $(x,0)$ and $(0,y)$ that are solutions of the given equation.

1. $x + y = 4$
   (4, 0) and (0, 4)
2. $x + y = -4$
   (−4, 0) and (0, −4)
3. $x - y = 3$
   (3, 0) and (0, −3)
4. $-x + y = 5$
   (−5, 0) and (0, 5)
5. $x + 2y = 6$
   (6, 0) and (0, 3)
6. $x + 2y = -4$
   (−4, 0) and (0, −2)
7. $-x + 5y = 5$
   (−5, 0) and (0, 1)
8. $3x + 2y = 12$
   (4, 0) and (0, 6)

Exercises

Graph each equation on a coordinate plane. Use a separate set of axes for each equation. Check students' graphs.

A **1– 8.** Graph each equation in Class Exercises 1–8.

9. $2x + y = 4$

10. $-2x + y = -2$

11. $3x - y = 6$

12. $x - 3y = -3$

13. $2x + 5y = 10$

14. $2x - 3y = 12$

Graph each equation on a coordinate plane. Use a separate set of axes for each equation. Check students' graphs.

Example | $y = 4$

Solution | Use $y = 0x + 4$. **Find three solutions.**

$x = 1, y = 0(1) + 4 = 4$
$x = 2, y = 0(2) + 4 = 4$
$x = 3, y = 0(3) + 4 \doteq 4$

Solutions are (1,4), (2,4), and (3,4). Graph these points and draw the line.

B 15. $y = 2$ 16. $x = -4$ 17. $y = 0$

18. $y = -3$ 19. $x = 6$ 20. $y = \frac{3}{2}$

Application

The Computer and Linear Equations

For a science project, Nina watered 6 identical lima bean gardens by different amounts. The crop yields were as shown below.

water applied daily in cm	0.5	1.0	1.5	2.0	2.5	3.0
yield of lima beans in kg	1.6	1.8	2.0	2.4	2.6	2.8

Can Nina use these data to predict crop yields for other amounts of daily watering? If we plot her findings as shown on the graph at the right, we see that the points are very nearly in line.

The computer program on page 351 gives the equation of the line which best fits the points on the graph. Notice how the coordinates of the points are entered in line 40. The sample run tells us that the equation of the desired line is $y = 0.5x + 1.32$. From its graph, shown beside the sample run, Nina could predict, for example, that about 3.0 kg of lima beans would grow if she applied 3.3 cm of water daily.

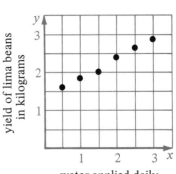

water applied daily in centimeters

Use graph paper and the computer to answer each question.

1. The chart below shows the distance a spring stretched when different masses were hung from it.

Mass in kg	0.3	0.6	0.9	1.2	1.5	1.8
Stretch in cm	2.1	4.9	6.0	7.1	8.9	10.8

← Spring

← Mass

About how much stretch would a 1 kg mass produce? 6.36 cm

2. The chart below shows temperature changes as a cold front approached.

Time in hours from 1st reading	0	1	2	3	4
Temperature in °C	21	17	14	12	7

Estimate the temperature 1.5 hours from the first reading. 15.85°C

```
10  DIM L[50]
20  PRINT "NUMBER OF POINTS ON GRAPH IS";
30  INPUT N
40  DATA .5,1.6,1,1.8,1.5,2,2,2.4,2.5,2.6,3,2.8
50  FOR I=1 TO N
60  READ X,Y
70  LET A=A+X
80  LET B=B+Y
90  LET C=C+X*X
100 LET D=D+X*Y
110 NEXT I
120 LET Q=N*C—A*A
130 LET R=INT(100*(N*D—A*B)/Q+.5)/100
140 LET S=INT(100*(B*C—A*D)/Q+.5)/100
145 PRINT
150 PRINT "EQUATION OF BEST FITTING LINE IS"
155 PRINT "Y=";R;"X";
160 IF S>=0 THEN 190
170 PRINT S
180 STOP
190 PRINT " ";"+";S
200 END
```

```
RUN
NUMBER OF POINTS ON GRAPH IS?6
EQUATION OF BEST FITTING LINE IS
Y= .5X + 1.32
END
```

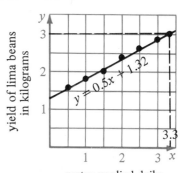

yield of lima beans in kilograms

$y = 0.5x + 1.32$

3.3

water applied daily in centimeters

11-5 Graphing a System of Equations

Objective To solve systems of equations using graphs.

The graphs of two linear equations

$$x - y = -3 \quad \text{and} \quad x + y = -1$$

are shown here. Their point of intersection appears to have coordinates $(-2,1)$. If this point lies on the graph of each of the equations, its coordinates must be a solution that both equations have in common. To check whether or not this is true, let us see if the ordered pair $(-2,1)$ satisfies both of the given equations. Since

$$-2 - 1 = -3 \quad \text{and} \quad -2 + 1 = -1$$

are both true, $(-2,1)$ is a solution of the given **system of equations.** Such a system is usually written as

$$x - y = -3$$
$$x + y = -1$$

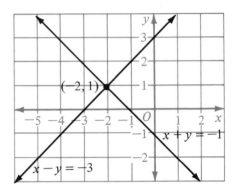

Example | Solve the system.

$$x + y = 6$$
$$x - y = 2$$

Solution | **Make tables of solutions and draw the graphs.**

$x + y = 6$			$x - y = 2$	
x	y		x	y
0	6		0	-2
6	0		2	0
3	3		6	4

Their intersection appears to be $(4,2)$. Check in both equations:

$$4 + 2 = 6 \quad \text{and} \quad 4 - 2 = 2.$$

Hence the solution is $(4,2)$.

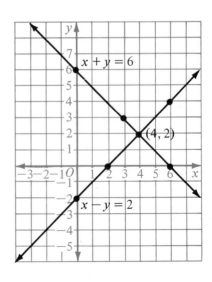

The graphs of the equations in the system

$$x - y = -3$$
$$x - y = -1$$

are parallel lines. Since they do not intersect, this system of equations has no solution.

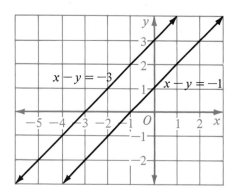

Class Exercises

Is the given ordered pair of numbers a solution of the given system of equations?

1. $(4, -1)$, $x + y = 3$ Yes
$x - y = 5$

2. $(5, -1)$, $x + y = 6$ No
$x - y = 4$

3. $(3,1)$, $x + 2y = 5$ Yes
$x - y = 2$

4. $(-4,4)$, $2x + y = 7$ No
$x - y = 5$

Exercises

Find the solution set of each system by graphing. Check your results. If no solution exists, so state.

A **1.** $x + y = 4$ $(3, 1)$
$x - y = 2$

2. $x + y = 6$ $(1, 5)$
$x - y = -4$

3. $2x + y = 5$ $(3, -1)$
$x - y = 4$

4. $-2x + y = 5$ $(-2, 1)$
$x + y = -1$

5. $3x + y = 6$ No solution
$3x + y = 3$

6. $2x - y = 4$ No solutic
$2x - y = 6$

7. $x + 2y = 6$ $(4, 1)$
$x - y = 3$

8. $x - 2y = 0$ $(2, 1)$
$2x + y = 5$

9. $3x + 2y = 12$ $(2, 3)$
$2x + y = 7$

B **10.** $2x + 3y = -10$ $(-2, -2)$
$2x + y = -6$

11. $4x + 3y = 7$ $(1, 1)$
$x - 2y = -1$

12. $2x + y = 4$ $(0, 4)$
$x - 2y = -8$

13. $3x + 2y = 7$ $(1, 2)$
$2x - y = 0$

14. $4x + 3y = 4$ $(1, 0)$
$x - 3y = 1$

15. $x - 3y = -5$ $(-2, 1)$
$x + 3y = 1$

16. $3x + 2y = -4$ $(0, -2)$
$2x + 3y = -6$

17. $2x + 3y = 4$ $(\frac{1}{2}, 1)$
$4x - 2y = 0$

18. $3x + 3y = 4$ $(1, \frac{1}{3})$
$x + 6y = 3$

A Surgeon

Daniel Hale Williams (1856–1931) was one of the pioneers in open-heart surgery. He received his medical degree from Chicago Medical College in 1883. After completing his internship at Mercy Hospital in Chicago, he was appointed surgeon to the South Side Dispensary. In 1891, he founded Provident Hospital in Chicago. Not long after, he performed a landmark operation in heart surgery. The patient reportedly was the victim of a knife wound in an artery lying a few centimeters from the heart. Dr. Williams repaired the wound with an innovative and delicate surgical procedure. Newspaper reports of the operation brought him instant fame.

Among his many achievements, Dr. Williams was chief surgeon of Freedman's Hospital in Washington, D.C., and organized the National Medical Association in Atlanta, Georgia. For twenty-five years he made annual visits to Meharry Medical College in Nashville, Tennessee, serving without salary as a visiting professor of surgery.

In 1931, Dr. Williams was inducted into the American College of Surgeons. He published nine scientific papers and was awarded honorary degrees by Wilberforce University and Howard University.

Research Activity Find a picture of the human heart in a biology book or encyclopedia. What are the main parts of the heart? What are their functions? How is blood pressure measured?

Career Activity Find out what different kinds of jobs are to be found in a hospital and what kinds of training they require.

Self-Test

Words to remember:
graph of the equation [p. 348]
linear equation in two variables [p. 348]
system of equations [p. 352]

State whether or not the given ordered pair is a solution of the equation.

1. $(3, -1)$, $2x - y = 7$ Yes

2. $(5, -3)$, $x - 2y = 11$ [11-3]
Yes

Give an equivalent equation for y in terms of x.

3. $2x - 3y = 6$
$y = \frac{2}{3}x - 2$

4. $-5x + 4y = -40$
$y = \frac{5}{4}x - 10$

Find ordered pairs of the form $(x,0)$ and $(0,y)$ that are solutions of the given equation.

5. $x + y = -7$
$(-7, 0)$ and $(0, -7)$

6. $-x + 2y = 11$ [11-4]
$(-11, 0)$ and $(0, \frac{11}{2})$

Graph each equation on a coordinate plane. Use a separate set of axes for each equation. Check students' graphs.

7. $3x - 4y = -12$

8. $y = -3$

Solve each system by graphing. If no solution exists, say so.

9. $x - y = -3$ $(-1, 2)$
$2x + y = 0$

10. $2x - 3y = 6$ $(-3, -4)$ [11-5]
$y = -4$

11. $2x - y = -2$ $(1, 4)$
$2x + y = 6$

12. $x + 2y = 4$ No solution
$x + 2y = 6$

Self-Test answers and Extra Practice are at the back of the book.

Chapter Review

Write the letter that labels the correct answer.

1. In a rectangular coordinate system, the axes intersect at the __?__. C **[11-1]**
   A. first component
   B. ordinate
   C. origin
   D. ordered pair

2. Each point in the plane is located by a(n) __?__. D
   A. pair of numbers
   B. number
   C. axis
   D. ordered pair of numbers

3. The graphs of $(-3,0)$, $(4,0)$, $(4,2)$, and $(1,2)$ are the vertexes of a __?__. B
   A. square
   B. trapezoid
   C. rectangle
   D. parallelogram

4. The triangle whose vertexes have coordinates $(-1,-3)$, $(-2,5)$ and $(6,5)$ has area equal to __?__. A
   A. 32
   B. 64
   C. 8
   D. 4

5. To move from the graph of $(-10,6)$ to the graph of $(-10,7)$ we move __?__. A
   A. up
   B. down
   C. left
   D. right

6. If (a,b) are the coordinates of a point in Quadrant I, then $(-a,-b)$ are the coordinates of a point in __?__. C **[11-2]**
   A. Quadrant I
   B. Quadrant II
   C. Quadrant III
   D. Quadrant IV

7. A point with negative abscissa and positive ordinate must lie in Quadrant __?__. B
   A. I
   B. II
   C. III
   D. IV

8. If points P and Q have equal ordinates they must lie __?__. A
   A. on the same horizontal line
   B. on the same vertical line
   C. in the same quadrant
   D. in different quadrants

9. The points for which the product of abscissa and ordinate is positive lie in __?__. B
   A. Quadrants I and II
   B. Quadrants I and III
   C. Quadrants I and IV
   D. Quadrant I only

10. A solution of the equation $2y = x + 1$ is __?__. A **[11-3]**
    A. $(-1,0)$
    B. $(0,1)$
    C. $(1,2)$
    D. $(2,1)$

11. If we solve for y in terms of x in the equation $x - 3y = 6$, we get __?__. D

   A. $x = 3y + 6$ B. $3y = x - 6$

   C. $y = \frac{1}{3}x - 6$ D. $y = \frac{1}{3}x - 2$

12. In the equation $2x - 7y = 4$, a solution with ordinate 6 is __?__. B

   A. $(6,23)$ B. $(23,6)$

   C. $\left(6,1\frac{1}{8}\right)$ D. $(-19,6)$

13. A solution of the equation $x - y = 3$ whose graph lies in Quadrant III is __?__. C

   A. $(7,4)$ B. $(-7,4)$
   C. $(-2,-5)$ D. $(-10,-7)$

14. The graph of $3x + 5y = 1$ does not pass through __?__. C   [11-4]

   A. Quadrant I B. Quadrant II
   C. Quadrant III D. Quadrant IV

15. The graph of $-3x + 8y = 12$ intersects the y-axis at the point whose coordinates are __?__. A

   A. $\left(0,1\frac{1}{2}\right)$ B. $(0,-4)$

   C. $(-4,0)$ D. $(0,12)$

16. The graph of $2x - y = 10$ intersects the x-axis at the point whose coordinates are __?__. D

   A. $(0,10)$ B. $(0,5)$ C. $(-10,0)$ D. $(5,0)$

17. The graph of $y = -2$ is __?__. C

   A. not a line B. a vertical line
   C. a horizontal line D. a line containing the origin

18. A solution of the system $\begin{array}{l} x + y = 1 \\ x - y = 1 \end{array}$ is __?__. D   [11-5]

   A. $(0,1)$ B. $(3,-2)$ C. $(0,-1)$ D. $(1,0)$

19. If a system of linear equations has no solution, then the graphs of the two equations __?__. C

   A. intersect at the origin B. lie in different quadrants
   C. are parallel D. are the same

Locating Points on Earth

Earth can be thought of as a sphere rotating about an *axis* that passes through two points at the ends of a diameter, the North Pole and the South Pole (N and S).

A **great circle** on a sphere is the intersection of the sphere with a plane passing through the center of the sphere. The great circle whose plane is perpendicular to the axis is called the **equator,** as shown in the figure at the right. The equator separates Earth into the Northern and Southern Hemispheres.

Any semicircle with endpoints at the poles is called a **meridian.** The meridian passing through the city of Greenwich, England, is the official **prime meridian.** Its intersection with the equator is labeled 0°.

Now we can set up a degree-coordinate system on Earth so that each point on the surface can be assigned an ordered pair of coordinates: (*longitude, latitude*).

The **longitude** is the number of degrees between 0° and 180° that any place is *east* or *west of the prime meridian.* The **latitude** is the number of degrees between 0° and 90° that the place is *north* or *south of the equator.* Thus, as indicated in the figure at the right, Vienna, Austria is about 20°E, 45°N.

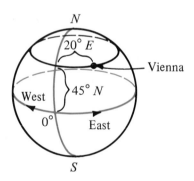

Globes or flat maps of Earth are often marked off in 10°-intervals of latitude (horizontal **parallels of latitude**) and longitude (vertical **meridians**).

A **Mercator Projection** is a flat map in which the meridians and parallels of latitude appear as lines crossing at right angles. Note in the Mercator Projection at the top of the next page that areas farther from the equator appear greater than they actually are.

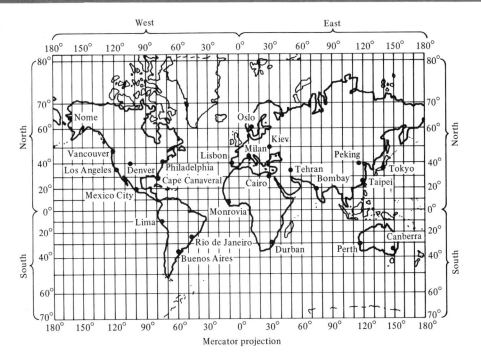

West   East

Mercator projection

From the map of the world shown above, find the cities near the following locations.

1. 100°W, 20°N
Mexico City

2. 30°E, 30°S
Durban

3. 120°W, 35°N
Los Angeles

4. 30°E, 30°N
Cairo

5. 75°W, 10°S
Lima

6. 30°E, 50°N
Kiev

7. 115°E, 30°S
Perth

8. 140°E, 35°N
Tokyo

9. 45°W, 25°S
Rio de Janeiro

10. 115°E, 40°N
Peking

11. 165°W, 65°N
Nome

12. 10°W, 5°N
Monrovia

From the map above, give the longitude and latitude of the following places to the nearest 5 degrees.

13. Oslo, Norway
10°E, 60°N

14. Cape Canaveral, Florida
80°W, 30°N

15. Peking, China
115°E, 40°N

16. Vancouver, British Columbia
120°W, 50°N

17. Buenos Aires, Argentina
60°W, 35°S

18. Philadelphia, Pennsylvania
75°W, 40°N

19. Bombay, India
75°E, 20°N

20. Lisbon, Portugal
10°W, 40°N

21. Denver, Colorado
105°W, 40°N

22. Milan, Italy
10°E, 45°N

23. Tehran, Iran
50°E, 35°N

24. Canberra, Australia
150°E, 35°S

Chapter Test

1. Each ordered pair of numbers locates a __?__ in the plane. point   **[11-1]**

2. The negative direction on the vertical axis is __?__. down

3. The vertexes of a triangle have coordinates $(-5, 2)$, $(-1, 7)$, and $(-1, -1)$. The area of this triangle equals __?__. 16

4. Every point in the plane is assigned a pair of __?__. coordinates   **[11-2]**

5. The x-coordinate of a point P is called the __?__ of P. abscissa

6. If $x > 0$ and $y < 0$ then (x,y) are the coordinates of a point in Quadrant __?__. IV

7. If a point has ordinate 0, it lies on the __?__. x-axis

8. Is $(4,3)$ a solution of the equation $y = x + 1$? No   **[11-3]**

9. A solution of the equation $3x - 5y = 14$ is $(-2,$ __?__$)$. -4

10. Does Quadrant I contain any points whose coordinates are solutions of the equation $x + y = -1$? No

11. A solution of the equation $4x - 3y = 12$ whose graph lies on the y-axis is __?__. $(0, -4)$

12. The equation $2x + 17y - 11 = 0$ is an example of a(n) __?__ equation in two variables. linear   **[11-4]**

13. The graph of an equation consists of all points whose coordinates are __?__ of the equation. solutions

14. Does the graph of $y = 5$ cross the x-axis? No

15. If the graph of $ax + by + c = 0$ passes through the origin, must $c = 0$? Yes

16. Is $(2,3)$ a solution of the following system? No   **[11-5]**
$$x + y = 5$$
$$x - y = 1$$

17. If the graphs of two linear equations are parallel lines, the system of equations has __?__ solution(s). No

18. A solution to a system of linear equations is found at the point of __?__ of the graphs of the equations. intersection

Skill Review

Convert the percent to a decimal.

1. 25% 0.25 **2.** 100% 1.00 **3.** 12.5% 0.125 **4.** 15% 0.15

5. 120% 1.20 **6.** 33% 0.33 **7.** 50% 0.50 **8.** 1550% 15.50

9. 200% 2.00 **10.** 11% 0.11 **11.** 47.32% 0.4732 **12.** 22.2% 0.222

Convert the decimal to a percent.

13. 0.15 15% **14.** 1.25 125% **15.** 1.61 161% **16.** 0.76 76%

17. 0.45 45% **18.** 0.782 78.2% **19.** 1.17 117% **20.** 0.21 21%

21. 10.15 1015% **22.** 0.133 13.3% **23.** 1.312 131.2% **24.** 5.5 550%

Convert the percent to a fraction or mixed number in lowest terms.

25. 25% $\frac{1}{4}$ **26.** 18% $\frac{9}{50}$ **27.** 80% $\frac{4}{5}$ **28.** 75% $\frac{3}{4}$

29. 12.5% $\frac{1}{8}$ **30.** 40% $\frac{2}{5}$ **31.** 115% $1\frac{3}{20}$ **32.** 16% $\frac{4}{25}$

33. 23% $\frac{23}{100}$ **34.** 71% $\frac{71}{100}$ **35.** 10% $\frac{1}{10}$ **36.** 212% $2\frac{3}{25}$

Convert to a decimal; then to a percent.

37. $1\frac{1}{4}$ 1.25; 125% **38.** $\frac{1}{2}$ 0.5; 50% **39.** $\frac{4}{5}$ 0.8; 80% **40.** $\frac{7}{10}$ 0.7; 70%

41. $2\frac{1}{8}$ 2.125; 212.5% **42.** $\frac{3}{4}$ 0.75; 75% **43.** $\frac{5}{2}$ 2.5; 250% **44.** $2\frac{3}{10}$ 2.3; 230%

45. $\frac{15}{16}$ 0.9375; 93.75% **46.** $\frac{7}{8}$ 0.875; 87.5% **47.** $1\frac{2}{5}$ 1.4; 140% **48.** $1\frac{1}{10}$ 1.1; 110%

49. $\frac{16}{5}$ 3.2; 230% **50.** $11\frac{1}{4}$ 11.25; 1125% **51.** $\frac{6}{5}$ 1.2; 120% **52.** $3\frac{5}{8}$ 3.625; 362.5%

Convert to a fraction or mixed number in lowest terms.

53. 0.12 $\frac{3}{25}$ **54.** 1.60 $1\frac{3}{5}$ **55.** 0.125 $\frac{1}{8}$ **56.** 0.8 $\frac{4}{5}$

57. 1.25 $1\frac{1}{4}$ **58.** 3.05 $3\frac{1}{20}$ **59.** 0.66 $\frac{33}{50}$ **60.** 0.45 $\frac{9}{20}$

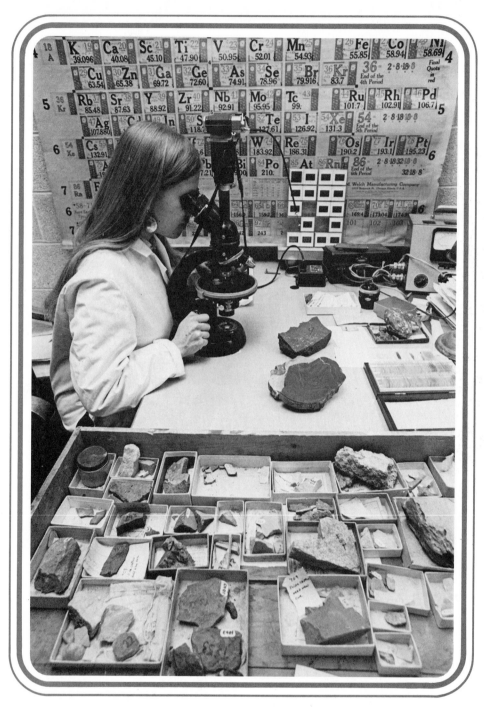

A geologist studying the structure of rock specimens.

12

Right Triangles

12-1 Square Roots

Objective To estimate the positive square root of a positive number by determining the two consecutive integers between which the square root lies.

Any nonnegative real number n can be expressed as the product of two equal factors, $a \times a$. Each factor a is called a **square root** of n. Thus, 2 is a square root of 4 because $2 \times 2 = 4$.

A positive number has two different square roots, which are opposites of each other. For example, the square roots of 4 are 2 and -2 because

$$2 \times 2 = 4 \quad \text{and also} \quad -2 \times -2 = 4.$$

The *positive* square root of a number n is denoted by the symbol $\sqrt{n}$, and the *negative* square root by the symbol $-\sqrt{n}$. Thus

$$\sqrt{4} = 2 \quad \text{and} \quad -\sqrt{4} = -2.$$

If $n = 0$, then $a \times a = 0$. The only value a can have is 0. Therefore,

$$\sqrt{0} = -\sqrt{0} = 0.$$

Suppose $n < 0$, say $n = -4$; then $a \times a = -4$. But there is no real number whose square is a negative number. Thus a negative number has no real-number square root.

In this chapter, we shall be concerned generally with the positive square roots of numbers.

If $\sqrt{n}$ is an integer, then n is called a **perfect square**. For example, $\sqrt{4}$ is the integer 2, so 4 is a perfect square. The positive square root of a number is often not an integer. We can estimate such a square root by finding the next smaller and the next larger integer between which the square root lies.

Example	Determine the two consecutive integers between which $\sqrt{52}$ lies.
Solution	Since $\sqrt{49} = 7$ and $\sqrt{64} = 8$, $$\sqrt{49} < \sqrt{52} < \sqrt{64}$$ $$7 < \sqrt{52} < \quad 8$$ Thus $\sqrt{52}$ lies between 7 and 8.

Class Exercises

Read each symbol.

1. $\sqrt{5}$ 2. $\sqrt{76}$ 3. $\sqrt{12}$ 4. $-\sqrt{10}$ 5. $-\sqrt{300}$

If the symbol names an integer, state the integer. If not, name the two consecutive integers between which the given number lies.

6. $\sqrt{9}$ 3 7. $\sqrt{30}$ 5 and 6 8. $\sqrt{36}$ 6 9. $\sqrt{85}$ 9 and 10 10. $\sqrt{11}$ 3 and 4

11. $\sqrt{1}$ 1 12. $\sqrt{81}$ 9 13. $\sqrt{23}$ 4 and 5 14. $-\sqrt{100}$ −10 15. $-\sqrt{0}$ 0

16. $\sqrt{55}$ 7 and 8 17. $\sqrt{6}$ 2 and 3 18. $\sqrt{74}$ 8 and 9 19. $\sqrt{4}$ 2 20. $\sqrt{47}$ 6 and 7

Exercises

If the symbol or expression names an integer, state the integer. If not, name the two consecutive integers between which the number lies.

A 1. $\sqrt{64}$ 8 2. $\sqrt{121}$ 11 3. $-\sqrt{400}$ −20 4. $-\sqrt{144}$ −12

5. $\sqrt{7^2}$ 7 6. $\sqrt{0^2}$ 0 7. $(\sqrt{3})^2$ 3 8. $(\sqrt{14})^2$ 14

9. $\sqrt{15}$ 3 and 4 10. $\sqrt{37}$ 6 and 7 11. $\sqrt{3}$ 1 and 2 12. $\sqrt{92}$ 9 and 10

13. $\sqrt{133}$ 11 and 12 14. $\sqrt{850}$ 29 and 30 15. $\sqrt{2600}$ 50 and 51 16. $\sqrt{146}$ 12 and 13

B 17. $\sqrt{25} - \sqrt{16}$ 1 18. $\sqrt{100} + \sqrt{49}$ 17 19. $\sqrt{144 + 25}$ 13

20. $\sqrt{25 - 16}$ 3 21. $\sqrt{45 + 45}$ 9 and 10 22. $\sqrt{13 + 68}$ 9

23. $\sqrt{200 - 164}$ 6 24. $\sqrt{79 - 61}$ 4 and 5 25. $\sqrt{49} + \sqrt{8}$ 9 and 10

26. $\sqrt{64} + \sqrt{2}$ 9 and 10 27. $\sqrt{121} - \sqrt{33}$ 5 and 6 28. $\sqrt{100} - \sqrt{10}$ 6 and 7

12-2 Approximating Square Roots

Objective To use the "divide and average" method to approximate square roots.

We can obtain a good approximation to the positive square root, a, of a number n by using this fact:

> If $\sqrt{n} = a$, then $n = a \times a$ and $\dfrac{n}{a} = a$.

That is, if we divide a number by its square root, the quotient is the same as the divisor. If we divide by an estimate for a that is *greater* than a, then the quotient will be *less* than a. The average of the divisor and quotient can be used as a new estimate for a.

Example | Approximate $\sqrt{11}$ to the tenths' place.

Solution | **Step 1. Estimate $\sqrt{11}$.**
$\sqrt{9} = 3$ and $\sqrt{16} = 4$. Thus $3 < \sqrt{11} < 4$. Since 11 is closer to 9 than to 16, you might try a number closer to 3 than to 4, say 3.2, as a first estimate of $\sqrt{11}$.

Step 2. Divide 11 by the estimate.
At each stage, the approximation is correct to the number of matching decimal places in the divisor and quotient. Here, $\sqrt{11} \approx 3$.

```
              3.43
    3.2 ) 11.0,00
           9 6
           1 4 0
           1 2 8
             1 20
               96
              240
```

Step 3. Find the average of the divisor and the quotient.

$$\frac{3.2 + 3.43}{2} = \frac{6.63}{2} \approx 3.32$$

Repeat Steps 1–2.
Step 1. $\sqrt{11} \approx 3.32$
Step 2. $11 \div 3.32 \approx 3.31$
Now the quotient and divisor agree to the tenths' place and $\sqrt{11} \approx 3.3$.

```
                3.31
    3.32 ) 11.00,00
            9 96
            1 04 0
              99 6
              4 40
              3 32
              1 08
```

Class Exercises

State the first digit of the indicated square root.

1. $\sqrt{14}$ 3
2. $\sqrt{40}$ 6
3. $\sqrt{68}$ 8
4. $\sqrt{22}$ 4
5. $\sqrt{89}$ 9

6. $\sqrt{3}$ 1
7. $\sqrt{63}$ 7
8. $\sqrt{300}$ 17
9. $\sqrt{905}$ 30
10. $\sqrt{721}$ 26

State the next estimate for the square root of the dividend.

Example

$$\begin{array}{r} 2.8 \\ 2\overline{)5.6} \end{array}$$

Solution The average of 2 and 2.8 is $\dfrac{2 + 2.8}{2} = 2.4$. Therefore, the next estimate is 2.4.

11. $\begin{array}{r} 4 \\ 3\overline{)12} \end{array}$ 3.5

12. $\begin{array}{r} 6.6 \\ 6\overline{)39.7} \end{array}$ 6.3

13. $\begin{array}{r} 4.18 \\ 4.3\overline{)18.0,00} \end{array}$ 4.24

14. $\begin{array}{r} 8.4 \\ 8.5\overline{)71.4,0} \end{array}$ 8.45

15. $\begin{array}{r} 16.0 \\ 14\overline{)225.0} \end{array}$ 15.0

16. $\begin{array}{r} 12.8 \\ 13\overline{)167.0} \end{array}$ 12.9

17. $\begin{array}{r} 20.3 \\ 20\overline{)406.0} \end{array}$ 20.2

18. $\begin{array}{r} 9.5 \\ 10\overline{)95.0} \end{array}$ 9.75

19. $\begin{array}{r} 5.38 \\ 5.2\overline{)28.00} \end{array}$ 5.29

20. $\begin{array}{r} 7.12 \\ 7.3\overline{)52.00} \end{array}$ 7.21

21. $\begin{array}{r} 0.50 \\ 0.6\overline{)0.30} \end{array}$ 0.55

22. $\begin{array}{r} 0.92 \\ 0.9\overline{)0.83} \end{array}$ 0.91

Exercises

Find each square root to the tenths' place.

A

1. $\sqrt{8}$ 2.8
2. $\sqrt{53}$ 7.2
3. $\sqrt{33}$ 5.7
4. $\sqrt{5}$ 2.2

5. $\sqrt{71}$ 8.4
6. $\sqrt{13}$ 3.6
7. $\sqrt{26}$ 5.1
8. $\sqrt{84}$ 9.1

9. $\sqrt{42}$ 6.4
10. $\sqrt{97}$ 9.8
11. $\sqrt{63}$ 7.9
12. $\sqrt{29}$ 5.3

13. $\sqrt{145}$ 12.0
14. $\sqrt{190}$ 13.7
15. $\sqrt{424}$ 20.5
16. $\sqrt{254}$ 15.9

B

17. $\sqrt{3.2}$ 1.7
18. $\sqrt{15.9}$ 3.9
19. $\sqrt{21.5}$ 4.6
20. $\sqrt{72.8}$ 8.5

21. $\sqrt{387.1}$ 19.6
22. $\sqrt{106.6}$ 10.3
23. $\sqrt{790.4}$ 28.1
24. $\sqrt{213.7}$ 14.6

25. $\sqrt{1058}$ 32.5
26. $\sqrt{4723}$ 68.7
27. $\sqrt{9999}$ 99.9
28. $\sqrt{5476}$ 74

29. $\sqrt{0.8}$ 0.8
30. $\sqrt{0.9}$ 0.9
31. $\sqrt{0.05}$ 0.2
32. $\sqrt{0.21}$ 0.4

12-3 Using a Square-Root Table

Objective To use a table to approximate square roots.

We can use the table on page 394 to approximate the square roots of integers from 1 to 100. For example, to find $\sqrt{63}$ using the table, we locate 63 in the column headed "Number." We read off the value beside 63 in the "Square Root" column: 7.937.

$$\sqrt{63} \approx 7.937$$

To find an approximation for the square root of a number which lies between two entries in the table, we can use a process called **interpolation**. For example, to approximate $\sqrt{5.5}$, we assume that $\sqrt{5.5}$ is 0.5 of the way from $\sqrt{5}$ to $\sqrt{6}$.

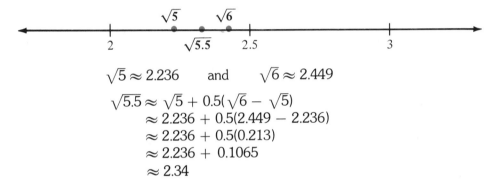

$$\sqrt{5} \approx 2.236 \quad \text{and} \quad \sqrt{6} \approx 2.449$$

$$\begin{aligned}
\sqrt{5.5} &\approx \sqrt{5} + 0.5(\sqrt{6} - \sqrt{5}) \\
&\approx 2.236 + 0.5(2.449 - 2.236) \\
&\approx 2.236 + 0.5(0.213) \\
&\approx 2.236 + 0.1065 \\
&\approx 2.34
\end{aligned}$$

Class Exercises

Use the table on page 394 to approximate each number.

1. $\sqrt{10}$ 3.162
2. $\sqrt{13}$ 3.606
3. $\sqrt{29}$ 5.385
4. $\sqrt{36}$ 6.000

5. $\sqrt{44}$ 6.633
6. $\sqrt{58}$ 7.616
7. $\sqrt{71}$ 8.426
8. $\sqrt{95}$ 9.747

9. $10\sqrt{2}$ 14.14
10. $10\sqrt{83}$ 91.10
11. $10\sqrt{64}$ 80.00
12. $10\sqrt{17}$ 41.23

13. $\frac{1}{10}\sqrt{55}$ 0.7416
14. $\frac{1}{10}\sqrt{34}$ 0.5831
15. $\frac{1}{10}\sqrt{79}$ 0.8888
16. $\frac{1}{10}\sqrt{49}$ 0.7000

State the two values between which the given square root lies.

17. $\sqrt{45.8}$
6.708,
6.782

18. $\sqrt{7.1}$
2.646,
2.828

19. $\sqrt{70.6}$
8.367,
8.426

20. $\sqrt{19.5}$
4.359,
4.472

21. $\sqrt{25.3}$
5.000,
5.099

22. $\sqrt{68.2}$
8.246,
8.307

Exercises

Use the table on page 394 to approximate each number to the nearest hundredth.

Example | $6\sqrt{48}$

Solution | From the square-root table, $\sqrt{48} \approx 6.928$.
Thus $6\sqrt{48} \approx 6(6.928) = 41.568$.
To the nearest hundredth, $6\sqrt{48} = 41.57$.

A 1. $\sqrt{39}$ 6.25

2. $\sqrt{43}$ 6.56

3. $\sqrt{60}$ 7.75

4. $\sqrt{18}$ 4.24

5. $10\sqrt{63}$ 79.37

6. $10\sqrt{77}$ 87.75

7. $\frac{1}{10}\sqrt{82}$ 0.91

8. $\frac{1}{10}\sqrt{90}$ 0.95

9. $7\sqrt{6}$ 17.14

10. $8\sqrt{88}$ 75.05

11. $3\sqrt{12}$ 10.39

12. $2\sqrt{31}$ 11.14

B 13. $\sqrt{53} + \sqrt{21}$ 11.86

14. $\sqrt{27} + \sqrt{81}$ 14.20

15. $\sqrt{72} - \sqrt{25}$ 3.49

16. $\sqrt{94} - \sqrt{48}$ 2.77

17. $2\sqrt{54} - \sqrt{5}$ 12.46

18. $\sqrt{66} - 4\sqrt{3}$ 1.20

Use interpolation to approximate each number to the nearest hundredth.

19. $\sqrt{18.3}$ 4.28

20. $\sqrt{97.5}$ 9.87

21. $\sqrt{47.1}$ 6.86

22. $\sqrt{61.8}$ 7.86

23. $\sqrt{32.4}$ 5.69

24. $\sqrt{8.7}$ 2.95

25. $\sqrt{92.6}$ 9.62

26. $\sqrt{50.9}$ 7.13

27. $2\sqrt{59.2}$ 15.39

28. $7\sqrt{23.4}$ 33.86

29. $5\sqrt{74.8}$ 43.24

30. $3\sqrt{86.1}$ 27.84

C 31. $\sqrt{340}$ (Hint: $\sqrt{340} = \sqrt{100 \times 3.4} = 10\sqrt{3.4}$) 18.39

32. $\sqrt{1120}$ 33.46

33. $\sqrt{0.77}$ $\left(\text{Hint: } \sqrt{0.77} = \sqrt{\frac{1}{100} \times 77} = \frac{1}{10}\sqrt{77}\right)$ 0.88

34. $\sqrt{0.051}$ 0.26

35. A square room has an area of 37.5 m². Find the length of a side of the room to the nearest 0.1 m. 6.1 m

36. The height of a triangle is the same as its base. If the area of the triangle is 45 cm², what is the length of the base to the nearest 0.1 cm? 9.5 cm

37. The height of a parallelogram is half the length of its base. The parallelogram has an area of 35 cm². Find the height to the nearest 0.1 cm. 4.2 cm

38. One base of a trapezoid is three times as long as the other base. The height of the trapezoid is the same as the shorter base. The trapezoid has an area of 72 cm². Find the height to the nearest 0.1 cm. 6 cm

EXTRA! More About Square Roots

Suppose we want to approximate $\sqrt{148}$. The table on page 394 does not list this square root. However, by using the following property we can approximate $\sqrt{148}$.

> ## property
>
> The positive square root of the product of two positive numbers is equal to the product of their positive square roots. That is, if $x > 0$ and $y > 0$, then
>
> $$\sqrt{xy} = \sqrt{x} \times \sqrt{y}$$

Thus, since

$$148 = 4 \times 37$$

we can write

$$\sqrt{148} = \sqrt{4 \times 37}$$
$$= \sqrt{4} \times \sqrt{37}$$
$$= 2\sqrt{37}$$

From the table, $\sqrt{37} \approx 6.083$. Thus:

$$\sqrt{148} = 2\sqrt{37} \approx 2(6.083)$$
$$\sqrt{148} \approx 12.166$$

Use the table on page 394 to approximate each square root to the nearest hundredth.

1. $\sqrt{1025}$ 32.02
2. $\sqrt{960}$ 30.98
3. $\sqrt{252}$ 15.88
4. $\sqrt{567}$ 23.81

5. $\sqrt{1078}$ 32.83
6. $\sqrt{468}$ 21.64
7. $\sqrt{1216}$ 34.87
8. $\sqrt{726}$ 26.94

9. $\sqrt{2900}$ 53.85
10. $\sqrt{270}$ 16.43
11. $\sqrt{1184}$ 34.41
12. $\sqrt{1836}$ 42.85

13. $\sqrt{185}$ 13.60
14. $\sqrt{341}$ 18.47
15. $\sqrt{161}$ 12.69
16. $\sqrt{249}$ 15.78

Calculator Activity Many calculators have a "square-root key" which enables you to find the square root of a number directly. If such a calculator is available, use it to check your answers to Exercises 1–16, above.

EXTRA! Square Root Symbol

When early mathematicians wrote about the square roots of numbers, they often used the term *radix,* which is the Latin word for root. The first widely used square root symbol, shown at the right, came from rx., an abbreviation of the word radix. The symbol $\sqrt{}$, which we use today, first appeared in print in 1525 in the book *Die Coss,* by Christoff Rudolff. The symbol may have come from a German form of the letter r. By the seventeenth century, $\sqrt{}$ became the standard symbol for square root.

Self-Test

Symbols and words to remember:

$\sqrt{n}$ [p. 363] $-\sqrt{n}$ [p. 363]

square root [p. 363]   positive square root [p. 363]

negative square root [p. 363]   interpolation [p. 367]

Name the two consecutive integers between which the square root lies.

1. $\sqrt{75}$ 8 and 9   2. $\sqrt{3}$ 1 and 2   [12-1]

3. $\sqrt{30}$ 5 and 6   4. $\sqrt{59}$ 7 and 8

State the next estimate for the square root of the dividend.

       3.5   7.8   8.2

5. $4\overline{)14.0}$ 3.75   6. $7\overline{)55.0}$ 7.4   8.6 7. $9\overline{)74.0}$ [12-2]

8. Find $\sqrt{68}$ to the tenths' place. (Do not use the square-root table.) 8.2

Use the table on page 394 to approximate each number to the nearest hundredth.

9. $\sqrt{84}$ 9.17   10. $3\sqrt{7}$ 7.94   [12-3]

11. $\sqrt{62} - \sqrt{60}$ 0.13   12. $\sqrt{24.6}$ 4.96

Self-Test answers and Extra Practice are at the back of the book.

12-4 The Pythagorean Property

Objective To apply the Pythagorean Property.

In a right triangle, the side opposite the right angle is called the **hypotenuse.** It is always the longest of the three sides. The other two sides are called the **legs.**

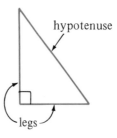

The following important property of right triangles was known to the Babylonians some 4000 years ago. However, it is believed that the first general proof of this property was given by the Greek mathematician, Pythagoras, about 2500 years ago. For this reason, it is usually named for him.

> ## property
>
> **For any right triangle, the square of the length of the hypotenuse is equal to the sum of the squares of the lengths of the other two sides. (The Pythagorean Property)**

For the right triangle at the right, this property states that

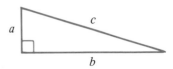

$$c^2 = a^2 + b^2.$$

You can verify the Pythagorean Property for the right triangle in the diagram below by counting the small unit squares within each of the large squares. The total number of unit squares in the squares on the legs $(9 + 16)$ is equal to the number in the square on the hypotenuse (25).

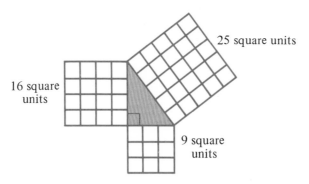

The following property is also true:

> ## property
>
> **If the square of the length of one side of a triangle is equal to the sum of the squares of the lengths of the other two sides, then the triangle is a right triangle.**

For the triangle at the right:
If $c^2 = a^2 + b^2$, then the triangle is a right triangle.

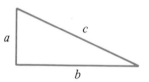

Class Exercises

Verify the Pythagorean Property by counting the small squares in the diagrams.

1.

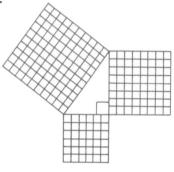

$$100 = 36 + 64$$

2.

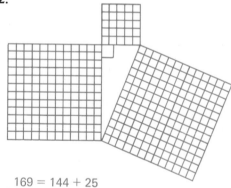

$$169 = 144 + 25$$

State whether or not a triangle with sides of the given lengths is a right triangle.

Example | **2, 3, 4**

Solution | $2^2 = 4$, $3^2 = 9$, and $4^2 = 16$.

Since $4 + 9 \neq 16$, **2, 3, 4** cannot be the lengths of the sides of a right triangle.

3. 3, 4, 5 Yes

4. 4, 5, 6 No

5. 8, 15, 17 Yes

6. 8, 10, 12 No

7. 5, 12, 13 Yes

8. 6, 8, 10 Yes

9. $1\frac{1}{2}$, 2, $2\frac{1}{2}$ Yes

10. 4, 5.5, 7 No

Exercises

State whether or not a triangle with sides of the given lengths is a right triangle.

A **1.** 3 m, 5 m, 7 m No

2. 10 m, 30 m, 32 m No

3. 9 cm, 12 cm, 15 cm Yes

4. 10 cm, 24 cm, 26 cm Yes

5. 20 mm, 21 mm, 29 mm Yes

6. 7 km, 11 km, 13 km No

7. 10 cm, 14 cm, 16 cm No

8. 11 mm, 20 mm, 21 mm No

The lengths of the legs of a right triangle are given. Find the area of the square on the hypotenuse of the triangle.

Example 1	**1 m, 3 m**

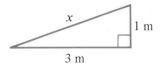

Solution **Make a sketch. By the Pythagorean Property:**

$$x^2 = 1^2 + 3^2 = 1 + 9 = 10$$

Thus the area of the square on the hypotenuse is 10 m².

9. 2 km, 2 km 8 km²

10. 2 m, 5 m 29 m²

11. 21 m, 30 m 1341 m²

12. 15 cm, 5 cm 250 cm²

13. 5.1 mm, 3.2 mm 36.25 mm²

14. 1.2 cm, 2.1 cm 5.85 cm²

A right triangle has sides of lengths *a*, *b*, and *c*, with *c* the length of the hypotenuse. Find the length of the missing side of the triangle to the nearest hundredth.

Example 2	**a. $a = 1, b = 5$**	**b. $a = 6, c = 8$**

Solution

a. $a^2 + b^2 = c^2$
 $1^2 + 5^2 = c^2$
 $1 + 25 = c^2$
 $26 = c^2$
 $c = \sqrt{26} \approx 5.099$
To the nearest hundredth,
$c = 5.10$

b. $a^2 + b^2 = c^2$
 $6^2 + b^2 = 8^2$
 $36 + b^2 = 64$
 $b^2 = 64 - 36 = 28$
 $b = \sqrt{28} \approx 5.292$
To the nearest hundredth,
$b = 5.29$

B **15.** $a = 2, b = 3$ $c \approx 3.61$

16. $a = 4, b = 7$ $c \approx 8.06$

17. $a = 5, c = 8$ $b \approx 6.25$

18. $a = 3, c = 6$ $b \approx 5.20$

19. $b = 2, c = 9$ $a \approx 8.78$

20. $b = 4, c = 7$ $a \approx 5.75$

C **21.** $a = 0.8, b = 2.6$ $c \approx 2.72$

22. $a = 5.3, b = 6.1$ $c \approx 8.08$

23. $a = 3.4, c = 6.4$ $b \approx 5.42$

24. $b = 2.9, c = 3$ (See Exercise 33, page 368) $a \approx 0.77$

Problems

Solve. Give your answer to the nearest tenth.

B   1. A rectangle is 6 cm long and 4 cm wide. How long is a diagonal of the rectangle? 7.2 cm

2. A square has sides that are 7 cm long. How long is a diagonal of the square? 9.9 cm

3. Suzanne and Barry hiked 8 km east and then 5 km north. How far were they then from their starting point? 9.4 km

4. A right triangle has two sides of lengths 2 and 3. Find two possible lengths for the third side. 3.6 or 2.2

C   5. Find the height of an isosceles triangle with two sides of length 37 cm and the base of length 24 cm. (*Hint:* The height is the length of the line segment that bisects the base and is perpendicular to it.) 35 cm

EXTRA! Pythagorean Triples

The Pythagorean Property relates the lengths, a and b, of the two legs of a right triangle with the length c of the hypotenuse by the equation: $a^2 + b^2 = c^2$.

Any three *counting numbers* (a, b, c) that satisfy this equation are called a **Pythagorean Triple.** For example, 3, 4, and 5 form such a triple because $3^2 + 4^2 = 5^2$.

Example	Show that (6, 8, 10) is a Pythagorean Triple.
Solution	$6^2 + 8^2 = 36 + 64 = 100$
	$10^2 = 100$
	Since $6^2 + 8^2 = 10^2$, (6, 8, 10) is a Pythagorean Triple.

Notice that $6 = 3 \times 2$, $8 = 4 \times 2$, and $10 = 5 \times 2$. In general, if (a, b, c) is a Pythagorean Triple and n is a counting number, then (an, bn, cn) is also a Pythagorean Triple

Find the missing number in the Pythagorean Triple (a, b, c).

1. 5, 12, __?__ 13	**2.** 20, 48, __?__ 52	**3.** 15, __?__, 17 8
4. 30, __?__, 34 16	**5.** __?__, 84, 85 13	**6.** 20, 21, __?__ 29
7. 60, 63, __?__ 87	**8.** __?__, 24, 25 7	**9.** __?__, 48, 50 14
10. 9, 40, __?__ 41	**11.** 45, 200, __?__ 205	**12.** 11, __?__, 61 60

12-5 Special Right Triangles

Objective To find the lengths of the other two sides of an isosceles right triangle, and of a 30°–60° right triangle, given the length of one side.

In an **isosceles right** triangle, the two legs are equal in length. Also, the angles opposite the legs are equal in measure. Since

$$\angle A + \angle B + \angle C = 180°$$

and

$$\angle C = 90°,$$

$$\angle A + \angle B = 90°.$$

Since $\angle A = \angle B$, each has the measure 45°.

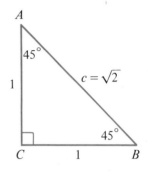

In the diagram, each leg is 1 unit long. From the Pythagorean Property:

$$c^2 = 1^2 + 1^2$$
$$c^2 = 2$$
$$c = \sqrt{2}$$

This suggests the following property:

property

If each leg of an isosceles right triangle has length a, then the hypotenuse has length $a\sqrt{2}$.

Example 1 | A square has sides of length 5 cm. Find the length of a diagonal of the square.

Solution | The shaded right triangle shown has legs of length 5 cm. Thus, the hypotenuse d has length $5\sqrt{2}$ cm.

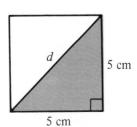

A **30°–60° right triangle,** such as $\triangle ABC$ at the right, can be thought of as half an equilateral triangle. Suppose $\overline{BC}$ is 1 unit long. Since $DB = AB$, $\overline{AB}$ is 2 units long. From the Pythagorean Property:

$$(AC)^2 + (BC)^2 = (AB)^2$$
$$(AC)^2 + 1^2 = 2^2$$
$$(AC)^2 = 3$$
$$AC = \sqrt{3} \approx 1.732$$

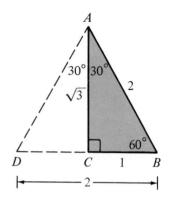

Notice that the larger the angle is, the longer its opposite side is. Thus in this 30°–60°–90° triangle, the sides have lengths 1, 1.7, and 2, respectively.

---

property

If the shorter leg of a **30°–60° right triangle** has length a, then the other leg has length $a\sqrt{3}$, and the hypotenuse has length **2a**.

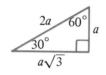

Example 2 | Find the values of x and y.

Solution | Since the hypotenuse is 8 cm long:

$$x = \frac{1}{2} \times 8 \text{ cm} = 4 \text{ cm} \quad \text{and}$$

$$y = 4\sqrt{3} \text{ cm}$$

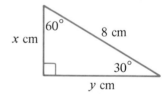

Class Exercises

State the values of x and y in each triangle. Explain how you determined the values.

1.

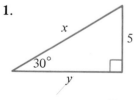

$x = 10; y = 5\sqrt{3}$

2.

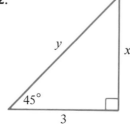

$x = 3; y = 3\sqrt{2}$

3.

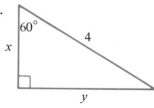

$x = 2; y = 2\sqrt{3}$

4.

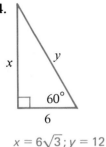

$x = 6\sqrt{3}; y = 12$

5.

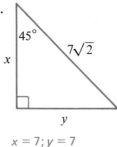

$x = 7; y = 7$

6.

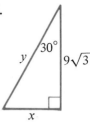

$x = 9; y = 18$

Exercises

Use the table on page 394 to approximate the value of x to the nearest hundredth.

A

1.

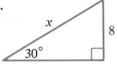

$x = 16$

2.

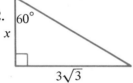

$x = 3$

3.

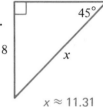

$x \approx 11.31$

4.

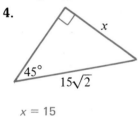

$x = 15$

5.

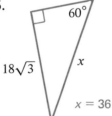

$x = 36$

6.

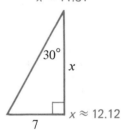

$x \approx 12.12$

7.

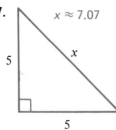

$x \approx 7.07$

$x = 5$

8.

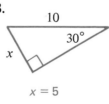

$x = 5$

9.

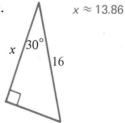

$x \approx 13.86$

10.

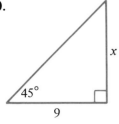

$x = 9$

11.

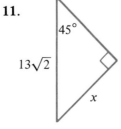

$x = 13$

12.

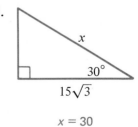

$x = 30$

Use the table on page 394 to approximate the values of x and y to the nearest tenth.

B **13.**

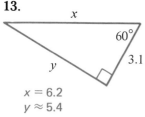

$x = 6.2$
$y \approx 5.4$

14.

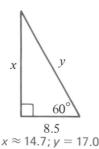

8.5
$x \approx 14.7; y = 17.0$

15.

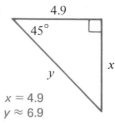

$x = 4.9$
$y \approx 6.9$

16.

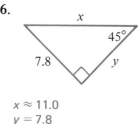

$x \approx 11.0$
$y = 7.8$

17.

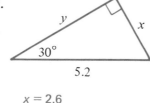

5.2
$x = 2.6$
$y \approx 4.5$

18.

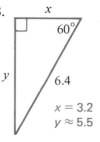

$x = 3.2$
$y \approx 5.5$

Problems

Solve.

A 1. How long is the hypotenuse of an isosceles right triangle which has legs 25 mm long? 35.35 mm

2. On a baseball field, the distance from home plate to first base is about 27 m. If the four bases are the vertexes of a square, what is the distance from second base to home plate? 38.178 m

B 3. A checkerboard has 8 squares on each side. If one side of each square is 5 cm long, how far is it from one corner of the board to the opposite corner? 56.56 cm

4. Find the length of an altitude of an equilateral triangle with sides of length 12 cm. 10.392 cm

C 5. The area of a square cake pan is 900 cm². What is the length of a diagonal of the pan? 42.42 cm

6. A 10 m pole is supported in a vertical position by three 6 m guy wires. If one end of each wire is fastened to the ground at a 60° angle, how high on the pole is the other end fastened? 5.196 m

7. A rhombus has angles of 60° and 120°. Each side of the rhombus is 8 cm long. What are the lengths of the diagonals? (*Hint:* The diagonals bisect the angles of the rhombus.) 8 cm and 13.856 cm

12-6 Trigonometric Ratios

Objective To calculate the sine, cosine, and tangent of an acute angle of a right triangle, given the lengths of two sides.

The left-hand diagram below shows three similar right triangles with an acute angle in common, $\angle A$. Since they are all $30°$–$60°$ right triangles, in each triangle the ratio of the length of the leg opposite $\angle A$ to the length of the hypotenuse is

$$\frac{a}{2a}, \quad \text{or} \quad \frac{1}{2}.$$

This ratio is called the **sine of $\angle A$,** usually abbreviated as **sin A.**

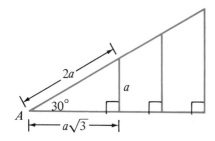

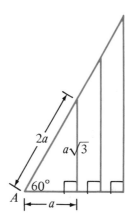

For each of the three similar right triangles in the right-hand diagram, where $\angle A = 60°$, you can see that

$$\sin A = \frac{a\sqrt{3}}{2a} = \frac{\sqrt{3}}{2}.$$

Thus you see that this **trigonometric ratio** depends on the measure of the angle A, but not on the size of the triangle. For a given length of hypotenuse, the greater the measure of $\angle A$, the greater the length of the leg opposite it. You can see from the diagram at the right that as the measure of $\angle A$ approaches $90°$, the leg opposite $\angle A$ approaches the same length as the hypotenuse. Thus, the sine ratio approaches $\frac{c}{c}$, or 1. As the measure of $\angle A$ approaches 0,

$\sin A$ approaches $\frac{0}{c}$, or 0.

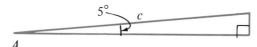

Two other important trigonometric ratios that depend on the measure of an acute angle in a right triangle are the **cosine of $\angle A$,** written **cos A,** and the **tangent of $\angle A$,** written **tan A.**

Refer to the diagram at the right, in which the lengths of the opposite side, adjacent side, and hypotenuse are denoted by the abbreviations "opp.," "adj.," and hyp." Then:

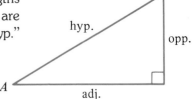

$$\sin A = \frac{\text{length of side opposite } \angle A}{\text{length of hypotenuse}} = \frac{\text{opp.}}{\text{hyp.}}$$

$$\cos A = \frac{\text{length of side adjacent to } \angle A}{\text{length of hypotenuse}} = \frac{\text{adj.}}{\text{hyp.}}$$

$$\tan A = \frac{\text{length of side opposite } \angle A}{\text{length of side adjacent to } \angle A} = \frac{\text{opp.}}{\text{adj.}}$$

Example | In the triangle at the right, give the value of:

a. sin A   b. cos A   c. tan B

Solution | **Find the value of x.**

$$4^2 + x^2 = 7^2$$
$$x^2 = 49 - 16 = 33$$
$$x = \sqrt{33}$$

a. $\sin A = \dfrac{\text{opp.}}{\text{hyp.}} = \dfrac{\sqrt{33}}{7}$ **b.** $\cos A = \dfrac{\text{adj.}}{\text{hyp.}} = \dfrac{4}{7}$

c. $\tan B = \dfrac{\text{opp.}}{\text{adj.}} = \dfrac{4}{\sqrt{33}}$

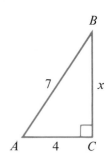

Class Exercises

Use the diagram at the right to state the value of each ratio.

1. $\sin A \quad \frac{3}{5}$ **2.** $\cos A \quad \frac{4}{5}$

3. $\tan A \quad \frac{3}{4}$ **4.** $\sin B \quad \frac{4}{5}$

5. $\cos B \quad \frac{3}{5}$ **6.** $\tan B \quad \frac{4}{3}$

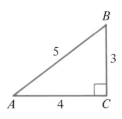

Use the 30°–60° right triangle shown to state the value of each ratio. Square-root signs are to be retained.

7. $\sin 60° \quad \frac{\sqrt{3}}{2}$ **8.** $\cos 30° \quad \frac{\sqrt{3}}{2}$

9. $\cos 60° \quad \frac{1}{2}$ **10.** $\sin 30° \quad \frac{1}{2}$

11. $\tan 30° \quad \frac{1}{\sqrt{3}}$ **12.** $\tan 60° \quad \sqrt{3}$

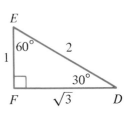

Exercises

In each exercise, write the value of sin A, cos A, tan A, sin B, cos B, and tan B. Give each ratio in lowest terms. Square-root signs are to be retained.

A **1.** $\frac{5}{13}, \frac{12}{13}, \frac{5}{12}, \frac{12}{13}, \frac{5}{13}, \frac{12}{5}$

2.

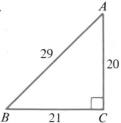

3.

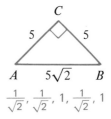

$\frac{1}{\sqrt{2}}, \frac{1}{\sqrt{2}}, 1, \frac{1}{\sqrt{2}}, 1$

4.

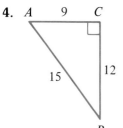

5.

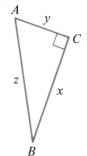

6. $\frac{r}{t}, \frac{s}{t}, \frac{r}{s}, \frac{s}{t}, \frac{r}{t}, \frac{s}{r}$

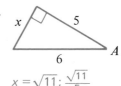

2. $\frac{21}{29}, \frac{20}{29}; \frac{21}{20}, \frac{20}{29}, \frac{21}{29}, \frac{20}{21}$ **4.** $\frac{4}{5}, \frac{3}{5}, \frac{4}{3}, \frac{3}{5}, \frac{4}{5}, \frac{3}{4}$ **5.** $\frac{x}{z}, \frac{y}{z}, \frac{x}{y}, \frac{y}{z}, \frac{x}{z}, \frac{y}{x}$

Find the value of x. Then find tan A.

B **7.**

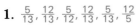

$x = 8; \frac{8}{15}$

8.

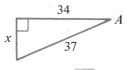

$x = \sqrt{213}; \frac{\sqrt{213}}{37}$

9.

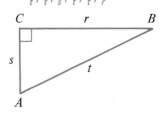

$x = \sqrt{11}; \frac{\sqrt{11}}{5}$

Find the value of x. Then find sin A.

10.

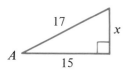

$x = \sqrt{29}; \frac{5}{\sqrt{29}}$

11.

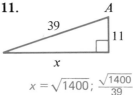

$x = \sqrt{1400}; \frac{\sqrt{1400}}{39}$

12.

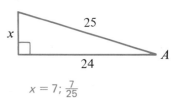

$x = 7; \frac{7}{25}$

Find the value of x. Then find cos A.

13.

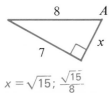

$x = \sqrt{15}; \frac{\sqrt{15}}{8}$

14.

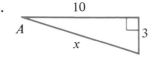

$x = \sqrt{109}; \frac{10}{\sqrt{109}}$

15.

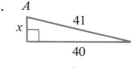

$x = 9; \frac{9}{41}$

12-7 Using a Trigonometric Table

Objective To solve a right triangle by using a trigonometric table.

The table on page 395 gives the *approximate* values of the sine, cosine, and tangent of angles with measure $1°, 2°, 3°, \ldots, 90°$. For example, to find $\sin 30°$, look down the "$\angle A$" column to $30°$. To the right of it in the "$\sin A$" column, you see the value 0.5000. This agrees with the value $\frac{1}{2}$ that we determined for $\sin 30°$ on page 379. Do you see that

$$\tan 67° \approx 2.3559?$$

You can use the trigonometric table to **solve right triangles,** that is, to find approximations for the measures of the sides and angles.

Example 1 Solve $\triangle ABC$ by finding:
  a. the value of x **b.** $\angle A$ **c.** $\angle B$

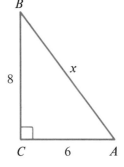

Solution **a.** $x^2 = 6^2 + 8^2 = 36 + 64 = 100$
  $x = \sqrt{100} = 10$
  b. $\tan A = \frac{8}{6} \approx 1.3333$. From the tangent column in the table, you find that the angle whose tangent is closest to 1.3333 has measure $53°$. Thus $\angle A \approx 53°$.
  c. $\angle A + \angle B + \angle C = 180°$
  $53° + \angle B + 90° = 180°$
  $\angle B = 180° - 143° = 37°$
  Thus $\angle B \approx 37°$.

Example 2 Solve $\triangle ABC$ by finding:
  a. $\angle B$; **b.** the value of x;
  c. the value of y.

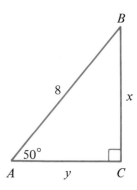

Solution **a.** $\angle A + \angle B + \angle C = 180°$
  $50° + \angle B + 90° = 180°$
  $\angle B = 180° - 140° = 40°$
  b. Use one of the following equations:
  $$\sin 50° = \frac{x}{8} \qquad \cos 40° = \frac{x}{8}$$
  $$0.7660 \approx \frac{x}{8}$$
  $x \approx 8 \times 0.7660 \approx 6.128$
  To the nearest whole number, $x = 6$.

c. *Method 1*

$\sin 40° = \frac{y}{8}$

$0.6428 \approx \frac{y}{8}$

$y \approx 8 \times 0.6428$

$y \approx 5.1$

Method 2

$6^2 + y^2 \approx 8^2$

$y^2 \approx 64 - 36$

$y^2 \approx 28$

$y \approx \sqrt{28} \approx 5.3$

To the nearest whole number, $y = 5$.

Class Exercises

Use the table on page 395 to state an approximation for each ratio.

1. $\sin 23°$ 0.3907 **2.** $\cos 81°$ 0.1564 **3.** $\tan 57°$ 1.5399 **4.** $\sin 35°$ 0.5736 **5.** $\cos 76°$ 0.2419

6. $\tan 19°$ 0.3443 **7.** $\sin 68°$ 0.9272 **8.** $\cos 4°$ 0.9976 **9.** $\tan 87°$ 19.0811 **10.** $\sin 42°$ 0.6691

Use the table on page 395 to find the measure of the angle to the nearest degree.

11. $\sin A = 0.9$ 64° **12.** $\cos A = 0.5$ 60° **13.** $\tan A = 1.3$ 52° **14.** $\sin A = 0.36$ 21°

15. $\cos A = 0.84$ 33° **16.** $\tan A = 1$ 45° **17.** $\cos A = 0.22$ 77° **18.** $\tan A = 0.6$ 31°

Find angle measures to the nearest degree and lengths to the nearest whole number.

19. $\angle A$ 26° **20.** x 8 **21.** $\angle B$ 64°

22. $\angle B$ 35° **23.** x 6 **24.** y 7

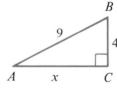

Exs. 19–21

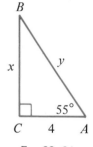

Exs. 22–24

Exercises

Use the table on page 395 to solve $\triangle ABC$ with $\angle C = 90°$.
Find lengths to the nearest whole number and angle measures
to the nearest degree.

B

1. $x = 2, y = 5$ 5, 22°, 68° **2.** $x = 4, y = 6$ 52, 34°, 56°

3. $x = 3, z = 6$ 5, 30°, 60° **4.** $y = 7, z = 9$ 6, 39°, 51°

5. $\angle A = 47°, z = 5$ **6.** $\angle B = 62°, z = 8$

7. $\angle B = 15°, x = 20$ **8.** $\angle A = 70°, y = 30$

9. $\angle A = 28°, x = 10$ **10.** $\angle B = 39°, y = 9$

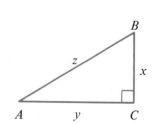

5. 4, 3, 43°   6. 4, 7, 28°   7. 5, 21, 75°

8. 82, 88, 20°   9. 10, 21, 62°   10. 11, 14, 51°

A Sculptor

Is there a statue or piece of sculpture in your town hall or in your neighborhood park? Did you ever wonder who made it?

Edmonia Lewis was a famous sculptor of the nineteenth century. For the most part, she chose people as her subjects. One of her first works was a medallion of abolitionist John Brown, published in 1865. She also sculptured a bust of U.S. Civil War hero Colonel Robert Gould Shaw which she modeled from photographs, and busts of Abraham Lincoln and U.S. Senator Charles Sumner. At times, Edmonia Lewis would get ideas from abstract themes or from poetry. Henry Wadsworth Longfellow's poem *Hiawatha* was the inspiration for her sculpture of an American Indian group. Her most important early work was a composition of two emancipated slaves entitled *Forever Free*. In 1865, she went to Rome to study sculpture. She completed a sculpture of Longfellow in 1869, which she modeled mainly from observing him in the streets of Rome. The sculpture was later placed in the Harvard College Library.

Edmonia Lewis exhibited six of her works at the Philadelphia Centennial Exposition of 1876. Collections of her works have been placed on display at the Frederick Douglass Institute of Negro Arts and History, the Fogg Museum at Harvard University, the California Public Library at San José, and the Kennedy Gallery in New York.

Research Activity Find out about the different styles of sculpture. Can you find any examples of realistic or neo-classic sculpture? Try to find some photos in an art book of the works of Edmonia Lewis. Can you tell which style she used?

Career Activity Find out what types of materials and tools sculptors work with. Before they begin, do sculptors usually draw sketches of what they intend to make?

Henry Wadsworth Longfellow

Self-Test

Symbols and words to remember:

$c^2 = a^2 + b^2$ [p. 371]
$\cos A$ [p. 380]
Pythagorean Property [p. 371]
30°–60° right triangle [p. 376]
sine [p. 379]
tangent [p. 380]

$\sin A$ [p. 379]
$\tan A$ [p. 380]
isosceles right triangle [p. 375]
trigonometric ratio [p. 379]
cosine [p. 380]
solving a right triangle [p. 382]

State whether or not a triangle with sides of the given lengths is a right triangle.

1. 7, 8, 10 No

2. 5, 12, 13 Yes

[12-4]

A right triangle has sides of lengths a, b, and c, with c the length of the hypotenuse. Find the length of the third side.

3. $a = 5$, $b = 7$ $\sqrt{74}$

4. $a = 9$, $c = 15$ 12

Find an approximation to the nearest hundredth for x.

[12-5]

5.

10
x
10

14.14

6.

x
45°
$9\sqrt{2}$

9

7.

7
30°
x

12.12

8.

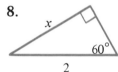

x
60°
2

1.73

Use the diagram to name each ratio.

9. $\sin A$ $\frac{x}{z}$

10. $\sin B$ $\frac{y}{z}$

11. $\cos A$ $\frac{y}{z}$

12. $\tan B$ $\frac{y}{x}$

[12-6]

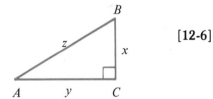

B
z
x
A y C

Refer to the diagram at the right.

13. Which equation is *incorrectly* written? b

[12-7]

    a. $\cos E = \frac{8}{17}$ **b.** $\cos D = \frac{8}{17}$ **c.** $\sin D = \frac{8}{17}$

14. Find the value of x. 15

15. Find $\angle D$ to the nearest degree. 28°

16. Find $\angle E$ to the nearest degree. 62°

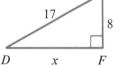

E
17
8
D x F

Self-Test answers and Extra Practice are at the back of the book.

Chapter Review

Write the letter that labels the correct answer.

1. A positive number has exactly __?__ different square root(s). B   [12-1]
   A. 1   B. 2   C. 3   D. 4

2. Which number lies between 8 and 9? C
   A. $\sqrt{54}$ B. $\sqrt{85}$ C. $\sqrt{72}$ D. $\sqrt{61}$

3. If we divide a positive number n by an estimate for $\sqrt{n}$, the __?__   [12-2]
   of the divisor and quotient can be used as a better estimate for
   $\sqrt{n}$. C
   A. sum   B. product   C. average   D. greater

4. If 7.4 is used as an estimate for $\sqrt{55.5}$, the next estimate will be B
   A. 7.5   B. 7.45   C. 7.42   D. 7.4

5. To find an approximation for the square root of a number which   [12-3]
   lies between two entries in a table, we use the process called
   __?__. D
   A. divide and average   B. estimation
   C. trigonometry   D. interpolation

6. To approximate $\sqrt{18.3}$ using a table, we assume that $\sqrt{18.3}$ is
   __?__ of the way from $\sqrt{18}$ to $\sqrt{19}$. A
   A. 0.3   B. 0.5   C. one third   D. $\sqrt{0.3}$

7. A table gives $\sqrt{10} \approx 3.162$ and $\sqrt{11} \approx 3.316$, so a good approximation for $\sqrt{10.5}$ is __?__. A
   A. 3.239   B. 3.162   C. 3.316   D. 3.2

8. The side opposite the right angle of a right triangle is the __?__. B   [12-4]
   A. leg   B. hypotenuse   C. vertex   D. shortest side

9. For a right triangle with hypotenuse of length c and legs of lengths
   a and b, the Pythagorean Property states that __?__. C
   A. $c^2 = a^2 - b^2$ B. $c^2 + a^2 = b^2$
   C. $c^2 = a^2 + b^2$ D. $c^2 + b^2 = a^2$

10. A right triangle with legs of lengths 7 and 24 has hypotenuse of
    length __?__. B
    A. 625   B. 25   C. $\sqrt{527}$ D. $\sqrt{31}$

11. In an isosceles right triangle, each acute angle has measure __?__ . B [12-5]

    A. 30°   **B.** 45°   **C.** 60°   **D.** 90°

12. If each leg of an isosceles right triangle has length a, then the hypotenuse has length __?__ . A

    A. $a\sqrt{2}$ **B.** $a\sqrt{3}$ **C.** $\dfrac{a}{2}$ **D.** $2a$

13. In a 30°–60°–90° triangle with hypotenuse of length 2, the side opposite the 60° angle has length __?__ . C

    A. 1   **B.** $2\sqrt{2}$ **C.** $\sqrt{3}$ **D.** $\dfrac{\sqrt{3}}{2}$

14. Refer to the diagram. $\sin A =$ __?__ C [12-6]

    A. $\dfrac{a}{b}$ **B.** $\dfrac{b}{c}$ **C.** $\dfrac{a}{c}$ **D.** $\dfrac{c}{a}$

15. $\cos A =$ __?__ B

    A. $\dfrac{a}{b}$ **B.** $\dfrac{b}{c}$ **C.** $\dfrac{a}{c}$ **D.** $\dfrac{c}{a}$

Exs. 14, 15

16. A right triangle has legs of lengths 3 and 4. The tangent of the angle opposite the leg of length 3 is __?__ . A

    A. $\dfrac{3}{4}$ **B.** $\dfrac{3}{5}$ **C.** $\dfrac{4}{5}$ **D.** 5

17. The table on page 395 indicates that $\sin 57° \approx$ __?__ . A [12-7]

    A. 0.8387   **B.** 0.5446   **C.** 1.5399   **D.** 0.8290

18. If $\tan A \approx 0.5317$, the table on page 395 indicates that $\angle A =$ __?__ , to the nearest degree. D

    A. 29°   **B.** 32°   **C.** 58°   **D.** 28°

19. Refer to the diagram. Use the table on page 395 to determine that $x =$ __?__ , to the nearest integer. A

    A. 3   **B.** 4   **C.** 8   **D.** 5

John Napier and Napier's Bones

The earliest known computing device was the ancient abacus, a plate of metal with grooves, and pebbles in the grooves. To perform a computation, the pebbles were moved and the answer was read from the final position of the pebbles.

An interesting second computing aid came centuries later when John Napier (1550–1617) made **Napier's rods,** or as they came to be known, **Napier's bones.** Workers in his father's business made frequent errors in arithmetic, so Napier, a Scottish mathematician, invented his computing rods to be used by the workers to multiply integers.

To make Napier's bones, start with ten long rectangular strips of paper or cardboard. Put a different digit at the top of each. The nine multiples of the digit are written in a column in nine squares below, with the tens' digit of each multiple in the top half of the square and the ones' digit in the bottom half.

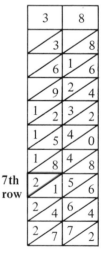

Example | Use Napier's bones to multiply 7×38.

Solution | Place the three rod next to the eight rod as shown. The answer is then read from the seventh row.

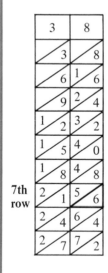

Ones' place = 6

Tens' place =
5 + 1 = 6

Hundreds' place = 2

Answer: 266

The example below (6 × 239) shows how sometimes you must carry a digit from one place to the next.

Ones' place = 4
8 + 5 = 13, Tens' place = 3

Hundreds'.place = 1 + 1 + 2 = 4
Thousands' place = 1
Answer: 1434

Inspecting the multiplication problem below suggests how you might use Napier's bones to carry out a multiplication when both factors have two digits or more.

$$
\begin{array}{r}
62 \\
\times 21 \\
\hline
62 \quad\longleftarrow 1 \times 62 \\
124 \quad\longleftarrow 2 \times 62 \\
\hline
1302
\end{array}
$$

In today's age of hand calculators, Napier's bones may seem cumbersome and impractical; but they remained popular for centuries following their invention.

Make a set of Napier's bones and use them to multiply the following numbers. (You may need to make duplicate rods for some digits.)

1. 6 × 27   **2.** 8 × 68   **3.** 5 × 623

4. 22 × 83   **5.** 47 × 79   **6.** 683 × 562

Research Activity Go to your library and find out more about John Napier. Find out what *logarithms* are.

Chapter Test

1. $-\sqrt{49} = $ __?__ -7   [12-1]

2. $\sqrt{29}$ lies between what two consecutive integers? 5 and 6

3. Using the divide and average method, $\sqrt{35} = $ __?__ to the tenths' [12-2]
   place. 5.9

4. Using the divide and average method, $\sqrt{127.5} = $ __?__ to the
   nearest tenth. 11.2

5. Using the table on page 394, $3\sqrt{12} = $ __?__ to the nearest [12-3]
   hundredth. 10.39

6. Using interpolation, $\sqrt{15.3} = $ __?__ to the nearest hundredth. 3.91

7. A triangle has sides of lengths 6, 12, and 13. Is it a right triangle? No [12-4]

8. A rectangle has sides of lengths 7 m and 24 m. How long is its
   diagonal? 25 m

9. How long is the hypotenuse of an isosceles right triangle which [12-5]
   has legs 6 cm long? $6\sqrt{2}$ cm

10. Use the table on page 394 to approximate
    the value of x to the nearest tenth. 6.1

    7

    30°

    x

11. $\sin A = $ __?__ $\frac{3}{5}$ [12-6]

    3

    A 4

12. $\tan 60° = $ __?__ $\sqrt{3}$

13. Use the table on page 395 to find that [12-7]
    $x = $ __?__ to the nearest whole number. 3

    6

    x

    34°

    y

14. To the nearest degree, $\angle A = $ __?__ 51°

    8

    x

    A 5

Cumulative Review

[Ch. 10]

1. A quarter, a nickel, and a dime are tossed. How many possible outcomes are there? What is the probability of each outcome?

   8 possible outcomes: $\frac{1}{8}$

Each of ten cards contains one of the letters from the word *straw-berry*. A card is drawn at random. Find:
   a. the probability of each event
   b. the odds against each event

2. $p(a)$ 3. $p(t \text{ or } b)$ 4. $p(r)$ 5. $p(a \text{ or } r)$

   $\frac{1}{10}$; 9 to 1 $\frac{1}{5}$; 4 to 1 $\frac{3}{10}$; 7 to 3 $\frac{2}{5}$; 3 to 2

6. Of 120 students polled at Middlebury School, 54 owned skateboards. About how many of the 1800 students at the school own skateboards? 810 students

[Ch. 11]

7. Graph and label the ordered pairs $(0, 5)$, $(3, 2)$, $(-1, 1)$ and $(3\frac{1}{2}, -4)$ on a coordinate plane. Check students' graphs.

8. Find the solutions of $6x - 5y = 2$ with abscissas 0, -2, and 3.

9. Graph $-2x + 3y = 5$. Check students' graphs. $(0, -\frac{2}{5})$ $(-2, -\frac{14}{5})$ $(3, \frac{16}{5})$

10. Solve the system $x - y = 2$
    $2x + y = 4$ $(2, 0)$

[Ch. 12]

11. $\sqrt{57}$ is between what two consecutive integers? 7 and 8

12. Approximate $\sqrt{63}$ to the tenths' place. 7.9

13. A right triangle has hypotenuse of length 13 cm and one leg of length 12 cm. How long is the other leg? 5 cm

14. In a 30°–60°–90° triangle, the shorter leg has length 5.5 m. How long is the other leg? $5.5\sqrt{3}$ m

15. Find   a. $\sin A$ $\frac{2}{\sqrt{53}}$
       b. $\tan B$ $\frac{7}{2}$

Final Review

1. Draw a histogram for the data at the right.
   Check students' papers.
2. Find the range, median, mean, and all the modes for
   the data at the right. 6; 8; 7.6; 7 and 8

Quiz scores of 25 students				
6	7	8	7	8
6	8	7	8	6
10	9	9	6	8
7	9	10	8	7
6	7	4	9	10

Exs. 1–2

3. According to the graph below, the greatest temperature change was recorded between which two consecutive hours? 11 P.M. and 12 P.M.

4. What is the range of recorded temperatures? 14

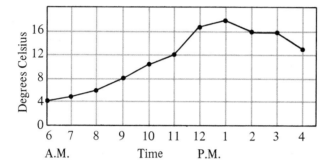

Exs. 3–4

5. Write 20.903 in expanded form.
   $(2 \times 10^1) + (9 \times 0.1) + (3 \times 0.001)$

Compute.

6. $(6.3 - 8.7)(2.5 - 4.9)$
   5.76

7. $(-3.1)(1.5)(-0.7)$
   3.255

8. $1.7057 \div (-0.037)$
   -46.1

Express as a fraction in lowest terms.

9. $-4\frac{2}{3} \div \left(-3\frac{4}{15}\right)$ $1\frac{3}{7}$

10. $1\frac{7}{9} + \left(\frac{5}{6} - \frac{8}{3}\right)$ $-\frac{1}{18}$

11. $12\frac{5}{6} - 6\frac{3}{7} \div 1\frac{2}{7}$ $7\frac{5}{6}$

Express each ratio as a fraction in lowest terms.

12. 3 min to 40 s $\frac{9}{2}$

13. 130 cm to 2.6 m $\frac{1}{2}$

14. 1.2 kg to 880 g $\frac{15}{11}$

15. A distance of 2 cm on a map represents an actual distance of 75 km. What
    distance on this map would represent 525 km? 14 cm

Complete.

16. 12% of 145 = $\underline{?}$ 17.4

17. 37 = $\underline{?}$ % of 40 92.5

18. 51 = 30% of $\underline{?}$ 170

19. A $40 clock is on sale for $32.80. Find the percent of discount. 18%

392

20. A salesperson who is paid a 9% commission earned $270. What was the amount of sales? $3000

Refer to the diagram at the right.

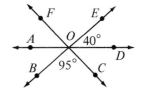

21. $\angle EOC =$ __?__ ° 85   **22.** $\angle FOE =$ __?__ ° 95

23. $\angle FOB =$ __?__ ° 85

24. In the diagram at the right, $\overline{ST} \parallel \overline{RU}$. Find the values of x and y.

Which pairs of triangles must be congruent? Give your reason (SAS, ASA, or SSS).

25.

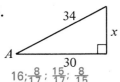

Congruent; ASA

26.

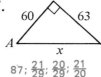

Not Congruent

27.

Congruent; SAS

28. Solve: $\frac{5}{6}(2x + 1) = \frac{3}{2}x + 2.$ {7}   **29.** For which values of t is $-3t + 5 > 2$? {..., $-2, -1, 0$}

30. $800 is invested in an account which pays 5% compounded quarterly. What will be the value of the account in 6 months? $820.13

31. Find the area of a triangle with base 3 m and height 2.5 m. $3.75m^2$

32. Find the volume of a cone with height 9 cm and radius 2 cm. (Use $\pi \approx 3.14$.)
37.68 cm^3

A coin is chosen at random from 3 pennies, 5 dimes, and 2 quarters.

33. What is the probability of choosing a quarter? $\frac{1}{5}$

34. What are the odds against choosing a dime? 1 to 1

35. Graph (3,2), ($-1,0$), and (7,-5) on a coordinate plane.
Check students' papers.

Find the value of x. Then find $\sin A$, $\cos A$, and $\tan A$.

36.

$16; \frac{8}{17}; \frac{15}{17}; \frac{8}{15}$

37.

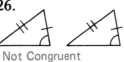

$87; \frac{21}{29}; \frac{20}{29}; \frac{21}{20}$

38.

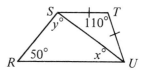

$4; \frac{4}{5}; \frac{3}{5}; \frac{4}{3}$

Number	Square Root	Number	Square Root	Number	Square Root	Number	Square Root
1	1.000	26	5.099	51	7.141	76	8.718
2	1.414	27	5.196	52	7.211	77	8.775
3	1.732	28	5.292	53	7.280	78	8.832
4	2.000	29	5.385	54	7.348	79	8.888
5	2.236	30	5.477	55	7.416	80	8.944
6	2.449	31	5.568	56	7.483	81	9.000
7	2.646	32	5.657	57	7.550	82	9.055
8	2.828	33	5.745	58	7.616	83	9.110
9	3.000	34	5.831	59	7.681	84	9.165
10	3.162	35	5.916	60	7.746	85	9.220
11	3.317	36	6.000	61	7.810	86	9.274
12	3.464	37	6.083	62	7.874	87	9.327
13	3.606	38	6.164	63	7.937	88	9.381
14	3.742	39	6.245	64	8.000	89	9.434
15	3.873	40	6.325	65	8.062	90	9.487
16	4.000	41	6.403	66	8.124	91	9.539
17	4.123	42	6.481	67	8.185	92	9.592
18	4.243	43	6.557	68	8.246	93	9.644
19	4.359	44	6.633	69	8.307	94	9.695
20	4.472	45	6.708	70	8.367	95	9.747
21	4.583	46	6.782	71	8.426	96	9.798
22	4.690	47	6.856	72	8.485	97	9.849
23	4.796	48	6.928	73	8.544	98	9.899
24	4.899	49	7.000	74	8.602	99	9.950
25	5.000	50	7.071	75	8.660	100	10.000

SINES, COSINES, AND TANGENTS OF ANGLES
FROM 1 TO 90 DEGREES

Angle	Sine	Cosine	Tangent	Angle	Sine	Cosine	Tangent
1°	.0175	.9998	.0175	46°	.7193	.6947	1.0355
2°	.0349	.9994	.0349	47°	.7314	.6820	1.0724
3°	.0523	.9986	.0524	48°	.7431	.6691	1.1106
4°	.0698	.9976	.0699	49°	.7547	.6561	1.1504
5°	.0872	.9962	.0875	50°	.7660	.6428	1.1918
6°	.1045	.9945	.1051	51°	.7771	.6293	1.2349
7°	.1219	.9925	.1228	52°	.7880	.6157	1.2799
8°	.1392	.9903	.1405	53°	.7986	.6018	1.3270
9°	.1564	.9877	.1584	54°	.8090	.5878	1.3764
10°	.1736	.9848	.1763	55°	.8192	.5736	1.4281
11°	.1908	.9816	.1944	56°	.8290	.5592	1.4826
12°	.2079	.9781	.2126	57°	.8387	.5446	1.5399
13°	.2250	.9744	.2309	58°	.8480	.5299	1.6003
14°	.2419	.9703	.2493	59°	.8572	.5150	1.6643
15°	.2588	.9659	.2679	60°	.8660	.5000	1.7321
16°	.2756	.9613	.2867	61°	.8746	.4848	1.8040
17°	.2924	.9563	.3057	62°	.8829	.4695	1.8807
18°	.3090	.9511	.3249	63°	.8910	.4540	1.9626
19°	.3256	.9455	.3443	64°	.8988	.4384	2.0503
20°	.3420	.9397	.3640	65°	.9063	.4226	2.1445
21°	.3584	.9336	.3839	66°	.9135	.4067	2.2460
22°	.3746	.9272	.4040	67°	.9205	.3907	2.3559
23°	.3907	.9205	.4245	68°	.9272	.3746	2.4751
24°	.4067	.9135	.4452	69°	.9336	.3584	2.6051
25°	.4226	.9063	.4663	70°	.9397	.3420	2.7475
26°	.4384	.8988	.4877	71°	.9455	.3256	2.9042
27°	.4540	.8910	.5095	72°	.9511	.3090	3.0777
28°	.4695	.8829	.5317	73°	.9563	.2924	3.2709
29°	.4848	.8746	.5543	74°	.9613	.2756	3.4874
30°	.5000	.8660	.5774	75°	.9659	.2588	3.7321
31°	.5150	.8572	.6009	76°	.9703	.2419	4.0108
32°	.5299	.8480	.6249	77°	.9744	.2250	4.3315
33°	.5446	.8387	.6494	78°	.9781	.2079	4.7046
34°	.5592	.8290	.6745	79°	.9816	.1908	5.1446
35°	.5736	.8192	.7002	80°	.9848	.1736	5.6713
36°	.5878	.8090	.7265	81°	.9877	.1564	6.3138
37°	.6018	.7986	.7536	82°	.9903	.1392	7.1154
38°	.6157	.7880	.7813	83°	.9925	.1219	8.1443
39°	.6293	.7771	.8098	84°	.9945	.1045	9.5144
40°	.6428	.7660	.8391	85°	.9962	.0872	11.4301
41°	.6561	.7547	.8693	86°	.9976	.0698	14.3007
42°	.6691	.7431	.9004	87°	.9986	.0523	19.0811
43°	.6820	.7314	.9325	88°	.9994	.0349	28.6363
44°	.6947	.7193	.9657	89°	.9998	.0175	57.2900
45°	.7071	.7071	1.0000	90°	1.0000	.0000	

Answers to Self-Tests

CHAPTER 1

Page 13 **1.** Crust **2.** Inner Core **3.** 3000–5000 km **4.** Mantle **5.** 110 kg **6.** About 70 kg **7.** About 45 kg **8.** About 90 kg **9.** About 6° to 22° **10.** February and March **11.** June **12.** May

Page 19 **1.**

Interval	Frequency
140–144	3
145–149	4
150–154	3
155–159	5
160–164	3
165–169	2

2.

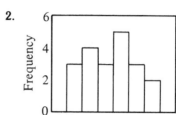

Students' heights in cm

3. 155–159 cm **4.** 5 **5.** 3 **6.** 9 **7.** 9 **8.** 10

CHAPTER 2

Page 39 **1.** 10^4 **2.** 10^8 **3.** $(4 \times 10^4) + (8 \times 10^2) + 7$ **4.** 8691 **5.** hundredths **6.** $7 + (2 \times 0.1) + (4 \times 0.01) + (6 \times 0.001)$ **7.** 403.62 **8.** 65.009

9. 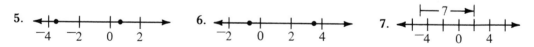 **10.** > **11.** < **12.** <, < **13.** 8; 17; 31 **14.** 4; 40; 95

15. 36; 96; 480 **16.** 4; 10; 17

Page 57 **1.** $n = 21 - 14$; $n = 7$ **2.** $n = 32 - 8$; $n = 24$ **3.** $n = 48 \div 6$; $n = 8$ **4.** $n = 24 \div 8$; $n = 3$ **5.** $x = 2$ **6.** $n = 0$ **7.** $t = 3$ **8.** $y = 8$ **9.** 49.646 **10.** 18.041 **11.** 8.494 **12.** 0.003 **13.** 15.504 **14.** 0.612 **15.** 0.01488 **16.** 13.2374 **17.** 4.1 **18.** 49.5625 **19.** 8.33 **20.** 0.27 **21.** 72 m **22.** 5.5 km **23.** 6.9 km **24.** $2.25

CHAPTER 3

Page 81 **1.** [number line: −40 −20 0 20 40] **2.** [number line: −800 −400 0 400 800] **3.** 6 **4.** 1.2

5. [number line: −4 −2 0 2] **6.** [number line: −2 0 2 4] **7.** [number line with 7: −4 0 4]

8. [number line with 4: −2 0 2] **9.** −5.4 **10.** 7.5 **11.** −0.6 **12.** 1.99 **13.** −3 **14.** 3.8 **15.** −4

16. −3.71

Page 97 **1.** −8 **2.** 4 **3.** $11 - 9 = 2$; $11 - 2 = 9$ **4.** $-8 - 29 = -37$; $-8 - -37 = 29$ **5.** $1.2 + (-3.6)$ **6.** $8.72 + 13.51$ **7.** 4.7 **8.** 14.59 **9.** −6.2 **10.** 405 **11.** 0 **12.** 51 **13.** −0.9 **14.** 18.87 **15.** −21.9648 **16.** 70.097 **17.** −4 **18.** −7.6 **19.** 4.5 **20.** −0.6

CHAPTER 4

Page 120 **1.** $\frac{1}{9}$ **2.** 35 **3.** $\frac{4}{7}$ **4.** $7\frac{4}{5}$ **5.** $\frac{-2}{7}$; $\frac{2}{-7}$ **6.** $\frac{1}{-4}$; $-\frac{1}{4}$ **7.** −37 **8.** $-6\frac{2}{5}$ **9.** Answers will vary. **10.** $\frac{25}{17}$ **11.** $-\frac{2}{9}$ **12.** $\frac{5}{9}$ **13.** $-\frac{3}{10}$ **14.** $\frac{3}{10}$ **15.** $-\frac{2}{7}$ **16.** 26

Page 137 **1.** $\frac{36}{45}$; $\frac{10}{45}$ **2.** $-\frac{25}{30}$; $-\frac{2}{30}$ **3.** $\frac{70}{180}$; $\frac{75}{180}$; $\frac{12}{180}$ **4.** $\frac{63}{84}$; $\frac{48}{84}$; $-\frac{14}{84}$ **5.** $\frac{11}{24}$ **6.** $-\frac{17}{24}$ **7.** $1\frac{2}{15}$ **8.** $5\frac{8}{9}$ **9.** $1\frac{1}{2}$ **10.** $1\frac{1}{21}$ **11.** $-3\frac{1}{5}$ **12.** $-\frac{1}{4}$ **13.** $\frac{89}{50}$ **14.** $-\frac{13}{40}$ **15.** $0.958\overline{3}$ **16.** −2.064

CHAPTER 5

Page 152 **1.** $\frac{1}{4}$ **2.** $\frac{3}{5}$ **3.** $\frac{3}{50}$ **4.** $\frac{200}{1}$ **5.** Incorrect **6.** Correct **7.** $n = 4$ **8.** $n = 28$

Page 165 **1.** 1.6 **2.** 18 **3.** 20% **4.** $26\frac{2}{3}$ **5.** 20% **6.** 12% **7.** 20 **8.** 5% **9.** $1928.40
10. $70.50 **11.** $1116 **12.** $24,220

CHAPTER 6

Page 189 **1.** one **2.** no **3.** 3500 **4.** 0.2 **5.** diameter **6.** 251.2 cm **7.** 132 **8.** 2.8 **9.** 45°
10. protractor **11.** 90° **12.** 50° **13.** parallel **14.** equal **15.** 138°, 42°, and 138° **16.** parallel
Page 209 **1.** 90°, 60° **2.** Yes **3.** 60° **4.** Two **5.** T **6.** F **7.** F **8.** T **9.** F **10.** F **11.** 24 cm
12. 3.8 cm, 4 cm, 4 cm **13.** No **14.** $\triangle EFD$ **15.** $\angle C$ **16.** CB **17.** Yes **18.** Yes **19.** No
20. No

CHAPTER 7

Page 230 **1.** $1 + x$ **2.** $12n$ **3.** The difference when z is subtracted from thirteen
4. The sum when the product of five and z is added to twelve **5.** -12 **6.** -3 **7.** 14 **8.** $\frac{1}{5}$ **9.** $\frac{y}{3} = -9$
10. $2(n + 4) = 14$
11. The product when three is multiplied by nine less than y is less than or equal to negative eight.
12. The sum when the product of negative seven and x is added to thirty-one is greater than four.
13. No **14.** No **15.** Yes **16.** Yes

17.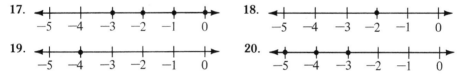

18.

19.

20.

Page 239 **1.** 12 **2.** 8 **3.** -54 **4.** -35 **5.** $\{-3\}$ **6.** $\{22\}$ **7.** $\{8\}$ **8.** $\{-6\}$
9. $\{-1, 0, 1, 2, \ldots\}$ **10.** $\{-10, -11, -12, \ldots\}$ **11.** $\{-5, -4, -3, -2, \ldots\}$
12. $\{3, 2, 1, 0, \ldots\}$

CHAPTER 8

Page 254 Answers will vary. **1. a.** The sum of 20 and what number gives 39? **b.** There are 2 algebra
classes and a total of 39 students. If there are 20 students in one class, how many students are in the
other class?
2. a. What number multiplied by 12 gives 84? **b.** A small auditorium has 12 seats in each row. If the
auditorium holds 84 seats, how many rows are there?
3. a. Seven less than what number is 23? **b.** Joanne needed 23 minutes to solve a difficult science
problem. This was 7 minutes less than Sarah needed to solve the problem. How long did Sarah need to
solve the problem?
4. a. One third of what number is 9? **b.** One third of the students in the glee club were chosen to be in
the concert choir. If 9 students from the glee club were accepted in the concert choir, how many students
are in the glee club? **5.** $18 + n = 51$ **6.** $n - 13 = 44$ **7.** $n + (n + 16) = 71$
8. $w + w + (w + 20) + (w + 20) = 120$ **9.** 19 **10.** 31 **11.** 19, 20 **12.** 12, 39

Page 263 **1.** 1.625 km **2.** 6 cm by 7.5 cm **3.** $82.50 **4.** $5600 **5.** 359,970,000 km²
6. About $2,300,000

CHAPTER 9

Page 281 **1.** 36 cm² **2.** 12 cm² **3.** 12 cm **4.** 32 cm **5.** 442 mm² **6.** 10.75 m² **7.** 12 cm²
8. 10 m **9.** $\frac{77}{8}$ square units **10.** 78.5 cm² **11.** 225π square units **12.** 12π m²
Page 299 **1.** 471 cubic units **2.** 540 cubic units **3.** 3375 mm³ **4.** 1056 cm³ **5.** $26\frac{2}{3}$ cubic units
6. 140 cm³ **7.** 117.23 cubic units **8.** 2616.67 cubic units **9.** 135 square units
10. 48 square units; 72 square units **11.** 502.4 square units; 602.88 square units
12. 36 square units; 72 square units **13.** 5500 **14.** 0.4 **15.** 440 g **16.** 0.0899 g

17. 4π square units; $\frac{4}{3}\pi$ cubic units   **18.** 1256 square units   **19.** $14\frac{1}{7}$ cubic units

20. $\frac{32}{3}\pi$ cubic units

CHAPTER 10

Page 317   1. No   **2.** Yes   **3.** 5; $\frac{1}{5}$ **4.** 4; $\frac{1}{4}$ **5.** $\frac{4}{7}$ **6.** 0   **7.** $\frac{1}{7}$ **8.** 1   **9.** 1 to 2   **10.** 2 to 1   **11.** 1 to 5
12. 1 to 1

Page 327   1. $\frac{14}{15}$ **2.** $\frac{2}{5}$ **3.** $\frac{1}{2}$ **4.** $\frac{2}{3}$ **5.** $\frac{3}{400}$ **6.** $\frac{397}{400}$ **7.** 45 fuses   **8.** 397 to 3   **9.** 5.51   **10.** 12.6¢
11. 2.5   **12.** 5

CHAPTER 11

Page 343   1. a. *J* **b.** *F* **c.** *A* **d.** *D* **e.** *B* **f.** *C*

2. a.

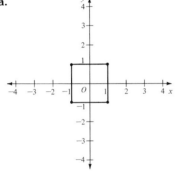

3. a.

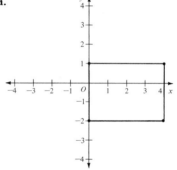

b. $A = 2 \times 2 = 4$

b. $A = 3 \times 4 = 12$

4. a.

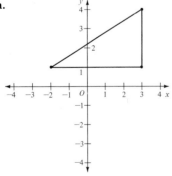

b. $A = \frac{1}{2} \times 5 \times 3 = 7.5$

5. a. $(-5, -4)$ **b.** $(-2, 3)$ **c.** $(-2, -3)$ **d.** $(3, 3)$ **e.** $(0, 4)$ **f.** $(2, -2)$ **6.** *A, K, M*
7. *A, C, D, E, F, H, K, M*   **8.** *D, O, J*

Page 355 **1.** Yes **2.** Yes **3.** $y = \frac{2}{3}x - 2$ **4.** $y = \frac{5}{4}x - 10$ **5.** $(-7, 0)$, $(0, -7)$
6. $(-11, 0)$, $(0, 5\frac{1}{2})$

7.

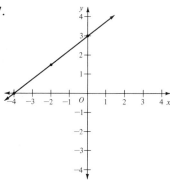

8.

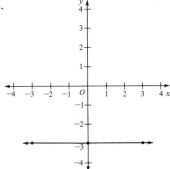

9.

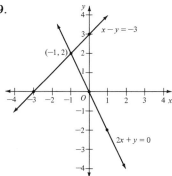

10.

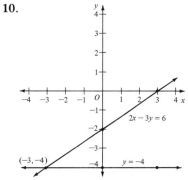

11.

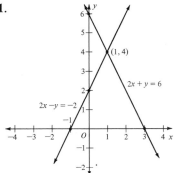

12.

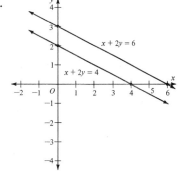

CHAPTER 12

Page 370 **1.** 8 and 9 **2.** 1 and 2 **3.** 5 and 6 **4.** 7 and 8 **5.** 3.75 **6.** 7.4 **7.** 8.6 **8.** 8.2 **9.** 9.17
10. 7.94 **11.** 0.13 **12.** 4.96
Page 385 **1.** No **2.** Yes **3.** $c = \sqrt{74}$ **4.** $b = 12$ **5.** $x = 14.14$ **6.** $x = 9$ **7.** $x = 12.12$
8. $x = 1.73$ **9.** $\frac{x}{z}$ **10.** $\frac{y}{z}$ **11.** $\frac{y}{z}$ **12.** $\frac{y}{x}$ **13.** b **14.** $x = 15$ **15.** $28°$ **16.** $62°$

Extra Practice—Chapter 1

For use after page 13:

Use the table below for Exercises 1–6.

Money Earned Annually by Student Organizations in Hopeville

Year	Food Sales	Dances	Yard Care	Washing Cars	Craft Sales
1974	$150	$240	$120	$60	$30
1975	$163	$260	$130	$65	$33
1976	$175	$280	$140	$70	$35
1977	$188	$300	$150	$75	$38
1978	$200	$320	$160	$80	$41

1. How much more did the student organizations earn in 1978 than in 1974? $201   **[1-1]**
2. Which activity earned the least amount from 1974 to 1978? Craft Sales
3. Which activity earned the greatest amount from 1974 to 1978? Dances
4. Which activity earned twice as much as washing cars? Yard Care
5. Which activities earned more than yard care? Food Sales, Dances
6. How much more did the students earn with dances than with food sales from 1974 to 1978? $524

Use the bar graph at the right for Exercises 7–9.

7. What were the sales in produce? $550   **[1-2]**
8. What were the sales in all four departments? $2600
9. What was the difference between meat sales and dairy sales? $550

Supermarket sales for one day

Use the line graph at the right for Exercises 10–12. It shows the amount collected from a clothes washer each day of one week.

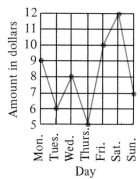

Money collected from washer

10. How much was collected on the busiest day? $12

11. How much was collected on the slowest day? $5

12. What is the range of the amounts collected? $7

[1-3]

For use after page 19:

In Exercises 1 and 2, check students' papers.

For Exercises 1–8, use the data given at the right.

Mass in Kilograms of 31 Students

50	52	52	51	53	52	50
52	54	50	52	50	53	
51	52	54	51	52	53	
53	50	54	53	54	52	
52	54	54	50	54	51	

[1-4]

1. Make a frequency table.
2. Make a histogram.
3. Which mass has the greatest number of students? 52 kg
4. How many students are at least 52 kg? 21 students

5. Find the range. 4
7. Find the median. 52

6. Find the mean. $52\frac{3}{31}$
8. Find the mode. 52

[1-5]

DATA

For Exercises 9–12, use the data given at the right.

75	75	85	95	80
70	70	75	90	75
65	85	45	80	70
50	80	55	75	55
75	75	60	70	60
70	80	60	65	65

9. Find the range. 50
10. Find the mean. 71
11. Find the median. 70 and 75
12. Find the mode. 75

Extra Practice—Chapter 2

For use after page 39:

Write each numeral as a power of 10.

1. 100,000 10^5

2. 100,000,000 10^8

Express as a single power of 10.

3. $10^3 \times 10^1$ 10^4

4. $10^2 \times 10^7$ 10^9

Write in expanded form, using exponents.

5. 9164
$(9 \times 10^3) + (1 \times 10^2) + (6 \times 10^1) + 4$

6. 51,008
$(5 \times 10^4) \times (1 \times 10^3) + 8$

Write the decimal numeral for the following:

7. $(6 \times 1000) + (4 \times 100) + (1 \times 10) + 8$ 6418

8. $(5 \times 10,000) + (3 \times 1000) + (0 \times 100) + (5 \times 10) + 1$ 53, 051

Name the place value of the underlined digit.

9. 656.03<u>8</u> thousandths

10. <u>7</u>00.5608 hundreds

11. 0.<u>9</u>84 tenths

12. <u>1</u>0.00579 tens

Write in expanded form.

13. 92.46
$(9 \times 10) + 2 + (4 \times 0.1) + (6 \times 0.01)$

14. 3.0468
$3 + (4 \times 0.01) + (6 \times 0.001) + (8 \times 0.0001)$

Write the decimal for the expanded form.

15. $(6 \times 100) + (4 \times 10) + 8 + (3 \times 0.1) + (5 \times 0.01)$ 648.35

16. $(4 \times 100) + (3 \times 10) + 1 + (5 \times 0.1) + (9 \times 0.01)$ 431.59

17. $(5 \times 10) + (4 \times 0.01) + (3 \times 0.0001)$ 50.0403

Write the decimal.

18. twenty-four and five hundredths 24.05

19. three hundred and sixteen thousandths 0.316

For each exercise, draw a number line and graph the given numbers.

20. 0.6, 0.2, 1, 1.8

21. 2.4, 0, 1.7, 2.6, 0.8

In Exercises 20 and 21, check students' papers.

Replace each __?__ with < or > to make a true statement.

22. 1.011 __?__ 1.101 <

23. 9.202 __?__ 9.22 >

24. 0.75 __?__ 0.55 __?__ 0.35 >;>

25. 8 __?__ 8.17 __?__ 8.2 <;<

402

Evaluate the given expression for the given replacement set.

26. $7 + n$; $\{0, 3, 24\}$ 7; 10; 31   **27.** $y - 19$; $\{23, 56, 101\}$ 4; 37; 82

28. $13x$; $\{4, 9, 50\}$ 52; 117; 650   **29.** $t \div 27$; $\{0, 81, 135\}$ 0; 3; 5

30. $n + 9$; $\{16, 36, 96\}$ 25; 45; 105   **31.** $15k$; $\{4, 40, 400\}$ 60; 600; 6000

32. $x \div 32$; $\{32, 256, 352\}$ 1; 8; 11   **33.** $93 - z$; $\{55, 93, 0\}$ 38; 0; 93

For use after page 57:

Write a related equation. Find the value of n.

1. $16 + n = 23$ $n = 23 - 16$; $n = 7$ **2.** $n - 9 = 35$ $n = 35 + 9$; $n = 44$ [2-5]

3. $7n = 56$ $n = 56 \div 7$; $n = 8$ **4.** $n \div 6 = 6$ $n = 6 \times 6$; $n = 36$

5. $n + 28 = 51$ $n = 51 - 28$; $n = 23$ **6.** $9n = 99$ $n = 99 \div 9$; $n = 11$

Replace each variable to make a true equation.

7. $16 \times n = 0$ $n = 0$ **8.** $9 \times (4 + 8) = (9 \times 4) + (9 \times l)$ $l = 8$ [2-6]

9. $(8 + 9) + 6 = 8 + (p + 6)$ $p = 9$ **10.** $z \times 8 = 8 \times 7$ $z = 7$

Add or subtract.

11. $63.583 + 8.09$ 71.673   **12.** $27.0564 + 281.4$ 308.4564   [2-7]

13. $2.4 + 3.183 + 9$ 14.583   **14.** $4 + 0.463 + 43.4$ 47.863

15. $83.407 - 29.49$ 53.917   **16.** $85.604 - 85.2$ 0.404

Multiply.

17. 9.1×2.94 26.754   **18.** 0.73×0.18 0.1314   [2-8]

19. 8.74×0.008 0.06992   **20.** 7.029×3.3 23.1957

21. 0.364×0.29 0.10556   **22.** 2.47×24.7 61.009

Divide. If the division is not exact, round the quotient to the nearest hundredth.

23. $130.24 \div 16.2$ 8.04   **24.** $151.29 \div 12.3$ 12.3   [2-9]

25. $2.0506 \div 0.24$ 8.54   **26.** $328.865 \div 9.01$ 36.5

27. $25.26 \div 1.06$ 23.83   **28.** $2.8927 \div 2.8$ 1.03

Solve.

29. A bicycle trail is 19.2 km long. Last week Frank rode the trail 6 times. How far   [2-10]
did he ride? 115.2 km

30. Joy commutes to work by bus at a cost of $1.50 and goes home by train at a cost
of $2.10. How much does it cost for her to commute each day? $3.60

31. During a flood, a river reached a high-water mark of 4.8 m. This was 1.9 m higher
than the record high. What was the record high before this flood? 2.9 m

32. Fran Johnson bought a camera for $17.65. She gave the clerk a $20 bill. How
much did she receive in change? $2.35

Extra Practice—Chapter 3

For use after page 81:

Show the graph of the opposite of the given number on a number line.

1. 5 $^-5$ **2.** $^-8$ 8   **[3-1]**

3. $^-6$ 6   **4.** 1 $^-1$

5. 10 $^-10$ **6.** $^-10$ 10

Write the missing number.

7. If $n = {}^-5$,   **8.** If $-n = 8$,

     then $-n = \underline{\quad?\quad}$ 5   then $n = \underline{\quad?\quad}$ $^-8$

9. If $-n = 4$,   **10.** If $-n = {}^-7$,

     then $n = \underline{\quad?\quad}$ $^-4$ then $n = \underline{\quad?\quad}$ 7

11. If $n = 0$,   **12.** If $-n = {}^-1$,

     then $-n = \underline{\quad?\quad}$ 0   then $n = \underline{\quad?\quad}$ 1

For Exercises 13-22, check students' papers.
Graph the numbers on a number line.

13. 1, $^-1.5$, 0, $^-2$ **14.** $^-5$, 5, $^-4$, 1   **[3-2]**

15. $^-2$, $^-4$, 2.5, $^-3.5$ **16.** 3, 1.2, $^-2.4$, $^-5$

17. 1, 1.5, $^-2$, 2.5   **18.** 0, $^-1$, 2, $^-3$

19. Draw an arrow starting at the graph of $^-4$ to represent the number 5.

20. Draw an arrow starting at the graph of 6 to represent the number $^-8$.

21. Draw an arrow with endpoint at the graph of $^-5$ to represent the number 6.

22. Draw an arrow with endpoint at the graph of 3 to represent the number $^-6$.

Complete.

23. $2.7 + 7.2 = \underline{\quad?\quad}$ 9.9   **24.** $^-4.9 + {}^-7.6 = \underline{\quad?\quad}$ $^-12.5$ **[3-3]**

25. $^-3.8 + {}^-3.8 = \underline{\quad?\quad}$ $^-7.6$ **26.** $^-148 + {}^-256 = \underline{\quad?\quad}$ $^-404$

27. $\underline{\quad?\quad} + {}^-2.3 = {}^-2.9$ $^-0.6$ **28.** $18.12 + \underline{\quad?\quad} = 19.78$ 1.66

29. $^-2.8 + {}^-3.9 = \underline{\quad?\quad}$ $^-6.7$ **30.** $1.9 + 19 = \underline{\quad?\quad}$ 20.9

31. $^-3.7 + \underline{\quad?\quad} = {}^-5$ $^-1.3$ **32.** $\underline{\quad?\quad} + 19.14 = 19.45$ 0.31

33. $^-8 + 6 = \underline{\quad?\quad}$ $^-2$ **34.** $6.1 + {}^-2.3 = \underline{\quad?\quad}$ 3.8   **[3-4]**

35. $9 + {}^-12 = \underline{\quad?\quad}$ $^-3$ **36.** $^-5.2 + 2.9 = \underline{\quad?\quad}$ $^-2.3$

37. $4.1 + \underline{\quad?\quad} = 0$ $^-4.1$ **38.** $^-14.75 + 9.94 = \underline{\quad?\quad}$ $^-4.81$

39. $^-2.25 + \underline{\quad?\quad} = 0$ 2.25   **40.** $4.038 + {}^-6 = \underline{\quad?\quad}$ $^-1.962$

For use after page 97:

Find the given difference.

1. $30 - 18$ 12
2. $16 - 24$ ⁻8 [3-5]
3. ⁻$3 - $⁻$8$ 5
4. ⁻$42 - $⁻$31$ ⁻11
5. $43 - $⁻$136$ 179
6. ⁻$17 - 49$ ⁻66
7. $15 - $⁻$3$ 18
8. $12 - 48$ ⁻36

Write two differences corresponding to each equation.

9. $12 + $⁻$8 = 4$
10. $7 + 14 = 21$ $21 - 7 = 14; 21 - 14 = 7$
11. ⁻$27 + 19 = $⁻$8$
12. ⁻$33 + 49 = 16$ $16 - $⁻$33 = 49; 16 - 49 = $⁻$33$
13. ⁻$30 + $⁻$12 = $⁻$42$ See below.
14. ⁻$57 + 49 = $⁻$8$ ⁻$8 - $⁻$57 = 49; $⁻$8 - 49 = $⁻$57$
15. ⁻$26 + 16 = $⁻$10$ See below
16. ⁻$46 + 19 = $⁻$27$ ⁻$27 - $⁻$46 = 19; $⁻$27 - 19 = $⁻$46$

9. $4 - 12 = $⁻$8; 4 - $⁻$8 = 12$
11. ⁻$8 - $⁻$27 = 19; $⁻$8 - 19 = $⁻$27$

Write the given difference as a sum.

17. $6.5 - $⁻$1.3$ $6.5 + 1.3$
18. ⁻$3.1 - $⁻$6.1$ ⁻$3.1 + 6.1$ [3-6]
19. $2.3 - 4.4$ $2.3 + $⁻$4.4$
20. ⁻$31.2 - 4.3$ ⁻$31.2 + $⁻$4.3$
21. $44.4 - 57.6$ $44.4 + $⁻$57.6$
22. $28.9 - $⁻$35.6$ $28.9 + 35.6$
23. $2.56 - $⁻$9.83$ $2.56 + 9.83$
24. $30.01 - 18.87$ $30.01 + $⁻$18.87$

Find the given difference.

25. $33 - 19$ 14
26. $13 - 18$ ⁻5
27. ⁻$43 - 17$ ⁻60
28. $25 - $⁻$38$ 63
29. $6.4 - $⁻$2.9$ 9.3
30. ⁻$2.6 - 4.9$ ⁻7.5
31. ⁻$8.7 - $⁻$9.8$ 1.1
32. $0.25 - 18.13$ ⁻17.88

Find the product.

33. $4(-26)$ −104
34. $-76(-6)$ 456 [3-7]
35. $-12(12)$ −144
36. $19(-8)$ −152
37. $-10(-13 + 13)$ 0
38. $-8(-6 - 8)$ 112
39. $(-12 + 4)(7 - 8)$ 8
40. $(8 - 10)(10 - 8)$ −4

41. $2.4(-1.2)$ −2.88
42. $-1.4(3.4)$ −4.76 [3-8]
43. $-8.2(-6.1)$ 50.02
44. $-4.3(-3.4)$ 14.62
45. $-2.15(1.15)$ −2.4725
46. $3.14(-8.1)$ −25.434
47. $-4.25(-3.14)$ 13.345
48. $1.11(-1.11)$ −1.2321

Simplify.

49. $-36 \div 9$ −4
50. $-65 \div (-5)$ 13 [3-9]
51. $3.06 \div (-1.7)$ −1.8
52. $-6.4 \div 0.8$ −8
53. $(-43.16 \div 8.3) \div (-5.2)$ 1
54. $(0.5)(-4.3) \div (-5)(-0.8)$ −0.5375
55. $(2.5)(-3.4) \div (-2.5)(1.7)$ 2
56. $(-1.01)(-1.11) \div (1.25)(-0.1)$ 9.688

13. ⁻$42 - $⁻$30 = $⁻$12; $⁻$42 - $⁻$12 = $⁻$30$
15. ⁻$10 - $⁻$26 = 16; $⁻$10 - 16 = $⁻$26$

Extra Practice—Chapter 4

For use after page 120.

Complete.

1. $6 \times \underline{\ ?\ } = 1\frac{1}{6}$

2. $11 \div 12 = \underline{\ ?\ } \quad \frac{11}{12}$ **[4-1]**

3. $3\frac{8}{9} = \frac{?}{9} \quad 35$

4. $5\frac{7}{16} = \frac{?}{16} \quad 87$

5. $\frac{19}{8} = \underline{\ ?\ } \quad 2\frac{3}{8}$

6. $\frac{44}{7} = \underline{\ ?\ } \quad 6\frac{2}{7}$

Write each fraction in two other ways.

7. $-\frac{4}{9} \quad \frac{-4}{9}; \frac{4}{-9}$

8. $-\frac{3}{7} \quad \frac{-3}{7}; \frac{3}{-7}$ **[4-2]**

9. $-\frac{5}{6} \quad \frac{-5}{6}; \frac{5}{-6}$

10. $-\frac{9}{10} \quad \frac{-9}{10}; \frac{9}{-10}$

Complete.

11. $-3\frac{3}{8} = \frac{?}{8} \quad -27$

12. $-\frac{25}{4} = \underline{\ ?\ } \quad -6\frac{1}{4}$

13. $-10\frac{7}{9} = \frac{?}{9} \quad -97$

14. $-\frac{35}{6} = \underline{\ ?\ } \quad -5\frac{5}{6}$

In Ex. 15–16, answers may vary.

15. Give two fractions equal to $-\frac{11}{13}$. $-\frac{22}{26}; -\frac{33}{39}$ **[4-3]**

16. Give two fractions equal to $-\frac{15}{16}$. $-\frac{30}{32}; -\frac{45}{48}$

Reduce each fraction to lowest terms.

17. $\frac{32}{28} \quad \frac{8}{7}$

18. $-\frac{48}{54} \quad -\frac{8}{9}$

19. $\frac{95}{145} \quad \frac{19}{29}$

20. $-\frac{63}{81} \quad -\frac{7}{9}$

21. $\frac{180}{210} \quad \frac{6}{7}$

22. $-\frac{144}{96} \quad -\frac{3}{2}$

Compute each product.

23. $-\frac{3}{4} \times \frac{4}{5} \quad -\frac{3}{5}$

24. $\frac{21}{56} \times \frac{20}{25} \quad \frac{3}{10}$ **[4-4]**

25. $\frac{7}{8} \times \left(-\frac{4}{7}\right) \quad -\frac{1}{2}$

26. $\frac{13}{27} \times \frac{81}{78} \quad \frac{1}{2}$

27. $-\frac{9}{14} \times \frac{18}{45} \quad -\frac{9}{35}$

28. $-\frac{15}{16} \times \left(-\frac{56}{25}\right) \quad \frac{21}{10}$

29. $-\frac{21}{52} \times \left(-\frac{13}{15}\right) \quad \frac{7}{20}$

30. $6\frac{3}{4} \times \left(-1\frac{1}{3}\right) \quad -9$

31. $-2\frac{5}{8} \times \left(-\frac{16}{19}\right) \quad 3\frac{15}{128}$

32. $-4\frac{2}{7} \times 2\frac{1}{4} \quad -9\frac{9}{14}$

For use after page 137.

Replace each set of fractions with equal fractions having the least common denominator (LCD).

1. $\frac{3}{4}, \frac{5}{6}$ $\frac{9}{12}, \frac{10}{12}$

2. $-\frac{8}{9}, \frac{7}{15}$ $-\frac{40}{45}, \frac{21}{45}$ [4-5]

3. $-\frac{3}{7}, -\frac{2}{5}$ $-\frac{15}{35}, -\frac{14}{35}$

4. $\frac{5}{8}, -\frac{1}{6}$ $\frac{15}{24}, -\frac{4}{24}$

5. $\frac{1}{4}, \frac{7}{12}, \frac{5}{18}$ $\frac{9}{36}, \frac{21}{36}, \frac{10}{36}$

6. $-\frac{1}{2}, \frac{3}{7}, \frac{5}{6}$ $-\frac{21}{42}, \frac{18}{42}, \frac{35}{42}$

Express the sum or difference as a proper fraction in lowest terms or as a mixed number with fraction in lowest terms.

7. $\frac{5}{9} + \left(-\frac{1}{4}\right)$ $\frac{11}{36}$

8. $-\frac{9}{14} - \frac{5}{32}$ $-\frac{179}{224}$ [4-6]

9. $3\frac{5}{7} - 2\frac{1}{3}$ $1\frac{8}{21}$

10. $-4\frac{7}{12} + 2\frac{4}{5}$ $-1\frac{47}{60}$

11. $2\frac{7}{12} + \frac{1}{5} + 1\frac{1}{6}$ $3\frac{19}{20}$

12. $3\frac{9}{14} - \left(-\frac{19}{21}\right)$ $4\frac{23}{42}$

Divide. Express all results in lowest terms.

13. $\frac{4}{7} \div \frac{9}{20}$ $1\frac{17}{63}$

14. $-\frac{11}{13} \div \left(-\frac{23}{52}\right)$ $1\frac{21}{23}$ [4-7]

15. $-16 \div 2\frac{10}{11}$ $-5\frac{1}{2}$

16. $3\frac{5}{6} \div \left(-1\frac{5}{12}\right)$ $-2\frac{12}{17}$

17. $-\frac{5}{6} \div 4\frac{1}{2}$ $-\frac{5}{27}$

18. $3\frac{5}{9} \div (-32)$ $-\frac{1}{9}$

Express as a fraction or a mixed number in lowest terms.

19. 1.34 $1\frac{17}{50}$

20. -1.75 $-1\frac{3}{4}$ [4-8]

21. 2.05 $2\frac{1}{20}$

22. -3.025 $-3\frac{1}{40}$

23. 0.008 $\frac{1}{125}$

24. -0.048 $-\frac{6}{125}$

Express as a terminating or repeating decimal.

25. $\frac{7}{4}$ 1.75

26. $\frac{9}{2}$ 4.5

27. $-\frac{21}{22}$ $-0.9\overline{54}$

28. $\frac{7}{16}$ 0.4375

29. $-\frac{161}{189}$ $-0.8\overline{51}$

30. $-\frac{287}{385}$ $-0.7\overline{45}$

Extra Practice—Chapter 5

For use after page 152.

Write each ratio as a fraction in lowest terms.

[5-1]

1. 4 to 12 $\frac{1}{3}$
2. 9 to 6 $\frac{3}{2}$
3. $\frac{12}{27}$ $\frac{4}{9}$
4. $\frac{175}{75}$ $\frac{7}{3}$
5. 10 months to 2 years $\frac{5}{12}$
6. 3 kL to 90 L $\frac{100}{3}$
7. 45 min to 2 h $\frac{3}{8}$
8. 55 cm to 1 m $\frac{11}{20}$

Tell whether or not the proportion is correct.

[5-2]

9. $\frac{2}{3} = \frac{6}{7}$ No
10. $\frac{4}{9} = \frac{24}{54}$ Yes
11. $\frac{10}{11} = \frac{20}{21}$ No
12. $\frac{8}{7} = \frac{9}{8}$ No
13. $\frac{11}{22} = \frac{44}{88}$ Yes
14. $\frac{21}{27} = \frac{7}{9}$ Yes

Solve the proportion.

15. $\frac{n}{4} = \frac{12}{9}$ $5\frac{1}{3}$
16. $\frac{21}{n} = \frac{7}{12}$ 36
17. $\frac{15}{2} = \frac{n}{5}$ $37\frac{1}{2}$
18. $\frac{60}{48} = \frac{10}{n}$ 8
19. $\frac{6}{5} = \frac{n}{35}$ 42
20. $\frac{25}{31} = \frac{n}{100}$ $80\frac{20}{31}$
21. $\frac{n}{90} = \frac{53}{100}$ $47\frac{7}{10}$
22. $\frac{75}{3} = \frac{n}{2}$ 50
23. $\frac{40}{n} = \frac{2}{3}$ 60
24. $\frac{39}{2} = \frac{n}{3}$ $58\frac{1}{2}$

For use after page 165.

Find the percent or number.

[5-3]

1. What is 5% of 60?   3
2. What is 15% of 84?   12.6
3. What is 8% of 24?   1.92
4. What is 40% of 75?   30
5. 12 is what percent of 14? $85\frac{5}{7}\%$
6. 11 is what percent of 11?   100%
7. 16 is what percent of 20?   80%
8. 20 is what percent of 15? $133\frac{1}{3}\%$
9. 30 is 25% of what number?   120
10. 8 is 45% of what number? $17\frac{7}{9}$
11. 36 is 200% of what number?   18
12. 29 is 87% of what number? $33\frac{1}{3}$

Solve.

13. The Sport Shoppe sold 25 sets of golf clubs in May and 20% more than that in June. How many did they sell in June? 30 sets   **[5-4]**

14. Maggie planted 15 tomato plants. Six of the plants died. What percent of the plants lived? 60%

15. The band concert drew an audience of 275 persons on Thursday. On Friday, 50 more persons attended than on Thursday. What was the percent of increase? $18\frac{2}{11}$%

16. Of 512 entrants in a walkathon, 32 did not 'finish. What was the percent of decrease between the number of entrants and the number who finished the walkathon? 93.75%

17. The Flower Pot Place sold 115 lilac plants in May and 20% fewer than that in June. How many did they sell in June? 92 plants

18. An auctioneer receives a commission of 15% of sales. At one auction, the sales were $13,750. What was the commission? $2062.50   **[5-5]**

19. A potter pays an art store 25% commission to sell his pots. One set of pots sold for $225. What commission did he pay the art store? $56.25

20. A real-estate agency pays its salespersons a commission of 6%. What would the commission be on a house selling for $44,500? $2670

21. A jewelry store pays its salespersons a commission of 14%. During one week the total sales were $175,000. How much did the store pay in commissions? $24,500

22. A salesperson earns 8% commission on all furniture she sells. How much would she earn for selling $2500 worth of furniture? $200

23. Tim Mayes earns a commission of 12% on what he sells. Last month his sales totaled $14,500. How much did he earn? $1740

24. A yacht salesperson worked for a commission rate of 7% of sales. Last year, total sales were $350,000. Find the salesperson's commission for the year. $24,500

Extra Practice—Chapter 6

For use after page 189.

Complete.

1. A segment has __?__ endpoint(s).   2   [6-1]
2. 25 m = __?__ cm   2500   3. 5300 m = __?__ km   5.3
4. 160 cm = __?__ mm   1600   5. 400 mm = __?__ m   0.4

6. A chord of a circle is a __?__ joining two points of the circle.   segment   [6-2]
7. If the radius of a circle is 30 cm, then the circumference of the circle is __?__.
   (Use $\pi \approx 3.14$.)   188.4 cm
8. If the radius of a circle is 3.5 cm, then the circumference of the circle is __?__.
   (Use $\pi \approx \frac{22}{7}$.)   22 cm
9. If the diameter of a circle is 24, then the circumference of the circle is __?__.
   (Use $\pi \approx 3.14$.)   75.36
10. If the diameter of a circle is 14.7, then the circumference of the circle is __?__.
    (Use $\pi \approx \frac{22}{7}$.)   46.2
11. If the circumference of a circle is 26.4, then the radius of the circle is __?__.
    (Use $\pi \approx \frac{22}{7}$.)   4.2
12. If the circumference of a circle is 18.84, then the radius of the circle is __?__.
    (Use $\pi \approx 3.14$.)   3

13. The angles formed by bisecting an obtuse angle are both __?__.   acute   [6-3]
14. You can bisect an angle by using a __?__ or by using a __?__.   protractor; compass
15. Angles with the same measure are called __?__ angles.   equal
16. Two complementary equal angles each have measure __?__. 45°
17. If one of two vertical angles has measure 65°, the other angle has measure __?__. 65°
18. Perpendicular lines form __?__ angles.   right

19. Whenever two parallel lines are cut by a transversal, the alternate interior angles   [6-4]
    are __?__.   equal
20. If two lines are cut by a transversal so that the corresponding angles are equal,
    then the two lines are __?__.   parallel
21. If two of the angles formed when two lines intersect have a sum of 95°, the
    sum of the measures of the other two angles is __?__. 265°

For use after page 209.

1. If the measure of one angle of a right triangle is 60°, what are the measures of **[6-5]**
   the other two angles?   30°; 90°
2. What is true about the sides of an equilateral triangle?   All sides are equal.
3. What kind of triangle has sides 3 cm, 6 cm, and 5 cm?   scalene
4. How many acute angles does a right triangle have?   2

True or false? Write T or F.

5. Every rectangle is a parallelogram.   T **[6-6]**
6. Every parallelogram is a quadrilateral.   T
7. Opposite sides of a trapezoid are parallel.   F
8. Opposite angles of a parallelogram are equal.   T

9. Octagons have fewer sides than hexagons.   F **[6-7]**
10. Equilateral hexagons have seven equal sides.   F
11. What is the perimeter of a square with one side 65 mm?   260 mm
12. If one side of a rectangle is 2.8 m long and the perimeter of the rectangle is
    17.2 m, how long are the other sides of the rectangle?   2.8 m; 5.8 m; 5.8 m
13. The perimeter of an equilateral pentagon is 131.5 cm. Find the length of each
    side.   26.3 cm

14. If three angles of one triangle are equal to three angles of another triangle, must **[6-8]**
    the triangles be congruent?   No
15. If two sides and the included angle of one triangle are equal to two sides and the
    included angle of another triangle, must the triangles be congruent?   Yes

Complete each statement for the
congruent triangles.

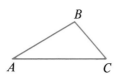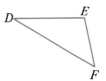

16. $\triangle DEF \cong$ __?__ $\triangle ABC$
17. $\angle C =$ __?__ $\angle F$
18. $DE =$ __?__ AB

19. Are all similar polygons also congruent polygons?   No **[6-9]**
20. Are all triangles similar?   No;
21. Are all equilateral triangles similar?   Yes
22. Are all quadrilaterals similar?   No
23. Are all isosceles trapezoids similar?   No
24. Are all equilateral octagons similar?   Yes

Extra Practice—Chapter 7

For use after page 230.

Write a variable expression.

1. Nine less than x $x - 9$ **2.** n divided by six $\frac{n}{6}$ **[7-1]**

3. The difference when z is subtracted from t $t - z$

4. The quotient when eight is divided by w $\frac{8}{w}$

Write a word phrase. Check students' papers.

5. $x + 9$ **6.** $3y - 1$ **7.** $\frac{n}{4}$ -4 **8.** $-2z - 3$

9. $\frac{3}{7}t$ **10.** $-x + 2$ **11.** $\frac{6}{z}$ 9 **12.** $a - 1$

Evaluate the expressions if $z = -2$.

13. $-3 - z$ -1 **14.** $\frac{1}{6}z$ $-\frac{1}{3}$ **15.** $-2 + z$ -4 **16.** $-\frac{2}{z}$ 1 **[7-2]**

17. $2z - 2$ -6 **18.** $-\frac{z}{4}$ $\frac{1}{2}$ **19.** $-3z + 3$ 9 **20.** $2 + z$ 0

Translate into a number sentence.

21. A number x is the product of negative two and one half. $x = -2 \times \frac{1}{2}$ **[7-3]**

22. The quotient when t is divided by negative one is twelve. $\frac{t}{-1} = 12$

23. The sum of y and four is less than seven. $y + 4 < 7$

Write a word sentence. Check students' papers.

24. $2(x - 3) < 4$ **25.** $3(y - 1) \geq -4$

26. $5(1 - z) > -2$ **27.** $-4w - 3 \leq -5$

Tell whether or not the given number is a solution of the given open sentence.

28. $x - 10 > -2; 9$ Yes   **29.** $y + 1 \leq 0; -1$ Yes   **[7-4]**

30. $8 - 3x = -4; -4$ No   **31.** $-y - 1 = -1; 0$ Yes

32. $-2x \geq 12; -3$ No   **33.** $-3y \leq 0; 10$ Yes

34. $x < -6 - 2x; -1$ No   **35.** $x - 3 = 12 - 4x; 3$ Yes

Graph the solution set. The replacement set for x is $\{-5, -4, -3, -2, -1, 0\}$.

36. $x + 2 < 0$ $\{-5, -4, -3\}$ **37.** $x \leq 1$ $\{-5, -4, -3, -2, -1, 0\}$ **[7-5]**

38. $x = -4$ $\{-4\}$ **39.** $x + 1 = -6$ No solution

40. $3 - x \leq 2$ No solution   **41.** $3x - 1 \geq 5$ No solution

42. $-x \geq -3$ $\{-5, -4, -3, -2, -1, 0\}$ **43.** $4x + 2 = -6$ $\{-2\}$

44. $-x + 8 \leq 8$ $\{0\}$ **45.** $x - 4 \geq -6$ $\{-2, -1, 0\}$

For use after page 239.

Use one of the properties of equality to form a true sentence.

1. If $x = 13$, then $x - 7 =$ ___?___ 6   2. If $y = -3$, then $4y =$ ___?___ -12 [7-6]
3. If $z = 5$, then $z + 3 =$ ___?___ 8   4. If $t = 4$, then $-3t =$ ___?___ -12
5. If $x = -11$, then $x + 4 =$ ___?___ -7 6. If $y = 5$, then $3y =$ ___?___ 15
7. If $z = -5$, then $-2z =$ ___?___ 10   8. If $t = -4$, then $3 - t =$ ___?___ 7

Solve each equation and check your solution.

9. $x + 3 = 5$ $\{2\}$ 10. $y + 2 = 1$ $\{-1\}$ [7-7]
11. $z - 3 = 6$ $\{9\}$ 12. $t - 4 = 3$ $\{7\}$
13. $x - 5 = -7$ $\{-2\}$ 14. $y + 8 = -3$ $\{-11\}$
15. $z + 7 = 9$ $\{2\}$ 16. $t + \frac{1}{2} = \frac{1}{2}$ $\{0\}$
17. $6x = 36$ $\{6\}$ 18. $-4x = 12$ $\{-3\}$
19. $7z = -28$ $\{-4\}$ 20. $-9t = -45$ $\{5\}$
21. $\frac{2}{3}x = 4$ $\{6\}$ 22. $-\frac{1}{2}x = 5$ $\{-10\}$
23. $-\frac{3}{4}z = -9$ $\{12\}$ 24. $\frac{5}{6}t = -15$ $\{-18\}$
25. $-2 + 8 = x + 2$ $\{4\}$ 26. $y + 6 = 3 - 7$ $\{-10\}$
27. $z(5 - 3) = -20$ $\{-10\}$ 28. $t(3 - 5) = 10$ $\{-5\}$
29. $x + 3x = 12$ $\{3\}$ 30. $-y - 2y = -6$ $\{2\}$
31. $6z - 2z = 24$ $\{6\}$ 32. $-t - (-3t) = -2$ $\{-1\}$
33. $4x = 1 - 5$ $\{-1\}$ 34. $-3y = -2 - 13$ $\{5\}$
35. $\frac{1}{2}x = \frac{2}{3} + \frac{1}{6}$ $\{\frac{5}{3}\}$ 36. $-\frac{1}{4}x = \frac{3}{4} - \frac{7}{8}$ $\{\frac{1}{2}\}$

Solve. The replacement set for x is {the integers}.

37. $x - 2 > -4$ $\{-1, 0, 1, \ldots\}$ 38. $x + 3 \le -4$ $\{\ldots, -9, -8, -7\}$ [7-8]
39. $x + 1 \ge -3$ $\{-4, -3, -2, \ldots\}$ 40. $x - 4 < -2$ $\{\ldots, -1, 0, 1\}$
41. $x + 4 \le 4$ $\{\ldots, -2, -1, 0\}$ 42. $x - 5 \ge 3$ $\{8, 9, 10, \ldots\}$
43. $x + 5 > -1$ $\{-5, -4, -3, \ldots\}$ 44. $x - 1 \le 0$ $\{\ldots, -1, 0, 1\}$
45. $x - 9 > 2$ $\{12, 13, 14, \ldots\}$ 46. $x + 6 \ge 7$ $\{1, 2, 3, \ldots\}$
47. $20 < 4x$ $\{6, 7, 8, \ldots\}$ 48. $-25 > 5x$ $\{\ldots, -8, -7, -6\}$
49. $24 > -2x$ $\{-11, -10, -9, \ldots\}$ 50. $-24 > -2x$ $\{13, 14, 15, \ldots\}$
51. $-3x \ge 0$ $\{\ldots, -2, -1, 0\}$ 52. $6x < -6$ $\{\ldots, -4, -3, -2\}$
53. $-7x \ge -70$ $\{\ldots, 8, 9, 10\}$ 54. $-8x \le -80$ $\{10, 11, 12, \ldots\}$
55. $-4 < -x$ $\{\ldots, 1, 2, 3\}$ 56. $4 - 10 \le x$ $\{-6, -5, -4, \ldots\}$
57. $-x \le 1$ $\{-1, 0, 1, \ldots\}$ 58. $9x > -18$ $\{-1, 0, 1, \ldots\}$
59. $-2x < -5$ $\{3, 4, 5, \ldots\}$ 60. $-3x > 7$ $\{\ldots, -5, -4, -3\}$

Extra Practice—Chapter 8

For use after page 254.

Write two practical problems that are expressed by the given equation. For the first problem, use a simple direct question. Check students' papers. Answers will vary.

1. $x + 16 = 43$ 2. $25 + x = 33$ [8-1]
3. $5x = 80$ 4. $13x = 91$
5. $x - 9 = 36$ 6. $17 - x = 11$
7. $\frac{1}{5}x = 7$ 8. $\frac{2}{3}x = 12$

Write an equation that could be used to solve the problem.

9. The sum of 19 and some number is 41. Find the number. $19 + n = 41$ [8-2]

10. When a certain number is subtracted from 32, the difference is 15. Find the number. $32 - n = 15$

11. When 9 is subtracted from some number, the difference is 25. Find the number. $n - 9 = 25$

12. The sum of two numbers is 64, and one of the numbers is 14 greater than the other number. Find the numbers. $n + (n + 14) = 64$

13. The sum of two numbers is 97, and one of the numbers is 56 greater than the other number. Find the numbers. $n + (n + 56) = 97$

14. The perimeter of a square is 36 m. Find the length of each side. $4s = 36$

15. The length of a rectangle is 35 cm greater than the width. The perimeter of the rectangle is 150 cm. Find the dimensions of the rectangle.
$w + (w + 35) + w + (w + 35) = 150$

Write an equation and use it to solve the problem.

16. The sum of two numbers is 75, and one number is 17 greater than the other number. Find the lesser number. 29   [8-3]

17. The sum of two numbers is 53, and one number is 21 greater than the other number. Find the greater number. 37

18. The difference when 27 is subtracted from a number is 18. Find the number. 45

19. Find two consecutive integers whose sum is 43. 21 and 22

20. The sum of two numbers is 31, and one number is 17 greater than the other. One number is even and is 3 more than 3 times the other number, which is odd. Find the numbers. 7 and 24

21. Find two consecutive even integers whose sum is 46. 22 and 24

For use after page 263.

Solve.

1. The scale on a map of Vega City is 1 cm → 0.25 km. The distance on the map from the Largo Building to the Königsberg Bridge is 5.5 cm. What actual distance does this represent? 1.375 km   [8-4]

2. The scale on a map of Lasco Desert is 1 cm → 0.25 km. The distance on the map from Point Peso to the Verdi Oasis is 12.6 cm. What actual distance does this represent? 3.15 km

3. On a scale drawing of a rectangular-shaped table top the scale is 1 cm → 0.5 m. The table top measures 4 m by 3 m. What size is the drawing? 8 cm by 6 cm

4. A map of Dallas, Texas, has the scale of 1 cm → 0.5 km. What map distance corresponds to an actual distance of 11 km? 22 cm

5. Lee Chung deposited $800 at 6.5% interest for 2 years. What was the simple interest earned? $104   [8-5]

6. Gerda Cohen deposited $900 at 5.5% interest for 3 years. What was the simple interest earned? $148.50

7. Bob Esposito invested $2000 at 6% annual interest. How much must he invest at 7% annual interest for his annual income from both investments to total $470? $5000

8. Althea Washington invested $2000 at 7% annual interest. How much must she invest at 6% annual interest for her annual income from both investments to total $410? $4500

9. What is the annual cost of operating a 300 W color television if it is used an average of 4 h per day and the electric rate is 6.5¢ per kW · h? 26¢   [8-6]

10. A helium balloon, Double Eagle II, made the first successful balloon crossing of the Atlantic Ocean in August, 1978. It traveled approximately 5200 km in 137 h. What was its average rate of speed in kilometers per hour? About 38 km/h

11. A pipeline system is about 1500 km long and about 0.75 m in diameter. The pipeline cost $3.1 billion to complete. What is the cost per kilometer? About $2,066, 667

Extra Practice—Chapter 9

For use after page 281.

For use after page 281.

1. A square has side 8 cm. What is its area? 64 cm^2 [9-1]
2. A rectangle has base 15 cm and height 7 cm. What is its area? 105 cm^2
3. A parallelogram has base 7 m and height 8 m. What is its area? 56 m^2
4. A rectangle has area 96 cm² and base 16 cm. What is its height? 6 cm
5. A square has area 64 cm². What is its perimeter? 32 cm
6. A square has perimeter 32 cm. What is its area? 64 cm^2
7. A parallelogram has area 48 cm² and height 6 cm. What is its base? 8 cm
8. A rhombus has perimeter 36 m. What is its base? 9 m

9. The base of a triangle is 28 mm long and its height is 36 mm. What is its area? 504 mm^2 [9-2]
10. The base of a triangle is 31 cm long and its height is 22 cm. What is its area? 341 cm^2
11. A trapezoid has bases 4 m and 5 m, and height 4.6 m. What is its area? 20.7 m^2
12. A trapezoid has bases 12 cm and 19 cm, and height 14 cm. What is its area? 217 cm^2
13. The legs of a right triangle are 6 m and 8 m. What is its area? 24 m^2
14. The legs of a right triangle are 21 mm and 28 mm. What is its area? 294 mm^2
15. The area of a triangle is 18 m². The base is 6 m. What is the height? 6 m
16. The area of a triangle is 35 cm². The base is 20 cm. What is the height? 3.5 cm

Find the area of the circle described.
17. radius = $1\frac{1}{2}$ units (Use $\pi \approx \frac{22}{7}$.) $7\frac{1}{14}$ sq. units   [9-3]
18. diameter = 12 cm (Use $\pi \approx 3.14$.) 113.04 cm^2
19. circumference = $20\,\pi$ (Leave answer in terms of π.) 100π
20. Find the area of the shaded part of the figure. (Use $\pi \approx 3.14$.) 50.24 m^2

For use after page 299.

For use after page 299.

Find the volume. In Exercises 1 and 2, leave answers in terms of π.
1. A cylinder with base diameter 12 and height 8. 288π [9-4]
2. A cylinder with base radius 4 and height 14. 224π
3. A prism whose base is a triangle with area 32 and whose height is 9. 288
4. A prism whose base is a square with area 35 and whose height is 10. 350
5. A cube with edge 12 cm. 1728 cm^3
6. A rectangular prism which measures 14 m by 11 m by 6 m. 924 m^3

7. A hexagonal pyramid with base 40 cm² and height 8 cm. about 106.67 cm³   **[9-5]**

8. A pentagonal pyramid with base 48 cm² and height 9 cm. 144 cm³

9. A cone whose base has radius 5 and which has height 11. (Leave answer in terms of π.) $91\frac{2}{3}\pi$

10. A cone whose base has diameter 18 mm and whose height is 21 mm. (Leave answer in terms of π.) 56.7π mm³

11. The base of a prism is a triangle with sides 4, 6, and 7. The height of the prism   **[9-6]**
170

Find the lateral area and the total surface area of each solid below.
Leave your answer to Exercise 14 in terms of π.

12.

6

6

6

144; 216

13.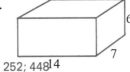

6

7

252; 448

14.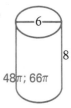

6

8

48π; 66π

Complete.

15. $4.6\,L = \underline{\ ?\ }\ cm^3$ 4600   **16.** $600\,g = \underline{\ ?\ }\ kg$ 0.6   **[9-7]**

17. $850\,mL = \underline{\ ?\ }\ L$ 0.85   **18.** $800\,cm^3 = \underline{\ ?\ }\ L$ 0.8

19. If the mass of 1 cm³ of gold is 19.3 g and of 1 cm³ of aluminum is 2.7 g, what is the difference in mass of 25 cm³ of each? 4.15 g

20. If the mass of 1 cm³ of milk is 1.03 g, what is the mass of 1 L of milk? 1030 g

21. If the mass of 1 cm³ of gasoline is 0.66 g, what is the mass of 1 L of gasoline? 660 g

22. Find the area and volume (in terms of π) of a sphere with radius 2. 16π; $\frac{32}{3}\pi$ **[9-8]**

23. What is the area of a sphere with radius 20 cm? (Use $\pi \approx 3.14$.) 5024 cm²

24. What is the volume of a sphere with radius $\frac{3}{4}$? (Leave answer in terms of π.) $\frac{9}{16}\pi$

25. If the area of a sphere is 16π, what is its volume in terms of π? $\frac{32}{3}\pi$

Extra Practice—Chapter 10

For use after page 317.

Is the experiment random?

1. You choose a frozen yogurt from 6 flavors available at a shop.   No   [10-1]
2. You spin a spinner with 8 possible outcomes.   Yes
3. You choose a one-dollar bill, a five-dollar bill, a ten-dollar bill, and a twenty-dollar bill from a container.   No
4. You choose 1 card from a standard deck of 52 cards.   Yes

Write the number of possible outcomes and the probability of each outcome.

5. A letter is randomly chosen from a, b, c, d, e, and f. $6; \frac{1}{6}$
6. A marble is drawn at random from a hat containing a red, a green, a white, and a yellow marble. $4; \frac{1}{4}$
7. A tag is drawn at random from a bag containing a white, a blue, an orange, a black, a red, and a green tag. $6; \frac{1}{6}$
8. One card is drawn at random from a standard deck of 52 cards. $52; \frac{1}{52}$
9. One marble is chosen at random from a bag with 8 differently colored marbles. $8; \frac{1}{8}$
10. One team is selected from 14 teams. $14; \frac{1}{14}$
11. The pointer on a spinner stops at one of the numbers 1, 2, 3, or 4. $4; \frac{1}{4}$
12. The pointer on a spinner stops at one of the letters a, b, c, or d. $4; \frac{1}{4}$
13. A penny and a nickel are tossed. $4; \frac{1}{4}$
14. A penny, a nickel, and a dime are tossed. $8; \frac{1}{8}$

The letters of the word *mathematics* are written on 11 cards. If a card is drawn at random, what is the probability of drawing:

15. an m? $\frac{2}{11}$ 16. an a? $\frac{2}{11}$ [10-2]
17. an i? $\frac{1}{11}$ 18. a t? $\frac{2}{11}$
19. an e? $\frac{1}{11}$ 20. a d?   0
21. an a or an e? $\frac{3}{11}$ 22. an m or a t? $\frac{4}{11}$
23. an a, an s, or an m? $\frac{5}{11}$ 24. an a, an m, or a t? $\frac{6}{11}$

Jesse has 8 tape cassettes in her car. One cassette stars the Rockers, 2 star Clarence Klow, 3 star Sue and the Stompers, and 2 star Queen Quigley. Jesse selects a cassette at random. What are the odds:

25. in favor of selecting a cassette by Clarence?   1 to 3   [10-3]
26. against selecting a cassette by Sue and the Stompers?   5 to 3

For use after page 327.

Answer the question asked. Base your answer on the evidence given in the problem.

1. Ray has 15 hits out of 27 times at bat. What is the probability that he will make a hit the next time at bat? $\frac{5}{9}$ **[10-4]**

2. A train arrived on time 4 out of the past 7 days. What is the probability that it will not be on time tomorrow? $\frac{3}{7}$

3. A wrestler has won 7 of his last 8 matches. What is the probability that he will win his next match? $\frac{7}{8}$

4. A meteorologist has predicted the weather correctly 865 times out of 1000. What is the probability she will be correct the next time? $\frac{173}{200}$

An inspector on an assembly line chose 200 condensers at random out of 4000 condensers. He found that 6 of the 200 condensers were defective.

5. What is the probability that a consumer will get a defective condenser? $\frac{3}{100}$ **[10-5]**

6. What is the probability that a consumer will not get a defective condenser? $\frac{97}{100}$

7. About how many defective condensers would you expect to find in the 4000 condensers? 120 condensers

8. If you randomly choose a condenser, what are the odds against your getting a defective condenser? 97 to 3

9. If you randomly choose a condenser, what are the odds for your getting a condenser that is not defective? 97 to 3

Extra Practice—Chapter 11

For use after page 343.

1. Name the graph of each ordered pair:

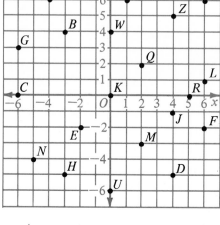

   a. (1,6) *P*   **b.** (5,0) *R*   **[11-1]**
   c. (6,1) *L*   **d.** (−3,4) *B*
   e. (−3,−5) *H* **f.** (4,−5) *D*
   g. (0,0) *K*   **h.** (−5,−4) *N*
   i. (−6,3) *G*   **j. (6,6)** *I*
   k. (2,−3) *M* **l.** (2,2) *Q*
   m. (−6,0) *C*   **n.** (0,−6) *U*
   o. (6,−2) *F*   **p.** (−2,−2) *E*
   q. (0,4) *W*   **r.** (4,5) *Z*
   s. (4,−1) *J*   **t.** (−4,6) *V*

On a coordinate plane, **a.** sketch the polygon whose vertexes are given and **b.** find the area of the polygon.

  2. (3,3), (−3,−3), (3,−3), (−3,3) *A = 36*   **3.** (3,4), (−2,4), (−2,−2), (3,−2) *A = 30*
  4. (0,5), (3,−3), (−3,−3) *A = 24*   **5.** (−3,−1), (−3,2), (6,2), (6,−1) *A = 27*
  6. (0,5), (5,5), (5,−2) *A = 17$\frac{1}{2}$*   **7.** (1,1), (4,4), (1,4), (4,1) *A = 9*
  8. (2,−1), (6,−1), (6,−5), (2,−5) *A = 16*

Exercises 9–14 refer to the figure at the right.

  9. Write the coordinates of the   **[11-2]**
     given points:
     a. *M*   **b.** *A*   **c.** *S*
     d. *C*   **e.** *P*   **f.** *R*
     g. *B*   **h.** *Q*   **i.** *E*
     j. *H*   **k.** *D*   **l.** *G*

  10. Name all points shown whose ordinates
     are negative. *B, C, P, Q, R*

  11. Name all points shown whose abscissas
     are 0. *R*
                                   B, Q
  12. Name all points shown in Quadrant IV.

  13. Name all points shown on the x-axis. *E, S*

  14. Name all points shown whose ordinates are 0. *E, S*

      a. (−4, 1)   **b.** (−2, 4)   **c.** (−1, 0)   **420**   **g.** (2, −4)   **h.** (5, −6)   **i.** (3, 0)
      d. (−5, −2)   **e.** (−2, −4)   **f.** (0, −2)   **j.** (1, 3)   **k.** (6, 4)   **l.** (2, 1)

For use after page 355.

State whether or not the given ordered pair is a solution of the equation.

1. $(3,1)$, $2x + y = 5$ No
2. $(-1,4)$, $x - 2y = -9$ Yes [11-3]
3. $(-2,-2)$, $y - 3x = 4$ Yes
4. $(-2,-2)$, $y - 3x = -4$ No
5. $(0,-3)$, $3x - 2y = -6$ No
6. $(-3,0)$, $2x - 3y = -6$ Yes
7. $(5,-2)$, $x - y = 3$ No
8. $(-4,-3)$, $3x + 2y = 18$ No

Give an equivalent equation for y in terms of x.

9. $3x + 2y = 6$ $y = 3 - \frac{3}{2}x$
10. $2y - 3x = -6$ $y = \frac{3}{2}x - 3$
11. $2x - 4y = 2$ $y = \frac{1}{2}x - \frac{1}{2}$
12. $3y - 5x = 10$ $y = \frac{5}{3}x + \frac{10}{3}$
13. $4x + 5y = 0$ $y = -\frac{4}{5}x$
14. $y - 6x = 12$ $y = 6x + 12$
15. $3y = 4x - 12$ $y = \frac{4}{3}x - 4$
16. $3x + 9 = 6y$ $y = \frac{1}{2}x + \frac{3}{2}$

Find ordered pairs of the form $(x,0)$ and $(0,y)$ that are solutions of the given equation.

17. $x + y = 2$ $(2, 0)$; $(0, 2)$
18. $x + y = -3$ $(-3, 0)$; $(0, -3)$ [11-4]
19. $x - y = -1$ $(-1, 0)$; $(0, 1)$
20. $x + 2y = 4$ $(4, 0)$; $(0, 2)$
21. $2x - y = 4$ $(2, 0)$; $(0, -4)$
22. $3x + 2y = 6$ $(2, 0)$; $(0, 3)$
23. $-3x + 4y = 12$ $(-4, 0)$; $(0, 3)$
24. $4y - 5x = -10$ $(2, 0)$; $(0, -\frac{5}{2})$

Graph each equation on a coordinate plane. Use a separate set of axes for each equation. Check students' papers.

25. $x - 2y = 3$
26. $x = -2$
27. $2x + 3y = 6$
28. $y = 1$
29. $3x - y = 3$
30. $y = x - 3$
31. $2x = y - 2$
32. $4x - 2y = 3$

Solve each system by graphing. If no solution exists, say so.

33. $2x - y = 0$
$x + y = -3$ $(-1, -2)$
34. $2x - 4y = 6$
$x = 5$ $(5, 1)$ [11-5]
35. $y - 2x = -3$
$y - 2x = 3$ No solution
36. $2x + 3y = 5$
$x - y = 0$ $(1, 1)$
37. $2x + y = 4$
$x - 2y = -8$ $(0, 4)$
38. $3x + 3y = 4$
$x + 6y = 3$ $(1, \frac{1}{3})$
39. $3x + y = 2$
$3x + y = 3$ No solution
40. $-2x + y = 5$
$x + y = -1$ $(-2, 1)$
41. $2x + 3y = -10$
$2x + y = -6$ $(-2, -2)$
42. $x - 2y = 0$
$2x + y = 5$ $(2, 1)$

Extra Practice—Chapter 12

For use after page 370.

Name the two consecutive integers between which the square root lies.

1. $\sqrt{8}$ 2 and 3
2. $\sqrt{2}$ 1 and 2
3. $\sqrt{7}$ 2 and 3   [12-1]

4. $\sqrt{12}$ 3 and 4
5. $\sqrt{19}$ 4 and 5
6. $\sqrt{22}$ 4 and 5

7. $\sqrt{27}$ 5 and 6
8. $\sqrt{33}$ 5 and 6
9. $\sqrt{39}$ 6 and 7

10. $\sqrt{60}$ 7 and 8
11. $\sqrt{43}$ 6 and 7
12. $\sqrt{73}$ 8 and 9

13. $\sqrt{58}$ 7 and 8
14. $\sqrt{80}$ 8 and 9
15. $\sqrt{78}$ 8 and 9

State the next estimate for the square root of the dividend.

$$\begin{array}{c}5.6\\16.\ 5\overline{)28.0}\end{array}\ 5.3 \qquad \begin{array}{c}7.1\\17.\ 8\overline{)57.0}\end{array}\ 7.6 \qquad \begin{array}{c}6.1\\18.\ 6\overline{)37.0}\end{array}\ 6.1$$ [12-2]

$$\begin{array}{c}3.7\\19.\ 4\overline{)15.0}\end{array}\ 3.9 \qquad \begin{array}{c}7.7\\20.\ 9\overline{)70.0}\end{array}\ 8.4 \qquad \begin{array}{c}8.4\\21.\ 7\overline{)59.0}\end{array}\ 7.7$$

$$\begin{array}{c}8.8\\22.\ 8\overline{)71.0}\end{array}\ 8.4 \qquad \begin{array}{c}6.6\\23.\ 6\overline{)40.0}\end{array}\ 6.3 \qquad \begin{array}{c}6.6\\24.\ 5\overline{)33.0}\end{array}\ 5.8$$

$$\begin{array}{c}9.2\\25.\ 9\overline{)83.0}\end{array}\ 9.1 \qquad \begin{array}{c}8.8\\26.\ 7\overline{)62.0}\end{array}\ 7.9 \qquad \begin{array}{c}7.3\\27.\ 6\overline{)44.0}\end{array}\ 6.7$$

Find each square root to the tenths' place. (Do not use the square-root table.)

28. $\sqrt{10}$ 3.2
29. $\sqrt{17}$ 4.1
30. $\sqrt{24}$ 4.9

31. $\sqrt{29}$ 5.4
32. $\sqrt{31}$ 5.6
33. $\sqrt{82}$ 9.1

34. $\sqrt{90}$ 9.5
35. $\sqrt{37}$ 6.1
36. $\sqrt{48}$ 6.9

37. $\sqrt{87}$ 9.3
38. $\sqrt{75}$ 8.7
39. $\sqrt{94}$ 9.7

Use the table on page 394 to approximate each number to the nearest hundredth.

40. $\sqrt{14}$ 3.74
41. $\sqrt{28}$ 5.29
42. $\sqrt{35}$ 5.92   [12-3]

43. $\sqrt{43}$ 6.56
44. $\sqrt{53}$ 7.28
45. $\sqrt{83}$ 9.11

46. $2\sqrt{6}$ 4.90
47. $3\sqrt{5}$ 6.71
48. $4\sqrt{3}$ 6.93

49. $8\sqrt{19}$ 34.87
50. $6\sqrt{41}$ 38.42
51. $10\sqrt{8}$ 28.28

52. $2\sqrt{76}$ 17.44
53. $10\sqrt{3}$ 17.32
54. $10\sqrt{24}$ 48.99

55. $\sqrt{21} - \sqrt{19}$ 0.22
56. $\sqrt{37} - \sqrt{27}$ 0.89
57. $\sqrt{8} - \sqrt{3}$ 1.10

58. $\sqrt{10} + \sqrt{10}$ 6.32
59. $\sqrt{13} + \sqrt{8}$ 6.43
60. $\sqrt{27} + \sqrt{2}$ 6.61

For use after page 385.

State whether or not a triangle with sides of the given lengths is a right triangle.

1. $3, 4, 5$ Yes
2. $30, 40, 50$ Yes   [12-4]
3. $8, 15, 17$ Yes
4. $8, 10, 12$ No
5. $6, 8, 10$ Yes
6. $10, 20, 24$ No

A right triangle has sides of lengths a, b, and c, with c the length of the hypotenuse. Find the length of the missing side to the nearest hundredth.

7. $a = 3, b = 4$ 5
8. $a = 10, b = 20$ 22.36
9. $a = 5, b = 5$ 7.07
10. $a = 3, b = 6$ 6.71
11. $a = 6, b = 8$ 10
12. $a = 7, b = 9$ 11.40

Find an approximation to the nearest hundredth for x.

13.

$x \approx 8.48$

14.

$x \approx 3.32$

15.
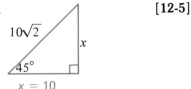
$x = 10$ [12-5]

16.

$x = 4$

17.
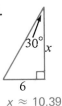
$x \approx 10.39$

18.
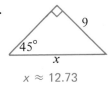
$x \approx 12.73$

Use the diagram to name each ratio.

19. $\tan A$ $\frac{x}{y}$
20. $\sin A$ $\frac{x}{z}$ [12-6]
21. $\cos A$ $\frac{y}{z}$
22. $\tan B$ $\frac{y}{x}$
23. $\cos B$ $\frac{x}{z}$
24. $\sin B$ $\frac{y}{z}$

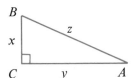

Refer to the diagram at the right.

25. Which equation is incorrectly written?   C   [12-7]
  a. $\cos D = \frac{12}{13}$
  b. $\sin E = \frac{12}{13}$
  c. $\cos E = \frac{13}{12}$
  d. $\tan E = \frac{12}{x}$
  e. $\sin D = \frac{x}{13}$

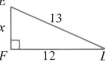

26. Find the value of x.   5
27. Find $\angle D$ to the nearest degree. $23°$
28. Find $\angle E$ to the nearest degree. $67°$

Glossary

Abscissa (p. 340) The first number in an ordered pair of numbers which are coordinates assigned to a point in the plane. Also called the *x*-coordinate.

Absolute value of a number (p. 80) The absolute value of a number *a*, denoted by $|a|$, equals *a* if *a* is positive or zero and equals the opposite of *a* if *a* is negative.

Acute angle (p. 180) An angle between 0° and 90°.

Acute triangle (p. 190) A triangle with three acute angles.

Alternate interior angles (p. 185) One of two pairs of angles formed when two lines are cut by a transversal. In the diagram, $\angle 3$ and $\angle 6$ are one such pair, and $\angle 4$ and $\angle 6$ the other.

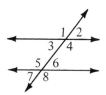

Angle (p. 180) A figure formed by two rays with a common endpoint.

Arc of a circle (p. 177) A portion of a circle which joins two points on the circle.

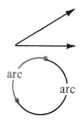

arc   arc

Area (p. 271) Amount of interior surface of a geometric figure.

Associative property (p. 42) Changing the grouping of addends in a sum, or of factors in a product, does not change the sum or product. For any numbers *a, b,* and *c,*
$a + (b + c) = (a + b) + c$ and $a \times (b \times c) = (a \times b) \times c$.

Bar graph (p. 5) A graph in which the length of each bar is proportional to the number it represents.

Base, numerical (p. 27) A number which is raised to some power. In 10^4, 10 is the base.

Base of a geometric figure *See* names of specific figures.

424

Beaufort scale (p. 12) A scale which assigns the numbers 0 through 12 to various ranges of wind speeds.

Binary system (p. 60) The place value numeration system in which a number is expressed as a sum of powers of two.

Bisector of an angle (p. 181) A ray which divides an angle into two equal angles.

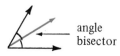

angle bisector

Chord (p. 177) A line segment joining two points of a circle.

chord

Circle (p. 177) A plane figure whose points are all the same distance from a given point in the plane, called the *center* of the circle.

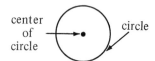

center of circle   circle

Circumference (p. 177) The distance around a circle.

Common denominator (p. 121) A common multiple of the denominators of two or more fractions.

Commutative property (p. 42) Changing the order of addends in a sum, or of factors in a product, does not change the sum or product. For any numbers *a* and *b*,
$a + b = b + a$ and $a \times b = b \times a$.

Complementary angles (p. 180) Two angles whose sum is 90°.

Cone (p. 286) If a point which is not in the plane of a circle is joined by line segments to each point of the circle, the solid formed is a cone.

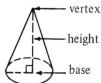

vertex
height
base

Congruent polygons (p. 201) Polygons which have the same size and same shape.

Coordinate (p. 34) The number paired with a point on the number line.

Coordinates of a point (p. 340) An ordered pair of numbers assigned to a given point in the plane.

Corresponding angles (p. 185) One of four pairs of angles formed when two lines are cut by a transversal. The corresponding angles shown are $\angle 1$ and $\angle 5$, $\angle 2$ and $\angle 6$, $\angle 3$ and $\angle 7$, and $\angle 4$ and $\angle 8$.

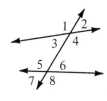

Corresponding parts (p. 201) Parts of two polygons which are paired in a one-to-one correspondence.

Cosine of an angle (p. 380) If $\angle A$ is one of the acute angles in a right triangle, the cosine of $\angle A$ ($\cos A$) is the ratio of the length of the side adjacent to $\angle A$ to the length of the hypotenuse.

$$\cos A = \frac{b}{c}$$

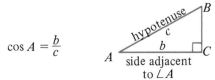

Cylinder (p. 282) A solid with two parallel bases that are congruent circles and one curved lateral face.

base

Decibel (p. 23) A unit for measuring the level of sound.

Decimal system (p. 28) The number system which is based on powers of ten.

Diameter of a circle (p. 177) A chord which contains the center of a circle. Also, the length of such a chord.

diameter

Distributive property (p. 43) For any numbers $a, b, c, a(b + c) = ab + ac$.

Endpoint (p. 173) A boundary point of a line segment or ray.

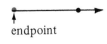

endpoint

Equation (p. 222) A number sentence in which two number phrases are connected by the symbol "$=$."

Equator (p. 358) The great circle on Earth whose plane is perpendicular to the axis. It separates Earth into the Northern and Southern Hemispheres.

Equilateral polygon (p. 197) A polygon with all sides equal.

Equilateral triangle (p. 190) A triangle with all sides equal.

Equivalent equations (p. 234) Equations which have the same solutions.

Equivalent inequalities (p. 237) Inequalities which have the same solutions.

Evaluate an expression (p. 37) To replace a variable in an expression with one of its values and then complete the indicated arithmetic.

Expected value of a random variable (p. 324) To find the expected value of a random variable, multiply each possible value by its probability and add the results.

Experimental probability of an event (p. 318) The number of times an event occurs divided by the number of times the experiment is performed.

Event (p. 310) A collection of possible outcomes of an experiment.

Exponent (p. 27) A number which indicates the power to which the base is to be raised.

Factor (p. 27) Any of two or more numbers which are multiplied to form a product.

Frequency polygon (p. 16) A broken-line graph connecting the midpoints of the tops of the bars of a histogram.

Graph of an equation (p. 348) The line consisting of all points whose coordinates satisfy the equation.

Graph of a number (p. 34) The point on the number line paired with the number.

Great circle (p. 358) The intersection of a sphere with a plane passing through the center of the sphere.

Histogram (p. 14) A bar graph that displays numerical data which have been organized into equal intervals.

Hypotenuse (p. 371) The side of a right triangle which is opposite the right angle.

Improper fraction (p. 105) A fraction in which the numerator is greater than or equal to the denominator.

Inequality (p. 222) A number sentence in which the two phrases are connected by an inequality symbol.

Integers (p. 66) The set of whole numbers and their opposites: $\{\ldots, -2, -1, 0, 1, 2, \ldots\}$

Inverse operations (p. 40) Adding a number and subtracting the same number are inverse operations. Similarly, multiplying by a number and dividing by the same number are inverse operations.

Isosceles right triangle (p. 375) A right triangle in which the two legs are equal.

Isosceles trapezoid (p. 194) A trapezoid in which the non-parallel sides are equal.

Isosceles triangle (p. 190) A triangle having at least two sides equal.

Lateral area of a prism or cylinder (pp. 290, 291) The surface area not including the bases.

Lateral edge of a pyramid (p. 285) An edge of a pyramid which does not belong to the base.

Lateral face of a pyramid or prism (pp. 285, 290) A face which is not a base.

Latitude (p. 358) The number of degrees between $0°$ and $90°$ that a place is north or south of the equator.

Least common denominator (LCD) (p. 121) The least common multiple of the denominators of two or more fractions.

Least common multiple (LCM) (p. 121) The least positive number that is a multiple of two or more given numbers.

Line (p. 173) A line segment extended without end in both directions.

Line segment (p. 173) A part of a line consisting of two endpoints and all the points between them.

Longitude (p. 358) The number of degrees between $0°$ and $180°$ that any place is east or west of the prime meridian.

Lowest terms (for fractions) (p. 113) A fraction whose numerator and denominator have no common factor except 1 is said to be in lowest terms.

Mass (p. 293) Quantity of matter.

Mean (p. 17) The sum of a set of numbers divided by the number of numbers in the set.

Median (p. 17) In a set of numbers arranged in increasing order, the median is the middle number if there is an odd number of numbers. If the number of numbers is even, the median is the mean of the two middle numbers.

Metric system (p. 54) A system of measurement based on powers of ten.

Mixed number (p. 105) A whole number and a fraction.

Mode (p. 17) The number which occurs most frequently in a set of data.

Multiple (p. 121) The product of a given number and a whole number.

Napier's rods (p. 388) A computing device invented by John Napier (1550–1617) in order to calculate products.

Negative numbers (p. 66) Numbers graphed to the left of the zero reference point on a horizontal number line.

Nonrepeating decimal (p. 140) A decimal in which the digits do not exhibit a repeating pattern.

Obtuse angle (p. 180) An angle between $90°$ and $180°$.

Obtuse triangle (p. 190) A triangle which contains an obtuse angle.

Odds (p. 313) The ratio of the probability of an event occurring to the probability of the event not occurring is called the odds in favor of the event.

Open sentence (p. 222) A sentence that contains a variable, such as $2p - 1 > 5$.

Opposite numbers (p. 66) Two numbers whose sum is zero are opposites.

Ordered pair of numbers (p. 335) A pair of numbers in which one is designated as the first component and the other as the second component.

Ordinate (p. 340) The second number in an ordered pair of numbers which are coordinates assigned to a point in the plane. Also called the y- coordinate.

Origin (p. 34) The point on the number line whose coordinate is zero. The zero reference point.

Parallel lines (p. 185) Lines in the same plane which do not intersect.

Parallelogram (p. 194) A quadrilateral with opposite sides parallel. Any side of a parallelogram can be considered as the *base*.

Perfect square (p. 364) An integer whose square root is also an integer.

Perimeter of a polygon (p. 197) The sum of the lengths of the sides of the polygon.

Perpendicular lines (p. 180) Lines that intersect to form $90°$ angles.

Pictograph (p. 22) A graph which uses visual symbols to represent data.

Point (p. 173) A mathematical figure which can be pictured by a dot.

Polygon (p. 197) A closed plane figure formed by connecting line segments at their endpoints.

Positive numbers (p. 66) Numbers graphed to the right of the zero reference point on a horizontal number line.

Power (p. 27) A product of equal factors.

Prime factorization (p. 122) The representation of a whole number greater than zero as the product of primes.

Prime number (p. 122) A whole number greater than 0 having only two factors, itself and one.

Prism (p. 282) If two congruent polygons are placed in parallel planes so that their corresponding sides are parallel and their corresponding vertexes are joined by line segments, the solid formed is called a prism.

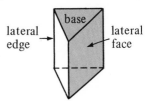

Probability (p. 307) A number between, and including, 0 and 1 which measures how likely it is that an event will occur.

Proper fraction (p. 105) A fraction in which the numerator is smaller than the denominator.

Property of one in multiplication (p. 43) The product of one and any given number a is the given number. That is, $1 \times a = a$.

Property of zero in addition (p. 42) Zero added to any number a gives the sum a; that is, $0 + a = a$.

Property of zero in multiplication (p. 42) The product of any number and zero is zero. That is, $0 \times a = 0$.

Protractor (p. 180) An angle-measuring device.

Pythagorean property (p. 371) For any right triangle, the square of the length of the hypotenuse is equal to the sum of the squares of the lengths of the other two sides.

Pythagorean triples (p. 374) Any three counting numbers a, b, c which satisfy the equation $a^2 + b^2 = c^2$.

Quadrant (p. 340) One of the four regions into which the axes of rectangular coordinate system separate the plane.

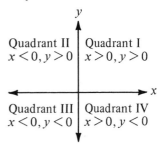

Quadrilateral (p. 194) A four-sided polygon.

Radius (p. 177) A segment joining any point of a circle to the center of the circle. Also, the length of such a segment.

radius

Random experiment (p. 307) An experiment in which the outcome is a matter of chance.

Random sample (p. 320) A sample in which each member of the population is equally likely to appear.

Random variable (p. 323) A variable whose value is determined by the outcome of a random experiment.

Range of a set of data (p. 17) The difference between the greatest number and the least number in the set.

Rational number (p. 110) Any number which can be represented as the quotient $\frac{a}{b}$ of two integers a and b, b not zero.

Ray (p. 173) A line segment extended without end in one direction only.

Reciprocal (p. 130) For all rational numbers a and b, if $ab = 1$, then a is the reciprocal of b and b is the reciprocal of a.

Rectangle (p. 194) A parallelogram which contains a right angle.

Rectangular coordinate system (p. 335) A system which consists of two number lines perpendicular to each other and intersecting at the origin of each. It is used to assign coordinates to points in the plane.

Repeating decimal (p. 133) A decimal in which a pattern of digits repeats without end.

Repetend (p. 133) The repeating part of a repeating decimal.

Replacement set for a variable (p. 37) The set of numbers which may be used to replace the variable.

Rhombus (p. 194) A quadrilateral with all sides equal.

Right angle (p. 180) A 90° angle.

Right triangle (p. 190) A triangle which contains a right angle.

Scalene triangle (p. 190) A triangle in which no two sides are equal.

Scientific notation (p. 100) The system of representing any positive number as the product of a power of ten and a number between 1 and 10.

Semicircle (p. 177) An arc which joins the endpoints of a diameter.

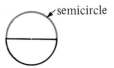
semicircle

Side of an angle (p. 180) One of the rays of an angle.

Similar polygons (p. 206) Polygons which have the same shape.

Sine of $\angle A$ (p. 379) For an acute angle A of a right triangle, the sine of $\angle A$ (sin A) is the ratio of the length of the side opposite $\angle A$ to the length of the hypotenuse.

Solution of an open sentence (p. 225) A value of a variable which yields a true statement.

Solution set of an open sentence (p. 227) The set of all values in the replacement set which are solutions.

Solving a right triangle (p. 382) The process of finding the measures of the sides and angles of a right triangle.

Sphere (p. 296) The set of all points in space which are a given distance from a fixed point.

Square (p. 194) A parallelogram with all sides equal and four right angles.

Square root (pp. 128, 363) If $a^2 = n$, then a is a square root of n.

Supplementary angles (p. 180) Two angles whose sum is 180°.

System of equations (p. 352) A set of two or more equations in the same variables.

Tangent of $\angle A$ (p. 380) For an acute angle A of a right triangle, the tangent of A (tan A) is the ratio of the length of the side opposite angle A to the length of the side adjacent to angle A.

30°–60° right triangle (p. 376) A right triangle in which the two acute angles are 30° and 60°.

Transversal (p. 185) A line which intersects two other lines.

Trapezoid (p. 194) A quadrilateral with exactly two sides parallel. The parallel sides are the *bases* of the trapezoid.

Trigonometric ratio (p. 379) Any of the sine, cosine, or tangent ratios.

Unit fraction (p. 106) A fraction whose numerator is one.

Value of a variable (p. 37) Any member of the replacement set of a variable.

Variable (p. 37) A letter which is used to stand for any member of a given set of numbers.

Vertex of an angle (p. 180) The common endpoint of the sides of an angle.

Vertex of a polygon (p. 197) A common endpoint of two sides.

Vertical angles (p. 181) Two angles whose sides form two pairs of opposite rays. In the diagram, ∠ 1 and ∠ 2 are vertical angles as are ∠ 3 and ∠ 4.

Volume (p. 282) The amount of space occupied by a solid.

x-axis (p. 335) The horizontal axis of a rectangular coordinate system.

y-axis (p. 335) The vertical axis of a rectangular coordinate system.

Zero meridian (p. 358) The semicircle on Earth which has end points at the poles and passes through Greenwich, England.

Zero reference point (p. 65) The point on the number line whose coordinate is zero.

Index

431

Credits

Creative art: Terry Presnall
Mechanical art: ANCO

Transforming Formulas

For use after page 297.

A formula, such as $A = bh$, is an equation involving more than one variable. A is expressed in terms of b and h. When we transform the formula into an equivalent formula expressing h in terms of A and b, we say that we are *solving the formula for h*.

Example 1 | Solve the formula $A = bh$ for h.

Solution | Write the formula. $\qquad\qquad\qquad\qquad A = bh$

Divide both members by b to get $\qquad \frac{A}{b} = h$

h by itself as one member.

Example 2 | Solve $v = w + at$ for t.

Solution | Write the equation. $\qquad\qquad\qquad v = w + at$

Subtract w from both members to

get the term with t by itself. $\qquad v - w = at$

Divide both members by a $\qquad\qquad \dfrac{v = w}{a} = t$

to get t by itself.

Solve each formula for the variable indicated in parentheses.

1. $A = lw$ (l) $\frac{A}{w} = l$

2. $C = \pi d$ (d) $\frac{C}{\pi} = d$

3. $\pi = \frac{C}{2r}$ (C) $2\pi r = C$

4. $I = Prt$ (t) $\frac{I}{Pr} = t$

5. $A = \frac{1}{2}bh$ (b) $\frac{2A}{h} = b$

6. $V = \frac{Bh}{3}$ (h) $\frac{3V}{B} = h$

7. $m = \frac{a + b}{2}$ (a) $2m - b = a$

8. $t = S - Sr$ (r) $\frac{t - S}{-S} = r$

9. $L = 2\pi rh + 2\pi r^2$ (h) $\frac{L - 2\pi r^2}{2\pi r} = h$

10. $2A = h(a + b)$ (b) $\frac{2A}{h} - a = b$

Surface Areas of Pyramids and Cones

For use after page 297.

The base of a regular pyramid is an equilateral polygon, and the lateral faces are congruent triangles. The *slant height* is the height of a triangular face.

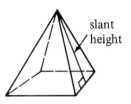
slant height

To find the lateral surface area of a regular pyramid, multiply the perimeter of the base by the slant height and divide by 2.

Lateral area $= \frac{1}{2} \times$ (perimeter of base) $\times$ slant height

The same formula is used to find the lateral surface area of a right cone. In the case of a cone, the perimeter is the circumference of the circular base ($2\pi r$).

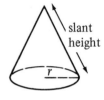
slant height

r

Example	**Find the lateral surface area of a right cone with slant height 10 and radius 5.**
Solution	**Lateral area $= \frac{1}{2} \times (2 \times \pi \times 5) \times 10$** **$= 50\pi$**

Right Cone

Solve.

1. A fluorite crystal is in the shape of two regular square pyramids base to base. The slant height is 2.0 cm, and the base is 3.2 cm on a side. Find the surface area of the crystal. 25.6 cm²

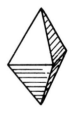

2. Some campers at Camp Millstream want to make a canvas shelter in the shape of a right cone with slant height 5 m and radius 3 m. How many square meters of material will they need? (Use $\pi \approx$ 3.14.) 47.1 m²

3. A right cone fits snugly inside a regular square pyramid. Both have a slant height of 14 cm. The square base is 10 cm on a side. Find the lateral surface areas of the pyramid and the cone. (Use $\pi \approx$ 3.14.) 280 cm² and 219.8 cm²

4. Two regular pyramids are such that the dimensions of one are twice those of the other. The larger pyramid has a square base 20 cm on a side and a slant height of 12 cm. Find the lateral surface area of each pyramid. What is the ratio of the two surface areas? 480 cm²; 120 cm²; 4 to 1

The Graph of a Linear Inequality

For use after page 350.

The graph of an equation in two variables such as $y = x + 1$ is a straight line drawn on a coordinate plane. The line separates the plane into three sets of points:

> the points *on* the line,
> the points *above* the line,
> the points *below* the line.

For points not on the line, $y > x + 1$ or $y < x + 1$.

Example	Graph the inequality $y < x + 1$.
Solution	First graph the related equation $y = x + 1$ as a dashed line. All points in the region below this line have coordinates (x,y) such that $y < x + 1$. Shade this region to show the graph of $y < x + 1$. Check: Choose a point in the shaded region, say $(2,1)$. See whether it is a solution of the given inequality.

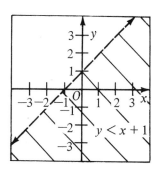

$$y \underset{?}{\overset{<}{}} x + 1$$
$$1 < 2 + 1$$
$$1 < 3 \ \checkmark$$

Graph each inequality. The graph is the region described below.

1. $y < x + 3$ **2.** $y < 2x + 3$ **3.** $y < 2 - x$

4. $y < 3 - x$ **5.** $y > x + 5$ **6.** $y > x - 3$

7. $y < 4 - x$ **8.** $y > 1 - 2x$ **9.** $2x + 5y > 10$

1. Below the line through $(0,3)$ and $(-3,0)$ 2. Below the line through $(-1\frac{1}{2},0)$ and $(0,3)$ 3. Below the line through $(0,2)$ and $(2,0)$ 4. Below the line through $(0,3)$ and $(3,0)$ 5. Above the line through $(-5,0)$ and $(0,5)$ 6. Above the line through $(0,-3)$ and $(3,0)$ 7. Below the line through $(0,4)$ and $(4,0)$ 8. Above the line through $(0,1)$ and $(\frac{1}{2},0)$ 9. Above the line through $(0,2)$ and $(5,0)$

Answers to Selected Exercises

Chapter 1   Statistics and Graphs

PAGES 3–4 EXERCISES   **1.** 17,938 km   **3.** 17,384 km
5. 23,002 km   **7.** 689 km   **9.** 2320 km   **11.** 18,430 km
13. 23,018 km   **15.** 79,130 km   **17.** 162,684 km
19. 115 min   **21.** 115 min   **23.** 45 h 59 min

PAGE 7 EXERCISES   **1.** About 17 m   **3.** About 50 m
5. About 90 m   **7.** About 72 m

9.
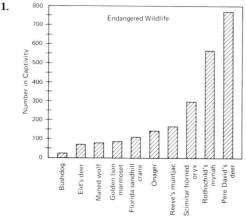

11.
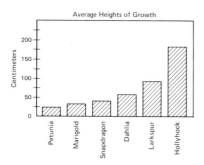

9.
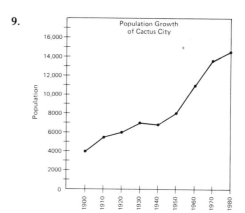

PAGE 15 EXERCISES   **1.** 53, 56, 61, 65, 65, 73, 74,
77, 80, 82, 82, 87, 88, 89, 90, 91, 94, 94, 94, 96,
99, 102, 106, 106, 107

3.
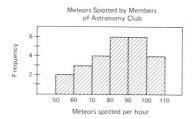

5.
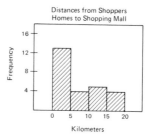

PAGE 18 EXERCISES   **1.** 65 cm   **3.** 64 cm   **5.** 6   **7.** 5

PAGE 11 EXERCISES   **1.** About 800 km
3. About 2600 km   **5.** 125 min   **7.** 11:30 A.M.

PAGES 20–21 CHAPTER REVIEW   **1.** D   **3.** B   **5.** B
7. C   **9.** C   **11.** C   **13.** D   **15.** A

PAGES 22–23 4°C = lowest average monthly temperature. This occurred in April. July had the greatest monthly precipitation. This was 9 cm.

In 1979 a total eclipse of the sun and moon occurs. No total eclipse of sun in 1982.

One record and four books cost $18. One record and 3 books cost $15. $13 remaining. With $13 the following combinations of books and records can be bought: 1 record, 1 book; 1 record, 2 books; 2 records, no books; no records and 1, 2, 3, or 4 books.

PAGE 25 SKILL REVIEW 1. 81 **3.** 107 **5.** 154
7. 96 **9.** 83 **11.** 966 **13.** 1048 **15.** 1427 **17.** 149
19. 1650 **21.** 12 **23.** 37 **25.** 35 **27.** 32 **29.** 68
31. 475 **33.** 372 **35.** 194 **37.** 654 **39.** 121 **41.** 72
43. 265 **45.** 480 **47.** 1380 **49.** 819 **51.** 1182
53. 27,840 **55.** 86,086 **57.** 24 **59.** 65 **61.** 379
63. 78 **65.** 94 **67.** 27 **69.** 32

Chapter 2   The Decimal System

PAGES 29–30 EXERCISES 1. 10^3 **3.** 10^4 **5.** 10^9
7. $(5 \times 10^3) + (8 \times 10^2) + (1 \times 10^1) + 2$
9. $(8 \times 10^3) + (4 \times 10^1) + 9$ **11.** 7475 **13.** 631,060
15. 2573 **17.** 8,407,459 **19.** 87 **21.** 875 **23.** 785
25. 8754 **27.** 2457

PAGE 30 EXTRA 1. a. 87 **b.** 87 **3. a.** 568 **b.** 568

PAGE 33 EXERCISES 1. $(6 \times 10) + 5 + (4 \times 0.1)$
3. $(6 \times 0.1) + (5 \times 0.01) + (4 \times 0.001)$
5. $(2 \times 10) + 9 + (8 \times 0.1) + (6 \times 0.01)$
7. $(4 \times 0.01) + (7 \times 0.001) + (2 \times 0.0001)$
9. 0.479 **11.** 572.08 **13.** 13.0052 **15.** 604.307
17. Twenty-six and five tenths
19. Three hundred and twelve thousandths
21. Eleven ten-thousandths
23. Four hundred one and nine hundred six ten-thousandths **25.** 14.7 **27.** 500.015 **29.** 250.346

PAGE 36 EXERCISES

1.

3.

5.

7. $9 < 15$ **9.** $21 > 19$ **11.** $0.8 > 0.5$
13. $6.1 < 6.3 < 6.5$ **15.** $1.53 > 1.43$ **17.** $>$
19. $<, <$ **21.** $>$ **23.** $<$ **25.** $<, <$ **27.** $>$
29. $<$ **31.** $>$

PAGE 38 EXERCISES 1. 18 **3.** 13 **5.** 315 **7.** 92
9. 2262 **11.** {12, 14, 25} **13.** {32, 256, 288}
15. {26, 22, 13} **17.** {1701, 693, 1260}

19. {30, 12, 2} **21.** {7, 15, 21} **23.** {16, 31, 41}
25. {24, 48, 108}

PAGE 41 EXERCISES 1. $x = 102 - 83$, $x = 19$
3. $x = 242 - 116$, $x = 126$
5. $x = 200 - 133$, $x = 67$
7. $x = 235 - 135$, $x = 100$
9. $x = 316 - 299$, $x = 17$
11. $x = 48$ **13.** $y = 156 \div 12$, $y = 13$
15. $y = 600 \div 25$, $y = 24$ **17.** $y = 132 \div 44$, $y = 3$
19. $y = 1818 \div 18$, $y = 101$
21. $y = 781 \div 71$, $y = 11$
23. $y = 2000 \div 16$, $y = 125$

PAGE 41 APPLICATION 3. 103,389
5. 723 Explanation: $7 \times 11 \times 13 = 1001$, and any 3-digit number times 1001 gives a 6-digit number which is just the same sequence of digits repeated once.

PAGE 44 EXERCISES 1. 47 **3.** 107 **5.** 40 **7.** 120
9. 520 **11.** 18,000 **13.** 60 **15.** 540 **17.** 344
19. 228 **21.** 220 **23.** 247 **25.** 850 **27.** 6500

PAGE 46 EXERCISES 1. 87.8 **3.** 14.731 **5.** 44.164
7. 53.3249 **9.** 24.201 **11.** 12.0216 **13.** 11.626
15. 9.53 **17.** 20.378 **19.** 40.6 **21.** 1.534 **23.** 11.757
25. 7.5666 **27.** 42.98 **29.** 70.33 **31.** 194.4154

PAGE 47 PROBLEMS 1. $7.16

PAGE 49 EXERCISES 1. 41.3 **3.** 55.44 **5.** 14.884
7. 285.5265 **9.** 1.3344 **11.** 0.42966 **13.** 0.001505
15. 370.44 **17.** 1.0759

PAGE 50 PROBLEMS 1. $136.50

PAGE 50 APPLICATION 1. B + W television is cheaper to operate; 45 cents to run the B + W television, $1.65 to run the color television **3.** $14.16

PAGE 53 EXERCISES 1. 2.9 **3.** 7.5 **5.** 4.36
7. 0.209 **9.** 31.27 **11.** 41.8 **13.** 51.7 **15.** 50.6
17. 201.6 **19.** 9.16 **21.** 17.52 **23.** 83.03 **25.** 542.80
27. 7.63

PAGE 53 PROBLEMS 1. $8.43 **3.** 4 kg with 24¢ left over

PAGE 54 APPLICATION 1. 3 **3.** 0.764 **5.** 100
7. 0.5 **9.** 1000 **11.** 0.375

PAGE 56 PROBLEMS 1. 1.5 kg **3.** 0.5 kW·h
5. $830.58 **7.** 4.8 cm by 6.4 cm

PAGE 58 CHAPTER REVIEW 1. D **3.** A **5.** D **7.** A
9. B **11.** D **13.** B **15.** A **17.** C **19.** C

PAGE 61 EXERCISES 1. 5 **3.** 11 **5.** 23 **7.** 44
9. 63 **11.** 111_{two} **13.** 1010_{two} **15.** 10100_{two}
17. 11111_{two} **19.** 110100_{two} **21.** 1010111_{two}

23.

+	0	1
0	0	1
1	1	10_{two}

25. 100001_{two}

27. 10000010_{two} **29.** 110_{two} **31.** 110111010_{two}

PAGE 63 SKILL REVIEW **1.** 8.5 **3.** 48.037 **5.** 5.588
7. 367.41 **9.** 6785.813 **11.** 0.0984 **13.** 2.6 **15.** 2.51
17. 5.05 **19.** 3.042 **21.** 0.977 **23.** 8184.5765
25. 32.76 **27.** 8.37624 **29.** 0.001615 **31.** 0.0733
33. 4531.868 **35.** 71.08374 **37.** 9.55 **39.** 3.20
41. 44.44 **43.** 0.92 **45.** 0.37 **47.** 0.13

Chapter 3 Positive and Negative Numbers

PAGE 69 EXERCISES **1.** -4 **3.** 46

5. (number line) -4, -3, 2, 5

7. (number line)

9. (number line)

11. (number line)

13. (number line)

15. (number line)

17. (number line) **19.** -6 **21.** 1

23. 19 **25.** -118 **27.** -1776 **29.** 26,475

31. (number line)

33. (number line)

35. (number line)

37. Positive **39.** -6 **41.** -0.75 **43.** 0 **45.** -4

PAGE 72 EXERCISES

1. (number line: -1.5, -1.2, 5.4)

3. (number line: -12.4, -3.0, -1.1, 0.7)

5. (number line) 2

7. (number line) -3

9. (number line) -3.5

11. (number line) 4

13. (number line) -2

15. (number line) 2.5

17. (number line) 3

19. (number line) -2.5

21. (number line) -3

23. 5; 3 **25.** 3; -3 **27.** -2.5; 3.75 **29.** -9.875; 6.75
31. -0.4; 0.35

PAGE 73 EXTRA

1. (number line) -5, -1, 0, 3, 4

3. (number line) -4, -3, -1, 0, 2

5. (number line: -2.5, 3.5) -4, -2.5, 0, 3.5

7. > **9.** > **11.** > **13.** >

PAGE 76 EXERCISES

1. (number line) 9, 4, 5

3.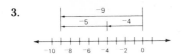

5.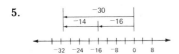

7.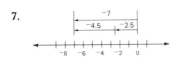

9. 12.1 **11.** ⁻3.4 **13.** ⁻147 **15.** ⁻404 **17.** ⁻1.3
19. ⁻7.0 **21.** ⁻8 **23.** ⁻46 **25.** ⁻17

PAGE 77 PROBLEMS **1.** 298 m

PAGE 77 APPLICATION **1.** 4 **3.** ⁻6

PAGE 79 EXERCISES

1.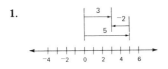

$5 + {}^-2 = 3$

3.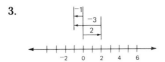

$2 + {}^-3 = {}^-1$

5.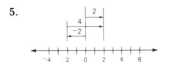

$^-2 + 4 = 2$

7.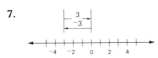

$^-3 + 3 = 0$

9. 5 **11.** ⁻3.5 **13.** 3 **15.** ⁻12 **17.** ⁻2.7 **19.** ⁻0.6
21. 2 **23.** ⁻7 **25.** ⁻5.9

PAGE 80 PROBLEMS **1.** 10,000 m

PAGE 81 EXTRA **1.** 1 **3.** 0 **5.** 8

PAGES 83–84 EXERCISES

1.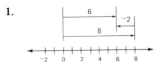

$6 - 8 = {}^-2$

3.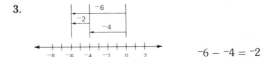

$^-6 - {}^-4 = {}^-2$

5.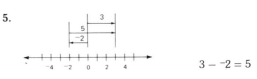

$3 - {}^-2 = 5$

7.

$^-5 - 0 = {}^-5$

9. 6 **11.** ⁻37 **13.** ⁻8 **15.** 33 **17.** 8 **19.** 34 **21.** ⁻8
23. 160 **25.** 22 = 16 − ⁻6; ⁻6 = 16 − 22
27. 29 = 16 − ⁻13; ⁻13 = 16 − 29
29. 15 = ⁻21 − ⁻36; ⁻36 = ⁻21 − 15

PAGE 84 PROBLEMS **1.** 27.7 s

PAGE 86 EXERCISES **1.** 7.5 + 1.3 **3.** 1.3 + ⁻3.4
5. 33.3 + ⁻46.5 **7.** 3.56 + 8.73 **9.** 5 **11.** ⁻20
13. 9.9 **15.** ⁻4.5 **17.** 10.88 **19.** ⁻13 **21.** 98.42
23. 99.45

PAGE 87 PROBLEMS **1.** $81.42

PAGE 90 EXERCISES **1.** $3(-4) = -12$

3. $4(-6) = -24$

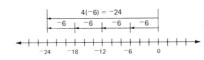

5. -72 **7.** -48 **9.** -132 **11.** -256 **13.** 12
15. 600 **17.** 1680 **19.** -2400 **21.** -1890
23. -1400 **25.** -294 **27.** 231 **29.** 460

PAGE 92 EXERCISES **1.** -9.62 **3.** -39.76
5. 19.04 **7.** 89.89 **9.** -15.96 **11.** -14.31
13. 17.92 **15.** 2.1 **17.** -4.7725 **19.** 7.02 **21.** -36
23. 17.856 **25.** -1.3455 **27.** 2.412 **29.** 9.89

PAGE 92 PROBLEMS **1.** 10¢

PAGE 93 EXTRA **1.** 10^{-1} **3.** $\frac{1}{1000}$; 10^{-3}

PAGE 96 EXERCISES **1.** -6 **3.** 17 **5.** -0.8
7. -0.9 **9.** 0 **11.** -1.7 **13.** 26.6 **15.** -34.7
17. -21.23 (rounded to hundredths)

19. 3.41 (rounded to hundredths)
21. -0.001 (rounded to thousandths)

PAGE 96 PROBLEMS **1.** 50¢

PAGE 98 CHAPTER REVIEW **1.** C **3.** C **5.** B
7. A **9.** D **11.** A **13.** D **15.** C **17.** D

PAGE 101 **1.** 3.460092×10^6 **3.** 6.59×10^{-3}
5. 3.85×10^2
7. White blood cell is about twice as large as the red blood cell. **9.** 300,000 km/s
11. 5,980,000,000,000,000,000,000,000,000 g
13. 6.96×10^{10} cm

PAGE 103 CUMULATIVE REVIEW **1.** Wednesday
3. Friday **5.** 10 **7.** 162 cm **9.** $>$ **11.** $>$ **13.** $>$
15. 83.11 **17.** 73.09 **19.** 22.812
21. 2.98 (rounded to hundredths) **23.** -1.6 **25.** 9.1
27. 66.42

Chapter 4　Rational Numbers

PAGE 108 EXERCISES **1.** 6 **3.** 3 **5.** $\frac{1}{3}$ **7.** $\frac{3}{7}$ **9.** 7
11. 19 **13.** 17 **15.** $5\frac{2}{3}$ **17.** $3\frac{1}{10}$ **19.** 14

PAGE 111 EXERCISES **1.** $\frac{1}{-8}; \frac{-1}{8}$ **3.** $-\frac{1}{12}; \frac{1}{-12}$
5. $-\frac{1}{3}; \frac{-1}{3}$ **7.** $\frac{-1}{25}; \frac{1}{-25}$ **9.** $\frac{-3}{17}; -\frac{3}{17}$ **11.** $\frac{-7}{17}; \frac{7}{-17}$
13. $-\frac{11}{3}; \frac{11}{-3}$ **15.** $-\frac{5}{2}; \frac{5}{-2}$ **17.** $-\frac{10}{3}$ **19.** $-\frac{47}{8}$
21. $-\frac{59}{7}$ **23.** $-\frac{93}{8}$ **25.** $-1\frac{3}{5}$ **27.** $-1\frac{7}{11}$ **29.** $-4\frac{4}{7}$
31. $-10\frac{3}{5}$

PAGE 114 EXERCISES (For Exs. 1–9, answers may vary.)
1. $\frac{26}{28}; \frac{39}{42}$ **3.** $-\frac{22}{14}; -\frac{33}{21}$ **5.** $\frac{24}{50}; \frac{36}{75}$ **7.** $-\frac{42}{58}; -\frac{63}{87}$
9. $\frac{74}{20}; \frac{111}{30}$ **11.** $\frac{1}{3}$ **13.** $-\frac{1}{2}$ **15.** $\frac{1}{2}$ **17.** $\frac{2}{3}$ **19.** $-\frac{3}{2}$
21. $-\frac{4}{3}$ **23.** $\frac{3}{4}$ **25.** $\frac{2}{3}$ **27.** $-\frac{2}{3}$ **29.** $\frac{5}{7}$

PAGE 118 EXERCISES **1.** $\frac{8}{35}$ **3.** $-\frac{1}{6}$ **5.** $\frac{7}{18}$ **7.** 0
9. -4 **11.** $-\frac{1}{12}$ **13.** -6 **15.** -14 **17.** $13\frac{1}{8}$ **19.** $\frac{1}{30}$
21. $18\frac{3}{64}$

PAGE 119 PROBLEMS **1.** 6 **3.** 60

PAGE 123 EXERCISES **1.** $\frac{6}{9}; \frac{5}{9}$ **3.** $-\frac{12}{20}; \frac{7}{20}$ **5.** $\frac{25}{15}; \frac{21}{15}$
7. $-\frac{88}{33}; \frac{27}{33}$ **9.** $\frac{28}{12}; \frac{9}{12}$ **11.** $-\frac{117}{1326}; -\frac{170}{1326}$ **13.** $\frac{35}{378}; \frac{153}{378}$
15. $\frac{68}{128}; -\frac{3}{128}$ **17.** $\frac{4}{8}; \frac{6}{8}; \frac{7}{8}$ **19.** $\frac{27}{144}; \frac{40}{144}; -\frac{6}{144}$
21. $\frac{56}{420}; -\frac{175}{420}; -\frac{48}{420}$ **23.** $\frac{98}{420}; \frac{36}{420}; \frac{203}{420}$

PAGE 126 EXERCISES **1.** $1\frac{5}{24}$ **3.** $\frac{23}{24}$ **5.** $\frac{11}{48}$ **7.** $\frac{17}{240}$
9. $-\frac{5}{96}$ **11.** $-\frac{87}{160}$ **13.** $1\frac{5}{8}$ **15.** $-1\frac{23}{48}$ **17.** $1\frac{23}{48}$
19. $1\frac{19}{24}$ **21.** $4\frac{1}{6}$ **23.** $9\frac{3}{16}$ **25.** $1\frac{11}{12}$ **27.** $9\frac{11}{12}$

PAGE 127 PROBLEMS **1.** $\frac{4}{5}$ h **3.** $\frac{59}{150}$ **5.** $\frac{8}{63}$

PAGE 128 EXTRA **1.** 10 **3.** 5 **5.** 9 **7.** $8^{\frac{1}{3}} = 2$

PAGE 130 EXERCISES **1.** $1\frac{1}{2}$ **3.** 1 **5.** $-\frac{5}{22}$
7. $-10\frac{1}{2}$ **9.** 1 **11.** $\frac{2}{5}$ **13.** $-\frac{22}{25}$ **15.** $\frac{1}{9}$ **17.** 5 **19.** 4
21. -6 **23.** $-1\frac{3}{11}$

PAGE 131 PROBLEMS **1.** $2\frac{2}{9}$

PAGE 134 EXERCISES **1.** $\frac{35}{10}$ **3.** $-\frac{45}{100}$ **5.** $\frac{2023}{1000}$
7. $\frac{2013}{10}$ **9.** $2\frac{3}{4}$ **11.** $-\frac{3}{8}$ **13.** $1\frac{5}{8}$ **15.** $\frac{52}{125}$ **17.** 1.25
19. 0.15 **21.** 0.375 **23.** 0.3125 **25.** -0.0125
27. $1.\overline{3}$ **29.** $0.\overline{5}$ **31.** $-0.\overline{63}$ **33.** $0.7\overline{14285}$
35. $0.\overline{076923}$ **37.** Repeating **39.** Repeating
41. Terminating
43. a. One **b.** Two **c.** Three **d.** Four
45. Since $0.\overline{9} = 1$, any terminating decimal can be represented as a repeating decimal.

PAGE 136 EXTRA **1.** $\frac{4}{9}$ **3.** $\frac{19}{9}$ **5.** $\frac{40}{33}$ **7.** $\frac{7}{30}$

PAGE 138–139 CHAPTER REVIEW **1.** D **3.** C
5. C **7.** B **9.** D **11.** A **13.** B **15.** D

PAGE 143 SKILL REVIEW **1.** $\frac{4}{5}$ **3.** $\frac{3}{5}$ **5.** $5\frac{2}{3}$ **7.** $4\frac{1}{6}$
9. $1\frac{1}{6}$ **11.** $-\frac{9}{10}$ **13.** $9\frac{3}{4}$ **15.** $-3\frac{1}{15}$ **17.** $\frac{1}{2}$ **19.** $\frac{3}{10}$
21. $3\frac{1}{2}$ **23.** $-1\frac{1}{3}$ **25.** $\frac{5}{8}$ **27.** $1\frac{2}{9}$ **29.** 3 **31.** -3
33. $\frac{1}{12}$ **35.** $-\frac{2}{3}$ **37.** $25\frac{2}{3}$ **39.** $-66\frac{2}{3}$ **41.** 6 **43.** -40
45. $1\frac{1}{3}$ **47.** $-\frac{1}{3}$ **49.** $2\frac{1}{2}$ **51.** $-2\frac{2}{7}$

Chapter 5　Ratio and Proportion

PAGE 148 EXERCISES **1.** $\frac{5}{2}$ **3.** $\frac{4}{15}$ **5.** $\frac{1}{3}$ **7.** 5 **9.** $\frac{50}{13}$
11. $\frac{5}{6}$ **13.** $\frac{7}{50}$ **15.** $\frac{3}{2}$ **17.** $\frac{5}{3}$

PAGE 148 PROBLEMS **1.** 22:1; \$1.32, \$1.98, \$4.62
3. $\frac{4}{7}, \frac{3}{7}$, yes

PAGE 151 EXERCISES **1.** Correct **3.** Incorrect
5. Incorrect **7.** Incorrect **9.** $n = 5$ **11.** $n = \frac{1}{2}$
13. $n = 7$ **15.** $n = 64$ **17.** $n = 210$ **19.** $n = 20$
21. $n = \frac{2}{5}$ **23.** $n = 1.5$

PAGE 151 EXTRA **1.** If the 9 sopranos sang the 3 songs simultaneously, then it would still take 12 minutes for 27 sopranos to sing those 3 songs. However, if the 27 sopranos sing in 3 sets of 9 each, consecutively, it would take 36 minutes.

PAGE 152 PROBLEMS **1.** 450 words **3.** 4.8 L
5. \$135

PAGE 156 EXERCISES **1.** 0.36 **3.** 90%, $\frac{9}{10}$
5. 150%, 1.5 **7.** 37.5%, 0.375 **9.** 248%, $\frac{62}{25}$ **11.** 92%
13. 350% **15.** 33% **17.** 4.03% **19.** 12 **21.** 3.84
23. 80 **25.** 70 **27.** 306

PAGE 156 PROBLEMS **1.** 90% **3.** 31.25% **5.** $4.48
7. 625 cartons **9.** 54% **11.** 99 students

PAGES 159–160 PROBLEMS **1.** 36.58 million
passengers **3.** 50% increase **5.** 25% decrease
7. 14% increase **9.** 10% decrease **11.** 4% **13.** 76

PAGES 162–164 PROBLEMS **1.** $1848 **3.** 2.5%
5. $2.48, $2.47 **7.** 35%, $44.85 **9.** $39.90
11. $15.63
13. Insurance = $1365, total income = $10,920
15. 44.8% **17.** 40% **19.** $9000

PAGE 166 CHAPTER REVIEW **1.** C **3.** A **5.** D
7. D **9.** C **11.** A **13.** D **15.** A **17.** D **19.** A

PAGE 171 SKILL REVIEW **1.** $\frac{3}{4}$ **3.** $\frac{3}{8}$ **5.** $\frac{1}{3}$ **7.** $\frac{3}{11}$
9. $\frac{3}{5}$ **11.** $\frac{3}{10}$ **13.** $n = 22$ **15.** $n = 55$ **17.** $n = 96$
19. $n = 45$ **21.** $n = 2.2$ **23.** $n = 31.5$ **25.** $n = 4.42$
27. $n = 16.8$ **29.** 40% **31.** 70% **33.** $66\frac{2}{3}$%
35. 37.5% **37.** 60% **39.** 180% **41.** 15% **43.** 8%
45. 375% **47.** $33\frac{1}{3}$% **49.** 4.5% **51.** 37.5% **53.** 1.28
55. 12.5% **57.** 175 **59.** 4.25 **61.** 25% **63.** $18\frac{2}{3}$

Chapter 6 Measurement and Geometry

PAGE 175 EXERCISES **1.** 48.6, 4.86, 0.486
3. 580, 5.8, 0.58 **5.** 9300, 930, 9.3
7. 3500, 350, 35 **9.** 400, 40, 4 **11.** 100, 1, 0.1
13. 10,000, 1000, 10 **15.** 5900, 590, 59
17. 6 cm, 60 mm **19.** 7.5 cm, 75 mm **21.** 7.5
23. 2000 **25.** 0.00001 **27.** Answers may vary.

PAGE 179 EXERCISES **1.** 314 cm **3.** 238.64 mm
5. 10.676 m **7.** 660 **9.** 154 **11.** $\frac{33}{125}$ **13.** 70
15. $\frac{49}{200}$ **17.** $26\frac{3}{5}$ **19.** The perimeters are equal.

PAGE 183 EXERCISES **1.** ∠COB and ∠COE
3. ∠AOC and ∠BOD **5.** ∠COD and ∠DOE **7.** 62°
9. 45° **11.** 35° **13.** a. 60° b. 30° c. 50°

15. **17.**

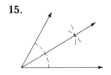

19. a.

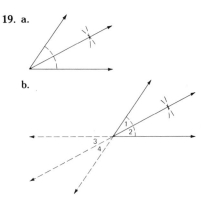

b.

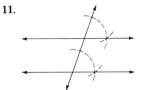

c. Yes; by vertical angles, ∠3 = ∠2 and ∠4 = ∠1.
So ∠3 = ∠4 since ∠1 = ∠2.

PAGE 184 EXTRA (bottom) ∠ACB is a right angle in
each case.

PAGES 186–187 EXERCISES **1.** ∠3 and ∠5;
∠2 and ∠8 **3.** ∠6, ∠4, ∠2
5. ∠2 = 68°, ∠3 = 112°, ∠4 = 68°, ∠5 = 112°,
∠6 = 68°, ∠7 = 112°, ∠8 = 68° **7.** ∠5
9. ∠2 = 35°, ∠3 = 90°, ∠4 = 55°, ∠5 = 35°,
∠6 = 145°, ∠7 = 35°, ∠8 = 90°

11.

13. ∠ADB = 110°

PAGE 188 EXTRA **1.** The distances between parallel
lines along each ray are the same. The distances along the
first ray are not equal to the distances along the second ray.
3. $\overline{CX}$ and $\overline{DY}$ are parallel to $\overline{EB}$ because corresponding
angles are equal. $AX = XY = YB$ from Exs. 1 and 2.

PAGES 191–193 EXERCISES **1.** D, H, I
3. A, B, E, G **5.** C, D, G, H, I **7.** G **9.** 130°
11. 70° **13.** 60° **15.** $x° = 65°, y° = 25°$
17. $x° = 35°, y° = 130°$ **19.** $x° = 30°, y° = 90°$
21. 125°

PAGE 193 EXTRA **1.** ∠A + ∠B + ∠C = 180°

PAGES 195–196 EXERCISES **1.** 7 **3.** 130° **5.** 3
7. 125° **9.** 3 **11.** 90° **13.** 15 **15.** 19° **17.** 70°
19. 13.5 **21.** 148° **23.** 14 cm **25.** Yes **27.** 12 cm

PAGES 199–200 EXERCISES **1.** 149.2 mm **3.** 18
5. Octagon **7.** 20 cm **9.** $37\frac{1}{2}$ **11.** 6.7 cm **13.** 40
15. 13 cm **17.** 56 cm

PAGE 200 EXTRA **1.** 6. A hexagon. Perimeter of hexagon = 3 × diameter

PAGES 203–204 EXERCISES **1.** *KERM* **3.** *RM*
5. 2 **7.** 90° **9.** 150° **11.** Congruent; ASA
13. Not congruent **15.** Congruent; SAS
17. $\triangle AOB \cong \triangle COD$; $\triangle AOD \cong \triangle COB$
19. $\triangle MIL$, $\triangle IMS$, $\triangle LSM$
21. $\triangle EBD$, $\triangle EDF$, $\triangle EBF$, $\triangle GDF$

PAGE 208 EXERCISES **1.** *DEF* **3.** $\angle B$ **5.** 4
7. 26° **9.** 4 **11.** Yes **13.** Yes **15.** Yes **17.** 9
19. Only if the rectangles are squares
21. $a = 18$, $b = 12$, $c = 17\frac{1}{3}$

PAGES 210–211 CHAPTER REVIEW **1.** C **3.** B
5. C **7.** B **9.** C **11.** C **13.** D **15.** C **17.** B **19.** C
21. B **23.** C **25.** C

PAGE 215 CUMULATIVE REVIEW **1.** $A = -1\frac{3}{5}$,
$B = -\frac{2}{5}$, $C = \frac{4}{5}$ **3.** $6\frac{3}{8}$ **5.** $11\frac{1}{9}$ **7.** $-1\frac{7}{24}$ **9.** 12
11. 0 **13.** 30 **15.** $12\frac{1}{2}\%$ **17.** 3.72 **19.** 100.48 cm
21. Bisect a right angle, then bisect one of the angles just constructed. **23.** Yes. $\triangle AOB \cong \triangle DOC$ by SAS.

Chapter 7 Equations and Inequalities

PAGES 218–219 EXERCISES **1.** $q + 8$ **3.** $7z$
5. $9(a + 3)$ **7.** c **9.** e **11.** d
13. The sum of negative five and t
15. The sum when the product of two and x is added to seven **17.** The quotient when a is divided by two
19. One half the product of a and b

PAGE 221 EXERCISES **1.** 13 **3.** 5 **5.** 4 **7.** -1
9. 0 **11.** 1 **13.** 0 **15.** 1 **17.** 30 **19.** 0 **21.** -2
23. 0 **25.** $\frac{1}{4}$ **27.** 3

PAGES 223–224 EXERCISES **1.** e **3.** d **5.** a
7. $t - 7 = 2$ **9.** $\frac{t}{8} < \frac{1}{2}$ **11.** $\frac{1}{2}t \le 8$ **13.** $-\frac{1}{2} > 2(-1)$

PAGE 226 EXERCISES **1.** True **3.** False **5.** True
7. True **9.** True **11.** True **13.** True **15.** True
17. No **19.** No **21.** Yes **23.** Yes **25.** Yes
27. Yes **29.** Yes

PAGES 228–229 EXERCISES

1.

3.

5.

7.  **9.** ∅

11.

13. $\{\ldots, -3, -2, -1\}$
15. $\{4, 5, 6, \ldots\}$ **17.** $\{3, 4, 5, \ldots\}$ **19.** $\{3\}$
21. $\{0, 1\}$ **23.** $\{-5, -4, -3, -2, -1, 0\}$

PAGE 229 EXTRA

1.

3.

5. The solution sets are the same.

PAGES 232–233 EXERCISES **1.** 18 **3.** -4 **5.** 4
7. -6 **9.** 33 **11.** 6 **13.** 4 **15.** -7 **17.** -11
19. 14 **21.** 6 **23.** 24 **25.** 8 **27.** -9

PAGE 233 APPLICATION **1.** 17¢ **3.** 19¢
5. 15 for $35.98

PAGE 236 EXERCISES **1.** $\{12\}$ **3.** $\{5\}$ **5.** $\{-20\}$
7. $\{-11\}$ **9.** $\{20\}$ **11.** $\{-1\}$ **13.** $\{3\}$ **15.** $\{-9\}$
17. $\{1\}$ **19.** $\{4\}$ **21.** $\{-4\}$

PAGE 236 EXTRA **1.** -4, -3 **3.** 9, 13

PAGES 238–239 EXERCISES

1.

3.

5. $\{-4, -5, -6, \ldots\}$ **7.** $\{8, 9, 10, \ldots\}$
9. $\{-5, -6, -7, \ldots\}$ **11.** $\{-11, -12, -13, \ldots\}$
13. $\{6, 5, 4, \ldots\}$ **15.** $\{4, 3, 2, \ldots\}$
17. $\{4, 3, 2, \ldots\}$ **19.** $\{10, 9, 8, \ldots\}$
21. $\{-9, -10, -11, \ldots\}$

PAGES 240–241 CHAPTER REVIEW **1.** C **3.** C
5. B **7.** D **9.** B **11.** C **13.** D **15.** C

PAGES 242–243 **1.** Skateboarders have more fun.
3. I accept your proposal for marriage. Love, Dora.

PAGE 245 SKILL REVIEW **1.** 20 **3.** -10 **5.** -28
7. 7 **9.** -11 **11.** -22 **13.** 37 **15.** -19 **17.** 42
19. 71 **21.** -17 **23.** 36 **25.** 24 **27.** 111 **29.** 21
31. -60 **33.** -34 **35.** 1 **37.** -8 **39.** 42
41. -56 **43.** -12 **45.** 2 **47.** 3 **49.** 60 **51.** 3
53. 144 **55.** 66 **57.** 7 **59.** -49 **61.** 18 **63.** 8
65. -6 **67.** -48

Chapter 8 Problem Solving

PAGE 248 EXERCISES Answers will vary. **1. a.** What number added to 24 gives 80? **b.** A city park department bought 80 trees to be planted. One day a crew planted 24 of the trees. How many more trees needed to be planted? **3. a.** Ten less than what number is 35? **b.** Irene and Tom drove to Pine Park. The return trip of 35 km was 10 km shorter than their trip to Pine Park. How far was their trip to Pine Park? **5. a.** What number multiplied by 12 gives 36? **b.** A dietician is preparing 36 school lunches. Each lunch includes an orange, and there are a dozen oranges in a bag. How many bags of oranges are needed? **7. a.** What number divided by 36 gives 50? **b.** Thirty-six clubs raised money to buy new equipment for the hospital. If each club raised an average of $50, how much money was raised in all? **9. a.** One quarter of what number is 20? **b.** Luis is 20 years old, one fourth as old as his grandfather. How old is Luis's grandfather? **11. a.** The sum of a number and three times the number is 16. What is the number? **b.** Cathy jogged 3 times as far Saturday afternoon as she did Saturday morning. If she jogged 16 km in all that day, how far did she jog in the morning? **13. a.** When you add the product of 2 and a number and the product of 2 and 4 more than the number, you get 28. What is the number? **b.** The length of a rectangle is 4 cm greater than the width. If the perimeter is 28 cm, what is the width of the rectangle? **15.** A carpenter had a box of 60 nails. After building 3 shelves there were 5 nails left. What was the average number of nails needed for each shelf? **17.** A veterinarian administers 3 doses of a medication from a full 60 mL bottle. If 5 mL of medication remained in the bottle, how many milliliters are there in one dose? **19.** A football coach spent 5 minutes of an hour's practice explaining a new play to the team. The remaining part of the hour was divided equally among three types of exercises. How much time was spent on each exercise?

PAGES 250–251 EXERCISES Answers may vary. **1.** $23 + n = 40$ **3.** $n + 21 = 33$ **5.** $n + (n + 8) = 19$ **7.** $n + (n + 1) = 101$ **9.** $l + (l + 45) = 120$ **11.** $x + x + x = 45$ **13.** $(x + 11) + (x + 11) + x = 58$ **15.** $2s + (s + 40) = 136$ **17.** $5n + 10(2n + 3) = 355$

PAGE 253 PROBLEMS 1. 17 **3.** 12 **5.** $5\frac{1}{2}$, $13\frac{1}{2}$ **7.** 50, 51 **9.** 37.5 m, 82.5 m **11.** 15 cm **13.** 23 cm, 23 cm, 12 cm **15.** Regular price, $8.50; championship-game price, $12 **17.** $12, $14, $21

19. He borrowed $30 and had saved $60. **21.** 9 years old **23.** 2 and $\frac{1}{2}$

PAGES 256–257 PROBLEMS 1. 2.25 m **3.** 18 cm **5.** Length, 9 cm; width, 6 cm **7.** About 90 km **9.** 1 cm → 60 km **11.** 300 km **13.** About 9.6 km

PAGES 259–260 PROBLEMS 1. $31.50 **3.** $2070 **5.** $8640 **7.** $3210 **9.** $15,300 **11.** $662.29 **13.** $1210; 13%; 15% **15.** Plan 2 **17.** $131.25

PAGES 261–262 PROBLEMS 1. No **3.** About 95 species **5.** 1,250,000 persons **7.** 22.08 W **9.** About 35 years

PAGE 263 APPLICATION 1. A monthly ticket **3.** Two 12-ride tickets and two one-way tickets

PAGES 264–265 CHAPTER REVIEW 1. C **3.** B **5.** C **7.** A **9.** B **11.** C

PAGES 266–267 1. 6 **3.** 3 **1.** 20 minutes **3.** 4 pieces; $\frac{1}{15}$ of the whole melon **5.** 65 geese **7.** $687,194,767.35

PAGE 269 SKILL REVIEW 1. 11.8 **3.** 1.23 **5.** −0.6 **7.** −5.5 **9.** 2.9 **11.** 3.523 **13.** 18.69 **15.** −1.65 **17.** 10.888 **19.** 16.112 **21.** −18.4 **23.** 18.124 **25.** 14.07 **27.** 4.03 **29.** 51.26 **31.** −51.3 **33.** −0.204 **35.** 8.2 **37.** −1305.5 **39.** −170.8 **41.** 996.2 **43.** 2.34 **45.** 13.13 **47.** 11.076 **49.** 95.7563

Chapter 9 Areas and Volumes

PAGES 273–274 EXERCISES 1. $b = 8$ cm, $h = 4$ cm, $A = 32$ cm² **3.** $b = 11.4$ m, $h = 5.2$ m, $A = 59.28$ m² **5.** $b = 7.3$, $h = 7.3$, $A = 53.29$ **7.** $A = 36$, $p = 30$ **9.** $A = 36$, $p = 26$ **11.** $A = 36$, $p = 28$ **13.** perimeters, area

PAGE 274 PROBLEMS 1. 25 cm² **3.** 7.5 cm **5.** 36 m **7.** Yes; no

PAGES 277–278 EXERCISES 1. 5 **3.** 7.5 **5.** 9 **7.** 12 **9.** 22 **11.** $A = 84$ cm², $p = 42$ cm **13.** $A = 36$, $p = 36$ **15.** 88 **17.** Each triangle has area 5 cm² **19.** 10 m **21.** 13.5 cm² **23.** 78

PAGES 279–281 EXERCISES 1. 50.24 km² **3.** 1.1304 cm² **5.** 803.84 cm² **7.** $3\frac{1}{7}$ square units **9.** $24\frac{16}{25}$ square units **11.** $\frac{7}{88}$ square units **13.** 3; 9 **15.** 5; 25 **17.** 36.75π cm² **19.** $(288 − 48\pi)$ cm² **21.** $1\frac{3}{4}$ units **23.** $(144 + 72\pi)$ square units; 24π units **25.** 32π square units; 16π units **27.** 16π square units; $16 + 4\pi$ units

PAGES 283–284 EXERCISES   **1.** 84 cm³   **3.** 36 cm³
5. 88 cubic units   **7.** 640π cm³   **9.** 5000π cm³
11. $\frac{529}{2}\pi$ mm³, or 264.5π mm³   **13.** 300 cubic units
15. 3140 cm³   **17.** No; the volume is multiplied by 27.

PAGES 287–288 EXERCISES   **1.** T
3. $\triangle PTQ$, $\triangle QTR$, $\triangle RTS$, $\triangle STP$ **5.** 9   **7.** 20 cm³
9. 1130.4 cm³   **11.** 960 cubic units   **13.** 924 cubic units
15. 0.60288 L   **17.** $\frac{1}{6}$ **19.** 500 cubic units   **21.** 4 units

PAGE 292 EXERCISES   **1. a.** 32 square units;   **b.** 62
square units
3. a. 640 square units;   **b.** 736 square units
5. a. 120π square units;   **b.** 192π square units

PAGE 292 PROBLEMS   **1.** 120 square units
3. a. 150 cm²;   **b.** 125 cm³   **5.** 54 cm²

PAGE 295 EXERCISES   **1.** 1000   **3.** 1.2   **5.** 3890
7. 2000   **9.** 700   **11.** 4.5 kg   **13.** 0.715 kg   **15.** 3.4 kg
17. 8.1 kg   **19.** 84 g   **21.** 1.43 kg   **23.** 8 kg

PAGE 295 PROBLEMS   **1.** No   **3.** 780 g
5. Answers may vary. For example, about twice as large

PAGE 297 EXERCISES   **1.** 36π square units;   36π
cubic units   **3.** 324π square units;   972π cubic units
5. 3813$\frac{1}{3}$ cubic units   **7.** The aluminum ball   **9.** No

PAGE 298 EXTRA   **1.** $\frac{256}{3}\pi$ cubic units   **3.** $\frac{2}{3}$ **5.** Yes

PAGES 300–301 CHAPTER REVIEW   **1.** D   **3.** C
5. C   **7.** A   **9.** B   **11.** C   **13.** C   **15.** D   **17.** C   **19.** B
21. D   **23.** D

PAGE 303 EXERCISES   **1.** 7.5 cm;   8.5 cm
3. 20.5 cm;   21.5 cm   **5.** 2   **7.** 4   **9.** 1   **11.** 2
13. 50 mm   **15.** 2000 m   **17.** 8.5;   2.5   **19.** 21.25
21. 18

PAGE 303 CALCULATOR ACTIVITY   57,705.375 mm³;
62,960.625 mm³

PAGE 305 CUMULATIVE REVIEW   **1.** 2 + 2x
3. Answers may vary. For example, seven more than the
product of 3 and x is less than or equal to 22.   **5.** 4
7. {−5}   **9.** {. . . , −3, −2, −1, 0, 1}   **11.** 23 h
13. 5 cm   **15.** 5 cm²   **17.** 64π square units
19. $\frac{20}{3}\pi$ cubic units

Chapter 10   Probability

PAGES 308–309 EXERCISES   **1.** The number of dots
on top is one of the following: 1, 2, 3, 4, 5, or 6.   **3.** $\frac{1}{6}$
5. 52; $\frac{1}{52}$ **7.** 20; $\frac{1}{20}$ **9.** 4; $\frac{1}{4}$

11. Answers may vary. For example, each person's name
could be written on a card. One name could then be
drawn from a bag.
13. Yes; since half the cards are red and half are black,
these outcomes are equally likely to occur.
15. Yes; out of four possible outcomes, one is "2 heads"
and one is "2 tails." Since the outcomes have equal
probabilities, they are equally likely to occur.

PAGE 309 EXTRA   The first situation involves matching
the birthdays of any two people in the class; the second
situation involves matching the birthday of one person with
that of a specific person in the class.

PAGE 312 EXERCISES   **1.** $\frac{1}{13}$ **3.** $\frac{1}{4}$ **5.** $\frac{1}{2}$ **7.** $\frac{2}{13}$ **9.** $\frac{1}{52}$
11. $\frac{5}{13}$ **13.** $\frac{3}{26}$ **15.** $\frac{1}{6}$ **17.** $\frac{1}{3}$ **19.** 0   **21.** 1   **23.** $\frac{1}{7}$
25. $\frac{3}{14}$ **27.** $\frac{5}{14}$ **29.** $\frac{1}{2}$ **31.** $\frac{5}{7}$ **33.** 1   **35.** $\frac{1}{6}$ **37.** $\frac{1}{3}$

PAGE 315 EXERCISES   **1.** 1 to 12   **3.** 1 to 1
5. 2 to 11   **7.** 1 to 51   **9.** 1 to 25   **11.** 3 to 1
13. The white die can show any number of dots from 1 to
6. For each number of dots on the white die, the yellow
die can have any number of dots from 1 to 6.
15. a. $\frac{1}{6}$ **b.** 5 to 1   **17. a.** $\frac{1}{18}$ **b.** 17 to 1
19. a. $\frac{4}{9}$ **b.** 5 to 4   **21. a.** $\frac{1}{12}$ **b.** 11 to 1
23. a. $\frac{5}{18}$ **b.** 13 to 5   **25. a.** $\frac{3}{4}$ **b.** 1 to 3
27. a. $\frac{5}{36}$ **b.** 31 to 5   **29. a.** $\frac{1}{36}$ **b.** 35 to 1
31. a. $\frac{1}{9}$ **b.** 8 to 1

PAGES 316–317 EXTRA   **1.** Yes; $\frac{1}{9}$ **3.** Yes; $\frac{1}{8}$

PAGE 319 EXERCISES   **1.** $\frac{7}{30}$ **3.** $\frac{3}{10}$ **5.** $\frac{8}{15}$ **7.** 0
9. 1 to 2   **11.** 11 to 19   **13.** 7 to 3   **15.** $\frac{2}{25}$ **17.** $\frac{139}{200}$
19. $\frac{31}{40}$

PAGES 321–322 EXERCISES   **1.** $\frac{2}{125}$ **3.** 42%
5. 1274 trees   **7.** $\frac{33}{200}$ **9.** 4080 people
11. Yes; about 62.4% favor passage.

PAGES 325–326 EXERCISES   **1.** Walking down the
stairs   **3.** 128 g
5. No; since the expected value is $-\frac{1}{2}$, the player is likely
to have a negative score.

PAGE 328–329 CHAPTER REVIEW   **1.** D   **3.** B   **5.** D
7. C   **9.** A   **11.** A   **13.** A   **15.** C

PAGES 330–331   **1.** $p(2) = \frac{1}{36}$, $p(3) = \frac{1}{18}$, $p(4) = \frac{1}{12}$,
$p(5) = \frac{1}{9}$, $p(6) = \frac{5}{36}$, $p(7) = \frac{1}{6}$, $p(8) = \frac{5}{36}$, $p(9) = \frac{1}{9}$,
$p(10) = \frac{1}{12}$, $p(11) = \frac{1}{18}$, $p(12) = \frac{1}{36}$
3. The bar for the value 7 is the longest.

PAGE 333 SKILL REVIEW **1.** $3\frac{3}{4}$ **3.** $6\frac{2}{3}$ **5.** -3
7. $11\frac{3}{10}$ **9.** $-7\frac{23}{24}$ **11.** $33\frac{11}{21}$ **13.** $44\frac{2}{3}$ **15.** $11\frac{3}{8}$
17. $-16\frac{20}{21}$ **19.** $-3\frac{31}{33}$ **21.** $6\frac{1}{3}$ **23.** $7\frac{2}{3}$ **25.** $\frac{1}{8}$ **27.** $3\frac{3}{8}$
29. $11\frac{1}{4}$ **31.** -11 **33.** 3 **35.** $-\frac{11}{49}$ **37.** $-\frac{4}{17}$
39. $-6\frac{17}{18}$ **41.** $\frac{7}{8}$

23. a.

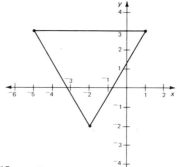

b. 15 sq. units

Chapter 11   The Number Plane

PAGES 338–339 EXERCISES **1.** H **3.** B **5.** J **7.** T
9. D **11.** K **13.** I **15.** E

17.

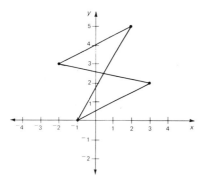

25. a.

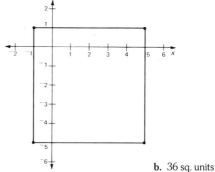

b. 36 sq. units

19.

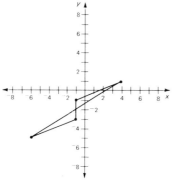

27. a.

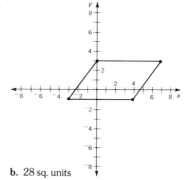

b. 28 sq. units

PAGE 339 APPLICATION **1.** Go one block east and
one block north. **3.** Go 3 blocks south.
5. Go 2 blocks west and 2 blocks north.

PAGES 341–342 EXERCISES **1.** (5,0) **3.** (4,−4)
5. (2,2) **7.** (0,0) **9.** (1,3) **11.** (−4,4) **13.** (4,4)
15. (1,−3) **17.** B, C, D, E **19.** K, L
21. A, B, C, D, E, N, P, Q **23.** K, L, M, N, P, Q
25. A, J **27.** F, M **29.** B, D, K, L

PAGES 346–347 EXERCISES **1. a.** Yes **b.** No
c. Yes **3. a.** No **b.** Yes **c.** Yes
5. a. Yes **b.** No **c.** Yes **7. a.** Yes **b.** No **c.** Yes
9. a. Yes **b.** No **c.** No

21.

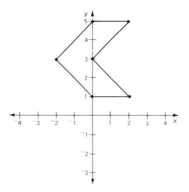

11. a. $y = 7 - x$ **b.** $(-5,12)$, $(-1,8)$, $(0,7)$, $(1,6)$, $(5,2)$

13. a. $y = x + 2$ **b.** $(-5,-3)$, $(-1,1)$, $(0,2)$, $(1,3)$, $(5,7)$

15. a. $y = x - 6$ **b.** $(-5,-11)$, $(-1,-7)$, $(0,-6)$, $(1,-5)$, $(5,-1)$

17. a. $y = \frac{5}{2} - \frac{x}{2}$ **b.** $(-5,5)$, $(-1,3)$, $(0,\frac{5}{2})$, $(1,2)$, $(5,0)$

19. a. $y = \frac{-x}{2} - 2$ **b.** $(-5,\frac{1}{2})$, $(-1,\frac{-3}{2})$, $(0,-2)$, $(1,\frac{-5}{2})$, $(5,\frac{-9}{2})$

21. a. $y = \frac{5}{2} - 2x$ **b.** $(-5,\frac{25}{2})$, $(-1,\frac{9}{2})$, $(0,\frac{5}{2})$, $(1,\frac{1}{2})$, $(5,\frac{-15}{2})$

23. a. $y = 4 - 6x$ **b.** $(-5,34)$, $(-1,10)$, $(0,4)$, $(1,-2)$, $(5,-26)$

25. a. $y = \frac{6x}{5} - 12$ **b.** $(-5,-18)$, $(-1,\frac{-66}{5})$, $(0,-12)$, $(1,\frac{-54}{5})$, $(5,-6)$ **27, 29, 31, 33.** Answers will vary.

5.

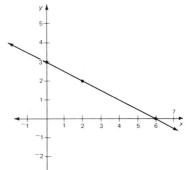

7.

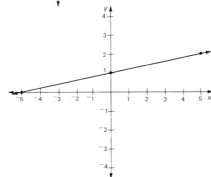

PAGES 349–350 EXERCISES

1.

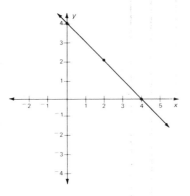

9.

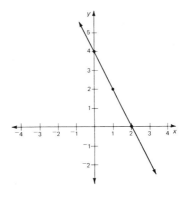

3.

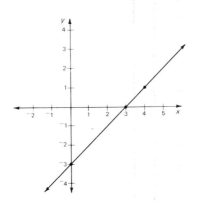

11.

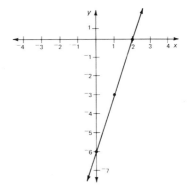

13.

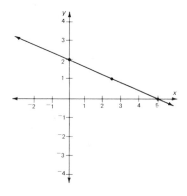

15.

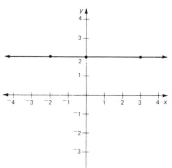

17.

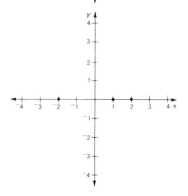

19.

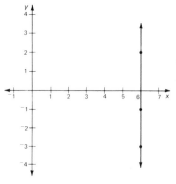

PAGE 351 APPLICATION

1. 1 kg mass produces 6.36 cm of stretch.

PAGE 353 EXERCISES

1.

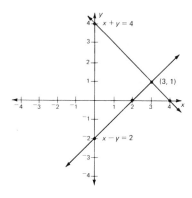

3.

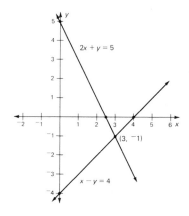

5.

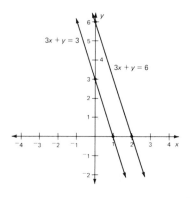

7.

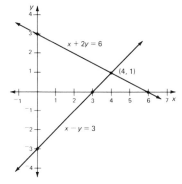

9.

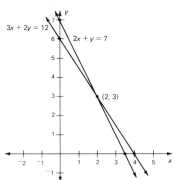

11.

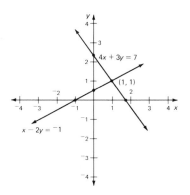

13.

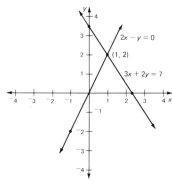

15.

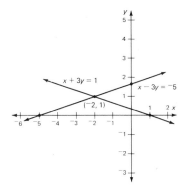

17.

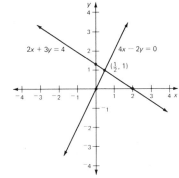

PAGES 356–357 CHAPTER REVIEW **1.** C **3.** B
5. A **7.** B **9.** B **11.** D **13.** C **15.** A **17.** C
19. C

PAGE 359 EXERCISES
1. Mexico City **3.** Los Angeles **5.** Lima **7.** Perth
9. Rio de Janeiro **11.** Nome **13.** 10°E, 60°N
15. 115°E, 40°N **17.** 60°W, 35°S **19.** 75°E, 20°N
21. 105°W, 40°N **23.** 50°E, 35°N

PAGE 361 SKILL REVIEW **1.** 0.25 **3.** 0.125 **5.** 1.20
7. 0.50 **9.** 2.00 **11.** 0.4732 **13.** 15% **15.** 161%
17. 45% **19.** 117% **21.** 1015% **23.** 131.2% **25.** $\frac{1}{4}$
27. $\frac{4}{5}$ **29.** $\frac{1}{8}$ **31.** $1\frac{3}{20}$ **33.** $\frac{23}{100}$ **35.** $\frac{1}{10}$ **37.** 1.25; 125%
39. 0.8; 80% **41.** 2.125; 212.5% **43.** 2.5; 250%
45. 0.9375; 93.75% **47.** 1.4; 140% **49.** 3.2; 320%
51. 1.2; 120% **53.** $\frac{3}{25}$ **55.** $\frac{1}{8}$ **57.** $1\frac{1}{4}$ **59.** $\frac{33}{50}$

Chapter 12   Right Triangles

PAGE 364 EXERCISES **1.** 8 **3.** −20 **5.** 7 **7.** 3
9. 3 and 4 **11.** 1 and 2 **13.** 11 and 12 **15.** 50 and 51
17. 1 **19.** 13 **21.** 9 and 10 **23.** 6 **25.** 9 and 10
27. 5 and 6

PAGE 366 EXERCISES **1.** 2.8 **3.** 5.7 **5.** 8.4 **7.** 5.1
9. 6.4 **11.** 7.9 **13.** 12.0 **15.** 20.5 **17.** 1.7 **19.** 4.6
21. 19.6 **23.** 28.1 **25.** 32.5 **27.** 99.9

PAGE 368 EXERCISES **1.** 6.25 **3.** 7.75 **5.** 79.37
7. 0.91 **9.** 17.14 **11.** 11.14 **13.** 11.86 **15.** 3.49
17. 12.46 **19.** 4.28 **21.** 6.86 **23.** 5.69 **25.** 9.62
27. 15.39 **29.** 43.24 **31.** 18.39 **33.** 0.26 **35.** 6.1 m
37. 4.2 cm

PAGE 369 EXTRA **1.** 32.02 **3.** 15.88 **5.** 32.83
7. 34.87 **9.** 53.85 **11.** 34.41 **13.** 13.60 **15.** 12.69

PAGE 373 EXERCISES **1.** No **3.** Yes **5.** Yes
7. No **9.** 8 km² **11.** 1341 m² **13.** 36.25 mm²
15. 3.61 **17.** 6.25 **19.** 8.78 **21.** 2.72 **23.** 5.42

PAGE 374 PROBLEMS **1.** 7.2 cm **3.** 9.4 km
5. 35 cm

PAGE 374 EXTRA **1.** 13 **3.** 8 **5.** 13 **7.** 87 **9.** 14
11. 205

PAGES 377–378 EXERCISES **1.** 16 **3.** 11.31 **5.** 36
7. 7.07 **9.** 13.86 **11.** 13 **13.** $x = 6.2$; $y \approx 5.4$
15. $x = 4.9$; $y \approx 6.9$ **17.** $x = 2.6$; $y \approx 4.5$

PAGE 378 PROBLEMS **1.** 35.35 mm **3.** 56.56 cm
5. 42.42 cm **7.** 8 cm and 13.856 cm

PAGE 381 EXERCISES **1.** $\sin A = \frac{5}{13}$, $\cos A = \frac{12}{13}$,
$\tan A = \frac{5}{12}$; $\sin B = \frac{12}{13}$, $\cos B = \frac{5}{13}$, $\tan B = \frac{12}{5}$
3. $\sin A = \frac{1}{\sqrt{2}}$, $\cos A = \frac{1}{\sqrt{2}}$, $\tan A = 1$; $\sin B = \frac{1}{\sqrt{2}}$,
$\cos B = \frac{1}{\sqrt{2}}$, $\tan B = 1$
5. $\sin A = \frac{x}{z}$, $\cos A = \frac{y}{z}$, $\tan A = \frac{x}{y}$; $\sin B = \frac{y}{z}$, $\cos B = \frac{x}{z}$,
$\tan B = \frac{y}{x}$ **7.** $x = 8$; $\tan A = \frac{8}{15}$
9. $x = \sqrt{11}$; $\tan A = \frac{\sqrt{11}}{5}$
11. $x = \sqrt{1400}$; $\sin A = \frac{\sqrt{1400}}{39}$
13. $x = 15$; $\cos A = \frac{\sqrt{15}}{8}$ **15.** $x = 9$; $\cos A = \frac{9}{41}$

PAGE 383 EXERCISES **1.** $z = \sqrt{29}$; $\angle A = 22°$;
$\angle B = 68°$ **3.** $y = \sqrt{27}$; $\angle A = 30°$; $\angle B = 60°$
5. $x \approx 4$; $y \approx 3$; $\angle B = 43°$
7. $y \approx 5$; $z \approx 21$; $\angle A = 75°$
9. $\angle B = 62°$; $y \approx 19$; $z \approx 21$

PAGES 386–387 CHAPTER REVIEW **1.** B **3.** C
5. D **7.** A **9.** C **11.** B **13.** C **15.** B **17.** A **19.** A

PAGE 391 CUMULATIVE REVIEW **1.** 8; $\frac{1}{8}$
3. $\frac{1}{5}$; 4 to 1 **5.** $\frac{2}{5}$; 3 to 2

7.

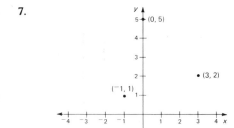

9.

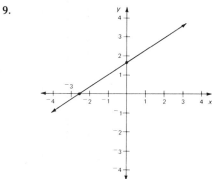

11. 7 and 8 **13.** 5 cm **15. a.** $\frac{5}{\sqrt{53}}$ **b.** $\frac{7}{2}$

Extra Practice

CHAPTER 1 PAGE 400 **1.** $201 **3.** Dances
5. Food sales and dances **7.** $550 **9.** $550 **11.** $5

PAGE 401

1.

Mass in kilograms	Frequency
50	6
51	4
52	9
53	5
54	7

3. 52 kg **5.** 4 kg
7. 52 kg **9.** 50
11. 72.5

CHAPTER 2 PAGE 402 **1.** 10^5 **3.** 10^4
5. $(9 \times 10^3) + (1 \times 10^2) + (6 \times 10^1) + 4$ **7.** 6418

9. thousandths **11.** tenths
13. $(9 \times 10) + 2 + (4 \times 0.1) + (6 \times 0.01)$ **15.** 648.35
17. 50.0403 **19.** 300.016

21.

23. < **25.** <, < **27.** 4; 37; 82 **29.** 0; 3; 5
31. 60; 600; 6000 **33.** 38; 0; 93

PAGE 403 **1.** $n = 23 - 16$; $n = 7$
3. $n = 56 \div 7$; $n = 8$ **5.** $n = 51 - 28$; $n = 23$
7. $n = 0$ **9.** $p = 9$ **11.** 71.673 **13.** 14.583
15. 53.917 **17.** 26.754 **19.** 0.06992 **21.** 0.10556
23. 8.04 **25.** 8.54 **27.** 23.83 **29.** 115.2 km
31. 2.9 m

CHAPTER 3 PAGE 404

1.

3.

5.

7. 5 **9.** -4 **11.** 0

13.

15.

17.

19.

21.

23. 9.9 **25.** -7.6 **27.** -0.6 **29.** -6.7 **31.** -1.3
33. -2 **35.** -3 **37.** -4.1 **39.** 2.25

PAGE 405 **1.** 12 **3.** 5 **5.** 179 **7.** 18
9. $4 - {}^-8 = 12$; $4 - 12 = {}^-8$
11. ${}^-8 - 19 = {}^-27$; ${}^-8 - {}^-27 = 19$
13. ${}^-42 - {}^-12 = {}^-30$; ${}^-42 - {}^-30 = {}^-12$
15. ${}^-10 - 16 = {}^-26$; ${}^-10 - {}^-26 = 16$
17. $6.5 + 1.3$ **19.** $2.3 + {}^-4.4$ **21.** $44.4 + {}^-57.6$
23. $2.56 + 9.83$ **25.** 14 **27.** -60 **29.** 9.3 **31.** 1.1
33. -104 **35.** -144 **37.** 0 **39.** 8 **41.** -2.88
43. 50.02 **45.** -2.4725 **47.** 13.345 **49.** -4
51. -1.8 **53.** 1 **55.** 2

CHAPTER 4 PAGE 406 **1.** $\frac{1}{6}$ **3.** 35 **5.** $2\frac{3}{8}$
7. $\frac{-4}{9}$, $\frac{4}{-9}$ **9.** $\frac{-5}{6}$, $\frac{5}{-6}$ **11.** -27 **13.** -97
15. Answers may vary; for example, $-\frac{22}{26}$, $-\frac{110}{130}$ **17.** $\frac{8}{7}$
19. $\frac{19}{29}$ **21.** $\frac{6}{7}$ **23.** $-\frac{3}{5}$ **25.** $-\frac{1}{2}$ **27.** $-\frac{9}{35}$ **29.** $\frac{7}{20}$
31. $2\frac{4}{19}$
PAGE 407 **1.** $\frac{9}{12}$, $\frac{10}{12}$ **3.** $-\frac{15}{35}$, $-\frac{14}{35}$ **5.** $\frac{9}{36}$, $\frac{21}{36}$, $\frac{10}{36}$ **7.** $\frac{11}{36}$
9. $1\frac{8}{21}$ **11.** $3\frac{19}{20}$ **13.** $1\frac{17}{63}$ **15.** $-5\frac{1}{2}$ **17.** $-\frac{5}{27}$ **19.** $1\frac{17}{50}$
21. $2\frac{1}{20}$ **23.** $\frac{1}{125}$ **25.** 1.75 **27.** $-0.95\overline{4}$ **29.** $-0.8\overline{51}$
CHAPTER 5 PAGE 408 **1.** $\frac{1}{3}$ **3.** $\frac{4}{9}$ **5.** $\frac{5}{12}$ **7.** $\frac{3}{8}$
9. Incorrect **11.** Incorrect **13.** Correct **15.** $n = 5\frac{1}{3}$
17. $n = 37\frac{1}{2}$ **19.** $n = 42$ **21.** $n = 47\frac{7}{10}$ **23.** $n = 60$
PAGE 409 **1.** 3 **3.** 1.92 **5.** $85\frac{5}{7}$% **7.** 80% **9.** 120
11. 18 **13.** 30 sets **15.** About 18% **17.** 92 pots
19. $56.25 **21.** $24,500 **23.** $1740

CHAPTER 6 PAGE 410 **1.** two **3.** 5.3 **5.** 0.4
7. 188.4 cm **9.** 75.36 units **11.** 4.2 units **13.** acute
15. equal **17.** 65° **19.** equal **21.** 265°

PAGE 411 **1.** 30°, 90° **3.** scalene **5.** T **7.** F **9.** F
11. 260 mm **13.** 26.3 cm **15.** Yes **17.** $\angle F$ **19.** No
21. Yes **23.** No

CHAPTER 7 PAGE 412 **1.** $x - 9$ **3.** $t - z$
5. The sum of x and nine
7. The quotient when n is divided by four
9. The product of three sevenths and t
11. The quotient when six is divided by z **13.** -1
15. -4 **17.** -6 **19.** 9 **21.** $x = -2\left(\frac{1}{2}\right)$ **23.** $y + 4 < 7$
25. The product of three and the difference when one is subtracted from y is greater than or equal to negative four.
27. The difference when three is subtracted from the product of negative four and w is less than or equal to negative five. **29.** Yes **31.** Yes **33.** Yes **35.** Yes

37.   **39.** No solution

41. No solution **43.**

45.

PAGE 413 **1.** 6 **3.** 8 **5.** -7 **7.** 10 **9.** $x = 2$
11. $z = 9$ **13.** $x = -2$ **15.** $z = 2$ **17.** $x = 6$
19. $z = -4$ **21.** $x = 6$ **23.** $z = 12$ **25.** $x = 4$
27. $z = -10$ **29.** $x = 3$ **31.** $z = 6$ **33.** $x = -1$
35. $x = 1\frac{2}{3}$ **37.** $\{-1, 0, 1, 2, 3, \ldots\}$
39. $\{-4, -3, -2, -1, 0, \ldots\}$

41. {0, −1, −2, −3, −4, . . .}
43. {−5, −4, −3, −2, −1, . . .}
45. {12, 13, 14, 15, 16, . . .}
47. {6, 7, 8, 9, 10, . . .}
49. {−13, −14, −15, −16, −17, . . .}
51. {0, −1, −2, −3, −4, . . .}
53. {10, 9, 8, 7, 6, . . .} **55.** {3, 2, 1, 0, −1, . . .}
57. {−1, 0, 1, 2, 3, . . .} **59.** {3, 4, 5, 6, 7, . . .}

CHAPTER 8 PAGE 414 1. a. What number plus 16 equals 43? **b.** Answers may vary.
3. a. Five times what number gives 80? **b.** Answers may vary.
5. a. The difference between what number and 9 is 36? **b.** Answers may vary.
7. a. One fifth of what number is 7? **b.** Answers may vary.
9. $19 + n = 41$ **11.** $n − 9 = 25$
13. $n + (n + 56) = 97$ **15.** $2l + 2(l + 35) = 150$
17. $n + (n − 21) = 53$; $n = 37$
19. $n + (n + 1) = 43$; $n = 21$; the numbers are 21 and 22.
21. $n + (n + 2) = 46$; $n = 22$; the numbers are 22 and 24.

PAGE 415 1. 1.375 km **3.** 8 cm by 6 cm **5.** $104
7. $5000 **9.** $28.47 **11.** About $2 million

CHAPTER 9 PAGE 416 1. 64 cm² **3.** 56 m²
5. 32 cm **7.** 8 cm **9.** 504 mm² **11.** 20.7 m²

13. 24 m² **15.** 6 m **17.** $7\frac{1}{14}$ square units

19. 100π square units

PAGE 416 1. 288π (cubic units) **3.** 288 (cubic units)
5. 1728 cm³ **7.** About 106.67 cm³
9. 91.67π (cubic units) **11.** 170 (square units)
13. 252 (square units); 448 (square units) **15.** 4600
17. 0.85 **19.** 415 g **21.** 660 g **23.** 5024 cm²

25. $10\frac{2}{3}\pi$ (cubic units)

CHAPTER 10 PAGE 418 1. No
3. No (It is random, however, if we are sure you do not look.) **5.** 6; $\frac{1}{6}$ **7.** 6; $\frac{1}{6}$ **9.** 8; $\frac{1}{8}$ **11.** 4; $\frac{1}{4}$ **13.** 4; $\frac{1}{4}$
15. $\frac{2}{11}$ **17.** $\frac{1}{11}$ **19.** $\frac{1}{11}$ **21.** $\frac{3}{11}$ **23.** $\frac{5}{11}$ **25.** 1 to 3
PAGE 419 1. $\frac{5}{9}$ **3.** $\frac{7}{8}$ **5.** $\frac{3}{100}$ **7.** 120 **9.** $\frac{97}{3}$

CHAPTER 11 PAGE 420 1. a. P **b.** R **c.** L **d.** B
e. H **f.** D **g.** K **h.** N **i.** G **j.** I **k.** M **l.** Q **m.** C
n. U **o.** F **p.** E **q.** W **r.** Z **s.** J **t.** V
3. Area of rectangle, 30 square units
5. Area of rectangle, 27 square units
7. Area of square, 9 square units
9. a. $(−4,1)$ **b.** $(−2,4)$ **c.** $(−1,0)$ **d.** $(−5,−2)$
e. $(−2,−4)$ **f.** $(0,−2)$ **g.** $(2,−4)$ **h.** $(5,−6)$ **i.** $(3,0)$
j. $(1,3)$ **k.** $(6,4)$ **l.** $(2,1)$ **11.** R **13.** E, S

PAGE 421 1. No **3.** Yes **5.** No **7.** No
9. $y = 3 − \frac{3}{2}x$ **11.** $y = −\frac{1}{2} + \frac{1}{2}x$ **13.** $y = −\frac{4}{5}x$
15. $y = \frac{4}{3}x − 4$ **17.** (2,0), (0,2) **19.** (−1,0) (0,1)
21. (2,0), (0,−4) **23.** (−4,0), (0,3)

25.

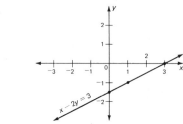

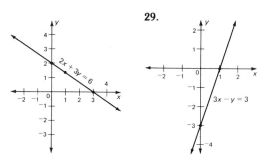

29.

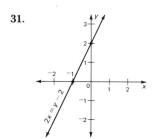

31.

33. $(−1,−2)$
35. No solution
37. (0,4)
39. No solution
41. $(−2,−2)$

CHAPTER 12 PAGE 422 1. 2 and 3 **3.** 2 and 3
5. 4 and 5 **7.** 5 and 6 **9.** 6 and 7 **11.** 6 and 7
13. 7 and 8 **15.** 8 and 9 **17.** 7.55 **19.** 3.85 **21.** 7.7
23. 6.3 **25.** 9.1 **27.** 6.65 **29.** 4.1 **31.** 5.4 **33.** 9.1
35. 6.1 **37.** 9.3 **39.** 9.7 **41.** 5.29 **43.** 6.56
45. 9.11 **47.** 6.71 **49.** 34.87 **51.** 28.28 **53.** 17.32
55. 0.22 **57.** 1.10 **59.** 6.43

PAGE 423 1. Yes **3.** Yes **5.** Yes **7.** $c = 5$
9. $c = 7.07$ **11.** $c = 10$ **13.** 8.49 **15.** 10 **17.** 10.39
19. $\frac{x}{y}$ **21.** $\frac{y}{z}$ **23.** $\frac{x}{z}$ **25.** c **27.** 23°